Innovation Diffusion and Political Control of Energy Technologies

Contributions to Economics

Christoph M. Schneider
Research and Development Management: From the Soviet Union to Russia
1994. ISBN 3-7908-0757-5

Bernhard Böhm/Lionello F. Punzo (Eds.)
Economic Performance
1994. ISBN 3-7908-0811-3

Lars Olof Persson/Ulf Wiberg
Microregional Fragmentation
1995. ISBN 3-7908-0855-5

Ernesto Felli/Furio C. Rosati/Giovanni Tria (Eds.)
The Service Sector: Productivity and Growth
1995. ISBN 3-7908-0875-X

Giuseppe Munda
Multicriteria Evaluation in Fuzzy Environment
1995. ISBN 3-7908-0892-X

Giovanni Galizzi/Luciano Venturini (Eds.)
Economics of Innovation: The Case of Food Industry
1996. ISBN 3-7908-0911-X

David T. Johnson
Poverty, Inequality and Social Welfare in Australia
1996. ISBN 3-7908-0942-X

Rongxing Guo
Border-Regional Economics
1996. ISBN 3-7908-0943-8

Oliver Fratzscher
The Political Economy of Trade Integration
1996. ISBN 3-7908-0945-4

Ulrich Landwehr
Industrial Mobility and Public Policy
1996. ISBN 3-7908-0949-7

Arnold Picot/Ekkehard Schlicht (Eds.)
Firms, Markets, and Contracts
1996. Corr. 2nd printing 1997.
ISBN 3-7908-0947-0

Thorsten Wichmann
Agricultural Technical Progress and the Development of a Dual Economy
1997. ISBN 3-7908-0960-8

Ulrich Woitek
Business Cycles
1997. ISBN 3-7908-0997-7

Michael Carlberg
International Economic Growth
1997. ISBN 3-7908-0995-0

Massimo Filippini
Elements of the Swiss Market for Electricity
1997. ISBN 3-7908-0996-9

Giuseppe Gaburro (Ed.)
Ethics and Economics
1997. ISBN 3-7908-0986-1

Frank Hoster/Heinz Welsch/Christoph Böhringer
CO_2 Abatement and Economic Structural Change in the European Internal Market
1997. ISBN 3-7908-1020-7

Christian M. Hafner
Nonlinear Time Series Analysis with Applications to Foreign Exchange Rate Volatility
1997. ISBN 3-7908-1041-X

Sardar M.N. Islam
Mathematical Economics of Multi-Level Optimisation
1998. ISBN 3-7908-1050-9

Sven-Morten Mentzel
Real Exchange Rate Movements
1998. ISBN 3-7908-1081-9

Lei Delsen/Eelke de Jong (Eds.)
The German and Dutch Economies
1998. ISBN 3-7908-1064-9

Mark Weder
Business Cycle Models with Indeterminacy
1998. ISBN 3-7908-1078-9

Tor Rødseth (Ed.)
Models for Multispecies Management
1998. ISBN 3-7908-1001-0

Michael Carlberg
Intertemporal Macroeconomics
1998. ISBN 3-7908-1096-7

Sabine Spangenberg
The Institutionalised Transformation of the East German Economy
1998. ISBN 3-7908-1103-3

continued on page 486

Karl Matthias Weber

Innovation Diffusion and Political Control of Energy Technologies

A Comparison of Combined Heat and Power Generation in the UK and Germany

With 27 Figures
and 42 Tables

Physica-Verlag
A Springer-Verlag Company

Series Editors
Werner A. Müller
Martina Bihn

Author
Karl Matthias Weber
Institute for Prospective Technological Studies
Joint Research Centre
European Commission
WTC Building
Isla de la Cartuja s/n
E-41092 Sevilla, Spain

ISBN 3-7908-1205-6 Physica-Verlag Heidelberg New York
D93 (1998)

Cataloging-in-Publication Data applied for
Die Deutsche Bibliothek – CIP-Einheitsaufnahme
Weber, Karl Matthias: Innovation diffusion and political control of energy technologies: a comparison of combined heat and power generation in the UK and Germany: with 42 tables / Karl Matthias Weber. – Heidelberg: Physica-Verl., 1999
(Contributions to economics)
Zugl.: Stuttgart, Univ., Diss., 1998
ISBN 3-7908-1205-6

Softcover Design: Erich Kirchner, Heidelberg

SPIN 10721292 88/2202-5 4 3 2 1 0 – Printed on acid-free paper

Preface

Two general questions stood at the beginning of this PhD-thesis, namely:

- What are the mechanisms which lead to the emergence and establishment of new technologies?
- How can this process of technological change be influenced politically?

In this sense, conceptual and theoretical interests were the early driving forces of the research work. This is also reflected in the considerable attention paid to the nature of technological change and political control. The result is an holistic perspective which builds on inputs from different disciplines and aims at dynamic interpretation. This, however, created a severe methodological problem: How could such a comprehensive perspective be used constructively? To develop this link between theory and forward-looking, policy-oriented analysis, and to devise a methodology which showed explicitly how this approach could be used in a constructive way were in fact the major challenges of this research project. The application to the example of combined heat and power generation, and the comparison of the developments in the UK and in Germany serve the purpose to demonstrate how this approach and methodology can be implemented in practice. These aspects were also of particular interest to the Institute for Prospective Technological Studies (IPTS), one of the institutes of the European Commission's Joint Research Centre, where most of the research work reported in this PhD-thesis was carried out.

In this context, the support given by the European Commission is particularly acknowledged because it financed the work at the Institute for Prospective Technological Studies (IPTS) from April 1993 to May 1996 by a grant of the Human Capital and Mobility Programme. The IPTS was located in Ispra (Italy) until September 1994, and since then in Sevilla (Spain); where I moved in April 1995. In between, I spent a period of six months as a visiting researcher at the Programme for Policy Research in Engineering, Science and Technology (PREST), University of Manchester, in order to carry out my empirical fieldwork in the UK. The supportive and inspiring atmosphere at PREST was without any doubt a very positive and creative source of ideas and contacts. The fieldwork on the German case was carried out while based in Sevilla, travelling several times to and across Germany in order to conduct interviews and collect data. Since June 1996, I had the opportunity to finish my PhD-thesis while working at IPTS on a number of other research projects. Most of the data and literature collection was finished by mid-1997, with only a few additions made since then.

The long and in-depth research reported here, which aimed at fulfilling the scientific requirements of a PhD as well as at providing a relevant contribution to actual policy issues, could not have been realised without the support of many of my colleagues. Both at IPTS and PREST I have received many useful inputs for my research work, and out of the numerous names which ought to be mentioned here, I would like to highlight in particular Patrick Canarelli, Carole Cohen, Ken Ducatel, Thomas Heinemeier, Nikos Kastrinos, Ian Miles and Fabiana Scapolo - be it for their scientific or their personal support. Among the many people with whom I got in contact during the last years, I particularly benefited from the scientific discussions I had with Ken Green, Stewart Russell, Remco Hoogma and Yvette Taminiau. Obviously, a word of thanks needs to be addressed to all the interviewees who provided essential pieces of empirical information to this work.

I suppose that my academic supervisor, Prof. Dr. Frank C. Englmann, must have been quite surprised when first addressed with the proposal to supervise a PhD-project by someone whom he hardly knew, who would be working at a remote place, and who had not even studied Economics. I have very much appreciated his open-minded spirit, which let me develop my ideas and research work while keeping me on the right track across a difficult terrain.

The role played by Prof. Dr. Ulrich Druwe for the realisation of this PhD-thesis went certainly beyond that of a normal co-referee. In fact, it was largely due to his influence that I decided not only to finalise my studies in political sciences, but also to take the risk of embarking on a more profound research project for a PhD in the social and economic sciences instead of following the more traditional professional trajectory of a process engineer. However, all remaining mistakes and shortcomings remain my own.

Several people could still be mentioned who have shared with me enjoyment and desperation during the more than four years of research, and who certainly had to bear a lack of personal attention from my side. However, I think that they are sufficiently close to me to know without specifically acknowledging how grateful I am for their support and patience.

Sevilla, November 1998

K. Matthias Weber

Table of Contents

PART IV: Concluding Remarks on Cogeneration, Political Control and the PET-Approach

List of Figures

List of Tables

Summary

This study analyses the processes of innovation and diffusion of cogeneration technology (also called combined heat and power - CHP) and the possibilities of steering and framing these processes by political means. The analysis is leant relevance by the ongoing liberalisation debate on energy supply systems in Europe. It also has as backdrop the desire for more sustainable energy technologies (of which CHP can be regarded as an example).

The underlying concern of the study is to support the development of forward-looking policy strategies. This concern implies that a wide range of factors have to be taken into account to capture the complex way in which policy affects the diffusion of new technology. The result is necessarily an interdisciplinary approach, because policy needs do not respect disciplinary boundaries. Perhaps it is not surprising that no conceptual or theoretical approach was available which provided a sufficiently broad foundation. Consequently, the study starts with a review of theoretical and conceptual approaches spanning relevant areas of economics, sociology and political sciences. In particular, elements of evolutionary economics, the literature on the social shaping of large technological systems, and self-referential concepts of political control are taken into account. The thesis is at this stage an attempt to synthesise these fields into a highly structured conceptual approach, here called the PET-system (Politics, Economics, Technology). It emphasises a thorough understanding of the technology in question as a basis for socio-economic analysis. In order to provide a guideline for applying this encompassing approach to specific studies, a methodology is devised which permits a multi-level picture of the technology in its context to be built up. This methodology is implemented in nine steps of technological and socio-economic analysis.

The second part of the thesis shows how the methodology might be implemented through the case of CHP. It starts by reviewing recent developments in the field of electricity supply and by introducing the technology of CHP. In this way the ground is laid for the empirical examples. A review of recent research findings on CHP provides a justification for selecting the UK and Germany as the two country case-studies to be examined. In this process of methodological specification it is also demonstrated where new empirical data needs to be generated, and by what means.

Part III of the thesis provides a systematic description of the processes under investigation in the UK and Germany. It is structured according to the elements of the PET-system, the future ramifications of which are explored through three scenarios for each country. By comparing the two case studies systematically it can be shown that the differences between them are the result of on the one hand path-dependent "lock-ins" between political, organisational and technological structures, and of corresponding actor strategies on the other. While Germany has maintained its monopolistic approach to energy supply, the British energy supply system has been liberalised and privatised in recent years. This structural change in the UK broke up several of the "lock-ins", finally triggering a process of substantial CHP-capacity expansion. However, the comparison indicates that the new liberalised British framework is not necessarily very conducive to CHP. It lacks several of the beneficial mechanisms of the monopolistic system in Germany and it has erected new, though different barriers to CHP-diffusion. In addition, liberalisation has favoured the uptake of advanced power-only technologies and created new path-dependencies which will remain effective for at least the next fifteen years. In the German case, it can be shown that in spite of the monopolistic structure of the electricity supply industry, some key elements have been crucial for the establishment of CHP, especially the strong position of horizontally integrated local utility companies and the existence of well-organised and influential industrial and engineering associations. A major reform of the energy supply system has only been initiated recently in view of the establishment of the Internal Energy Market in Europe. The modernisation of East German CHP-systems has been achieved within a completely restructured energy supply system, which replicated the monopolistic West German model and did not trigger a major structural reform in the reunified country.

The fourth and final part of the study formulates a number of generalised conclusions with regard to the political stimulation of CHP-diffusion in different regulatory frameworks. These focus on policies both at national and at European level. Firstly, the historical inertia of existing organisations and structures is underlined, as well as the importance of established technological paradigms. Secondly, it is highlighted that different policy areas such as technology policy, environmental policy and energy policy have a joint impact on both structural and behavioural determinants of CHP-diffusion. This shows that policy analysis can not be limited to policy objectives and actions, but needs to address also the internal processes of policy-making. Further methodological and theoretical lessons from the study are also drawn out. Although the framework developed was powerful and comprehensive, it was equally demanding to implement, especially in terms of the amount of data which needed to be collected and analysed. Nevertheless, experience shows that such an approach, although ambitious, is essential for understanding the complex mechanisms of change. Finally, some consideration is given to future research needs. Beyond the refinement and generalisation of the conceptual basis, further empirical extension of the PET-approach is needed on a country by country and technology by technology basis. They should aim at a full-scale prospective application of the PET-approach. In addition, it would be worth to extend the analysis of innovation diffusion processes beyond the needs of policy communities, for example by taking an industrialist's or technologist's control perspective.

Chapter 1: Introduction

1.1 The applied problem

The process of industrialisation in the last 200 years has transformed our societies to an extent and with a speed which has never been known in the history of humankind before. On the one hand, economic growth and wealth have been the outcome as well as the driving force of new products and production processes, but on the other hand the natural foundations of life are increasingly challenged.

Energy supply, environmental degradation and the diffusion of CHP

At least during the last twenty years it has become obvious that technological change and economic growth also have negative consequences which in the long term could endanger the sustainability of human life. The limitations of natural resources, environmental pollution, the potential of high-risk accidents as well as negative social impacts of new technologies have been recognised as being interwoven constraints on our present lifestyles. The logic of long-term environmental problems stands in sharp contrast to the prevailing short-term logic of market economies, which does not allow accounting for negative environmental and other social externalities.

In order to adjust these 'market failures', practically all industrialised countries have embarked on government policies to respond to environmental concerns,[1] using either market-based or regulative instruments. "Sustainable development" has become a guiding political concept in this context, implying a techno-economic development which can be reconciled with the requirements of a long-term persistence of our industrial economies and societies. It can be defined as "a pattern of social and structural economic transformations (i.e. "development")

---

1 This comprises environmental policy as well as technology and innovation policy, energy policy and tax policy.

which optimises the economic and other societal benefits available in the present, without jeopardising the likely potential for similar benefits in the future."[2]

One of the central ecological challenges is the increase of CO_2-emissions. It calls into question one of the foundations of industrialisation, namely the supply with energy from fossil resources. Taking the trends in global population growth and energy consumption, as described in several long-term scenario studies,[3] into consideration, the need for a change in global energy strategies is obvious. The reduction of per-capita primary energy consumption could in principle be achieved by either of two ways: demand-side measures, i.e. the reduction of energy consumption and the use of more efficient end-use technologies, or supply-side measures, i.e. the introduction and diffusion of new, cleaner and more efficient generation technologies.[4] The latter of these options is the focus of the following study.

Attempts to introduce non-fossil technologies have until now had only a limited success. This has been due either to technical restrictions (e.g. for hydropower, solar and wind energy), social resistance (nuclear energy) or high economic fears (as for photovoltaic and hydrogen technology). Only nuclear energy and hydropower have attained significant market shares in Europe. Fossil fuel-based technologies still dominate energy supply for the foreseeable future.[5] Therefore, this study focuses on some recent developments of cleaner and more efficient energy technologies based on fossil fuels.

Cogeneration technology (or combined heat and power, CHP) is one of the options about which wide consensus exists on its environmentally beneficial potential. It is an option which goes in the 'right' direction - even without defining in detail the concept of 'sustainable development'. Combined heat and power is not a new technology. Its principles have been known for several decades, but it was not before the oil price crisis of the Seventies that they attracted new interest for energy supply. Their diffusion can be seen as part of a general trend towards more efficient and therefore more environmentally compatible technologies. The associated debate can be labelled by headings like 'end-of-pipe' versus 'process-integrated' and 'clean' technologies, CHP being an example of a process-integrated technology.[6] A further important feature of cogeneration is its decentralised character. The difficulty of transporting heat over long distances without

---

2 Cited from (Goodland & Ledec, 1987, p. 20). For a more recent and comprehensive investigation of the concept of sustainable development see (Pearce & Warford, 1993).

3 See for example (Beaver, 1993; Eden, 1993; Starr, Searl, & Alpert, 1992).

4 See for example (Grubb, Walker, & al., 1992, p XIX-XX and parts II/III)

5 See footnote 3 and for more technology-related aspects (Nakicenovic & al., 1993).

6 'End-of-pipe' technologies add a clean-up system to a widely unaltered production process. This means that they represent additional environmental investment. 'Clean' or 'process integrated' technologies aim at modifying and restructuring the entire production process in order to keep the generation of emissions at a low level right from the beginning. This promises to be a less costly option in a long-term perspective. See for example (Kemp & Soete, 1992; Zimmermann, 1988)

excessive efficiency losses implies that an energy supply system which is largely based on CHP would be composed of decentralised generation plants.

Although at a first sight CHP may appear to be a very promising option, diffusion in the different EU Member States has been very uneven, raising the question why the development has been so variable. While in some cases like Norway with its abundant hydropower, the absence of CHP can be explained quite easily, in most of the others the situation is far more complex. Apparently, one cannot even speak of a fundamental lack of economic viability of CHP. Successful applications show that it is indeed economically and technically viable. Explanations for its limited diffusion must be sought elsewhere, taking into consideration factors such as structural and organisational settings of the energy supply system, or the imperfect decision behaviour of technology users and suppliers.

Energy supply, liberalisation and the role of public authorities

As indicated above, one of the reasons for studying CHP has been the insight that changes in the use of energy technologies are necessary for environmental reasons. Another concerns the role, the responsibilities and the possibilities of public authorities to influence these changes. Important decisions need to be taken now and over the next few years, which will determine the use of energy technologies for at least the coming fifty years. In order to manage this transformation process in a sustainable direction, there is a pressing need to improve our understanding of processes of technological change in the field of energy.

But public authorities are also directly involved in the energy sector through the energy utility companies which are often public or semi-public. In some countries even the major energy technology manufacturers are under public control. Public authorities therefore carry much responsibility for inducing and guiding change. The way in which the energy sectors have been managed in the last years, namely by public ownership and control, is nowadays confronted by increasing criticism. This has been paralleled by a trend towards a liberalisation of utility services, traditionally considered a public monopoly.[7]

The two issues mentioned above, namely the need for more sustainable energy technologies and the question of the right approach to managing future energy supply systems, merge in the case of cogeneration. The responses to these two challenges differ considerably among the Member States of the EU. Two very different examples have been chosen for an in-depth investigation in this study: the United Kingdom and Germany. While in the United Kingdom new liberal settings have been introduced, followed by a renaissance of CHP, Germany with its fairly high share of CHP-based power generation has maintained its traditional monopolistic supply system and even extended it to East Germany.

The question could be raised whether it makes much sense to study the diffusion of energy technologies solely at national level. Certainly, from the manufacturers'

---

7 Similar arguments in favour of liberalisation can be made with regard to other public utility services, such as transport or telecommunications, but also for the health sector and social security.

point of view energy technologies should be seen in a global context, because the market for large power plants is fully international. Cogeneration technology being a decentralised technology, the situation appears to be different. Also from the main customers' point of view it seems more appropriate to define the basic unit of analysis at country level, because the energy supply systems as well as energy policies are still highly country-specific. Global opportunities for manufacturers of CHP-technology should therefore be regarded as an additional, though not a decisive factor of their innovation and investment strategies.

A comparative perspective on the diffusion of energy technologies is regarded to be particularly useful for studying the effects of different types of national energy supply systems on innovation diffusion. This holds even more when focusing on the possibilities and limitations of political control. While the detailed study examines the United Kingdom and Germany since the early Eighties, a broader review of different factors affecting the diffusion of CHP (Chapter 6) has been carried out on the basis of a much wider selection of countries.

The political relevance of the study

Given the above mentioned arguments, the political relevance of the study is evident. CHP has been widely recognised as a benign option. At the political level this is reflected in the introduction of CHP-capacity targets and the formal intention to promote its wider use. However, the crucial question is how to meet these targets, and how existing obstacles to its wider diffusion could be removed.

Public authorities which attempt to exert influence on the innovation and diffusion of new technologies are confronted with a highly complex situation. It amalgamates aspects of technological feasibility, economic incentives, social acceptance, structural impacts and the multiple dynamic interactions between them. Focusing predominantly on innovation diffusion, and leaving out the preceding invention phase,[8] information is required on:

- The process of innovation diffusion itself which is mainly taking place inside the economic system,
- The impacts of possible political measures from different policy areas on the diffusion process,
- Impacts of possible innovations and policies on other parts of the social system, and
- The capabilities of the political system to formulate and implement adequate control measures.

Although the third point, i.e. the study of side-effects of policy measures, also represents an important subject for investigation it will play a minor role in the present work. With regard to the first two aspects, theory - and not only economic

8 Although inventive activities may also be affected to a certain extent by market demand, here they are regarded to be driven mainly by other factors. For the case of CHP this can be justified by its comparative technological maturity.

theory - is challenged by the need to provide an adequate explanation of how innovation diffusion works. Its driving forces and mechanisms are the basis for further impact and prospective studies during the process of political decision-making. In other words, innovation and diffusion theory is the key element for any further investigation to support policy-making. The fourth point raised will require some further consideration, because it is usually ignored in technology studies. In the majority of studies, the (techno-) economic system only is investigated, thereby neglecting the mutual links which exist between politics and economics. These links are particularly strong in areas like energy, telecommunication, or transport where traditionally government has been involved as a shareholder or an influential regulator of (sometimes) monopolistic companies. In other words, it is necessary to extend the empirical scope of the study and include the political realm. This has obvious conceptual consequences because it requires to integrate the issue of politics in the conceptual framework to be developed. Such a broad perspective on technological change and political control should also help the fictive clients of the study, i.e. political decision-makers, to get a more realistic and complete perspective on their own possibilities to exert influence on technological change by means of political measures. The introduction of such a fictive client is hence a useful device to facilitate and support the constructive orientation of the study and its underlying theoretical approach.

The following diffusion and innovation study should therefore

- Contribute to the ongoing discussion on adequate approaches to the political control of technological change in the context of the ongoing liberalisation debate;
- Provide insights into the possibilities for stimulating the diffusion of environmentally compatible energy technologies;
- Analyse in a comparative perspective recent experiences made with cogeneration technology, and explore them into the future.

A fundamental problem of policy-making consists of its long-term impacts. In order to make the right decisions today, it is desired and necessary to look ahead. Although this study is not intended to be a full prospective exercise, the underlying theoretical and methodological approach provides the foundation for future-oriented work. By broadening the range of arguments used to explain innovation diffusion the approach facilitates prospective analyses and therefore provides a basis for political decision-making.

Therefore, the central political interests of the study are the following:

- It is intended not just to describe the aggregate process of innovation, diffusion and political control of cogeneration technology, but to understand the possibilities and limitations which exist for governments to influence this process. There is a need to understand the processes and interrelationships which underlie the observable macroscopic patterns of change in order to be able to inform policy-makers why and how they could proceed and why certain seemingly helpful measures fail.

- Rather than a particular policy area, the interplay of different policies with regard to innovation diffusion is highlighted. In other words, the main interest is in the fate of the technology as a whole, subject as it is to a multitude of social, economic and technological forces, among which the co-ordinated effect of political forces plays a critical role.

1.2 The methodological problem

Given the brief outline of the problem area, it becomes obvious that the relationships and interdependencies in the field of innovation diffusion and political control are too complex to be dealt with using simple models, and that clear remedies are consequently not readily at hand. One has to deal with

- Technological developments,
- The emergence of new interests, values and pressures in society,
- Structural constraints on the economy and public authorities,
- Complex interactions of industry, research, government and interest groups,
- The application of a comparative and longitudinal research approach.

These requirements inevitably call for an interdisciplinary perspective and approach. With regard to this problem Krupp has formulated an interesting statement on the necessity of interdisciplinary approaches to energy research (Krupp, 1992, p. 11):

> *"Energy is the physical fuel of societal dynamics. Its form and use is an integral part of a society. Therefore, the politics and the economics of energy provision and consumption are too complex to be left to any particular discipline or elite. Also, there is no such discipline as energology. Therefore, an interdisciplinary approach has to be taken which methodologically has to muddle through."*

The methodological challenge of this study is to avoid 'muddling through' but to try to develop a straightforward and clear methodology. Basically, two alternative ideal types of approaches present themselves in this respect. Either one starts to collect different observable data on patterns related to the core issue, analyses these trends in a 'synthetic' way, and suggests an interpretation of the specific area in question ('*inductive*' approach); or one tries to build the study of the specific area explicitly on the generalised understanding of similar problem areas (theory-guided or '*deductive*' approach). For this study, the decision has been made to build the research work on existing conceptual and theoretical findings on technological change, i.e. to follow an approach which comes closer to the deductive than to the inductive model. This allows to be highly explicit and transparent about the research procedure followed, and it facilitates the transfer of the methodology to other cases beyond CHP in the future. In order to demonstrate its general applicability the generalised methodology to be developed will thus be tried out in a very explicit way for the case of CHP.

Obviously, the quality of such an approach calls for a thorough assessment of theoretical approaches and findings on technological change and political control. This will involve an attempt to make '*constructive*' use of theory, i.e. not only in the sense of generating some basic working hypotheses, but also to provide a methodological foundation for the research work itself.

1.3 The theoretical problem

The methodological research problem raises the question how constructive use can be made of theory and whether it is possible to 'use' theory as a foundation for applied research. Usually, the role of theory is limited either to an ex-post explanation of observed empirical patterns, or to the development of hypotheses which are empirically tested to enrich the body of academic and theoretical research. While the latter approach does not provide much help for specific policy problems which are concerned with a complex reality or even prospective issues, the former does not use theory in a constructive way. The challenge when developing a theoretical foundation for prospective research is that it must be sufficiently broad to capture all potential intervening factors and feedback mechanisms which can have an influence on the system evolution in the future. This requirement needs to be balanced with the need to be simple, explicit and understandable.

Political control as the frame of analysis

The broadest possible frame for the analysis of techno-economic change is the whole social system in which these changes occur. Cultural and social boundary conditions can play a decisive role for the success or the failure of new technologies. It is impossible to consider all these aspects simultaneously in a single diffusion study, although it has been shown that they can be very important in the case of energy technologies.[9] The focus of interest will therefore be on the political and techno-economic systems.[10] The latter is regarded as the immediate environment for emerging new technologies, while the former defines the wider the context and the embedding structures for technologies.

As already mentioned, efforts to exert political control on techno-economic change are directed towards a highly complex system. The theoretical understanding which prevails about the impacts of political tools on complex socio-technical systems is quite limited; theories of political control of technology are even at a rudimentary level.[11] Therefore the question *how* public policy could or

---

9 Nuclear energy is probably the best known example in this context (Keck, 1984). Another example for the dominance of socio-cultural factors over techno-economic and political ones is the introduction of information technologies, which was accepted by the employees much more slowly than expected.

10 For the fundamentals of the systems approach in political sciences see (Easton, 1965).

11 These views are supported e.g. by (Glagow & Willke, 1987; Görlitz, 1989; Görlitz, 1995; Keck, 1986; Meyer-Krahmer, 1989, p.6).

should intervene in a targeted way can not be answered fully, but will remain in the background of our analysis. The discussion of policy implications will be limited to the general possibilities and limitations of control, especially with regard to the autonomy of the techno-economic system from political intervention. The lessons which can be learnt from a comparison of past experience with energy technology policies in different countries should therefore be seen as a contribution to a more realistic assessment of what can be expected from public policy with regard to complex control problems in general.

The central role of theories of technological change

While being embedded inside the frame of political control, the diffusion of technological innovations represents the core of the study. Since the end of the Fifties efforts have been made to improve the understanding of technological change and its contribution to economic growth, based on the conviction that new technologies represent a main driving force for economic change and growth. Much empirical work has been carried out, and different models and various theories have been presented, attempting to explain why structural change in the techno-economic system takes place. Several theoretical approaches to innovation and diffusion, based on diverging assumptions, can be identified: equilibrium versus disequilibrium dynamics of diffusion, macroeconomic versus microeconomic orientation, uncertainty about new technologies versus perfect information, maximising versus institutionalised behaviour, heterogeneity versus uniformity of the agents, demand- versus supply-based driving forces of technical change,[12] to mention just a few.

In the context of political control, innovation and diffusion theory should not only explain how and why these changes take place, but also the motivations behind firms' behaviour and reactions to external impulses. This implies that it is not sufficient to describe innovation and diffusion processes at the macro-level. The underlying motivations, decision rules and regularities of firm behaviour must also be known in order to enable a realistic assessment of the reactions to control measures. In other words, inside any given context of political control it is necessary to go down to the micro-level of innovation diffusion.

The still dominant approach in economics - neo-classical theory - has been severely criticised for its deficiencies in dealing adequately with technical change. Although attempts have been made to improve the neo-classical approaches, the concepts of evolutionary economics will be preferred as background for this study.[13] First of all, an evolutionary approach appears to be more appropriate to cope with dynamic processes of change like innovation and diffusion. As a second argument, evolutionary economics is also more open to contributions from other research fields because it builds on conceptual elements which are shared with other disciplines. In addition to the economic lines of research, sociology has also

---

12 Cf. (Dosi, 1991, pp.189-192).

13 See (Witt, 1987, pp. 1-30) for a general introduction to evolutionary economics, and (Dosi, 1988c; Dosi, 1991a; Silverberg, Dosi, & Orsenigo, 1988) for evolutionary approaches in the context of innovation diffusion.

contributed considerably in the last fifteen years to the improvement of our understanding of technological change, mainly from a micro-perspective which, for this study, will help to complement the economic analyses. Particular attention will also be paid to the influence of changing technological characteristics on the success or failure of a technology. The underlying argument is that such technical issues have been neglected until recently in most economic and social sciences theories. This is not appropriate if, as in the case of energy supply, innovation diffusion takes place in complex technological systems. Broad classifications as the ones in process/product innovations or supplier dominated/production intensive/science based industries (Pavitt, 1984) are too simple for an adequate consideration of these systems' properties and their effects on the diffusion process.

1.4 The rationale of the study

From the perspective of the philosophy of science the study will proceed in a deductive rather than in an inductive way. The main steps to be made can be summarised as follows:

- Developing a generalised theoretical approach of innovation diffusion and political control;
- Deriving generalised conclusions or hypotheses on the processes of techno-economic and socio-political change;
- Formulating a general methodology for investigating empirical cases, dealing with all the different elements of the theoretical approach;
- Specifying the general methodology and hypotheses for the case of cogeneration; this will build on the available information on energy supply and cogeneration in order to identify gaps in the empirical information;
- Presenting, comparing and interpreting the empirical data along the lines of theory, allowing a critical assessment of the theoretical approach and the generalised hypotheses.

This is obviously not an ideal deductive procedure which aims at the testing of hypotheses, but an attempt to be as deductive as available theory permits. It is driven by the question whether it is possible to build on and make constructive use of the vast and diverse theoretical literature accumulated during the past decades. The traditional approach of developing first theoretical and then empirical hypotheses ('operationalisation') is challenged by the recognition that most social processes of change are highly complex and cannot be translated into simple testable hypotheses. This makes it increasingly difficult to use theory for constructive and prospective tasks and requires a novel approach to theory-based research. However, the preference for a quasi-deductive approach is based on the conviction that it is essential to be explicit about the researcher's guiding mental framework and to avoid the unconscious preconceptions of inductive research.

Therefore, the purposes and functions of the theoretical approach with regard to the comparative case study are to

- Provide a clear and coherent conceptual vocabulary as a tool to structure and debate research results,
- Formulate a basic understanding of innovation diffusion and its key determinants, hence a basic understanding of the systems dynamics,
- Delimit, map out and justify the aspects which are of central interest to the study and to situate them in a broader context,
- Structure the pieces of information which are needed to give a comprehensive and intelligible explanation of the observed processes.

In other words, the aim is to develop a kind of general methodology for policy-oriented diffusion studies which helps to structure the empirical work.[14] At the end of the study the question will be taken up again whether this theoretical and methodological approach has been useful or not.

1.5 Structure of the study

Consequently, the four main parts of the study will deal with the following aspects:

- Part I: Development of the theoretical model and methodological approach;
- Part II: Introductory analysis of the problem area and specification of the in-depth research methodology;
- Part III: Presentation, exploration and comparison of the empirical models;
- Part IV: Conclusions and recommendations with regard to policy-making, methodology and theory.

In the first part, a review of recent developments in economic and sociological innovation diffusion theory, and of theories of political control, will be presented in order to get an idea of the concepts and theories which represent the 'state of the art' (Chapter 2). As the review will conclude that the existing work does not provide an adequate approach to the problems addressed in this study, a new perspective will be introduced which attempts to bring together strands of theory from economics, political sciences and sociology in a coherent framework (Chapter 3). This new approach will be called PET ('Politics-Economics-

---

14 A similar argument has been presented by (Klein, 1983, p. 154, translation by MW), who underlines the necessity "to develop a theoretical frame of reference as a foundation for empirical research design, which enables the researcher to gain experimental knowledge in the empirical field, which in turn allows him to modify the holistically conceived frame of reference where necessary, or to make it more precise where possible."

Technology'), reflecting the importance of an interdisciplinary social sciences perspective which builds on a sound technological basis. Starting from this theoretical contribution, a basic methodology for studies of innovation diffusion and political control will be presented which provides the main steps for the rest of the study (Chapter 4). Both the methodology and the underlying theoretical approach are expected to be sufficiently general to find application to other examples of technological change and public policy.

The second part will include an introduction to the problem area of energy and electricity supply and to the technical issues involved in cogeneration technology (Chapter 5). A review of the diffusion and the research findings on cogeneration technology (Chapter 6) will be provided in order to be able to specify in detail the reasons and the methodology of the targeted empirical investigation (Chapter 7). The core empirical contribution will be based on an interview study in the two countries with technology suppliers, technology users, government representatives and representatives from associations. This will be complemented with information from a variety of other sources.

Using the pattern developed in the preceding chapters, in the third part the research findings from the UK and Germany will be presented, followed by a brief scenario-based outlook to the future (Chapters 8 & 9). This leads over to a comparative overview and analysis (Chapter 10).

The fourth part will be concerned with the analysis and interpretation of the results (Chapter 11). Beyond a discussion of the working hypotheses with regard to CHP-diffusion, some main policy conclusions will be drawn for the fictive client of the study, i.e. mainly policy-makers both at national and European level. Finally, the methodological and theoretical conclusions of the approach underlying the study will be discussed, along with an outlook on possibilities and necessities for further research.

A final note for the reader

Due to the interdisciplinary character of the study and hence the range of disciplines it is addressed to, several issues will have to be dealt with which - from a single-disciplinary perspective - would not need to be explained and might even appear to be evident. Other issues may not be discussed in sufficient depth to satisfy the specialist. However, it needs to be emphasised that this study intends to fulfil a bridging function between disciplines, and therefore needs to be accessible for readers with different backgrounds. Consequently, a highly sophisticated insider-language has been avoided in order to facilitate the comprehension by someone who is not familiar with the subtleties of economic, sociological or political vocabularies and ideas. The informed reader may hence skip certain chapters in which he regards himself to be already expert.

PART I:

The PET-System - A Theoretical and Methodological Approach to Innovation Diffusion and Political Control

In this first part, the theoretical and methodological foundation is laid for the comparative study of cogeneration technology. It becomes evident in the course of this part that there is no ready-made approach available which would fulfil the broad requirements of this study. A review of existing approaches to the analysis of technological change in different social science disciplines provides the knowledge base for developing the integrated approach needed, and a corresponding methodology. This approach has been called the *PET system approach* in order to reflect the importance assigned to politics, economics and technology. When reading the following part, it should be kept in mind that it is not the principal objective of this approach to derive testable hypotheses but to guide policy-oriented research on a sound theoretical basis. This procedural rationale is the key to obtain useful results which combine theoretical and empirical inputs. As this represents a rather unconventional perspective on the study of technological change and political control, it will be shown in Part II how it can actually be implemented for a practical case-study. Part I is therefore restricted to the following research issues:

- Chapter 2 summarises and reviews different approaches to the analysis of technological change and political control. The main focus is on recent theoretical lines of thinking in economics, sociology and political sciences.
- Building on these inputs, Chapter 3 introduces the PET system approach. It provides the conceptual language and the dynamic rationale underlying the evolution of a system which comprises behavioural and structural as well as technological elements.
- The methodology presented in Chapter 4 aims to make explicit how the PET perspective can be used constructively in order to facilitate and guide further applications. It outlines the major steps to be followed when implementing PET system approach for a particular study.

Chapter 2: Theoretical Approaches to Technological Change and Political Control Across Disciplines

2.1 Some basic concepts and definitions

A number of key concepts are frequently used in the context of research on technological change, but often with significantly different connotations. Meanings differ among disciplinary contexts, and also among different authors. The following brief discussion of such key terms aims to provide a foundation for the study on CHP and to avoid misunderstandings with regard to the various disciplines from which inputs will be used. At the end of Section 2.1, a general typology of models of theoretical approaches to technological change will be suggested. The discussion of different disciplinary ways of theorising about technological change will be discussed along the lines of this typology in the sections 2.2 to 2.5.

2.1.1 Technology and related definitions

The term *technology* is often used without questioning its precise meaning. Analytically, it can be characterised by at least two dimensions:[15]

- Artefact dimension: technologies are composed of physical hardware,
- Knowledge dimension: in order to make a technology work, different forms of knowledge are needed, both for its production and operation.

---

[15] In English and German, the meanings of apparently similar terms differ. While the German *Technik* mainly refers to the hardware dimension of realised artefacts, and *Technologie* is understood as the knowledge or science of realising such artefacts, the British notions of *technology* and *technique/technical* are different. 'Technology' encompasses both the artefact and the knowledge dimension, and *technique* is closely associated to a procedure or process for realising artefacts. In this study, the English notation will prevail unless specified differently.

The concept of *knowledge* needs to specified further. It comprises codified knowledge (documents, blueprints, etc.) as well as tacit elements, e.g. those embodied in the skills and capabilities of the workforce. Such a definition of technology may still be fairly simple, but it has been used frequently (Metcalfe & Boden, 1992). Other authors include further dimensions in their notions of technology. For example, Bijker distinguishes between processes, activities and artefacts (Bijker, Hughes, & Pinch, 1987, p. 4). Organisational aspects of importance for technologies to be operational have also been included in definitions of the term (Coombs, Green, Richards, & Walsh, 1998). *Skills*, i.e. practical and often tacit knowledge embodied in humans and acquired by learning can also be regarded as a separate feature of technology (Liso & Metcalfe, 1994). In the following, *technology* will be used in the first sense, i.e. comprising the dimensions of artefact and knowledge, while *technique* and *technical* will be used to describe the artefact and procedural dimensions of technology. Organisational aspects will be referred to separately.

The term *system* is applied to technologies which have a complex character and are composed of several interacting elements.[16] In social and economic sciences, the notion of technological systems usually highlights the interdependent relationships between technology and its social and economic context; therefore also the organisational issues mentioned before. Carlsson and Stankiewicz define a technological system as "a dynamic network of agents interacting in a specific economic/industrial area under a particular institutional infrastructure and involved in the generation, diffusion, and utilisation of technology" (Carlsson & Stankiewicz, 1991, p. 93). A similar understanding is supported by Hughes, but he argues against a too vague definition of what a technological system is, putting emphasis on the distinction between environment and system (Hughes, 1987, p. 52). Factors which may affect the operation of a system, but are not under the system's control, should not be regarded as components of the system (Hughes, 1987, p. 54). In other words, within a system the existence of feedbacks and interdependencies is taken into account, whereas the environment just exerts an influence on the system without being affected by it. Such an analytical simplification of a complex and interdependent reality is important, but needs to be justified in any specific case under study.

Innovation, *diffusion* and *technological change* are also used in a number of different ways. The aforementioned wider notions of technology or of technological systems already indicated the importance of non-technical implications of innovation, going beyond the artefact and knowledge dimensions. In principle, this was already recognised by (Schumpeter, 1964, p. 100, translation by MW) in his definition of innovation:

---

16 A term *system* is usually used to describe a set of elements and the relations between them. Unfortunately, this definition is very vague, and the difficulties to delimit a system and to identify its relations with the environment is problematic - unless one deals with a closed system (Rickards, 1985, p. 28). The distinction between a system and its environment is usually given by the lack of control of the system over its environment, i.e. the operation of the system does not affect substantially the state of the environment.

"1. Introduction of new or improvement of existing goods;

2. Introduction of new production methods;

3. Opening up of a new market;

4. Opening up of a new supply source;

5. Carrying out of a reorganisation, e.g. by creating or breaking up of a monopoly."

This definition of innovation obviously emphasises economic aspects, but similarly one can distinguish social and other types of innovations. In particular, the notion of *policy innovations* will become relevant later on, even if the techno-economic meaning will play the key role.

Innovation and diffusion were traditionally understood as two stages of the process of technological change. This distinction can still be maintained even if the underlying processes leading to technological change are interrelated and complex rather than a simple sequence of discrete steps. Innovation refers essentially to the introduction of novelties into the market or any other social environment. Diffusion denotes the spread and wider application of these novelties. Put in the terms of Meyer-Krahmer, innovation denotes "the process of development of technologically new or improved products and processes and their commercialisation on the market, respectively their integration within production" (Meyer-Krahmer, 1989, translation by MW). Diffusion is the process of the wider spread of a technology, and it can be distinguished with regard to different levels of diffusion: inside a company, between companies, between industries, and even between countries (Stoneman, 1983). The active promotion of these diffusion processes, especially inter-firm and inter-industry, is also called technology transfer. Diffusion has two further distinct components, namely *imitation* (i.e. using the same technology in the same industry, referring to the replication of the production process) and *adoption* (i.e. using the results or products of an innovative industrial process in another industry or sector, hence the wider acceptance of a new product). *Technological change* is a more encompassing concept which not only includes the wider use and application of new technologies, but also the transformation of knowledge and skills needed to make them operational. It represents the changes which in the end have been achieved in terms of both artefacts and knowledge. *Technical change* would then be limited to the artefact dimension of technological change.

Several taxonomies and classifications have been proposed for innovation and technological change in order to generate hypotheses for different types of technologies. Such an approach assumes that the content of technology itself matters for the analysis of innovation and diffusion process. The most well-known distinctions can be summarised as follows:

- <u>The content of an innovation:</u> The key issue differentiating issue is the actual substance and significance of an innovation. Schumpeter's definition is an example of this; a similar one has been given by Kline and Rosenberg (Kline & Rosenberg, 1989, p. 279).

- The radicality of an innovation: In most simple terms, incremental innovations are distinguished from radical ones. Another example is based on the distinction between *micro-inventions* and *macro-inventions* (Mokyr, 1990). Freeman and Perez have further added to incremental and radical innovations the concepts of new technology systems and changing techno-economic paradigms (Freeman & Perez, 1988).
- Two dimensional classifications: As an example, a matrix scheme which is based on the disruptive or reinforcing character of a technology on the one hand and its impacts on the market side on the other can be mentioned here (Abernathy & Clark, 1985). Henderson and Clark distinguish between the level at which an innovation occurs (component or architectural level) and its disruptive or reinforcing character. This leads to the identification of component, radical, architectural or modular innovations (Henderson & Clark, 1990).

Although the focus of this study is on the processes leading to technological transformations, the meaning of Technology Assessment (TA) needs to be explained as well - especially because it has become a major field of policy research. TA points to the important role which *expectations* of future impacts of new technologies have on technological choices to be made today. It is increasingly based on a multi-dimensional understanding of technological change and its impacts. These impacts can be on the technology itself, on economic developments, on social relationships, or on the natural environment, as well as on political agendas and decisions. Although TA may essentially be dealing with the final impacts and results of technological change, it should not be misunderstood as the anticipation of future developments. It is also an important determinant of ongoing decision-making processes shaping the future.

2.1.2 Definitions related to political control and regulation

The core interest of this study is to improve the understanding of processes of technological change, but close to it come the role, possibilities and limitations for government and society to steer technological change in a benign direction. *Political control* ('politische Steuerung') may not be the most common term to denote this political aspect of technological change, but it is more encompassing than *regulation*. Regulation refers to the political measures applied to an essentially market-based industry, especially in the sense of constraining the operation of market forces to protect citizens and to increase social benefits.[17] Examples are health and safety regulations in industry, licensing requirements, or emission limits. However, this definition already implies that technological change is regarded mainly as an economic phenomenon. Political control is a wider concept because it includes not only policies as outputs of the political process, but also the processes of policy formulation and implementation. In addition, a wider range

---

17 See e.g. the definition of Waterson: "Regulation (..) is taken as meaning control of an industrial activity by government, in the sense of actions such as restrictions on firms entering the industry, constraints on firms actually in the industry, or both." (Waterson, 1988, p. 2)

ofApolitical options is taken into account, based on the utilisation of the three basic forms of control: market, hierarchy/bureaucracy and solidarity/community. They can be characterised in terms of the dominance of one three main control measures (money, law and information).[18] Control is exerted not only through the targeted promotion of individual technological options, e.g. by means of technology policy, but also by addressing the systemic context of a technology ('context control'). In other words, political control comprises the entire interface between the political authorities and the object of control, as well as the processes within these authorities.

The interest in context control has been induced in recent years by the failure of targeted political control. This 'government failure' has been observed in the field of technological change as well as in other problem areas. It has led not only to a growing preference for market-based control mechanisms and the retreat of the state but also to a debate on whether political structures represent obstacles to efficient control. Already the formulation phase of policy making can be regarded as biased by political organisations because they represent an important but filter of both information and partial interests. On the other hand, examples of government failure can be easily contrasted with market failures. This has been particularly obvious in the context of negative, especially environmental, externalities of market systems, showing that the discussion concerning the 'right' control philosophy has not yet been concluded.

2.1.3 Multiple perspectives on technological change: A hierarchy of disciplines

The theoretical interest of this study lies in the combination of two areas of research: technological change and political control. A major issue is therefore how different disciplines have dealt with technological change. The research results over the past two decades have improved our understanding of technological change but remain scattered and unrelated. A multi-level system of disciplinary perspectives can be identified which shows the variety of relevant perspectives and the levels at which they apply.

In addition, historical studies fulfil a cross-cutting, horizontal function. They may be less systematic and theory-oriented in character, but provide rich information on the simultaneous effects of different factors on individual processes of technological change (Mokyr, 1990). The disciplines mentioned in Table 2.1 tend to study the systematic role of individual factors in these processes rather than their joint operation from an integrated perspective. The difficulty for policy-oriented studies is that they have to take the complexity and integrated character of real-world processes into consideration in order to provide sensible conclusions, while being sufficiently systematic and simple to be intelligible to decision-makers.

---

[18] The definition of political control used here is mainly based on (Mayntz, 1987) where the basic terminology is introduced and differentiated. She mentions "associations" or "private government" as a potential fourth form of control.

Level	Discipline	Research Areas
1. Individual level	Psychology Sociology	Individual interests, motivations and pre-conditions for behaviour; Cultural factors
2. Level of organisations such as companies, social groups, political organisations	Sociology Management Science Political Sciences	Motivations, conflicts and internal organisation of firms; The role of R&D
3. Level of relations between organisations such as markets, networks, hierarchies	Economics of Industrial Organisation Sociology	Economic processes and technological change; Economic organisation of sectors and industries
4. Level of national economies	Economics	Organisation and structures of national systems as factors of innovativeness, growth and competitiveness
5. Level of political authorities	Political Sciences	The role of policies and political processes for technological change and political control; Organisation of political processes
6. International level	Political Sciences Economics	Internationalisation of research and technological development, trade issues

Table 2.1: A hierarchy of research disciplines related to technological change

For the purpose of this study, the contributions from levels 3, 4 and 5 are regarded as particularly relevant. The detailed processes within organisations (level 2) are of less relevance than the interactions between them, even if a reliable conceptual understanding of intra-firm processes is needed as a theoretical foundation. Due to the fact that until now energy supply systems have shown strong national differences, the international level will be relevant to the study on CHP as an exogenous factor only. Anyway, the three remaining social science disciplines - economics, political sciences and sociology - already represent a challenging diversity of determinants and driving forces of technology-related change. The situation is further complicated by the necessity to integrate technical elements in the analysis. Obviously, it would be questionable to investigate technological change without having a good grasp and knowledge of the changing object itself. The difficulty will therefore consist of providing an integration without discontenting the individual contributing disciplines.

2.1.4 Multiple models of processes of change - a formal criterion

Different disciplines may be concerned with different types of factors, but the basic structures or models underlying disciplinary theories can nevertheless be similar. Examples of these basic models are structural/macro vs. behavioural/micro models, dynamic vs. static models, formal vs. appreciative models, or descriptive vs. explanatory models. In the following, a slightly different pattern for the categorisation of theories will be introduced which reflects their degree of complexity. In addition to purely descriptive approaches, there are basically three models for an explanation of processes of change, namely

- Linear models which attempt to identify the impact of individual 'factors' on a phenomenon, assuming that these factors are the dominant driving forces. This approach can be found in different disciplines, e.g. for the structure-conduct-performance model in industrial economics ('structural model'), for the role of sociological factors, for firm characteristics like size ('behavioural model'), or - with regard to technology - also for the stage model of invention, innovation and diffusion, reflecting a deterministic science- or technology-push perspective.[19] Approaches building on historical regularities reflect such linear thinking as well.
- Feedback models in which the feedback mechanisms between different phases of the process of technological change are taken into account. Simple examples are single self-reinforcing effects such as those induced by economies of scale and scope or by network externalities. More complicated feedback models interpret technological change as driven by the joint operation of technology-push and demand-pull factors and the interdependence between them.
- Systemic models which build on elements of feedback models but integrate them into a more comprehensive perspective. Basically they enable the analysis of complex real-world phenomena and their interrelatedness. Feedback mechanisms are a cornerstone of this interrelatedness, but also the distinction between different categories of "factors" (e.g. actors, structures, strategies, etc.). A crucial problem consists of the delimitation of the object of investigation, i.e. the definition of the 'inside' and the 'outside' of a system. It is essential to reduce the complexity of the problem to a manageable level. Three main tasks are implied by the systems approach, namely the identification of a reasonable delimitation, the identification of exogenous forces, and the description of the system's internal organisation and dynamics.

However, in feedback or systemic models it can still make sense to speak of phases (though not in the sense of causal steps) in order to describe the aggregate phases of change as observable macro-pattern, while keeping in mind the underlying interdependencies.

---

[19] There are several proposition for how to distinguish phases of technological change. The most well-known and simple one is probably the three step model of invention - innovation - diffusion. For further examples see(Hughes, 1987; Pinch & Bijker, 1987; Welfens, 1988).

2.1.5 Summary and synthesis

Based on the preceding analysis, a matrix of theories can be introduced which builds on two main dimensions:[20]

- Dimension 1: The disciplinary perspective, similar to the six levels identified in Table 2.1 which, in principle, could be refined in more detail.
- Dimension 2: The underlying model for the description and explanation of processes of change (i.e. linear, feedback or systemic), here applied to the interrelationship between technological change and the types of determinants considered (i.e. economic, sociological, political).

In principle, a systems approach would allow linking the different factors and components in a common and open modular framework. Its application would require a very explicit and precise delimitation of the system, and it would provide a comprehensive foundation for policy analysis which can still build on the existing contributions from the different disciplines. This issue will be taken up again later on, i.e. after the discussion of the current approaches.

The disciplines needed to understand the role of political control are in the first instance economics, political sciences and sociology. This does not mean that psychological or management aspects are not important, but only that they are not of central relevance for the perspective of this study. They may still be needed for the definition of concepts of firm or actor behaviour.

In the following section, the presentation and discussion of relevant theoretical approaches will proceed basically along the lines of the matrix suggested above. After a discussion of possibilities and limitations to measure technological change, the focus will be on the different disciplines at a first level and on the models used in these disciplines at a second level. However, as it is not always possible to identify corresponding models in the different disciplines and as individual approaches do not necessarily fit exactly in this ideal pattern, the matrix should be regarded as a general guide rather than as a rigid template. Basically, it will proceeded from simple descriptive and linear models, which might reflect some broad historical regularities, to the different, more sophisticated feedback and systems models. The simpler approaches often induced the search for better alternatives which looked at underlying behavioural or embedding structural determinants. In each discipline, systemic models, encompassing several levels of interdependent determinants, will be the final step. In such models the boundaries between disciplines already become fuzzy, indicating the need for an integration. By using some

---

[20] A possible third (and 'classical') dimension would be to distinguish between structural/macro models which focus on the role of structural settings for specific phenomena, and behavioural/micro models which mainly take the underlying mechanisms and strategies into account. However, this differentiation is already reflected to some extent in the disciplinary perspectives. For example, sociological approaches tend to have a micro view, whereas economic approaches are open to both structural and micro determinants, reflecting the traditional distinction in macro- and microeconomics.

of the approaches reviewed as building blocks, such an integration becomes an important ambition of the present study.

2.2 Descriptive models and the aggregate measurement of technological change, diffusion and innovation

Before being able to analyse the microscopic processes which underlie technological change, it is necessary to identify and measure the macroscopic evolution. This implies basically two questions:

- Which measures and indicators can be used to represent technological change, or patterns of invention, innovation and diffusion?
- How can these macroscopic patterns be described in conceptual or even mathematical terms?

2.2.1 Measurement of innovation and diffusion

Measuring the *diffusion* of a new technology is a fairly straightforward task once the technology itself has been identified and delimited. It can be achieved by counting and accumulating data about its use in terms of adoption and imitation. The *identification*, on the contrary, is far more difficult because technologies are not easy to define. An agreement needs to be found on the identity of a technology or an innovation. In practical terms, this means that technical specifications need to be established before statistical diffusion data can be collected.

Many innovations do not reach this level of detectability - either because they never diffuse on a market or because they just represent improvements of existing technologies. Moreover, technological innovations also comprise 'soft' elements (such as knowledge or organisation) which are difficult to measure. Technologies also change and are improved throughout their diffusion. This makes the separation of innovation and diffusion measurement problematic.

Traditional quantitative measures

A number of quantitative methods are available to measure innovative activity, e.g. R&D-expenditures, bibliometric analysis, patent data and technometrics (Archibugi, 1988; Archibugi, 1992; Grupp, 1987; Grupp, 1994; Sigurdson, 1990; Smith, 1992). Apart from problems of data availability, they are all confronted with severe shortcomings. For example, patent analysis is a reliable indicator only if inventions are in fact registered and protected by this means. Other ways for appropriating the benefits of innovative activities are more common in many industries, such as e.g. the exploitation of first-mover advantages, learning curve effects, secrecy, etc. Technometric analysis relies on the approximation of 'innovativeness' through technical performance data, but this is inappropriate to measure organisational innovations. Bibliometric analysis is only applicable in research areas, where the practice of scientific publications and citations is common, a condition which is often not fulfilled in applied R&D. Finally, R&D-expenditures are not an output measure, but an input measure, neglecting the varying efficiency across R&D activities.

Multi-dimensional representations and morphological analysis

If the intention is to measure innovation at a more disaggregate level, it is necessary to open up the 'black box' of technical systems and identify changes and improvements of their elements. In addition it would be useful to consider also the actual 'service quality' of a technology, i.e. in how far it can fulfil user requirements. A user is less interested in the detailed technical specifications and their improvements, but in the services which a technology delivers. Therefore, it has been suggested to map technological changes through a two-dimensional approach in terms of service and technical characteristics (Saviotti & Trickett, 1992; Saviotti & Metcalfe, 1984).

Foray and Grübler go even further by trying to identify the range of potential technologies in a given area (Foray & Grübler, 1990). They study the main functional units of a complex technology and their technical parameters. Their method of "morphological analysis" has been applied among other examples to technological paths of ferrous casting technology in different countries.[21] This approach helps to overcome the problem of measuring and identifying innovative improvements in differentiated technological systems, but requires an in-depth understanding of the technology in question.

2.2.2 Models for the representation of innovation and diffusion processes

Two basis models for representation of the macroscopic patterns of technological change can be distinguished. Stage models are a simple and straightforward way for differentiating the main phases of technological change while formalised mathematical descriptions can be quite sophisticated and aim to capture essential diffusion mechanisms.

Stage models

The sequential models of invention, innovation and diffusion (or others in a similar vein) have had a wide-ranging impact on policy-oriented research work. The idea that technological change emanates from inventive activities underlies the technology-push approach to technology and innovation policy, which was widespread at least until the late 1970s. Empirical research pointed to the importance of demand factors for inducing innovation, culminating in a similar uni-directional conceptualisation of demand-pull as the main driving force of technological change (Schmookler, 1966). Nowadays, both extremes are regarded as too mechanistic and simple, covering only very few facets of the complex process of technological change. They are too rough approximations to be useful as a foundation

---

21 For the application of this method to transport technology, see (Canarelli, 1994).

for targeted technology and innovation policy, even if they may still be interesting for classification purposes.[22]

Epidemic and psychological models

The key macroscopic features of diffusion processes can be captured by S-shaped curves reflecting both the time-consuming and competitive character of diffusion. In biological terms, this means that a technology competes for limited resources with another species and displaces it in an application niche. These patterns have been captured by a number of formal mathematical models.

Biological growth and substitution models

Models of S-shaped diffusion curves mainly draw on preceding work in population biology and on growth models developed in this context. Apart from their capacity to generate diffusion curves, the rationale which underlies the biological models is very appealing: the diffusion of a species by interaction with its environment results in a growth curve, which is limited by the scarcity of resources.

The Lotka-Volterra differential equation is probably the best known representation of competitive interactions in biological systems (Lotka, 1924; Volterra, 1931). Originally, it only allows for the representation of growth processes but it has been adapted for specific applications in economics.

$$\frac{dN_i}{dt} = a_i N_i - \frac{1}{g_i} \sum_{j=1}^{n} l_{ij} N_i N_j, i = 1, \ldots n$$

N_i = number of species N_i

l_{ij} = 'collision rate' between the species

a_i, g_i = constants

At a first glance the analogies with technology diffusion are striking: the species can be interpreted as different technologies, which diffuse in an economic market environment and are limited by potential demand. Obviously, these analogies remain at a very general level; economic reality is much more complex. In addition, the simple logistic models lack a realistic and differentiated micro foundation. Individual (economic or other) motivations and the processes underlying change can not be tackled, but such a micro-foundation would be essential for policy-oriented analysis. Neither can improvements to which technologies are subjected during diffusion (and which can change considerably their performance) be taken into account in the basic model.

---

22 Similar concepts have been proposed to describe the innovative process at the level of companies. They also started with a linear technology-push approach. Later, demand-oriented elements were added, leading to feedback models like the one proposed by Kline and Rosenberg which considered the phases of research, development, production and marketing (Kline & Rosenberg, 1989). In the 1990s, self-organisational approaches became a focus of interest also for R&D processes at the level of firms.

Economic applications

Several extensions of the original Lotka-Volterra equation have been developed in economics,[23] either as growth or as substitution models.[24] The more complicated the models are and the more parameters the models use, the less can they be linked to the initial biological analogy and therefore recur to an epidemic rationale. Consequently, the greater flexibility of multi-parameter models can only be justified in particular applications. Still, in many cases the standard logistic curve has revealed to be the most robust and reliable approximation.

In order to deal specifically with product innovations, marketing" models have been developed which distinguish between "innovative" and "imitating" adopters who are influenced in their decisions by mass-media on the one hand and word-of-mouth communication on the other (Bass, 1969). Basically these are extensions of the epidemic approach with a more sophisticated underlying rationale, and they have been further refined in recent years (Mahajan et al., 1990).

One of the main deficiencies of the growth and substitution models is that they are limited to binary processes. Marchetti and Nakicenovic developed a model which describes the simultaneous competition between more than two technologies (Marchetti & Nakicenovic, 1979). In the following years, they applied their model to a number of empirical case studies and it proved to be very useful for representing past diffusion processes.[25] The following major assumptions underlie their model:

- The diffusion of a technology can be subdivided into three distinct phases as measured by its market share: logistic growth, non-logistic saturation and logistic decline;
- One technology is in its saturation phase, and its market share is defined as the residual of the sum of the other (two or more) logistic market shares;
- It is the oldest technology which first enters the saturation phase, i.e. a technology more recently introduced cannot overtake an older one.

---

23 For a review of the growth and simple substitution models which have been formulated and extended on the basis of Lotka and Voterra's work, see e.g. (Grübler, 1990). Marketing models dealing with product innovations have been studied in (Mahajan, Muller, & Bass, 1990) and (Mahajan & Peterson, 1985).

24 Growth and substitution models differ in that the latter are just a normalised form of the former for competition between two species. In a growth model the environment is not just given by the one competitor like in a substitution model, and the upper growth limit needs to be defined. Three parameters are required in growth models: The upper growth limit or saturation level, the time when the curve reaches its point of inflection, and a characteristic growth time in relation to the saturation level (for example from 10 - 90% of saturation). A symmetric model is given when a saturation of 50 % is reached at the inflection point.

25 See the numerous publications of the International Institute for Applied Systems Analysis (IIASA), e.g. (Grübler & Nakicenovic, 1987), (Nakicenovic, 1986) and (Marchetti & Nakicenovic, 1979).

These models have also been applied prospectively, e.g. to the fields of transport and energy supply,[26] but their predictive power is limited by a number of severe shortcomings:

- They can only capture successful innovations, and can not take failures into account;
- They do not provide any underlying mechanisms and interdependencies which would allow drawing operational conclusions for diffusion policies.

A further shortcoming of most economic diffusion models consists of the disregard of technological improvements during diffusion by which technological performance can change considerably. While this interdependence of innovation and diffusion had already been recognised conceptually in the early Eighties, corresponding models have been developed only recently (Englmann, 1989).

2.2.3 Summary

The growth- or substitution-based diffusion models are useful for the rough representation of the S-shaped macro-patterns of successful innovations, but apart from the biological analogies to epidemic growth only a very limited rationale is provided which could explain how these processes come about. Complementary causal relationships at the micro-level have been investigated to check the validity of these growth models (Rogers, 1983). It has even been possible to derive logistic diffusion curves based on economic models which take certain micro-level factors into account.[27]

For forecasting analyses the models should be applied with much care due to their inherent limitations: pro-innovation bias, no qualitative changes and improvements of the technologies possible, no consideration of transformations of the technology's environment, and often an insufficient data quality.

From the perspective of political control, models which aim at describing the macro patterns of innovation diffusion are less useful because they cannot provide any point of access for the analysis of the effects of policy measures on technological changes.

Regarding the measurement of innovation and diffusion, the different methods available (R&D expenditures, patent analysis, etc.) should be treated with much care because they allow generating meaningful findings only under very specific circumstances. Multidimensional representations of technological changes appear

---

26 See for example the forecasting work in the field of energy in the late Seventies (Marchetti & Nakicenovic, 1979). Based on the extrapolation of historical regularities, nuclear power was predicted as the next major wave of energy supply. These predictions were not confirmed due to the nuclear crisis in the last fifteen years, showing how difficult it is to extrapolate from the past. More recent contributions (Nakicenovic & al., 1993) draw more careful conclusions.

27 For example, based on a lognormal distribution of firm size, a logistic diffusion curve was derived by Davies (Davies, 1979, as described in Grübler, 1990, p. 40).

to be very helpful to identify innovative elements within complex technological systems. They can be used to study systematically future opportunities of individual technological systems, but they require a good knowledge of the technologies in question. In this sense, they help to explain the diffusion of a technology in terms of its components' characteristics.

2.3 Economic approaches to technological change

Economic approaches attempt to explain technological change on the basis of economic forces, and highlight its repercussions on the economy. Since Solow's path-breaking contribution (Solow, 1957), technological change has been recognised as an essential factor for economic development and growth. In the last thirty years the relationships between technological change and economic factors have been extensively investigated. As expressed in numerous modelling attempts, technological change and especially innovation are now regarded as strongly determined and constrained by economic incentives and competitive pressures. They range from models to identify correlation between different factors (e.g. by means of econometrics[28]), to the integration of technological aspects into established and new schools of thought. Especially evolutionary economics has severely criticised neo-classical economics and of its inadequate ways of dealing with technological change and dynamics. In the following, several of the traditional economic approaches will be reviewed, starting with linear and feedback models. Due to its substantial role in introducing technological aspects into economic analysis, the emphasis will be put on concepts and findings from evolutionary economics. A crucial problem of traditional economics research on innovation diffusion has always been whether its findings could be generalised across industries and technologies. The traditional neglect of technological issues in economic research made it difficult to tackle this basic problem. Attempts were made to classify technologies in order to develop hypotheses which refer to specific classes only (Pavitt, 1984), but it was essentially due to evolutionary economics that technological aspects were brought back into the focus of attention of economics research.

2.3.1 Simple models - linear thinking

2.3.1.1 Economic cycles and historical regularities

The simplest type of models tries to establish causal relationships between certain parameters and the development of different phases of technological change. Based on historical records, regularities have been postulated for major phases of innovation and diffusion. The most important debates along such lines were con-

---

28 Econometrics is not considered in much detail here, because it is only a tool for identifying correlations, and does not have a theoretical and conceptual value in itself.

cerned with life-cycles and long-waves (either for industries or for entire economies), the role of large vs. small companies (Schumpeter hypothesis) and the role of demand pull vs. technology push factors.

The Schumpeter hypothesis

The probably most influential debate during the 1960s was concerned with the question whether firm size is conducive or detrimental to innovation. This debate goes back to the work of Schumpeter on the respective importance of the role of the entrepreneur and the economies of scale in R&D which can be achieved by large firms (Schumpeter, 1964; Schumpter, 1946). Later on, his arguments were simplified to a hypothesis about the general advantages of large firms with regard to innovation, and thus of concentrated industry and market structures. The empirical results on this issue were highly contradictory, and underlined first of all the limitations of highly simplified hypotheses about complex processes. More differentiated arguments have been proposed, taking into account differences between industries and technologies. They show that the Schumpeter-hypothesis seems to hold for industries like chemicals, which require high capital investment in research, but not necessarily for several other industries.

Life-cycle models

Based also on regular historical regularities, life-cycle models relate innovative performance to life-cycles of firms, industries or technologies. A technological life-cycle is paralleled by the establishment of a "dominant design configurations" as a result of testing, learning, and the establishment of widely shared mental frameworks and paradigms (Henderson & Clark, 1990; Utterback & Suarez, 1993). A straightforward reason for the establishment of a dominant design could be its 'superiority', but research in the sociology of technology has shown that beliefs and interests can have a pervasive influence on what is expected to be superior.

The concept of dominant designs refers to one type of life-cycles, namely product life-cycles. New technologies are regarded to pass different phases, from infant state via growth to maturation and stagnation. While the early stages are characterised by the establishment of the dominant design through product innovation, the later phases exhibit mainly incremental process improvements (Abernathy & Utterback, 1978).

These early life-cycle approaches have been refined in recent years (Tushman & Rosenkopf, 1992; Utterback, 1994; Utterback & Suarez, 1993), but nevertheless struggle with the problem of general applicability. In this regard, typologies of innovations allow differentiating between different types of life-cycles models.[29] In addition, technology life-cycles can be linked to industrial life cycles because

---

29 See e.g. (Henderson & Clark, 1990) who distinguishes between innovations at the component and the architectural level of technologies, each having either a reinforcing or an overturning impact on the dominant design. This model implies different types of changes needed for the firms in a changing industry.

the early introduction phase of a novel competing design offers opportunities for a start-up of new firms, whereas in the maturation phase economies of scale and the establishment of standards lead to a drop-out of many firms - until a new cycle starts.[30]

Long-waves

The debate on long waves as a major type of historical regularities in economic development has been introduced in the 1930s. With regard to technology, the main hypothesis consists of a correlation of these long waves of economic development with the emergence of some key technological paradigms. Building on earlier work of Kondratieff who formulated first hypotheses about the duration of these waves, several refinements have been proposed in the meantime.[31] With regard to the role of technology, recent work by Freeman and Perez aims to relate major changes of technological paradigms to waves of economic development (Freeman & Perez, 1988). This approach will be discussed later on.

Demand-pull vs. technology-push

Building on a linear stage model of technological change at the level of firms, the concepts of technology-push and of demand pull - at least in their pure forms - regard technological change as the result of forces operating on the scientific/R&D side or on the market-side respectively. These impulses would then induce innovative results in a quasi-automatic way. The major difference consists of the impact of technology-push and market-pull forces on the incremental or radical character of innovations. Market-pull forces tend to induce rather incremental changes, while technology-push appears to be more likely to initiate radical changes. Associated to this, the balance between market and scientific forces differs among industries and technological areas. Obviously, very different policy conclusions can be drawn from these conceptualisations with regard to the efficiency of technology policy and the stimulation of demand, e.g. by means of public procurement. This dualism led to attempts to reconcile them in combined models which take into account feedbacks between the different phases of technological change.[32]

---

30 For industry life cycles, the emergence of a new technology can be correlated with the number of firms (industry structure), c.f. (Utterback & Suarez, 1993). Technology life-cycles are based on the sequence of product and process innovation phases, see (Utterback, 1994) and (Utterback, 1979). In (Gort & Klepper, 1982) technology life-cycle issues are combined with entry and exit patterns of firms in an industry.

31 For more details about the long waves debate see especially (Vasko, 1987).

32 At the company-level, this has been conceptualised as a chain-linked model (Kline & Rosenberg, 1989). For the industrial level, Walsh has indicated for the case of the chemical industry that the role of demand and supply factors may vary over the industry's life cycle (Walsh, 1984).

2.3.1.2 Early micro-economic models

One of the first attempts to go beyond the simple description of macro-correlations and to provide a micro-founded economic explanation of diffusion was made by Mansfield (Mansfield, 1961; Mansfield, 1968). Although his early model is not particularly sophisticated, it has been very influential for later developments in diffusion theory. Many further improvements were made by making reference to the shortcomings in Mansfield's work.[33] A key element of Mansfield's first models (Mansfield, 1961) is the propagation of information about a new technology. His equation reflects a learning process based on accumulated knowledge, implying an epidemic propagation mechanism which leads to a logistic diffusion curve. More precisely, four key variables determine the decision of a firm to introduce a new process innovation:

- The risk of introducing a new technology, expressed by the size of investment.
- The expected profitability of the innovation as compared to alternative forms of investment.
- The number of potential users as compared to actual users, reflecting the importance of learning mechanism the knowledge already accumulated ('epidemic effect').
- Other unspecified variables.

Investment risk and profitability expectations are in this model related to the adoption and diffusion speed, a mechanism which Stoneman (Stoneman, 1983) has called "psychological", because it finally draws on sociological and psychological arguments of adoption behaviour rather than on economic calculus.[34] Empirical tests supported the correlation between profitability and diffusion, and between size of investment and diffusion. However, it should be taken into account that this was a purely micro-economic approach, structural variables have not been considered at all.

In the following years the model was extended in order to integrate several other intervening factors. This was necessary because the initial model proved to be insufficient to cope with other than the initial application cases. In one of Mansfield's later publications (Mansfield, 1968), seven main determinants of the diffusion rate are mentioned by which also product innovations and intra-firm diffusion could be tackled. They can be grouped in technology-related, adopter-related and structural factors:

1. Technology related factors

- Economic advantage of the new product/process as compared to the old one.

---

33 The progress of diffusion theory along the lines of critique of Mansfield's approach has been reviewed by Grübler (Grübler, 1990).

34 This psychological argument underlies different classification schemes of firms with regard to their eagerness to use new technologies, ranging from early adopters to laggards (Rogers, 1983). See also Chapter 2.4.1 .

2. Adopter-related factors

- Risk and uncertainty associated with the use of an innovation;
- Efforts made to try out and introduce an innovation;
- Internal learning effects and the extent to which they reduce the initial uncertainty;
- Skill-level and the scientific quality of an industry.

3. Structural factors

- Market and industry structure;
- Spin-off of external learning effects outside the industry.

Still, a number of shortcomings remained, like for example the separation of innovation and adoption, the application of a simple stage model of technological change, the static character of the technology's environment, and the neglect of improvements on the technology supply side.[35]

2.3.2 Explanatory models of the feedback and systemic type - neo-classical vs. evolutionary theorising

While the simple models aimed at identifying relationships between patterns of innovation diffusion and other macroeconomic factors, explanatory models build on a clearly specified micro-foundation, i.e. on a conceptual or formal model of economic behaviour and interaction. In this sense, Mansfield's models were just a first step towards a deeper understanding of innovation diffusion. In response to new empirical findings, it was also increasingly recognised that technological change is a very complex phenomenon which required to introduce more sophisticated conceptualisations of technological elements in economic models.[36] The debate between neo-classical and evolutionary economics builds essentially on these two dimensions: the different conceptualisation of micro-foundations on the one hand, and the degree to which technological aspects need to be represented in a differentiated way in models of techno-economic change.

2.3.2.1 Neo-classical models: the mechanistic analogy

According to Dosi (Dosi, 1988b), the basic features of the neo-classical approach to techno-economic change can be summarised in ten key points - even if it should

---

35 For further points of critique see (Coombs, Savioti, & Walsh, 1987, pp. 120-134) and (Grübler, 1990, Chapter 2).

36 See also the chain-linked model by Kline and Rosenberg (Kline & Rosenberg, 1989). It is comparatively narrow in its focus on firm-level feedbacks between different phases of technological change: invention/analytic design, detailed design/testing, production, marketing activities, market potentials and knowledge-generating research.

be admitted that since the early work of Solow (Solow, 1957) a number of improvements and extensions have been introduced:

- Scarcity is a key constraint for economic agents;
- Economic agents are regarded to decide in a rational and maximising way, leading to optimised behaviour;
- This leads to optimised and intentional behaviour; unintentional outcomes are not foreseen;
- Uncertainty can be reduced to a lack of information, still allowing for maximisation on the basis of probability distributions; there is no essential uncertainty in the sense of Knight, but only what could be called calculable risk;
- Agents are identical except of their preferences and endowments;
- The state in which an economy is or at least to which it converges is in general defined as an equilibrium, and therefore independent of preceding paths, behaviours, decisions and history;
- Non-economic institutions do not play a role for shaping economic behaviour and performance;
- Equilibrium states are inherently aspired in the economic system;
- Technology is basically a freely accessible and available good;
- Technical change can be captured in the concept of a shifting production function, relating output of production with the traditional production factors capital and labour (and potentially other inputs), giving rise to a limitation of technical change to process innovations.

In the course of the last two decades the standard neo-classical model of techno-economic change has been extended in several respects, relaxing some of the most restrictive assumptions and addressing also some of the shortcomings of Mansfield's model:

- Product innovations have been considered in an attempt to reduce the process innovation bias inherent to the production function approach (Bass, 1969; Mahajan et al., 1990).
- The time lag has been explained through learning and vintage effects (Cabe, 1991).
- Related to this is the problem of heterogeneity of firms, e.g. in terms of size, acceptable pay-off, or sensitivity to innovative impulses (David, 1975; Davies, 1979; Stoneman, 1983).
- Demand and supply elements have been joined in an integrated model in order to overcome the demand bias of neo-classical approaches (Cameron & Metcalfe, 1987; Metcalfe, 1981). Hence, both adoption and imitation as the two components of diffusion processes can be addressed.

- Disequilibrium and irreversibility are considered in order to be able to take e.g. the timing of innovation and adoption decisions into account;[37] however, the assumption of optimising behaviour is maintained (David, 1985).
- The importance of institutional settings for determining technological choices has also been addressed (Hodgson, 1988).
- Technological improvements during the diffusion process are considered (Metcalfe, 1981).

However, the basic assumption of maximising behaviour is maintained in these models. The main points of critique of neo-classical approaches focus on the unrealistic assumptions underlying the basic models, and - in spite of these unrealistic assumptions - their normative application. A circular argumentation is inherent to the *use* of these models: In order to be tractable, these models have to be simplified and abstracted from reality. Consequently, normative conclusions on desirable equilibrium states, promising socially optimal outcomes, etc. are valid only within these models. Unless it can be argued that the underlying assumptions are valid in specific empirical situations, it does not make any sense to use these normative conclusions as orientations for real-world economic systems. Unfortunately, this false inference is often made.[38] The conclusions drawn from these unrealistic models are not just regarded as analytical exercises, but also as normative guidelines for policy making. They are applied to the solution of applied problems although the underlying assumptions and premises do not hold.

A consequence of these shortcomings is a complete mis-specification of the problems to be addressed in a policy-oriented analysis. Empirically important aspects like e.g. institutional issues, fundamental uncertainty are excluded ex ante from the analysis, as well as many crucial technological determinants and constraints. In other words, although several important aspects have been addressed in recent neo-classical models, the treatment of technological change remains rather sketchy. The applicability of the neo-classical framework for analyses of complex problem areas such as technological change, innovation and public policies in thus severely limited.[39]

---

37 See e.g. Silverberg (Silverberg, 1991, p. 210) who states that "although the influence of profitability on adoption and diffusion rates has been extensively studied since Mansfield, the inverse - the effect of the timing of adoption on profitability, relative competitiveness and market structure - has surprisingly been largely neglected."

38 C.f. (Freeman, 1988, p.3): "(N)eo-classical economics elaborated and refined quantitative equilibrium analysis and mathematical models, which, although useful as a modelling exercise on highly restrictive assumptions, neglected some of the crucial elements involved in the long-term behaviour of the system. They therefore appeared to non-economists and to other social scientists to be concerned with the endless elaboration and refinement of assumptions which lacked both realism in relation to certain fundamental features of the system's behaviour and rigorous falsifiability of the predictions derived from the models."

39 For example, the neo-classical approach underlies the majority of studies and forecasts in the field of energy, where for a long time the correlation between economic growth and energy demand represented a key relationship. Innovative issues did not play an important role, apart from forecasts on larger units and higher plant efficien-

2.3.2.2 Evolutionary models: the biological analogy

Evolutionary ideas are not new in economics, the biological analogy has already been regarded as the "Mecca of economists" by Marshall, but he also admitted that it was too complex and difficult to be applied; consequently his contributions are known mainly in the mechanistic line of reasoning. Schumpeter's processes of "creative destruction" in which the entrepreneur is assigned a central role, has been the starting point for a revival of evolutionary ideas in the late 1970s. This is first of all associated with the names of Richard Nelson and Sydney Winter. They even adopted the label of "neo-Schumpeterian" economics for their work (Nelson & Winter, 1982, p. 39).

However, it is difficult to provide a good definition of what evolutionary economics really is. Four different definitions have been provided by Witt, Dosi, Ayres and Nelson,[40] which all have in common that they are sufficiently broad to encompass quite disparate lines of theorising under a common roof. They can be synthesised by the following four key characteristics of evolutionary theory:

- The theory formulates hypotheses about the temporal behaviour of systems and the emergence, selection and spread of novelties in them.

---

cies. The introduction of new technologies like renewable energies was only relevant with regard to their costs and environmental impacts.

40 Witt has proposed a general definition which is independent of biological analogies and which can be summarised by three criteria (Witt, 1987, translation by MW): "The theory is dynamic, i.e. it is concerned with a development which lapses over time, 2) The theory is based on the concept of irreversible time, i.e. it refers to developments which show an irreversible direction in time, 3) The theory explains, how it comes to novelties in the developments studied and which general impacts they have, i.e. it formulates hypotheses about the temporal behaviour of systems in which novelties occur and spread."
In Dosi's definition (Dosi, 1991b), the following characteristics can be singled out: 1. The world is full of opportunities of which only a very small share is exploited, 2. Agents are less maximising rather than varying in their behavioural norms, decision biases, inertial responses to environmental information, jointly with learning, creativity and discovery, 3. Interactions between economic agents are typically disequilibrium ones, and that very fact establishes a sort of "externality" by which individual participants learn, are selected , grow or die, history matters, 4. Institutional rules/learning/discovery and ex-post selection drive co-ordination and change, evolutionary means that the variety lacks of an ex-ante consistency, 5. History and institutions being integral parts of economic analysis, this analysis becomes less elegant and more context specific.
Ayres (Ayres, 1991, p. 256) stresses the non-equilibrium, non-linear, self-organisational and non-homothetic character of evolutionary models.
Finally, a fourth definition of evolutionary models has been given by Nelson (Nelson, 1995, p. 54), who points to three main elements: 1. Explanation of the movement of something over time, or why that something is what it is at a moment in time in terms of how it got there. 2. The analysis involves both random elements which generate or renew some variation in the variable in question, and mechanisms that systematically winnow on extant variation. 3. There are inertial forces that provide continuity of what survives the winnowing.

- It is a dynamic theory, allowing for temporally irreversible and history-dependent processes which are affected by random events, inertia and reinforcements.
- It is based on learning agents who interact in disequilibrium processes with each other and their environment on the basis of behavioural routines jointly with elements of creativity.
- Institutions and the specific context constrain development and need to be taken into account in the analysis.

This very broad definition sets the frame for more precise models of innovation, diffusion and structural change. It also reflects the criticism of the neo-classical way of modelling economic problems, and the critique of neo-classical economics has taken a lot of space especially in the early evolutionary contributions. A further characteristic is the distinction between macroscopic processes of change and the underlying microscopic mechanisms.

Inside evolutionary theory, two main strands can be further distinguished, namely formal models on the one hand, and "appreciative models" on the other.[41] Although appreciative models are usually not formulated in a mathematical language, they also represent a rigorous way of modelling by being precise in their definitions, delimitiations and conceptualisations, but allowing for a rich and realistic representation of complex processes which can not (yet) be tackled by formal approaches.

2.3.3 Main elements of evolutionary theories of techno-economic change

2.3.3.1 Basic assumptions and stylised empirical facts

The basic assumptions underlying evolutionary theory differ significantly from those of neo-classical theory. They are based on the recognition that the empirical findings of innovation and diffusion research in economics and economic history can not be reconciled with the prevailing neo-classical approaches. These empirical findings can be summarised in a number of key patterns ("stylised facts") which are regarded to be characteristic of techno-economic change and should therefore be taken into account in evolutionary modelling attempts (Arthur, 1989; Dosi, 1988c; Dosi, 1991a; Grübler, 1990; Silverberg et al., 1988). They will be presented in three main blocks, namely :

- The properties of technology,
- The behaviour of economic agents,
- Incentives, structures and mechanisms in the socio-economic environment.

---

41 Nelson and Winter introduced this expression, which reflects a rigid formulation of theory but does not involve mathematical formalisation (Nelson, 1987, pp. 19-20).

These stylised empirical facts on processes of innovation diffusion could also be regarded as a first step towards systematic appreciative theorising, i.e. the formulation of concepts which then allow analysing technological change by using a sophisticated vocabulary. A number of the more encompassing conceptual approaches which attempt to go a step further than these stylised facts of innovation and diffusion processes will be described later on.

Properties of technology[42]

Complexity and diversity of technologies

Technology is not a universal and free good with broadly uniform characteristics, but it is highly differentiated and heterogeneous with regard to both technical and economic features. In order to facilitate the analysis of different technologies, classifications of the main types of innovations can be useful. This indicates that technical issues can not be excluded from economic analysis.

Fundamental uncertainties and ex-ante inconsistency of technological choices

New technologies are inherently uncertain in both technological and economic terms, implying that their future evolution is open and unpredictable, and can thus not be captured by probabilistic approaches. Consequently, decisions about new technologies are subjected to qualitative uncertainty and the risk of inconsistency with future development.

Cumulativeness and the importance of learning for building up a knowledge-base

Technology is characterised by varying degrees of cumulativeness, i.e. present improvements build on preceding experience and achievements. Learning processes play a major role in this respect. Consequently, innovation and diffusion cannot be separated, they are intertwined and simultaneous processes. Based on learning, technologies which have been initially perceived as inferior can overtake competing options in the course of time.

Path-dependence, irreversibility and self-reinforcing effects

Irreversibilities occur in the course of technological progress due to several forms of increasing returns of adoption.[43] Technological change is therefore path-dependent; technological trajectories are an ex-post representation of this path-dependence.

---

[42] The following discussion draws on a number of sources which can not be discussed here individually. See e.g. (Dosi, 1982; Dosi, 1988c; Nelson, 1987; Nelson & Winter, 1982; Sahal, 1985).

[43] According to Arthur (Arthur, 1988a; Arthur, 1988b; Arthur, 1989), the main mechanisms for increasing returns to adoption are learning by using, network externalities, scale economies in production, informational increasing returns and technological interrelatedness.

The importance of scientific foundations and opportunities

While in former times new technological opportunities were created on the basis of experience, an important characteristic of recent technological development is its growing reliance on scientific knowledge. However, the degree of this science-based character may differ considerably among industries, as well as the degree of technological opportunity inherent to them.

Tacitness and embodied character of technology

In spite of this scientific foundation - which implies better documentation, understanding and reproducibility - technology is not an easily accessible free and public good, but is characterised by different forms of tacitness. Only parts of the technological information can be transferred easily as blueprints, others are embodied in the skills of people, in organisational practices and in the artefacts themselves.

Varying degrees of appropriability and costs

Technologies differ in the extent to which they allow appropriating the economic benefits derived from their application. From a different perspective, this means that the transfer of technologies with a large appropriability potential is usually very costly.

Specificity of technological knowledge

Tacitness and cumulativeness of technological information imply that technological knowledge is at least partly specific to individual firms.

Behaviour of economic agents

Strong uncertainty and diversity of behaviour

Due to technological and economic uncertainty, the behaviour of economic agents is diverse. They draw on different knowledge bases, therefore perceiving technological opportunities differently. One should even talk of "strong uncertainty" (Dosi, 1988c, p. 1134) because agents do not just lack complete information on a given number of possibilities but they do not even know which the different options exist.

Knowledge-bases and learning processes

Similar to technology, the knowledge base of actors depends on continuous learning processes, either through internal R&D or by acquisition of additional information from outside. Jointly with the specificity of the technological knowledge base, this implies also the specificity of this newly acquired knowledge.

The importance of expectations

Expectations about future technological developments are essential for the decision-making process of actors, possibly even more than their actual knowledge (vanLente, 1993).

Rule-guided behaviour and profitability orientation

Due to uncertainty it is unrealistic to assume optimising economic agents and perfect information. Broad rules of thumb and institutionalised patterns of behaviour guide decision-making processes, even if they may be profit-oriented.

Incentives, structures and mechanisms in the environment of firms

The importance of institutions

Institutions outside the narrowly defined economic system such as public agencies, universities and research centres as well as socio-cultural patterns of behaviour, influence significantly the process of technological and economic change.

Feedback processes and interrelatedness

The causal influence of some independent on other dependent variables is replaced by the notion of feedbacks between among the elements of a system. This is particularly important for the analysis of long-term developments and self-reinforcing mechanisms for which minor factors can be crucial.

Endogenising market structures and prices

Market shares and economic structures affect innovation and diffusion processes, but the reverse influence needs to be considered as well. This implies that it is necessary to endogenise factors in the analysis which have been regarded as exogenous in traditional analyses.

Interrelatedness of innovation and diffusion

The endogenisation of feedbacks implies that innovation and diffusion can not be separated because the interactions and mutual reinforcement between supply side (innovation, imitation) and demand side (adoption) are interconnected, e.g. by mechanisms of learning.

Dynamics of novelty generation and spread

Techno-economic change is regarded as a continuous process of change, i.e. a process in disequilibrium. The generation of novelties and the learning processes needed to use them successfully induce structural changes in the techno-economic environment. These changes, in turn, exert selection pressures on the firms and the technologies they apply. Adaptation to changing environmental conditions is therefore required from the firms.

Time-consuming and historical character of technological change

Given the above mentioned technological, behavioural and environmental characteristics, it is obvious that technological change is a dynamic and non-linear process. The interrelatedness with institutional changes, the cumulative aspects of technological change and the emergence of large technological systems make it an irreversible, historical and path-dependent process. In such a framework, the timing of decisions is an important determinant to be taken into account.

2.3.3.2 *Appreciative models of techno-economic change*

Building on these stylised facts, some important appreciative models will be presented in the following section. They address the most important of these facts at a conceptual level. Emphasis is given to those approaches which will be used in Chapter 3 to formulate the conceptual framework of this study.

The basic model of Nelson and Winter

The pioneering work of evolutionary economics is without doubt Nelson and Winters' "Evolutionary theory of economic change" (Nelson & Winter, 1982). It argues for behavioural approach to firm decision-making, which itself is based on a corresponding model of individual behaviour. "Routines" as contrasted to optimising behaviour of the neo-classical agent are seen as the normal operating mode of decision making in firms. Even the behaviour which is oriented towards the monitoring and introduction of novelties is guided by these "rules of thumb". Nelson and Winter distinguish three types of rules, namely operating characteristics for daily decisions, capital stock variations for structural decisions, and search routines to guide innovative activities.

Optimised rational choices are regarded to be impossible due to technological and economic uncertainties, and the limitations of information access and processing. Consequently, innovation and adoption decisions are not triggered by optimality rationales but by "satisficing" rationales, i.e. by vaguely defined criteria of performance and success. Once the perceived performance scores below a certain threshold level, search routines for better solutions are triggered. Moreover, routine and satisficing behaviour imply behavioural diversity among firms.

Biological analogies in terms of variation, selection and replication are drawn in their evolutionary model. Variation in technological terms (innovation) is not a purely random process but affected by a firm's existing routines and technologies in a path-dependent way. Basically, there are two stages of selection, one at the firm-level (innovation decision) and another at the market level (market success). Replication refers to firm growth as well as to adoption and imitation of technologies. The success of an innovation is then reflected in the growth of the market share of the carrying firm. However, it should be noted that their theory is not a full analogy to biological evolution, but rather a special case of general evolutionary theory. For example, in contrast to biological models, changes of routines are based on conscious individual decisions, not on random events.

The simulation studies which Nelson and Winter carried out focus on the evolution of market structure and on certain aspects of the Schumpeter-hypothesis.[44] They show that it is possible to carry out empirically meaningful calculations on the basis of evolutionary models which until then had been regarded as too complex.

---

44 For a detailed review of Nelson and Winter's models, see (Andersen, 1994, Ch. 4).

Trajectories, paradigms and mental frameworks: A macro representation of technological content

In the Eighties several appreciative and formal models were developed which dealt with the properties of technological change. One of the key problems behind these efforts was to understand why some technologies spread and advance, while others disappear or are not even considered at all. This work has been influenced by approaches in the sociology of science and technology (see also Section 2.4), where this "unevenness of the scientific frontier" had already been recognised earlier (Kuhn, 1970): Scientific research explores some pathways only and some approaches happen to dominate scientific thinking, even if their quality is already heavily questioned. The question for both science and technology therefore is, why certain opportunities are selected for further exploration, others not.

Dosi (Dosi, 1982), Sahal (Sahal, 1985), Nelson/Winter (Nelson & Winter, 1982) and other authors proposed similar concept to cope with these characteristics of technological development. Although their concepts carry different labels,[45] they stress similar ideas. They suppose the existence of a multitude of relatively independent potential technological paths. Only a few of them are further pursued, depending on the prevailing dominant cognitive framework or paradigm. The decision for one of the alternatives is guided by certain scientific and engineering convictions, shared tacitly by the scientific and technological communities. In the case of technology, these paradigms are strengthened by embodied or disembodied breakthroughs, and reinforced by structural interdependencies.

Probably the most influential of these approaches is Dosi's notion of "technological trajectories and technological paradigms" (Dosi, 1982; Dosi, 1988c). The approach is quite 'sociological' because it attempts to build upon the concepts of knowledge and the role of communities of practitioners as the main factors shaping the emergence and establishment of a general paradigm and of specific trajectories. His conceptual understanding of paradigms explicitly builds upon Kuhn's work on scientific paradigms. A technological paradigm is hence "a 'model' and a 'pattern' of solution of selected technological problems, based on selected principles derived from natural sciences and on selected material technologies" (Dosi, 1982, p. 152). However, the analogy with scientific paradigms is limited by the role of market forces for technological success, and by the fact that scientific paradigms need not necessarily be materialised in the form of working artefacts or "technological guideposts" (Sahal, 1985). "Technological trajectories" are then those potential paths of technological change which can be identified and which are believed to be feasible in agreement with the guiding principles of the technological paradigm.

Underlying the wider spread of a specific technology is then its compatibility with the interests on the technology supply and user side. Dosi (Dosi, 1988a) has proposed technological opportunity, market demand and appropriability conditions as

---

45 Just to mention some of them, see "technological trajectories and technological paradigms" (Dosi, 1982), "natural trajectories" (Nelson & Winter, 1982), "technological guideposts" (Sahal, 1985) and - at the level of specific technologies - "design configurations and dominant designs" (Henderson & Clark, 1990).

the key determinants of innovative behaviour on the supply side. It can be complemented for the user side by the right price and quality of an innovation, by an easy transfer of the relevant knowledge and information, and by a comparatively low risk and uncertainty associated with the technology (Kemp, Olsthoorn, Oosterhuis, & Verbruggen, 1991).

Basically, the key novelties of Dosi's approach are firstly the recognition that specific technological forces exist which determine the future direction of technological change, going hence far beyond the production function approach which dominated traditional economic analysis, and secondly the consideration of guiding cognitive patterns complementing or even replacing the standard economic maximisation assumption.

Information, knowledge and learning: A micro representation of technological content

In order to shape mental frameworks and to acquire the capabilities to use and develop technologies in line with a new paradigm, it is necessary to build up knowledge and skills related to this paradigm. Furthermore, information is needed to reduce at least some elements of uncertainty. Learning becomes a crucial element for technological change, as well as the processes by which firms become familiar with new developments. Clark and Juma (Clark, 1991; Clark & Juma, 1987) have introduced a very useful conceptualisation of this link. In contrast to the neo-classical understanding of information, they distinguish between "information" on the one hand and "knowledge" on the other. The transmission of information takes place - in abstract terms - in different steps involving processes of reception, transmission, and coding/decoding. Different types of communication channels are used in this process. While information is understood as "a dimensionless energetic category whose production and movement throughout the economic systems is pervasive and is accomplished in a variety of different ways, (..) it only becomes 'knowledge' having economic value to the extent that a particular context allows it to do so" (Clark & Juma, 1987, p. 95). Hence, knowledge is *useful* information in the sense that it is available and applicable by a company for economic purposes at a given time and place.

The research infrastructure is the location where most scientific and technological knowledge is generated. However, for companies, that want to exploit the potential of this knowledge it represents initially just information without any economic meaning to them. To describe the transformation of information into knowledge, Clark and Juma propose an "interactive model" which merges their concepts of information and knowledge with that of scientific and technological paradigms. Consequently, bottlenecks for the propagation of information can be found on both the user- and supply-side of technology.

Networks of innovators and user-producer relationships

The quality of user-supplier relationships in a specific technology field or industry are of major importance for the transfer of both tacit and formalised knowledge. The increasing importance of scientific and technological information for innovation and diffusion processes has even reinforced the role played by these relationships. The concept of networks has been used in economics as well as in sociology to describe and analyse them. Basically, networks provide an intermediate form of organisation between markets and hierarchies, involving both formal and informal arrangements (Freeman, 1991; Hippel, 1988) (Frances, Levacic, Mitchell, & Thompson, 1991). Typically, they comprise arrangements such as joint ventures, joint R&D, sub-contracting, publicly sponsored research programmes, and licensing agreements, but also personal contacts between individuals.[46] Network analysis can help to reveal the structure and membership of such arrangements.

Some rather general conclusions have been derived from these analyses. For example, successful innovation performance seems to depend on being well connected to both formal and informal networks. This implies that network-type relationships ought to be promoted. The concepts of networks also plays a key role in the literature on systems of innovation (Lundvall, 1992b).

Technological systems and technological regimes

It is not clear who first introduced the notion of technological systems to economics, but clearly both Rosenberg (Rosenberg, 1982) and Nelson/Winter (Nelson & Winter, 1982) were among the first to use it. The concept of technological systems acknowledges the heterogeneous and differentiated character of most technologies. They should not be regarded as "black boxes", but as composed of different interdependent subsystems, requiring complex organisational structures of companies, industries and economies to co-ordinate their operation.

Technological paradigms have been very useful to capture micro-level and cognitive factors as determinants of the directions of technological change. In other words, the attitudes and mental frameworks of scientists and engineers are taken into account. However, structural or macroscopic constraints as imposed e.g. by the interdependencies in a technological system are not considered. This need for a broader concept which comprises both micro-and macro-level elements has already been recognised by Nelson and Winter's discussion of "technological regimes" (Nelson & Winter, 1982). Along these lines, Georghiou et al. define a technology regime similarly as a "a set of design parameters which embody the principles which will generate both the physical configuration of the product and the process and materials from which it is to be constructed. The basic design parameters are the heart of the technological regime, and they constitute a frame-

---

46 Imai and Baba (Imai & Baba, 1991) developed a categorisation of networks which distinguishes ten different categories: 1. Joint ventures and Research Corporations, 2. Joint R&D agreements, 3. Technology exchange agreements, 4. Direct investment (minority holdings) motivated by technology factors, 5. Licensing and second-sourcing agreements, 6. Sub-contracting, production-sharing ad supplier networks, 7. Research Associations, 8. Government-sponsored joint research programmes, 9. Computerised data banks and value-added networks for technical and scientific interchange, 10. Other networks, including informal networks. Hagedoorn and Schakenraad (Hagedoorn & Schakenraad, 1990) distinguish six different modes of technology co-operation, comparable to networks, which cover similar types of relationships.

work of knowledge which is shared by the firms in industry (Georghiou & al., 1986, p. 32)." Still, this definition is very much geared towards design principles and practices. A broader suggestions has been made recently by Kemp et al. (Kemp, Miles, Smith, & al., 1994):

> *"A 'technological regime' refers to the whole complex of scientific knowledge, engineering practices, production process technologies, product characteristics, skills and procedures, and institutions and infrastructures which make up the totality of a technology."*

Such an encompassing and differentiated conceptualisation of technology as a system which needs to be embedded in a technological regime in order to be operational has wide-ranging implications. It helps to understand why it is so difficult to achieve the diffusion of an innovation: In addition to the cognitive obstacles, wide-ranging relationships have to 'match' the novelty before it can be successfully applied.

In other words, the systems perspective opens up the possibility to consider and analyse a large number of effects which contribute to the establishment and the stabilisation of technological developments. Reinforcing mechanisms between different elements of a system, the dependence on infrastructures, network externalities and phenomena like path-dependence and lock-in can be explained on the basis of a systems perspective.[47] Kemp et al. (Kemp et al., 1994) have identified six main types of "temporal increasing returns" which constrain future development paths: interactive learning, static economies of scale, dynamic economies and learning-by-doing, economies of scope, technological and pecuniary externalities, and network externalities. They are typical examples of 'systemic' phenomena affecting the direction of technological change.

Extending the system perspective: Systemic interdependencies and institutions

Wider systemic interdependencies

The concepts of technological regimes and of technological systems have already extended substantially the dominant micro-oriented perspectives by taking structural constraints into account. Freeman and Perez (Freeman & Perez, 1988) have gone another step further by pointing to constraints which are based on long-term economic developments associated to the pervasive spread of certain basic types of new technologies. Their starting point is a taxonomy of innovations which distinguishes between incremental and radical innovations, new technology systems and changes of techno-economic paradigms. This latter type implies changes which go beyond individual technological systems and their organisation in an industry. They comprise also the general economic structures throughout the entire economy. "Technological interrelatedness" is the key concept to explain this wide-ranging evolution, because developments in different industries and intercon-

---

47 See especially the contributions by Arthur (Arthur, 1988a; Arthur, 1988b; Arthur, 1989) on self-reinforcing effects, but also (David & Bunn, 1988; David, 1985) for some empirical cases, and (Katz & Shapiro, 1994; Katz & Shapiro, 1985) on network externalities.

nected by these paradigm-setting technologies. Their perspective puts technological change at the heart of long-term fluctuations of the economy, providing an underlying rationale for the emergence of Kondratieff's "long waves" (see also section 2.3.1.1). At present, information and telecommunication technology is regarded to be the new emerging techno-economic paradigm.

Their theory is also a good example of a general trend in economics to integrate processes of change at different levels. At a theoretical level, Freeman and Perez's arguments have received impulses from self-organisational approaches in other disciplines which are also concerned with non-linear dynamics and path-dependency. Typically, the emergence of regular macro-patterns and structures is interpreted as interrelated with autonomous processes of adjustment and co-operation at the micro-level.[48] In other words: "(A)ny world where structural conditions and innovation diffusion are dynamically coupled, that is wherein structural conditions influence innovation adoption and innovation adoption changes industrial structures, is going to exhibit non-linear dynamics and path-dependency" (Dosi, 1991a, p. 202).

The institutional element: National Systems of Innovation (NSI)

Another important structural and historically grown element which nevertheless merits some specific attention are institutions. Some researchers may claim that institutional economics does not represent a part of evolutionary economics but in the context of technological change, institutional elements have been integrated prominently in evolutionary approaches. This integration aims to endogenise institutional aspects into an evolutionary systems theory, not just at considering them as exogenous determinants of economic processes. The importance of institutions as a further type of structural determinants is based on the recognition that the direction of technological change can not only be guided by addressing the technological system itself, but also by manipulating the operation of its context, with institutions representing an important element of this context

Almost simultaneously, several ways of addressing this challenge have been suggested.[49] They all assume that differences in national institutions matter for the

---

48 For further contributions to self-organisation in innovation economics cf. (Bruckner & al., 1994; Silverberg, 1991; Silverberg et al., 1988), and for its application to social systems (Haken, 1983; Weidlich, 1991; Weidlich, 1992). While most of the formalised approaches deal with the general dynamics of socio-economic systems, a number of specific empirical examples have been studied recently (Nagel, Rasmussen, & Barrett, 1997; Schweitzer & Steinbrink, 1997).

49 See especially the following publications: (Carlsson & Stankiewicz, 1991; Edquist, 1997; Lundvall, 1992b; Nelson, 1993; Niosi & al., 1993; Patel & Pavitt, 1994). Carlsson and Stankiewicz speak of "technological systems" in this context but the basic ideas are very similar to national innovation systems. On institutional elements in evolutionary economic theory see (Hodgson, 1993). In their concept of Technology Support Systems (TSS), de Liso and Metcalfe avoid the limitations of national boundaries for conceptualising institutional elements. Furthermore, they do not refer to general institutions supporting the development and diffusion of new technology, but more specifically to certain areas of technological change (Liso & Metcalfe, 1994).

understanding of technological change, especially for the explanation of national differences. In the definition of Niosi et al. which builds to a large extent on Lundvall's work (Lundvall, 1992b), it is proposed to consider the following main elements in an National System of Innovation (Niosi & al., 1993, p. 211):

- Markets and natural resource determinants like consumers' incomes, their tastes and a country's resource base;
- User-producer relationships and other types of collaborative arrangements which are assumed to be stronger inside the national boundaries than outside;
- Technically-based interdependencies and areas of technological specialisation of countries;
- Politically driven linkages like those established through technology policy, military and civil procurement;
- The main links among the different elements of an NSI are constituted by financial flows, legal and policy links, the flow of technological and scientific information, and social flows (e.g. exchange of personnel).

Notwithstanding the recognition of institutional elements, the NSI-perspective regards the political realm still as an element external to the scope of its research. Consequently, from the perspective of this study on political control, there is still a missing link, namely the integration of politics into the analysis. This means that the full range of micro-level interactions is not yet captured, political interaction is not addressed. Finally, the focus of NSI is clearly on innovation and its determinants rather than on technological change as a whole and the diffusion of technologies throughout the economy.

2.3.3.3 Formal models

This study does not focus on formal modelling because the formal models proposed until now are still too crude to answer the research questions raised. Formal models can only capture in a highly structured way some selected mechanisms of the complex and multi-faceted real-world problems.

Nelson and Winter already modelled processes of economic change mathematically. They carried out simulation studies which built on the dynamic characteristics of Markov-chains. Later on, several other more sophisticated models have been developed. A very influential example has been presented by Silverberg et al. (Silverberg et al., 1988). It describes innovation and diffusion as self-organising processes. On the basis of a number of behavioural assumptions and a set of feedback loops between structural and behavioural level, it reflects several evolutionary mechanisms, but produces regular diffusion patterns. However, certain interesting evolutionary mechanisms can be described better by other than self-organisational models. For example, adaptation processes and random effects can be described by means of genetic algorithms (Canarelli, 1995b). However, as pointed out by Birchenhall (Birchenhall, Kastrinos, & Metcalfe, 1994), both genetic algorithms as well as self-organisational models may represent the random elements of evolution but they do not capture the narrowing impact of cognitive mechanisms. In other words, despite the progress made in recent years, formal evolutionary models are still in a rather early stage of development. For the moment, the appreciative modelling approaches appear to be more useful for the purpose of this study.

2.3.4 Summary and assessment

- It is obvious that the approaches presented above have become increasingly sophisticated and complex at the end of this section. This reflects the historical development of innovation and diffusion theory. Evolutionary approaches include several important factors in the analysis which were ignored in earlier economic analyses. In other words, "the development of innovation theory is one through which new elements of the firms' environment have been included in the theoretical systems" (Niosi & al., 1993). NSI and self-organisational models represent the culmination point of this development until now. Not only do they recognise the fundamentally dynamic and evolutionary character of technological change, but also take institutional and technological issues into account. Furthermore, the structure of the evolutionary approach facilitates the simultaneous consideration of macro and micro determinants.
- However, jointly with the neo-classical approaches, evolutionary economics suffers from a major shortcoming, namely the normative character of these theories with regard to policy-making. Government is regarded as an exogenous factor. The links between economic system and political system as well as the processes internal to the latter are not considered. In other words, those who are the main addressees of economic analysis do not play a central role in it. Governments are taken into account only as the outer shell of a selection environment which concentrates on economic players.
- Neo-classical approaches still suffer from two further severe shortcomings. First of all, they represent a model world built on unrealistic assumptions. This reduces the possibility to derive any relevant policy-oriented conclusions. Second, if it is intended to draw policy-conclusions, the models must be empirically meaningful. On the basis of neo-classical models the assumptions of the models have been prescribed as normative for real-world system, because within the (unrealistic) model the theoretical analysis led to beneficial outcomes. This abuse of the analytically very valuable results represents a severe logical error. A good example of this is the discussion on the advantages and disadvantages of government influence in the economy. In an ideal neo-classical model world government intervention ought to be reduced to the minimum, but if more realistic assumptions are taken into account like e.g. in NSI, then it becomes obvious that the co-ordinating function of government can be essential. Similar arguments hold with regard to large technical systems or infrastructures which can often not be built up over long periods without the support by government.
- A further general shortcoming of economic approaches is that they apply - by definition - only economic criteria for the evaluation of events and decisions, although other aspects should be taken into account as well. Especially the

analysis at the micro-level remains fairly superficial and restricted to markets and user-producer relationships.

- Hence, although the basic set-up of the recent evolutionary approaches seems to be very promising for the analysis of technological change and political control, the existing models still lack several essential elements. These are hoped to be found in sociological and political theory.

2.4 Sociological analysis of technological change

The economic approaches presented in the preceding section have focused on the interrelation between economic and technological change. From a sociological perspective quite different approaches to the emergence of new technologies have been developed which concentrate on micro-determinants of technological change. The New Sociology of Technology provides a perspective on technological change which concentrates on the underlying processes of shaping beliefs and expectations, learning, power relations and mutual influence among actors. Methodologically, the analysis of networks of interaction plays an important role. In addition, these sociological approaches have influenced much of the work on issues of political control. For example, network analysis can be used to reveal power relations and to understand the structural characteristics of different forms of political organisations. However, political control also needs to address structural aspects relating to the political context of technological change. Much technology-related sociological work has also dealt with the processes and organisational issues within firms or laboratories, but they are less relevant to the subject of this study.

2.4.1 Early sociological approaches - linear models and single factors

Quite at the same time as Mansfield, Rogers (Rogers, 1983, p. 5) laid the ground for sociologically inspired work on the diffusion and adoption of technologies. He defines diffusion as

> *"the process by which an innovation is communicated through certain channels over time among the members of a social system. It is a special type of communication, in that the messages are concerned with new ideas."*

This definition encompasses the four elements which he regards as essential for describing diffusion: innovation,[50] communication channels, time, and a social system. In his main book, he provides a critical synthesis and classification of

---

50 Innovation and innovators are in Roger's terminology defined differently than in economics. For example, in his typology of adopters (innovators, early adopters, early majority, late majority and laggards) the innovators are regarded to be the firms which *adopt* a new technology first, not the ones which introduce it to the market.

preceding research work on innovation diffusion across disciplines, but avoids to propose yet another simplified theory of diffusion. His review reflects very well the linear conceptualisation of the relationship between technological change and the determinants investigated at the time. Rogers distinguishes eight main areas of diffusion research (see Table 2.2), which focus on different dependent and independent variables (Rogers, 1983. p. 81-85).

He acknowledges several shortcomings of contemporary diffusion research, e.g. the bias in favour of successful innovations. This bias has often led to a harsh critique of the conservatism of non-adopters, but later studies of failed innovations have shown that this conservatism can be a very successful strategy. Rogers also indicates several of the methodological problems of diffusion research. Finally, his critique addresses the problem that the content of innovations is usually not questioned at all and not considered as the outcome of a social process. Together with the pro-innovation bias, this latter aspect has been one of the main starting points for the 'new' sociology of technological change.

Dependent Variables	Independent Variables
Earliness of knowing about innovation	Characteristics of members
Rate of adoption of different innovations in a social system	Attributes of innovations as perceived by members
Innovativeness and adopter categories	Characteristics of members, system-level variables
Opinion leadership	Characteristics of members, system norms and other variables, communication behaviour
Who interacts with whom in diffusion networks	Existence or not of a network link between members
Rate of adoption in different social systems	System norms, characteristics of the social system, type of innovation decision
Communication channel usage	Innovativeness and other characteristics of members, system norms, innovation attributes
Consequences of innovation	Characteristics of members, nature of the social system and of the innovation

Table 2.2: Dependent and independent variables in early diffusion research

2.4.2 The 'new' sociology of technological change - interdependencies, networks and systems

2.4.2.1 Roots and reasons

Findings in the field of sociology of science have been the starting point for sociological research dealing with processes of technological change. The basic hypothesis of the new sociology of technological change argues that the outcome of scientific research is to a large extent the result of social forces and conditions, rather than the discovery of objective truths. For example, research areas are selected through social mechanisms such as the incentive and peer review systems, financing, career opportunities, shared beliefs, etc.

However, the analogy from science to technology can not be easily drawn even if technology is nowadays increasingly science-based. While the insight that social relations and individuals' believes strongly affect the directions of scientific research can be easily confirmed for the *generation* of new technologies, the transfer of the analogy becomes more difficult for *adoption and diffusion*. The influence of market forces, and the necessity to provide working artefacts impose different acceptance mechanisms for innovation: scientific findings are basically accepted until they the opposite is proven, whereas technology has to prove its operational feasibility immediately. Building on the work of Merton (Merton, 1973) and Kuhn (Kuhn, 1970) some of the mechanisms which underlie the selection of scientific research fields have been developed further for technology. Most influential in this respect has been the work carried out under the headline of Sociology of Scientific Knowledge (SSK). It can be linked to the work of Dosi (Dosi, 1982) who also builds on Kuhn's notion of scientific paradigm, but introduces a sociological element in a predominantly economic perspective. In sociology, communities of practitioners, mental frameworks, constellations of interests and deliberate strategies of actors have been recognised as key forces for the selection of technological paths, shedding some light on the reasons why seemingly inferior technologies become established. Evidence for these theoretical generalisations is provided by a large number of historical studies. They complement, if not contradict, the economic determination of technological change, pointing to the importance of additional factors beyond economic ones. Three main insights can be identified as underlying most sociological work:

- The rejection of any kind of technological determinism or a linear determination of technological change as emanating from science;
- The conviction that essentially social forces shape the direction of technological change, both with regard to its content and its selection. It is driven by the interests and conscious or unconscious behaviour of actors in favour of certain options. This implies that it is possible to affect the direction of technological change by social forces, e.g. by means of political control. Still, the question remains open whether the complexity of the problems at stake allows to design targeted measures.
- Technological change is interpreted as a 'contingent' phenomenon, i.e. as a phenomenon which cannot be explained on the basis of some few single determinants, but heavily depends on its specific context and a broad set of interdependent factors.[51]

---

51 Cf. (Rickards, 1985) who points to the limitations of attempts to explain complex phenomena on the basis of single factor models. If one accepts the contingency-approach of 'it all depends', then this has a number of severe methodological implications. In particular, it does not make much sense to look for general correlations, but

Although the different sociological approaches may share the above mentioned basic insights, several different schools of thought need to be distinguished. One of the shared characteristics is the emphasis put on processes and relationships at the micro-level as determinants of technological change; structural aspects are not the focus of interest. This explains also the dominance of network or network-like concepts. Although the understanding of networks is similar to the use of the term in economics, it is generally used in a more encompassing way, including all types of relationships possible among actors.[52]

2.4.2.2 Actors-networks and the social construction of technology

Under the broad headline of 'Social Construction of Technology (SCOT)' a large number of studies has been carried out in the last two decades which focus not only on successful innovations - one of the main shortcomings of past research - but also on failures.[53] The approach is based on the conviction that not only the *use* of technologies is determined by social forces but also its *content* - a statement which shows the closeness to the Sociology of Scientific Knowledge (SSK). The central concerns of SCOT consists of the analysis of the influence which *relevant actors* and their interests have on both the choices and the content of technological developments. *Actors* can be understood both in the sense of individuals and of organisations, and are conceptually clearly distinguished from technical artefacts. This makes it easier to conceptualise the content of technology as being socially shaped.

In general, a rough variation/selection analogy is applied to technological change. The variability of an emerging technology is based on the different interpretations, meanings and interests which the actors assign to the technology in question ("interpretative flexibility"). The "stabilisation" of the technology is related to the achievement of a consensus among the actors about the pathway to follow. This formation of a consensus has also been labelled "closure". A further step in the analysis of the social construction of technology would be the investigation of the "wider context": the socio-cultural and political situation (i.e. a structural element) which shapes the norms and values of the actors, and hence underlies the meanings and interests assigned to a technology. However, this last step has rarely been implemented in any empirical study; they tend to focus on micro-sociological aspects. The majority of SCOT-studies are of a historical character, and aim at reconstructing stories of technological successes and failures along the conceptual patterns of the approach.

---

rather to develop general methodologies and approaches for studying specific phenomena.

52 For more details on the 'narrow' and the 'broad' definition of networks cf. (Frances & al., 1991) and (Thompson & al., 1991). In general, networks may be a frequently used conceptual term, but it is rarely well defined and far from being a homogeneous concept which could be clearly distinguished from hierarchies and markets as competing organisational forms.

53 See e.g. the studies on the bicycle (Pinch & Bijker, 1987) and on bakelite (Bijker, 1987). Some more recent examples can be found in (Bijker & Law, 1992b).

Closely related to SCOT is a second line of research under the label of "actor networks" (Callon, 1987). This name was chosen because an actor network is regarded to be less stable and consistent than an actor, but more stable and permanent than a network.[54] Actor networks include animate as well as inanimate elements, i.e. technological artefacts as well as organisations and individuals. These elements have essentially the same status in the actor network. The initiative for such an encompassing approach was based on the conviction that technical aspects ought to be incorporated in sociological analysis, and that technical, economic and sociological forces are interwoven. One of the ground-breaking applications of this approach was a study on the development of the electric vehicle in France by Electricité de France, which shows that they did not only try to imagine a new technological artefact but also a complete environment (or hence actor network) inside of which it could operate (Callon, 1980).

Later on, this approach has been modified under the headline of "techno-economic networks (TEN)" (Bell & Callon, 1994; Callon, 1992), opening up the sociological view to more techno-economic aspects. Technological change is conceptualised as being organised around three main "poles": a scientific pole, a technical pole and a market pole. Constituted around these poles, a techno-economic network can be defined as "a co-ordinated set of heterogeneous actors - for instance, public laboratories, centres for technical research, companies, financial organisations, users and the government - who participate collectively in the conception, development, production and distribution or diffusion of procedures for producing goods and services, some of which give rise to market transactions" (Callon, 1992, p. 73). Interactions take place by means of "intermediaries" which comprise texts, human beings and their skills, money, and technical artefacts.[55] Actors complement the notion of intermediaries in the sense that the latter can be attributed to the former. Similar to the SCOT approach, the crucial point is how a technology gets stabilised in the course of a process of "irreversibilisation", in which the processes of translation (between actors and by means of intermediaries) play the key role. This "convergence" of a TEN leads to the establishment of a consensus, in which the translation is facilitated. A further similarity with SCOT can be seen in the

---

54 See Callon's own definition (Callon, 1987, p. 93): "The actor network is reducible neither to an actor alone nor to a network. Like networks it is composed of a series of heterogeneous elements, animate and inanimate, that have been linked to one another for a certain period of time. The actor network can thus be distinguished from the traditional actors of sociology, a category generally excluding any non-human component and whose internal structure is rarely assimilated to that of a network. But the actor network should not, on the other hand, be confused with a network linking in some predictable fashion elements that are perfectly well defined and stable, for the entities it is composed of, whether natural or social, could at any moment redefine their identity and mutual relationships in some new way and bring new elements into the network. An actor network is simultaneously an actor whose activity is networking heterogeneous elements and a network that is able to redefine and transform what it is made of."

55 Compare Niosi et al. (Niosi & al., 1993, p. 211) and the links defined in National Systems of Innovation: financial flows, legal and policy links, technological, scientific, and informational flows, and social flows.

attempt to identify dependencies, interest structures and mutual influences among actors. While the SCOT-studies tend to be rather retrospective, actor network and TEN studies have been carried out on very recent cases and even in parallel with the actual implementation of a network. In this sense, the French actor network school has provided more 'constructive' contributions than SCOT. Both schools have shown an interest in successful innovations as well as in failures, helping to overcome the pro-innovation bias of research on technological change.

Methodologically, the identification of networks imposes severe difficulties which the constructivists as well as the actor network approach attempt to solve by working as closely as possible with the researcher and engineers under scrutiny. Close observation should allow to identify the networks of individuals or - at a more aggregate level - of organisations. Although some quantitative and more objective techniques have been developed to improve the reliability of these observations, an important subjective element remains and is recognised.

Other approaches which are closely linked to the ones mentioned before have been presented by Rammert (Rammert, 1988) who looks at innovation processes and their organisation in firms, and by MacKenzie (MacKenzie, 1987; MacKenzie, 1992). MacKenzie points especially to the role of “self-fulfilling prophecies” for the establishment of technologies and rejects the notion of their 'natural' unfolding of technological trajectories. Although being inspired by actor-networks, too, he predominantly builds on the systems approach of Hughes (Hughes, 1983; Hughes, 1987). A third interesting approach has been developed by Molina (Collinson & Molina, 1995; Molina, 1993). His “socio-technical constituencies” integrate social and technical elements from the micro to the macro (institutional) level.

2.4.2.3 Large technical systems

Hughes (Hughes, 1983; Hughes, 1987) introduced an approach to the study of technology which is broader than most sociological approaches, but it has played an important role for inspiring many of them. While especially social constructivism over-emphasises the role of social forces as compared to technical issues, the “technological systems” as defined by Hughes explicitly integrate technical elements in the analysis. His idea of “momentum” should certainly not be confused with technical determinism, but he argues that technical considerations can represent major constraints and incentives for further development. Related to this, his concept of “reverse salients” implies that technical bottlenecks represent areas on which R&D activities are concentrated. These technical elements are complemented by social forces, expressed especially in his notions of “systems builders” and “technological style”. They reflect the importance of individual actors and of cultural factors for the success and the emergence of technology.[56] Furthermore, structural settings of the system play a quite prominent role in his historical study

---

56 C.f. (Hughes, 1983, p. 462): “The evolving power systems were not, metaphorically speaking, driverless vehicles carrying society to destinations unknown and perhaps undesired. The systems did, however, have an internal drive and increasing momentum.”

on the emergence of electricity supply systems. The technical, economic and social organisation strongly constrains the possibilities of the actors to shape and affect the further development of the system. Hughes' approach thus combines technical, economic, political and social forces at structural and behavioural level in a single framework.

The consideration of these systemic interrelationships has been particularly fruitful for the analysis of "large technical systems (LTS)". These can be characterised as follows:

> *"(S)ome types of technical systems can be singled out as undisputedly large: those complex and heterogeneous systems of physical structures and complex machineries which (1) are materially integrated, or "coupled" over large spans of space and time, quite irrespective of their particular cultural, political, economic and corporate make-up, and (2) support or sustain the functioning of very large numbers of other technical systems, whose organization they thereby link."* (Joerges, 1988, p. 24)

The logic of development of a technological system consists of three main phases, namely first invention, development and innovation, second the phase of transfer, involving problems of adaptation to different environment ("technological styles") and finally the phase of consolidation and growth of a technology.

Hughes' own path-breaking comparative study on the emergence of electricity networks is first of all a very detailed historical account, but it exemplifies the meanings of his conceptual terms. Several studies of "large technical systems" have been carried out in the meantime.[57] Recently, the focus of interest has switched from the emergence of LTS to their reconfiguration (Summerton, 1994), i.e. on the question how these systems can change and adapt to different conditions and requirements. A number of case studies on the integration, evolution, merging or expansion of LTS have been carried out, but conceptually little progress has been made in the last years.

2.4.2.4 Constructive technology assessment

Constructive Technology Assessment (CTA) (Rip, Misa, & Schot, 1995; Schot, 1992) builds on an integration of some basic notions of evolutionary economics with those of sociological analysis. It concentrates on the roles different actors play in an evolutionary process of technological variation and selection. It is 'constructive' in the sense that it aims to identify actor strategies to affect and shape ongoing processes of change.

---

57 For a selection of studies see e.g.(Mayntz & Hughes, 1988). Examples contained in this volume are the American, German and French railway networks, telephone system in Germany, the U.S. air traffic system, the French power system, and as a more recent example the emergence of interactive videotext in Germany, France and the UK. The follow-up work on large-technical systems is collected in (Summerton, 1994).

Basically, the approach supports the concept of a co-evolution of technological systems and their environments (Schot, 1992), implying the necessity to operate at these two levels to promote technological change in a desirable direction. Co-evolution operates through the anticipation of future changes by actors, through institutional linkages and through strategies of active niche creation. CTA thus takes an intermediate path between an independent selection environment as used in the economic approaches and an environment which is socially shaped as assumed in the sociological perspectives. Actors can be located either inside the system, shaping the content of variations, or outside, hence affecting the selection environment, or even located at the interfaces between environment and system ("technological nexus").

Three basic mechanisms to affect technological change are distinguished: the development of alternative variations, the modification of the selection environment, and the creation or utilisation of the technological nexus, i.e. institutional interfaces between technology-based firms and their selection environment. One could argue that they corresponds roughly to three 'classical' types of policies, namely technology and innovation policies, structural and tax policies, and 'facilitating' policies like information supply, commitment campaigns etc. Consequently, CTA could be used in a fairly straightforward way as a framework for political control.

CTA has been developed as a reaction the more traditional TA approaches which focus primarily on the objective assessment of external effects of technology and assume the possibility to make clear choices between well-defined options. This view is rejected in CTA in favour of the idea that choices and changes take place permanently in a process which involves technical, economic, social and political factors. In this sense, it recommends a continuous and enduring commitment to shaping technology itself, underlining thus its constructive impetus.

CTA differs less from the other approaches mentioned in terms of the methods applied. All approaches have in common a reliance on network-based forms of analysis and on the mapping of interactions and strategies. The scope of analysis of CTA, however, is oriented towards more policy-oriented investigations, political authorities being one of the relevant actors exerting an influence on technology. Consequently, the aim of CTA is to develop actively strategies for the stimulation of specific technologies, e.g. through processes of strategic niche management (Kemp, Schot, & Hoogma, 1996).

2.4.3 Summary and assessment

- Sociological studies have pointed to the importance of determinants which are usually neglected in economic analysis: believes and mental framework, interests and coalitions of interest, the deliberate shaping of both the diffusion and the content of a technology.
- Furthermore, a broad range of technologies has been investigated. Failures and successes, "modest" artefacts and large systems have been subjected to detailed case-studies.

- Constructive Technology Assessment goes beyond sociological *analysis* by underlining the active role of the researcher in the process of social shaping of technology. Conceptually, it builds on both evolutionary and sociological concepts for the management of change. It also takes into account both structural and behavioural level.
- One of the main shortcomings of many sociological approaches consists of a neglect of structural determinants which - from a political perspective - leaves several potential points of access for political measures unexplored.[58] Notions such as "wider context" in SCOT (Pinch & Bijker, 1987, p. 46) and of "local and global networks" (Law & Callon, 1988) remain very vague and do not really capture structural aspects in a detailed way.[59]
- All approaches share a lack of analytical specificity; the concepts used remain vague and general. This is most pronounced in the case of actor-networks which can encompass everything from technological hardware to individual interests. This may be the result of the contingency which is underlined by all approaches, but a more differentiated vocabulary would nevertheless be helpful for guiding targeted empirical research work.[60]
- A further point of critique concerns the neglect of technical, physical and economic constraints to the establishment of technology.[61] Although social forces may co-shape the content of technology, they are not the only forces at work.
- Although the idea of social shaping of technology seems to be open to the consideration of political forces, this step has rarely been made. The micro-sociological focus of the majority of approaches does not allow considering several important political aspects which are often located at a structural level.

---

58 This critique has already been formulated by (Russell, 1986b) who pointed out that technological change is not just a "bottom-up" process but heavily constrained but structural and social settings which narrow down the autonomous decisions of actors.

59 Building on corporatist and sub-government-based concepts, Rüdig proposes a analogous structure of technology and socio-political realms according to different fields and levels: "(W)hat we are suggesting is that a systematisation of policy levels into sectors, sub-sectors and issues associated with policy communities and policy networks is matched by an equivalent systematisation of technology; different technology levels, technology systems and sub-systems, and individual technological artefacts forming one hierarchy matched by a system of technology actor-networks equivalent to policy communities and policy networks." (Rüdig, 1990, p. 24)

60 C.f. (Schneider, 1989, p. 28, translation by MW) who argues with regard to Callon's actor-networks as follows: "Important differences between relatively autonomous and independent areas of the social whole ("das soziale Ganze", MW) are levelled out - for example when science, technology, economy and policy-making are regarded without difference as social practice. In this connection, the special rationality and the respective modal logic governing each of these areas, is overlooked."

61 See e.g. (MacKenzie, 1992) who discusses what is 'natural' about Nelson and Winter's natural trajectories and Dosi's technological trajectories, i.e. in how far technical automatism should be accepted as operational once a paradigm is established. He prefers to interpret the trajectories as self-fulfilling prophecies, hence as socially determined and not obeying to physical necessities.

Neither is the problem addressed that both the processes leading to policies and those leading to new technologies matter from a policy-oriented perspective.

- The need to consider several levels of constraints (technical, economic and social ones) which jointly shape the pathway of technological change seems to be best reflected in Hughes differentiated view of the emergence of large technical systems. It maintains a distinction between technical hardware, the realm of markets, political influences, social forces and organisational issues, and shows their interrelatedness for reinforcing technological changes.

2.5 Theories of political control and regulation

The purpose of this study is - at the theoretical level - to link a conceptual understanding of technological change to that of political control. It has already been mentioned that such a synthesis is still missing but would be highly desirable as a foundation for policy-oriented technology analyses. Political control is more than the impact of different policies and measures on the economic and social behaviour of actors. It also comprises the organisational and operational aspects of the political process, how these affect the outcome of policy-making, and how therefore technological change is politically constrained. The following section deals therefore with the issue of how regulation and political control can be conceptualised and which elements could be useful for formulating and integrated conceptual framework for the political control of technological change. The review starts with simple models which build on a linear relationship between political aspects and technological change. Most of them assume that targeted political control of social systems is indeed feasible. While still maintaining a linear structure, several attempts have been made to address the shortcomings of the initial planning models (Sections 2.5.1.3 to 2.5.1.5). In a second step (2.5.2), network and self-organisational systems approaches will be revisited which take a more critical stance towards controllability, and represent a radical departure from the linear model.[62]

---

62 Some further distinctions could be easily made with regard to political approaches, namely 1) between structural models on the one hand and behavioural ones on the other, and 2) between approaches which focus on the internal capacities and processes of political systems to formulate policies and others which are concerned with the efficiency and implementation of political measures. This distinction between internal processes and efficiency/implementation results of policies, can be compared to the micro-/macro- dichotomy and as well to the relationship between subject and object of control. Görlitz (Görlitz, 1995) also indicates five different control concepts which reflect four different types of models of the political process, namely planning models (which are basically linear), policy-making models (which apply a mechanistic systems approach), policy-cycle models (which distinguish different phases of control), policy network models (which use a realistic representation of a wide range of stakeholders in political control), and self-organisational models (which are more critical of the targeted controllability of social systems and aim at taking into account their internal mechanisms).

As a general assessment, one can state first of all that the majority of policy-oriented approaches adopt a linear conceptualisation of technological change, raising questions such as "What is the impact of a political determinant on ..?". Other approaches focus on policy-making processes, feedbacks and systemic mechanisms in general terms but the link between policy-making and technological change is still relatively unexplored.[63] Secondly, in contrast to economics where the dynamic analysis of technological change plays an increasingly important role, the emphasis in political sciences is still on static analysis, although this seems to be changing slowly.

2.5.1 Linear relationships between the political and the techno-economic realm

The main common characteristic of the approaches presented in this paragraph consists of the exclusion of the interdependence of political and techno-economic changes, i.e. a linear relationship between political measures and techno-economic change is assumed. Furthermore, the first three groups (government measures affecting technological change, planning, and traditional TA/forecasting) do not question the internal constraints of the policy-making processes such self-interest of individual actors or the structural bias due to the organisational structure of the political process. The importance of behavioural factors and structural settings in the political system as determinants of policy is recognised is the next two groups (behavioural and structural approaches), but they still rely on a linear conceptualisation of the link between the political determinants and technology.

2.5.1.1 Government measures as factors affecting technological change

An enormous amount of literature is available on the effects of individual government measures on innovation, diffusion and technological change, including evaluation studies of government programmes. All these studies address predominantly specific empirical questions rather than conceptual issues, but implicitly they almost exclusively build on a linear understanding of the relationship between political and technological change. This type of research on the role of specific government initiatives has been started in the early Seventies with a number of large studies on the effects and spill-overs of military research.[64]

---

[63] This has been confirmed even in a very recent and encompassing publication on Science and Technology Studies (Jasanoff, Markle, Petersen, & Pinch, 1995, p. 528). According to the authors, this is due to the turn to the micro-level in the course of the constructivism debate in the Seventies which simplified and neglected the impact of structural parameters outside the laboratory level - elements which traditionally have played an important role in political sciences. Beyond these simplifications in the sociological realm, political scientists have been "reluctant to take on board the far-reaching implications of a constructivist approach that permanently problematized the science-state boundary".

[64] These are the TRACES, BATTELLE and Hindsight studies. Some of them were repeated in a modified way after some years.

The types of studies dealing with the impact of government measures on technological change are discussed below according to a scheme suggested by Bräunling (Bräunling, 1986). In general, they all build on a linear model of the impact of policy on technology. Government measures are regarded as external forces, impinging on different phases and aspects of technological change by means of technology policy, research policy, economic policy, information policy etc. The underlying model of technological change may be represented by one of the sophisticated models mentioned above, but the link with government policy is clearly linear:

The impact of context conditions

The organisational frameworks for industry, financial institutions, patent system, and general planning and regulation define the boundary conditions for any innovative activity. At a more specific level, such measures also comprise the support for the establishment of an innovation services infrastructure, e.g. business or technology parks. Finally, one can also include in this category studies which address the question of general government and/or market failure.

Supply-oriented measures

<u>Creation and finance of institutions, relevant for technological innovation and diffusion in a particular area/industry</u>

Examples of these measures are the funding for basic institutions like universities, publicly financed research institutes, technology assessment and transfer bodies. Also standardisation institutions and all other investment in the R&D-infrastructure make part of it. In a wider sense, other types of infrastructure investment needed for the successful spread of technologies, such as the transport and communication infrastructures, could be considered as well. With regard to energy and electricity supply, this refers also to the whole organisational set-up for dealing with the problems of natural monopolies. Several studies have been carried out which address the (dis-)advantages of regulated monopolistic structures vs. deregulation, liberalisation and privatisation (c.f. Part II).

<u>Targeted support of individual projects</u>

All studies which aim at the assessment of benefits emerging from large projects in selected areas such as nuclear energy, aeronautics, micro-electronics, or telecommunications fall under this category, as well as R&D programmes to support innovative activities of firms in general.

Demand-oriented measures

<u>Financial incentives and disincentives</u>

In contrast to the aforementioned category, financial (dis-)incentives can have a broad scope of application and can be directed to the potential users of new technologies. Their investment decisions are stimulated by taxes, grant schemes, etc.

Often they are also used to complement and replace institutional measures such as standards and limits.

Procurement

Especially in some certain areas (military, industries in public ownership, public buildings and equipment) the demand effect of public procurement is of great importance.[65]

Mediation and stimulation of actors

Measures in this context aim at inciting the active involvement of other social actors in R&D by means of persuasion rather than by financial or regulatory measures. Information is therefore an important element of such strategies.

2.5.1.2 Mechanistic top down approaches of policy implementation - linear thinking in planning

While the effectiveness of implemented political measures is mainly an object of study in economics and evaluation research, the creation of government measures (i.e. the process and determinants by which policies and regulations are formulated and implemented) is the focus of research in political sciences and political economy. The first type of approach in this realm is called 'planning' because it is characterised by the belief in an efficient, adequate and targeted policy reaction of government in response to shortcomings and failures in society, and especially in the techno-economic realm. In other words, it is assumed that

- Government can detect deviations of the actual state of affairs from the desired state,
- It disposes of the means to correct these deviations,
- It is able to formulate and implement specific measures for this purpose,
- The social realms react to these measures in a way which reflects the intentions of the planners.

This is a typical example of a top-down approach, which has also been termed "public interest theory" (Waterson, 1988, p. 6). The controlling authorities are regarded to be fully exogenous to the object of control. They dispose not only of complete information but also of effective and efficient tools to exert targeted influence. This may seem to be a very simplistic perspective on the operation of political control, but in fact it dominated the control debate in the 1970s. Energy planning was basically built on these assumptions, and also on the belief that future developments can be predicted quite accurately.

---

65 A good orientation for many countries is that about 50% of GNP is represented by public expenditure, even if this is only to a certain extent dedicated to procurement in the narrow sense.

Society and technological change are understood as mechanistic systems. The failure of this technocratic approach culminated in the debate about government or control failure. As a response, more differentiated approaches like the ones presented in the following sections were developed. They take into account the limitations of government and political decision making due to

- Informational and information processing restrictions (Section 2.5.1.3);
- Influence of internal interests (Section 2.5.1.4);
- Influence of external interests and the autonomy of the objects of control (Section 2.5.1.5).

2.5.1.3 The traditional forecasting and TA-approach - addressing the information barrier

Even in the most optimistic planning approaches, the capacity and information of government for the formulation of adequate policies is regarded to be limited by uncertainties and a lack of information about technologies and their future impacts. In order to reduce these informational deficits of government, technology assessment emerged in the 1970s. It was meant to provide the needed inputs to policy-makers about technologies and technological trends.[66] The early TA-attempts were still fuelled by a belief in the feasibility of targeted control, as expressed e.g. in the system dynamic models developed at the beginning of the 1970s, and the various economic, energy and political forecasting attempts related to them.[67] In other words, the mechanistic perspective is maintained in this approach because the control failures are regarded as a problem of information supply rather than of inherent limitations. The major shortcomings of this approach will become obvious when presenting the more recent interactive concepts of Technology Assessment (c.f. Section 2.5.2.4).

2.5.1.4 Understanding politics of technology - micro or behavioural explanations

In order to address the 'human factor' inside political authorities a number of theories have been proposed which explain why government measures are not necessarily developed according to the initial requirements. The problems posed are interpreted in terms of different types of (self-)interests of policy-makers. Consequently, the outcome of the political process and its impact on technological change also depends on these underlying motivations. The notion of interests points to the economic inspiration of these theories which build on the assumption

---

66 However, it should be taken into account that the foundation of the Office for Technology Assessment in the US was intended to provide information to parliament rather than to the executive branch, which had a quite good access to technological information through its own services.

67 A very interesting study was presented by Midttun and Baumgartner (Midttun & Baumgartner, 1986) who show how different energy forecasts have been used consciously in the public and political debates to manipulate decisions in energy policy.

that policy-makers initiate political activities predominantly in response to individual motivations and interests, and not for the sake of the public interests. Two major lines of thought can be distinguished in this context:

- The economic theory of party democracy (Downs, 1957; Lehner, 1981), which is essentially built on the argument that policy-makers in responsible positions try to maximise votes in elections and orient their decisions to this goal. The consideration of different interest groups as large reservoirs of votes is therefore of crucial importance.
- While the preceding argument can help to explain the behaviour of eligible politicians only, the more detailed formulation and implementation of measures in bureaucracy can be interpreted as governed by three major mechanisms: the influence of mental frameworks and formative elements which are specific for certain professions, the interest to minimise and reduce conflicts, in order to achieve at least a consensus at a minimum level, and the attempt to maximise the influence, competencies and budgets of individual policy departments (Müller & Vogelsang, 1979; Street, 1992; Walz, 1994).

These explanations of constraints on effective control at the micro-level are based on the traditional model of maximising agents, but in principle the routine and satisficing approach of Nelson and Winter could be applied as well, especially in order to explain the inertia of government and political administration.

2.5.1.5 Understanding politics of technology - macro or structural approaches

Research work in political sciences tends to focus on the impact of the organisation and operation of political processes on the political output. In other words, they tend to focus on structural aspects. A wide range of theoretical and approaches has been developed to explain the limits to control on this basis. Two main dimensions have been addressed concerning the organisational distribution of power and responsibilities, namely (Francis, 1993)

- Between different departments and levels of government, typically expressed in the problem of decentralisation of competencies;
- Between government and private sector.

Intra-government relationships

The distribution of competencies among different levels and among different departments of government are two typical sources of conflict within government. Centralised governments tend to introduce similarly centralised organisational structures for state-dominated industries. This excludes certain decentralised technological paths which would be feasible in a different organisational context. The French electricity supply system exemplifies this effect. On the contrary, the interwoven political structures in Germany ("Politikverflechtung") show how a decentralised system can look like, with corresponding impacts on the trajectories of local technology (Scharpf, 1978).

The conflicts between different departments of government and the lack of co-ordination between them are main sources of contradictory outcomes of the political process, of political standstill or of the dominance of certain policy areas over others. With regard to technological choices, control problems emerge if different policy departments counteract each others' policies. Examples of this can be found at all levels of government, including the EU and international organisations.[68] The establishment of inter-service working groups or task forces are typical attempts to build bridges between departments.

Government - group relationships[69]

In a pluralist model of control, the relationships between government and interest groups change with the electoral cycle. In a democratic system the influence of interests considered in the political process should change from time to time. Usually, it is implicitly assumed that the existing formal institutions guarantee that these changes of representation actually take place. Depending on how far this ideal model is distorted, a corporatist conceptualisation of government-group relationships may be more appropriate. In any case, one can argue that the possibilities to organise interests influences and constrains the outcomes of the political process (Olson, 1968; Schubert, 1992).

In contrast to the pluralist model, corporatist approaches assume more stable relationships between government and the private sector, which are not only defined in terms of the influence of interest groups in the political process, but also in terms of a delegation of responsibilities to industrial organisation. In other words, the tasks and responsibilities for political control are shared between government and private sector. One can distinguish different approaches dealing with these questions (Francis, 1993, Chapter 1; Lehmbruch, 1984):

- Under headlines such as "capture theory" and "iron triangle" a negative picture of corporatist relationships have been drawn up. Due to the dependence of government on the private sector, political decisions are in danger of being determined by the interest of the most powerful groups rather than by public interests. Critics of interventionist governance approaches favour this interpretation of government - private sector relationships because it supports their negative attitude towards the role of government. With regard to technology and innovation policy, capture theory helps to explain conservatism, but it is less useful for dealing with processes of change. In this latter respect it can only explain why certain pathways are selected by arguing that they have to match the existing political structural settings.

---

68 For a specific study dealing with technological transformation and the influence of government relations, see e.g. (Weber, 1986). Among other issues, he studies the positions and conflicts of German government departments in relation to a support programme for production technologies.

69 This issue can not be discussed in full detail here. For more details, see e.g. (Alemann, 1989; Goldthorpe, 1984; Hartmann, 1985; Schubert, 1989).

- Neo-corporatist approaches point to the constructive and positive role of the delegation of power to established organisations which represent social groups. They strongly support policies based on the decentralisation of power, subsidiarity, and the self-organisation and self-regulation of industry because they are regarded as more efficient ways to implement policies than the traditional top-down approach (Alemann, 1981; Streeck, 1984). Those who are most familiar with the situation in the different sub-systems should also be in charge of implementing the policies in detail.

The corporatist model offers a good, though highly simplified, representation of the situation in most European countries, whereas the pluralist model is better adapted to the situation in the US (Francis, 1993, p. 57). A far more complex concept is that of a 'regulatory regime'.[70] It integrates different types of structural settings and goes far beyond the distribution of power in formal political institutions. Regulatory regimes encompass also non-formalised principles, norms, values, procedures and channels of influence on policy-making.[71] Francis identifies two main dimensions which are included in the concept of regulatory regimes, namely the degree of centralisation of decision-making in government, and the degree of division of responsibility between the state and the private sector. The key to identifying the centres of influence lies in an analysis of the major groups of actors which participate in a regulatory regime. Typical such actors are (Francis, 1993, p. 65):

- State policy making authorities,
- Producer groups, industry, or service organisations, i.e. those actors who are subjected to regulation,
- Competitors or potential competitors of the regulated companies,
- Consumer organisations,
- Advocacy groups like e.g. environmental groups.

In order to characterise different types of regulatory regimes in more detail, additional concepts have been introduced, e.g. different national 'styles' of regulation (Vogel, 1986). A link can also be made to the approach of National Systems of Innovation where the notion of regulatory regime is also used. The concepts of

---

70 The regime-concept has given many impulses in the last years. It has been used in a number of different fields, e.g. as technological regimes (see above), international regimes (Kohler-Koch, 1989), or regulatory regime. For international relations, Krasner (Krasner, 1982) has proposed to define regimes as "a set of implicit or explicit principles, norms, rules and decision-making procedures around which actors' expectations converge in a given area of international relations", but the principles of this definition can also be applied to other fields of research.

71 It should be distinguished between a *socio-economic* understanding of regulatory regime, which refers to the rules and institutions relevant to an industry as boundary conditions of the industrial activity, and the *political* understanding of the concept (to which we refer here), which focuses on the political organisation of decision-making on specific rules and institutions. In other words, the two interpretations can be seen as the two sides of the same coin.

"governance" and "governance regime" are usually applied to economic systems as a whole while regulatory regimes refer to specific sectors (Campbell, Hollingsworth, & Lindberg, 1991; Hollingsworth, Schmitter, & Streeck, 1994).

2.5.1.6 Introducing dynamics

All the aforementioned approaches have a static character. They analyse and compare existing structures, but rarely look at processes of change. The inherently historical and path-dependent character of technological change as recognised in evolutionary economics is consequently not addressed. Investigations of shifts of regulatory regimes has been a first attempt to look at political control in a more dynamic way. However, in contrast to international politics, this type of research has not yet progressed very much in technology-related fields. Francis (Francis, 1993, p. 96) points out that "(c)hanges in regulatory regimes are brought about by the same factors that contributed to the creation of a regulatory regime in the first place." In practice, he refers to

- Ideological shifts about the role of the state in the economy, concerning the main assumptions and strategies underlying regulation,
- Changing assessments of the regulated industry itself,
- Alterations of the governing coalitions of influential actors,
- Changes of economic expectations toward the regulated industry,
- Technological innovation,
- Transfer of competencies to supra-national bodies.

However, regulatory regime shifts may allow describing a paradigmatic move from one 'equilibrium state' of regulation to another, but it does not really capture the permanent transformation processes to which regulatory regimes are subjected. The concept of regime shifts also allows to deal with the sources of changes, both in the environment of the regime (inducement factors such as technological changes) and within the regime (internal conflicts).

A very different line of dynamic thinking with regard to policy-making describes political control as a cyclical process, subdivided in different distinct phases which are linked to each other by feedback mechanisms. However, in relation to technological change it maintains the targeted and linear control model.[72] The major shortcoming of this approach consists of the neglect of overlaps between and deviations from the sequential heuristic.

---

[72] For a brief description of the policy cycle model see (Görlitz, 1995). Also in (Irwin & Vergragt, 1989) a feedback model of the relationship between regulation and innovation is proposed.

2.5.2 Feedback and system thinking from a political perspective

2.5.2.1 Policy research and networks

In parallel with developments in sociology and related to the corporatism/neo-corporatism debate, the network concept entered the field of political sciences, mainly due to the recognition of the importance of non-formal relationships as contrasted to the legally defined formalised relationships, both within government and between government and private sector. The analysis of networks addresses the meso-level of government - private sector relationships, in contrast to the macro-level corporatist arrangements. Basically, it is attempted to identify dependencies and power relationships which are not formally codified and institutionalised in order to point to the 'effective' dependencies at play. Marin and Mayntz have published a survey of research results (Marin & Mayntz, 1991b) which helps to understand why certain government decisions have been made, how government is subjected to biased influences from interest groups, and especially how coalitions between actors in the techno-economic and the political realm have determined technological choices. Policy networks can thus be characterised (Marin & Mayntz, 1991a)

- By their structure as inter-organisational arrangements,
- By their function which aims at formulating and implementing policies, mainly at the sectorial level.

The identification and evolution of membership in decision-networks and hence the representation of interests are therefore the keys for understanding the outcomes of the political process. Beyond the analysis of formal and informal links, network analysis also focuses on new emerging 'issues' and the formation of interests and networks related to these issues. They can easily cross the borders of existing policy and industrial areas, depending on whoever might have an interest in the issue at stake.

Network analysis in itself is not a theory, but rather a tool for identifying empirical relationships which might then lead to the formulation of theory. It has also been used for studying dynamic processes, e.g. in some empirical studies on the evolution of policy networks (Jansen, 1991; Kenis, 1991; Schneider & Werle, 1991). Network analysis allows to show at a very specific level that the formation of these informal relationships determines political and technological choices. It highlights detailed mechanisms and interdependencies, how they emerge and how they are perceived. To put it in the words of Schneider and Werle who refer to the emergence of telecommunications technology: "Thus, the perceived technical and organisational functions and anticipated externalities "generated" a network of actors who considered each other directly or indirectly relevant for the system" (Schneider & Werle, 1991, p. 110).

In contrast to the sociological network approaches which interpret networks as a model of reality, policy research underlines the role of network analysis as a tool. The level of aspiration is thus more modest than in the sociology of technology. However, the conceptualisation of policy-making in the form of networks has theoretical implications, because it indicates limitations and constraints to control inside the complex network of relationships.

A major shortcoming of earlier network analyses is their behavioural focus, but in recent years the concept has been further advanced. The approach of actor-focused institutionalism ("akteurszentrierter Institutionalismus") aims at combining structural and behavioural elements, and adopts also some self-organisational notions (Mayntz & Scharpf, 1995; Mayntz & Schneider, 1995).

2.5.2.2 Systems theory in political sciences

System thinking found its way to sociology and political sciences much earlier than to economics. Easton (Easton, 1965) introduced systems thinking to political sciences as a structured foundation for a theory of society and government. Although his line of reasoning had a considerable impact on research in political sciences, his framework remained predominantly descriptive and static. Furthermore, it is so general in character that much further elaboration is required to apply his principles of systems analysis to specific cases.

Basically, systems thinking should be seen as a methodology or as a model of theories rather than as a theory in itself. It is a way of structuring complex problems, a tool to narrow down specific research questions by specifying the coupling and linkages to related systems. Social system can typically be subdivided into economic, socio-cultural and political subsystems, which can then be further fine-structured (e.g. the political system into party parliamentary and political administrative part). The subsystems are coupled but for research-practical reasons they usually have to be studied separately. Although this basic subdivision of society can also be used as a framework for technology-related studies (Bräunling & Harmsen, 1975), other structures are possible as well. Depending on the research interest one can e.g. consider politics, economics and technology as the three constituting parts of society (Krupp, 1992).

The early systems models used a fairly complex representation of society, but maintained a mechanistic understanding of its operation. Consequently, the possibilities of targeted control of society were seen positively. They shared with most other approaches to the study of political control and regulation the control optimism of the 1970s.

2.5.2.3 Self-organisation and autopoiesis as models of political control

Self-organisation and self-reference are related but different concepts. Self-organisation stresses that the emergence of organisational structure of a system can be an autonomous process, triggered by certain boundary conditions, but in detail undetermined ex ante. In self-referential system, the focus of interest is the adaptation of an existing system to changes in the system's environment. This adaptation takes place in agreement with the internal mechanisms and logic of the system in question. Environmental changes can only trigger transformations of the system. How these changes look like is not determined by the exogenous impulse but by the internal structures and interests.

Formal models and their application to social system

In other disciplines (especially in physical systems theory) dynamic models have been developed which can at least in principle be applied to political and social issues. One of the key difficulties of political sciences with this type of systems theory has been its highly formalised (often mathematical) character. Social systems have long been regarded as too complex to be tractable within a mathematical framework, but recent attempts to model social systems by means of the master-equation approach or by applying the principle of synergetics have provided first promising results (Haken, 1983, Ch. 11; Schweitzer, 1997; Weidlich, 1992). It has been possible to show that these formal models cannot only be used to reproduce generalised dynamic patterns of social change but also to tackle specific empirical phenomena.

Qualitative models: autopoiesis

In general, autopoiesis is a specific example of a self-referential approach. In essence, the underlying idea is that systems react in agreement with their internal organisation, not necessarily in the way which was intended from outside. The theory of autopoiesis was suggested by Maturana and Varela as a qualitative approach to the study of living systems (Maturana & Varela, 1980). Initially it was meant for biological applications but later on formulated as a general model. It has been attempted frequently to translate their generalised systems model to the social sciences - with more or less success (Görlitz, 1989; Luhmann, 1984; Luhmann, 1988; Schimank, 1987). Especially the work of Luhmann has been criticised for being just a simple analogy rather than a accurate reconstruction of Maturana and Varela's approach (Druwe, 1989; Knie, 1994, footnote 1). However, the following main features can be identified as relevant for political control:[73]

- Autopoiesis is an attempt to formulate the ideas of self-organisation in a qualitative though formalised language.
- It can be applied to all 'living systems', 'living' being originally defined in the biological sense. All living systems are autopoietically organised. Social systems need hence to be understood as living systems.
- The basic idea of autopoiesis is that a system adapts to impulses from its environment, but the way it reacts depends predominantly on its internal structure ("Strukturdeterminiertheit"). More specifically, only the relations between its elements change ("operationale Geschlossenheit") but the basic organisation ("autopoiesis") remains the same. Given the fact that different autopoietic systems are structured in different ways, it is not possible to predict generally the reactions of autopoietic systems to external impulses.
- Evolution is therefore understood as a process of adaptation.

---

[73] It is not attempted to use the specific language developed by Maturana and Varela, but a simpler one which aims at capturing the main features of the model only. For more details on the application of Maturana and Varela' work to study the control of social systems, cf. (Druwe, 1989; Görlitz, 1994, Section 3.3-3.4).

- Society can be regarded as a multi-level autopoietic system.
- Control in social systems is hence confronted with the problem that the intentions of control measures are transformed into systemic adaptations according to the systems internal structure and relations, but not necessarily in line with the initial intentions.

A number of conclusions and hypotheses with regard to the political control of autopoietic social systems can be drawn:

- Successful control measures need to take the internal structure of the control objects into account. Detailed information about each individual system (having different internal structures) is needed in order to improve control.
- Consequently, political control should be interpreted in a non-deterministic, rather probabilistic sense.
- New types of control measures such as reflective law ("reflexives Recht"), or the decentralisation of society and social responsibilities are needed. In this regard, links with the corporatism/neo-corporatism debate can be identified because the advocates of neo-corporatist arrangements argue that the delegation and decentralisation of responsibilities would be an effective control approach.
- As the most prominent claim, the need for more market forms of control and less state or hierarchy has been expressed, in other words a stronger reliance on the self-organisation of the objects of control. It should be subjected to contextual (Glagow & Willke, 1987; Ulrich, 1994) or medial (Görlitz & Druwe, 1990) control only. This points to the need for decentralised control, or in more general terms for a non-hierarchical approach to control: object and subject of control are located at the same level and control each other mutually.

One of the basic problems of autopoiesis consists of its complex language and of the highly general character of the concepts used. This makes an operationalisation and application difficult. The general model needs to be translated into more specific theories (e.g. for social systems) in order to be able to derive relevant hypotheses to applied problems of policy-making.[74] In principle, the general conclusions of autopoietic thinking can also be expressed in far simpler terms. In this respect, one could raise the question whether it fulfils one of the criteria of a 'good' theory, namely that of being simple.

Autopoiesis as well as other self-organisational approaches can be applied both to the political system itself and to the social system controlled by it. In other words it applies to both elements of control: the subject, i.e. the political realm in which policies are formulated and implemented, and the object, i.e. the issue area which is subjected to political measures. Depending on the perspective, the question thus arises which of the social systems actually controls the other. This indicates that

74 For a detailed application of autopoiesis to political control, see (Görlitz, 1994). The policy approach which is developed for the political control of environmental issues is called "mediale Steuerung" and takes into account that the systems to be controlled can only be perturbed but hardly be controlled in detail .

there are clear links with other models discussed earlier, e.g. the notion of co-evolution of technological and social systems (Section 2.4.2.4).

A self-referential approach to economy, politics and technology

One interesting example could be identified for which the concepts of self-organisation and of self-reference are used in a framework for the study of technological change in the field of energy supply (Krupp, 1992). Luhmann's conceptual language represents the starting point for a systems theory of society which is combined with a theory of technological innovation (Luhmann, 1984; Luhmann, 1988). The main elements of Luhmann's approach as used in Krupp's "Schumpeter Dynamics" are the following:

- The differentiation of industrialised societies into sectors or subsystems, which are in his view essentially economy, politics, and technology;
- A fundamental lack of central controllability and the interdependence between all three subsystems underline the importance of interface organisations;
- Due to the operation of the subsystems according to their internal logic each subsystem generates major external costs;
- The conditions imposed by the self-referential character of the subsystems need to be fulfilled by every innovation in order to be introduced successfully.

The major problems and concerns of industrialised societies are then understood as intersystemic issues which are dealt with simultaneously by the different subsystems of society, each one applying its own internal logic and argumentation. Krupp proposes the concept of "intersystemic resonance" for issues which become important throughout society. Societal interactions on these issues lead finally to what can be termed societal learning and adaptation:

> *"Systems theory assumes that there is no extrasystemic center of societal control, so that issues are self-produced by the systems of society. A society is self-referencing in the sense that it can produce such issues and, at the same time receive them as targets and evolve in their direction - in a closed cycle of objectives and means."* (Krupp, 1992, p. 24)

In his systems theory of society Krupp introduces the notion of "Schumpeter Dynamics", i.e. the economically driven processes of technological change (building on profit orientation and competition). He argues that the dominant logic of economic growth and technological innovation is one of the issues which show intersystemic resonance, and which are widely shared throughout the different subsystems. In other words, the profit and competition logic has been agreed upon in all major subsystems, representing a kind of shared paradigm of change which enables positive feedbacks between them. Consequently, Schumpeter Dynamics has been highly stable and resistant to any type of critique, e.g. regarding the numerous negative externalities which is has created.

2.5.2.4 The 'new' technology assessment: mediation and socio-political process

Technology assessment was introduced at a time when several of the aforementioned limitations of control were not yet generally accepted. In addition, the systemic perspective on technological change and the democratic imperatives with regard to society's technological choices have modified the requirements of TA during the last decade. The recognition of fundamental limitations of determinism and hence of forecasting and control added to this problem. New approaches to technology assessment (Böhret & Franz, 1986; Elzen & Mulder, 1995; Janes, 1996; Leyten & Smits, 1996; Rip et al., 1995; Smits, Leyten, & Geurts, 1987) had to respond to several emerging problems:

- The information deficit, i.e. the need to provide rich and 'objective' data on technological developments, future trends and actual and potential risks/uncertainties involved;
- The implementation deficit, i.e. how can - in the terms of self-reference - the political steering of technological change be intentionally done, taking into account the internal structures and routines of the social systems to be controlled.
- The process deficit of earlier TA, i.e. the need to understand it as a permanent adjustment process during innovation and diffusion, instead of a single-step assessment at the beginning.
- The interaction deficit, i.e. the need to go beyond the provision of information and to integrate different relevant social actors into the decision processes. From a political point of view, this comprises also the facilitation of political control by involving the public, hence attempting to improve communication about technological challenges between the different centres of society.
- The constructive deficit, i.e. the need to achieve an active contribution of all relevant actors to the implementation of new technology in a benign way and the avoid negative side-effects.

This transformation of TA is obviously built on a holistic and contingent view of technological change as a phenomenon which concerns society as a whole. It acknowledges that the top-down planning approach is not feasible with regard to technological change. TA represents on the one hand a feedback to the political system, but on the other hand it also has a mediation function in order to achieve consensus and acceptance by the different (self-referential) subsystems of society. In this sense, TA is not a theory of political control, but a communication tool for establishing a societal dialogue about new technology. It helps to facilitate change and extend the knowledge base but it is also a means of political control.

2.5.3 Summary and assessment

- In general, it should be admitted that the conceptual understanding provided by political sciences for technology-related work is rather limited. Apart from TA, all the approaches are of a very general nature and need to be specified in order to be applicable to technological change. Technology policy and control studies predominantly assume a linear and mechanistic relationship between the political and the techno-economic system

- Corporatism, political economy and self-organisation have helped to shed light on the internal limitations of political control, and on the feedbacks from the techno-economic realm to the political system. By focusing on the political process and the constraints it imposes on control, they neglect the outcome of that process and the constraints imposed by other, e.g. technical, factors.
- Many of the theoretical approaches presented do not have a dynamic element, but are limited to a static description and comparison of structures and behavioural patterns. Dynamic elements are captured only in some of them: shifts of regulatory regimes, the evolution of network, new technology assessment and self-organisation clearly emphasise the process character of political control. Contrary to evolutionary economics, they do not provide the researcher with a dynamic rationale or explanation, but remain descriptive or - in the case of TA - action-oriented.
- Consequently, the simpler linear models should only be regarded as adequate simplifications in cases in which the feedbacks between economics and politics are very weak. For the purpose of this study which is dealing with strongly state-controlled technologies, this is certainly not the case.
- A promising attempt to integrate dynamic elements is the analysis of shifts of regulatory regimes which could be related to similar concepts in economics (e.g. National Systems of Innovation), but it is concerned only the shift of the regime itself and does not cover the dynamics of underlying processes.
- Self-organisational and self-referential systems approaches seem to be particularly useful for analysing political control and technological change because they provide a dynamic perspective and rationale on the problems involved in political control. Network analysis can be used as a helpful tool to reveal interdependencies but it does not cover the process as a whole.
- TA as it has developed in recent years should is an application of new ways of thinking about technological change and political control rather than a theoretical approach. It is a strategy for putting political control into practice, building on a pragmatic understanding of socio-technical change.
- A systems approach appears to be the adequate starting point for integrating different elements and levels of analysis, reflecting interdependencies between micro (e.g. motivations of decision-makers), meso- (e.g. networks) and macro- (e.g. neo-corporatist settings) levels.

2.6 Conclusions

2.6.1 Requirements of an integrated perspective

The preceding review of theoretical approaches from economics, political sciences and sociology to the study of processes of technological change has pointed to a number of important issues which need to be taken into account in a framework for the study of innovation diffusion and political control.

- Innovation diffusion and its political control are very complex phenomena, affected by a multitude of factors which operate at different levels. It is an inherently dynamic process which interconnects social, technological, economic and political driving forces. They bring about a diversity of technologies, and constrain their spread. Consequently, if it is intended to exert influence on this complex process by political means, it is inevitable to develop an approach which is able to capture this multifaceted character.
- It has become obvious that it is not appropriate to regard policies just as an exogenous force to technological change, but that it is necessary to take also the feedbacks to the political realm into account in order to explain technological change: policies are not just imposed on other actors but actively shaped by them, e.g. through pressure groups, formal corporatist arrangements and informal networks. For a policy-oriented analysis an approach is needed which integrates political actors into the system, as endogenous elements subjected to manifold influences and constraints on their control capabilities. This corresponds basically to a self-referential understanding of processes of change in society, with government being a part of the system to be controlled.
- Most of the existing approaches reviewed have emerged from social and economic sciences, not from technical and engineering disciplines. Technical constraints and interdependencies with social issues are not taken into account in a detailed way. However, the conceptual approaches related to evolutionary economics and sociology of technology admit the need to be more precise on the technical side, and they have opened up ways to take technical issues into account. In order to deal also with detailed technical information within an evolutionary or sociological study, it is necessary to provide adequate technological concepts as part of the framework of analysis.
- To move away from existing technological systems and controlling the transition to new ones requires the utilisation of mechanisms in the economic, political and social realm. For the purposes of this study, it is necessary to dispose of a coherent framework and vocabulary which allows considering a wide range of potential sources of change in a forward-looking way. A theoretical approach which would meet these requirements could not be identified. In other words, there is a clear need to integrate existing lines of thinking into a new and more comprehensive framework.
- The approaches reviewed provide many insights into different determinants of change but are, taken alone, too narrow to provide a foundation for construc-

tive policy-oriented analysis in an area such as energy supply. They either focus on the political process only or ignore this process and deal exclusively with the techno-economic and social issues. They either emphasise the structural elements of change or describe in detail the decision-making and behavioural regularities at the micro-level, but do not study them simultaneously. However, many of the approaches described above can be used as building blocks for the encompassing framework needed.

- Technological change is an inherently dynamic process, and it is essential to dispose of a conceptual tool to understand its operation in order to derive conclusions with regard to political control. An evolutionary logic as used in economics seems to be the most promising approach in this respect. As it is closely linked to notions of self-organisation and self-reference it seems possible to link it to concepts used in political sciences. An autopoietic understanding of the behaviour of actors and social systems could be particularly promising to capture the dynamic mechanisms at the micro level.
- A systems approach seems to be the best option to integrate the different contributions from the main disciplines into a coherent framework, and especially to take both structural and behavioural aspects into account. The different contributions discussed need to be joined as elements or modules in an overarching systemic framework.

2.6.2 Existing attempts to integrate perspectives

Until now, only very few attempts have been made to build explicitly a conceptual bridge between the different disciplines relevant to this study. A major convergence can be identified between evolutionary economics and the sociology of technology. Although they have very different roots, the basic understanding of the process of technological change is quite similar, and - even more important - sufficiently open to introduce elements of the other discipline. In several evolutionary approaches the importance of social institutions has been admitted, and the notion of selection environment not only comprises economic forces but also social ones. The concept of technological regimes is also used in both disciplines to capture interdependencies between technology and socio-economic elements. Still, communication across disciplinary boundaries is quite difficult, mainly due to differences in the definitions of concepts and underlying assumptions. As MacKenzie put it: "Economic analysis is often based upon assumptions sociologists regard as absurd, while sociological writing often almost ignores the dimension of cost and profit in its subject matter." (MacKenzie, 1992, p. 25). In this sense, the micro-sociological approaches should be seen as complementary to economic approaches because they are more explicit in describing the interactions and processes at the micro-level. In economics, this is often subsumed under the label of 'diversity', but without stating clearly how this diversity looks like. What is still missing is the actual integration in a single framework which would allow to investigate different cases from a wider perspective, and to bridge explicitly between economics and sociology with regard to technology studies (Blauwhof, 1994; Coombs, Saviotti, & Walsh, 1992; MacKenzie, 1992). This becomes obvious when looking at the convergence between the literature on national innovation systems in economics and on technology-based networks in sociology.

The necessity to bridge also between the political and the economic realm has already been recognised by Irwin and Vergragt in the context of environmental regulation (Irwin & Vergragt, 1989), but the conceptual shortcomings recognised have not been further explored yet. Present approaches to technology policy and innovation diffusion usually do not integrate the feedback links between political and economic realm, one of the few exceptions being the framework presented by Krupp (Krupp, 1992).

Krupp's framework also allows bridging between economic analysis and technical analysis. However, similar to most work on technological change in all social sciences, the link remains very general and superficial. The 'black box' of technology is not really opened, and there is still a clear lack of approaches which link a detailed technical analysis of new technology with a detailed analysis of the socio-economic relationships in which it is embedded.

2.6.3 Elements of an integrated approach

The overarching question of Part I has been to find out what is known about technological change and political control and how this might be integrated into a coherent research approach and methodology. It is suggested that a new framework should be formulated which builds on the following elements:

- A systems terminology as the means by which the different disciplines can be linked together. It will be applied both to the technology and the system of actors. The systems approach includes also the notion of systemic constraints which are based on the interdependencies among the system's elements.
- Conceptualisation of technology and technological change in a terminology which builds upon the concepts of regimes, paradigms, trajectories, systems, knowledge and learning. These are the richest available concepts for the study of technological change but they need to be complemented by in-depth technical information.
- A combination of self-organisational/self-referential concepts and Nelson and Winter's behavioural approach for modelling different actors. It will be complemented by elements of Clark and Juma's understanding of information processing and its transformation into knowledge.
- Interests of actors as the main driving forces of change, building on the conceptual understanding developed in evolutionary economics, but complemented by technological driving forces and constraints.
- The notion of socio-economic shaping as a key element but with a particular emphasis on political shaping. This would provide a solid basis for studying micro-level processes underlying change.
- Elements of the national systems of innovation concept which allows taking the wider structural and institutional context into account.

- A self-referential understanding at the level of systems and subsystems in order to complement the evolutionary and behavioural model elements. This allows to address the question how actors, organisations and subsystems react to impulses from their environment.
- System dynamics will be based on evolutionary ideas of variation and selection of technologies as constrained by political and economic forces. This is regarded to be at present the most sophisticated and useful option, even if the basic notion of autopoiesis may be quite similar.

Chapter 3: The PET- System: An Integrated Approach to Innovation Diffusion and Political Control

In the preceding chapter different approaches to technological change and political control have been reviewed. They allow gaining insights into different aspects of these phenomena but - taken individually - do not provide a comprehensive and well-structured foundation as needed for policy-oriented analysis. This critique is particularly relevant for an area such as energy supply where politics and economics closely interact to determine the technological choices of society. In order to address this deficit an integrated theoretical perspective is going to be developed: the PET-systems approach. PET stands for 'Politics, Economics and Technology'. In Chapter 4, the approach will be developed further into a methodology for the analysis of political control of technological change.

3.1 Introduction

Range of application and scope of the approach

Although the PET-system represents a generalised approach and methodology it has been developed for the specific context of modern industrialised societies. It is meant in particular for areas of research which focus on the following aspects:

- Technological areas, which are characterised by a superposition of market-driven and politically driven factors, e.g. infrastructure services in energy, telecommunications or transport.
- Processes of technological transformation, which are embedded in a network of interests of different actors. Issues such as product and process innovations or the roles of users and suppliers would be part of such an analysis.
- Technological systems in which institutional and organisational factors play an important role. System boundaries are often defined by such institutions, like e.g. in the case of national energy supply systems. For other examples it may

be more appropriate to start from a European perspective, e.g. in the case of telecommunications.

- Interactions among actors and stakeholders, especially if their internal processes and structures can be captured by a simplified actor model.
- Joint operation of techno-economic, social and political determinants as the basis for the study of political control. In such cases the consideration of elements of political self-control, politics, and societal self-reference plays an important role.

Actors and building blocks of systems

It has been argued that a systems perspective would provide a promising foundation for the integrated approach sought. However, 'system' is a highly multifaceted concept which needs to be given a more precise meaning in this study. Definitions of 'systems' are legion, and in the following - i.e. for structuring the variety of interrelated factors having an impact on innovation diffusion and political control - a fairly simple definition will be used. A system can be defined as a number of elements or sub-systems with specific interrelations.[75] Sub-systems (such as individual actors or organisations) can be defined in order to simplify the representation of the internal structure of the system without having to differentiate and describe in detail each individual element. A distinction can be made between 'structures', 'actors' and 'technology' as the main constituents of a system. While actors and technology represent its main sub-units, structures reflect the more stable constellations of a system's elements. In other words, by putting emphasis on the role of actors in a system, it is attempted to bridge between behavioural approaches on the one hand and traditional structural-functional perspectives on the other. Defining a system requires also to delimit it from its environment and to assess its autonomy and consistency. Formulating relationships between system and environment presupposes the existence of an open system. System behaviour can then be interpreted as a reaction to and impact on environmental changes. In general, the relations between environment and the system should be less close than the intra-systemic ones "for the system to have some level of coherence and persistence through time." (Niosi & al., 1993, p.210). Intra-systemic relations are characterised by interaction and interdependence, whereas extra-systemic relations can be defined by impulse and reaction, i.e. there are no feedbacks between environment and system.[76]

---

75 Cf. (Niosi & al., 1993, p. 210; Willke, 1987) for some main definitions on systems, environment and relations. A very good introduction to General Systems Theory, Cybernetics, Systems Analysis and Soft Systems Approaches is also provided in (Rickards, 1985).

76 There are different options about the issue of the relations between system and environment, c.f. (Willke, 1987). Some authors prefer to speak of 'blurred' boundaries between system and environment, thus taking account that it is very difficult to define empirically such boundaries. However, a researcher should make explicit what is included in his analytical perspective and what is not.

The choice and definition of system boundaries is difficult and depends on the perspective and research interest pursued. A very broad delimitation would obscure the most interesting relationships and lead to an unmanageable level of complexity, but a very narrow focus may lead to the omission of important aspects. Anyway, the definition of a system is always a simplification which aims to reduce its description to a manageable number of interdependencies.

In contrast to simpler definitions, which can be found in literature, this is a rather abstract definition of the term 'systems'. It is often just used to denote a relatively complex and interrelated set of factors, elements and relationships without delimiting it in detail or questioning its consistency. By being more explicit in this respect, an attempt is made to use the systems terminology in a more stringent way. Although in principle it must be admitted that innovation and political control are contingent phenomena, the scope of inquiry needs to be reduced and delimited for practical reasons. In this sense, PET is neither a reductionist nor a holistic approach. This is done by stating explicitly the interfaces between the system under study and its environment. It needs to be justified what is part of the system, and why it makes sense to consider it as such. In other words, the aim is not to formulate an all-encompassing theory of society, but to try to be precise in defining which elements of society are being investigated in detail and how they relate to their environment.

In principle, system theory allows formulating analytical statements only; it allows building models. The variables in these models are given meaning only insofar as the model is assumed to be a good representation of reality. This highlights the difference between a *model* (which has an abstract meaning only) and a *theory* (which makes statements on reality). It is thus of crucial importance to justify the applicability of the model to reality, in order to be able to formulate a theory which is based on the model.

In practice, this distinction is rarely made in soft systems approaches.[77] The terminology which has been defined for the PET-system is certainly not in perfect agreement with the aforementioned idealised considerations either. The soft systems approach fulfils the function of a 'tool' for the construction of a theory because it explicitly relates different conceptual issues to one another. It provides hence the 'organisational structure' or pattern for a theory, but the *content* of the theory needs to based on solid theoretical and empirical knowledge.

3.2 Delimitation of the system and exogenous forces

Constitution of the system

The delimitation and structure of the PET-system should not be misunderstood as an empirical reflection of the object of inquiry. Reality is contingent, and PET is a

---

[77] On the difference between systems theory and soft systems approaches, see (Rickards, 1985, p. 26).

partial model of this reality. The delimitation of the system is an analytical necessity in order to keep the research work manageable, and it must be done explicitly in order to show which assumptions underlie the research work.

The integrated perspective sought requires to define system borders which comprise both the techno-economic and the political realm. In addition to technology, the main actors are the users and suppliers of technologies, the different levels of policy-making and other relevant institutions which fulfil an intermediary or facilitating function. Different types of interactions can be identified among these actors. Due to the interest in *technological* change, emphasis is given to interactions which affect this key element of the PET-system. At an aggregate level, longer-term, structural transformations of the system are interpreted as a result of these interactions. The actors' behaviour in interaction processes is based on their perception of both system-endogenous and system-exogenous transformations. This means that a distinction is maintained between technology on the one hand and the social relationships by which it is shaped on the other. Elements of autonomous system dynamics, of socio-politico-economic shaping and of structural constraints are combined in this perspective. It is obvious that the technology as the object of interest must be understood in detail in order to assess its potential applications and the role of technical constraints on decision-making as opposed to socio-economic ones.

In more structured terms, the PET-system operates at five main levels:

- The level of technologies or technological options, which are subjected to permanent changes, transformations and improvements; substantially new options can emerge, too;
- The level of actors, who are characterised by interests, strategies, and different types of decisions;
- The level of interactions between actors: the generation and selection of technologies is based on their decisions and preceding decision-making processes;
- The level of the system-endogenous environment of the technology and the actors, constituted of the evolving structures of the system and the social institutions guiding the interactions. Basically, these structural elements provide the settings inside of which interactions take place and which are in turn the aggregate outcome of these interactions;
- The exogenous selection environment which is conceptualised as not being affected by the interactions in the system. The system as a whole has to adapt to exogenous developments but does not affect them significantly. It is evident that one cannot properly speak of 'selection' in this context because social systems evolve or adapt, but can hardly disappear.

The distinction between system and its (exogenous) environment is purely conceptual for the moment. In any empirical investigation, it is necessary to justify the exclusion of certain potential elements from the system. If it can be argued that - based on available empirical information - certain elements are not subjected to feedbacks, then the system's description can be simplified.

Depending on the research interest, further subsystems could be considered more prominently in an interdependent perspective such as the PET-system. For example, to investigate the impacts of education and training on innovation diffusion, the education system must be taken into account. Given the focus on the interactions between political and techno-economic realm, these other subsystems and inputs are considered here as exogenous factors. This shows that the system approach has disciplining function: it forces the researcher to formulate and structure his work area explicitly.

Exogenous structures and processes in the systemic selection environment

Based on the preceding paragraph, it is obvious that it is not possible to prescribe in general what is 'inside' and what is 'outside' the system boundaries. It always depends on the specific example. However, typical exogenous factors in the selection environment of a PET-system could be

- International resource prices and the resource situation of the system considered;
- For a national economic system, the patterns of economic growth and energy demand;
- Public opinion about new technologies, the emergence of new societal issues affecting the political agenda, or cultural frameworks and institutions;
- International policy decisions regarding e.g. the environment or trade;
- Technological developments outside the system, but affecting the opportunities which emerge inside of it (e.g. technology transfer, scientific advances);
- General economic, organisational and institutional settings of society which are not altered significantly by the evolution of the PET-system.

3.3 The central object of PET: technology

According to recent research findings on technological change, technology should not be conceptualised as a single entity or as a black box. The different hard and soft elements which make a technology operational are distributed among different actors, artefacts and locations.[78] Moreover, technology and technological change can be studied at two levels of analysis:

- The level of aggregate generalised concepts: shared cognitive patterns as described e.g. by technological paradigms (Dosi, 1982), or realised artefacts/technological guideposts (Sahal, 1985) by which the actual feasibility of a concept is symbolically expressed;

---

[78] Hard elements (artefacts) and soft elements (knowledge) can be further differentiated. For example, an artefacts consist of components and architectures, and knowledge has tacit/codified and public/private aspects.

- The micro level, which describes the processes involved in the information and artefact transformations needed at the level of the individual actor or firm (Clark, 1991; Clark & Juma, 1987).

Given this differentiated conceptualisation, it is obvious that a technology needs to be understood in detail in its different implications, constraints, dimensions and meanings before a sensible analysis of actors' interests and decisions with regard to this technology can be made. Although it may be useful to classify technologies into certain categories, e.g. as done in sectorial analyses (Pavitt, 1984), this would ignore the importance of technological specificities. The knowledge of such details can be decisive for the success or failure of control measures.[79] This becomes evident e.g. in the case of technological complementarities, which can only be identified on the basis of detailed technical understanding.

A technology and its evolution can be characterised by a number of main dimensions which are assessed by different actors in different ways. This comes quite close to the social constructivists' notion of "interpretative flexibility" which essentially says that a technology is perceived and interpreted in different ways by different people at different moments in time. The key problem is to identify these key dimensions in order to trace the qualitative changes of the technology in question. This would be a first step towards an analytical framework which captures both the technical and the social dimensions of technology. First steps in this direction have been made by Saviotti and Metcalfe with their concept of technical and service characteristics (Saviotti & Trickett, 1992; Saviotti & Metcalfe, 1984). However, given the broad system perspective outlined before and the range of actors and stakeholders considered, it seems to be necessary to include further dimensions. In a single study it will be impossible to study all possible dimensions, but only those of particular importance to the actors involved. This will be exemplified in Part II for the case of cogeneration. Still, there are a number of aggregate dimensions which are of general relevance to many new technologies:

- <u>Technological dimension</u> which is essential for assessing the technical feasibility and the related scientific and engineering perspectives. From an evolutionary and dynamic perspective, aspects such as the degree of complexity, technological uncertainty, cumulativeness and learning, tacitness, reinforcements and complementarities, etc. have been recognised as being of major importance for the assessment of technological trajectories.
- <u>Micro-economic dimension</u> for the different types of actors involved. Although this may be regarded as the major dimension of interest for technology users and suppliers, it is indirectly also highly relevant for other actors. It deals on the one hand with issues such as technological opportunities, appropriability, and demand expectations, and on the other with price and quality of innovations, with their feasibility, and with the associated risk and uncertainty.

---

79 This perspective implies that although technical characteristics may in the end have to be translated into economic assessments, the direct assessment of technology would allow to draw certain conclusions much earlier.

- Macro- or socio-economic dimension, which reflects the aggregate economic impact of a technology on society. This includes also questions of long-term development and the social aspects of sustainability such as employment, quality of life, or economic growth potential.
- Environmental and resource dimension, which represents in addition to socio-economic development a second long-term assessment element.
- Organisational and power dimension, which deals with the impacts of a technology on the strategic power position of different actors (including political actors). This goes beyond the direct economic impact and builds on the difficult question which long-term organisational impacts a new technology would have on the structure of the system.

Depending on the technology in question, the importance of each dimension can differ. Other dimensions might have to be added, e.g. in response to changing priorities on the political agendas. In addition, different actors in the system will assign different degrees of importance to the dimensions, but taken together they are all necessary to understand the range of interests and positions at stake.

In the context of diffusion of technologies it is evident that any analysis has to take the different alternatives to a technology into account. 'Alternatives' is a less clear-cut idea than it seems to be at first glance, because actors can have very different opinions about what an alternative might be.

3.4 Structures and structural changes

The term structures can have at least two different meanings, namely it can be used either to describe the internal differentiation of a unit of analysis, or represent the stable framework inside of which certain activities take place. In the following, both connotations will be relevant. Structural changes are interpreted as the transformations of the relatively stable settings of the system under study. These structures represent at the same time the context for interaction. It is important to recognise that structures emerge as a result of historical processes. They are the aggregate outcome of interactions which have taken place earlier on. In other words, structures and interactions are interconnected. Structures are comparatively stable, but nevertheless changing characteristics of a system, though changing with a considerably longer characteristic time than interactions.[80] This process of *structural evolution* implies that "the relative persistence of structures is no longer emphasised, new structures are considered as being continuously formed in the course of time, and complementarities over time do emerge as a result of historical

80 See on this aspect also the autopoiesis model of Maturana and Varela (Maturana & Varela, 1980). They make a distinction between "organisation" and "structure". Whereas the organisation of all autopoietic systems the same, they differ with regard to their structure and can change their structure in the course of adaptation processes while maintaining their organisation.

(or "evolutionary") linkages between events at different periods of time" (Arthur, Landesmann, & Scazzieri, 1991).

The notion of regimes comes close to this understanding of structures, in fact they overlap. A technological regime as defined by Kemp et al. comprises many different elements.[81] Their understanding of technological regimes comprises technological elements, micro-level elements such as behavioural practices, engineering styles etc., but also several structural elements. However, political structures are not considered in their definition. In order to avoid using a too comprehensive and undifferentiated concept which does not allow to make differentiated statements, the different elements (structures, behavioural practices, technological elements) will be kept separate.

In the following, the main types of structures to characterise the PET-system are introduced. They are regarded to co-evolve in response to the interactions in the system. Basically, this idea of co-evolution means that a broad range of structural conditions needs to fit to each other as well as to the technological changes considered in order to enable technology diffusion.[82]

3.4.1 Changes of technological structures

The key changes of the technological structures are reflected in the diffusion patterns of the technology under study. During the diffusion process the technologies is modified and improved. The diffusion potential can be defined either technically, socio-economically, institutionally or even spatially. It changes with the changing characteristics of the technology. In other words, early diffusion phases can be compared to niche development processes, by which the technological context for diffusion is transformed. Other characteristics of technological structures can be defined in terms of:

- Diffusion patterns of alternative and competing options;
- Qualitative changes of the entire technological system considered, e.g. its technical organisation in terms of architecture and components;

---

81 They define a technological regime as "the overall complex of scientific knowledge, engineering practices, production process technologies, product characteristics, skills and procedures, institutions and infrastructures which make up the totality of a technology." (Kemp et al., 1994, p. 7).

82 In order to make this concept of structures more operational, it would be useful to introduce classifications of the different structures considered. They could then be used to formulate a typology of 'PET-regimes'. One could start with existing typologies in political sciences and economics, like, for example, notions of open/closed, centralised/decentralised, and weak/strong political systems (Kitschelt, 1983), or centralised/decentralised, vertically/horizontally integrated, market-based/interventionist economic structures. Technological structures can be characterised as centralised/decentralised, homogeneous/heterogeneous or independent/interdependent. Information supply structures can be characterised as closed/open, formal/informal, or by intensity of informational flows.

- The evolution of technical infrastructures on which a technology relies; this is particularly relevant in large infrastructure systems like energy, transport or telecommunications which represent complex sets of interdependent elements, but it can also be applied to technical complementarities in a company;
- Interdependencies with other technologies.

While the technological structures refer to the main *object* of interest, i.e. to technology, the next three types of structures characterise the relationships between the different subjects of interest, i.e. the actors.

3.4.2 Changes of economic structures

Economic structures should not be misunderstood in the sense of market structures only. The term is used in a far more comprehensive sense which includes

- The type of economic organisation of the industry considered (e.g. regulated/liberalised);
- The evolution of the market forms (monopolistic, oligopolistic, competitive) and market shares;
- The creation, disappearance, fusion and separation of firms;
- Changes of regulatory settings and their intensity (regulatory regime);
- Distribution of public and private ownership;
- The extent and characteristics of competition and co-operation.

Economic structures need to be investigated both for the industries on the technology supply side and on the technology demand side. General economic settings such as the patent system, general taxation etc. are interpreted as exogenous conditions. Anyway, the elements of economic structures mentioned above should not be interpreted as a rigid pattern of analysis. Depending on the specific sector under study it might be necessary to consider additional aspects.

Given the fact that supply and demand side of technology are studied simultaneously, their respective economic context needs to be considered, too. In addition to the technology-related user-supplier link, each firm is embedded in a web of other market relationships which can be interpreted as its 'secondary markets' as compared to its main technology market. For instance, the purchase of certain technological components takes place on an important secondary market for a firm.[83] Changes on this secondary markets can induce changes on the main technology market, and vice-versa.

---

83 This concept of 'secondary markets' obviously comes close to a kind of network notion, but - see the definition of networks - they are explicitly dealing with market relationships in the economic context and not with any kind of informal links as networks.

3.4.3 Changes of information structures

Information and knowledge about technology are key elements to understand decisions and decision-making processes in the context of innovation, adoption and policy-making. Knowledge has been interpreted as one of the two main dimension of technology, in addition to the artifact dimensions. Learning depends on the access to information sources. Much effort has been put into investigating how R&D could be efficiently organised to learn better and generate knowledge. In Section 3.5, the question how actors or firms actually get access to information from their environment and transform it into knowledge will be discussed. In the following, the main concern is how technological information, its generation and distribution are organised and structured within the system. It has been recognised that these structures differ between countries. They have an important impact on the success of industries and on the performance of national economies. Some of the elements which need to be taken into consideration concern

- Research infrastructures and public R&D organisations in a country,
- Industrial organisations involved in the provision of technological information,
- Government bodies involved in similar activities, especially with regard to the promotion of technologies and the dissemination of technological information,
- Organisations which facilitate the transfer of technologies, e.g. from basic to applied research and to the market, from public to private firms, and also among private firms,
- Facilities and interfaces to connect firms to these different information sources.

Technological knowledge is a typical example of a good which is not only supplied through market relationships but probably to an even larger extent through informal channels. It is transferred by the employment of highly skilled employees, by network-like relationships with other public and private organisations, or by co-operation between firms in joint projects.

3.4.4 Changes of political structures

Regulatory regimes and specific regulations have already been integrated as a first political element into the analysis of economic structures. Structures of political control go beyond this and looks also at other policies and at determinants of technological change which are internal to the political system itself. The structures of the political system and the interfaces with other social systems need to be taken into account because the settings inside of which policies and regulations are formulated and implemented affect and shape technological change. These political structures are composed of elements such as

- Organisational integration of different interests in the political process,
- Dependence of policy-makers on external sources of information,
- Distribution of competencies between different levels of government,

- Distribution of competencies between different policy departments,
- Overall power of the political authorities to intervene in industrial affairs,
- Formal political institutions.

In other words, political structures represent the formal *and* informal settings for technology-related policy making and action. Similar to all other structures of the PET-system, they are regarded to be shaped by decisions at the micro-level.

3.5 Actors and motivations

3.5.1 The underlying actor model

The actor model which is used for the PET-system is based on three main sources of inspiration:

- A behavioural element, especially the contributions of Nelsons and Winter (Nelson & Winter, 1982). They build on earlier works by Simon (Simon, 1955; Simon, 1959) and Cyert and March (Cyert & March, 1963).
- An autopoietic element, which builds on a self-referential understanding of social systems. It is based on the work of Maturana and Varela (Maturana & Varela, 1980), which has penetrated the social sciences in the last decade.
- An information-theoretical element, using elements of Clark and Juma's contributions (Clark, 1991; Clark & Juma, 1987). They deal with the question how information from the environment can be absorbed by companies and transformed into operational knowledge.

Firms, routines and satisficing behaviour

The traditional and standard economic models of actor and firm behaviour is based on notions of maximisation, complete information and rationality. Although this traditional view has been improved and adapted to the needs of technological change (e.g. by relaxing some of the restrictive assumptions on complete information and homogeneity of actors), the behavioural concept as proposed by Nelson and Winter is regarded to provide a more realistic foundation for studying decision situations under uncertainty, especially technological uncertainty (Nelson & Winter, 1982). It can be summarised by the following characteristics:

- The firm model is based on a model of individual "skills", which are characterised by certain "programmes" for standard tasks, being executed in an automatic way. The capability to make "choices" is limited to variations of existing programmes. Individuals are able to learn from past experience in order to improve their skills. Skills are consequently specific to individuals.
- Similarly, one speaks of "routines" at firm level. They are differentiated into short-term "operating characteristics", medium-term "capital stock variations",

and longer-term, change-oriented "search routines". Routines are the main means for storing knowledge in stable and repetitive procedures. They are the outcome of preceding learning processes and thus contribute to the diversity of firms. By way of search routines, innovations can be introduced into firms. This process is subjected to the problem of bounded rationality with respect to new technologies. The dominant search mode is "local search" which means that it mainly takes place close to existing routines and knowledge, thereby reducing the uncertainty associated to search processes. Search routines are not constantly activated but only if the basic performance criteria of the firm are not achieved to a "satisficing" extent. The concept of "satisficing behaviour" is a core element of Nelson and Winter's thinking because it reflects the departure from the standard maximisation model. Decisions on certain non-routine types of investment such as strategic R&D are made only if the organisation falls short of its minimum level of satisfaction.

- Due to uncertainty and the local character of search procedures, there is a historical element in their model. It implies that the capabilities of firms depend on the routines which have been built up in the past. Knowledge and technology are not regarded as easily transferable goods any more.
- In principle, the model can be applied to private and public firms as well as to any other type of organisation (even if Nelson and Winter applied it predominantly to industrial firms). What differs is the content of the routines, the goals and the criteria for search and decision-making. For private firms the expectations to appropriate profits could be regarded as a criterion to innovate or to adopt a technology, for public organisations the fulfilment of some kind of public task or possibly also the self-interest of its members.
- These generalisations of the original firm model require to have a broader understanding of what "behaviour" means. It clearly goes beyond typical economic activities like investment and technological choices, and comprises also other types of activities such as political interactions, decisions to co-operate with other actors, etc.

Self-reference and autopoiesis

Self-reference and autopoiesis imply an understanding of behaviour which rejects the simple mechanistic stimulus-response model. They assume that "living systems"[84] react to external impulses in a way which is predetermined by their internal structure, and not necessarily according to the original intentions of the external impulse. As an example of autopoiesis, the operation of internal incentive mechanisms in large organisations can be mentioned. They can induce opportun-

---

84 Maturana and Varela refer explicitly to living systems which are subjected to autopoiesis. However, this should not be misunderstood in a purely biological sense, although they started to study biological systems. In their interpretation, all systems which obey to the mechanisms of autopoiesis are by definition living systems (Maturana & Varela, 1980).

istic behaviour like career orientation rather than behaviour which is in the interest of the organisation.

An autopoietic understanding of systems has major implications for control. For example, a black box model of control could certainly not be justified. Targeted control may be possible in principle, but only if the internal processing of these control impulses by the object of control is understood in detail. In practice, the only resort is seen in a probabilistic contextual control approach. In this case, and due to the recognition that control is subjected to fundamental limitations, the objects of control are considered as quite autonomous. Systems can be "perturbed", i.e. reactions be stimulated, but it is not possible to predict these reactions in detail. These limits to control are very important from a political perspective. While targeted control would not be feasible, possibilities for more modest measures remain. Their quality depends to a large extent on a good understanding of how the object of control operates. Control analysis therefore needs to concentrate on the object of control.

Neither can it be assumed from the perspective of autopoiesis that the objects of control show maximising behaviour, because in this case it would have to be perfectly informed about its environment and - even more important - be able to orient its decisions "optimally" in response to the state of the environment. In contrast to this, self-reference implies an orientation according to internal characteristics and objectives. This appears to be compatible with several of the aforementioned sociological findings on technological change, e.g. concerning the role of paradigms, shared beliefs and expectations.

Information theory

While neo-classical economics does not make a clear distinction between information and knowledge, information theory has focused exactly on the difficulties to transform the former into the latter. According to Clark and Juma (Clark & Juma, 1987), information needs to be interpreted as an intangible pervasive flow throughout economic systems, which has not yet been absorbed and integrated into a specific actor's "knowledge-base". It can not yet be used in an operational way to the benefit of this actors; it is not yet 'fixed'. In order to achieve this integration into a knowledge-base or a routine, several steps are needed between source of information and receiver: encoding, transfer, decoding and reception. Coding and decoding require usually the understanding of a specific language. Selecting and filtering relevant pieces of information is part of the reception process, and it is crucially affected by technological paradigms, orienting theories and mental frameworks. In other words, the transfer process of information into knowledge is far from being simple and objective.

These translation problems concerns all stages and facets of technological change. Research work is as much affected by it as the interactions of a firm with its outside world, or its internal information processing procedures. The perspective of information theory underlines that *perception* is the key to understanding actor behaviour showing that any apparently objective piece of information is evaluated by this filtering mechanism.

3.5.2 Actors' decisions and links to their environment

Based on the preceding arguments, the question of how decisions are actually made is going to be addressed now. Economic decisions (e.g. to develop or adopt a particular technology) as well as political actions (e.g. to promote a certain technology programme) will have to be taken into account. This model of decision-making uses the following main elements (c.f. also Figure 3.1):

1. Inputs to decision-making:
    - Perception of information;
    - Transformation into knowledge.
2. Transformation processes:
    - Goals, interests, strategies;
    - Resources in terms of production factors and access to finance;
    - Anticipations and expectations of future changes;
    - Decision-making processes, internal interests and conflicts.
3. Outputs of decision-making:
    - Final decisions;
    - Interactions.

Actors' interfaces with their environment - inputs

Actors receive a great variety of information inputs from their environment, both on structural changes and on the interactions and decisions of other actors. Their own interactions with others provide an additional source of information. The information flows through these different channels are neither completely perceived (quantitative limitations) nor objectively interpreted (qualitative limitations). Firm routines for the perception of information, governed by the dominant mental frameworks and paradigms of actors, fulfil the role of filters and give way to local search procedures.

The transformation of information into knowledge through these channels and organisational routines builds up the knowledge base of a firm. It play a crucial role for the quality of decision-making. There are routines for different monitoring activities e.g. concerning the technological developments, market relations, relevant public policies and other boundary conditions.

This process of building up a knowledge-base is further complicated by the existence of the above mentioned search paradigms. Not only do they filter the information which is going to be stored in the knowledge base, but they are also shaped by what is regarded as important in the search routines, i.e. by the content of the knowledge base itself. This is a typical example of a self-reinforcing feedback. However, information collection is not solely determined by these rather unconscious paradigms, but also by the consciously defined goals and strategies.

Consequently, the ability of a firm to learn (i.e. to transform information into knowledge, to store it in the knowledge-base and to apply it in the form of new routines) is constrained by many factors. This applies not only to innovations, but also to any other type of decision. In the case of innovation, the constraints are particularly strong due to the fundamental uncertainty involved.[85] These constraints on learning inhibit on the one hand innovation and diffusion, but on the other hand they are essential for maintaining the incentives to innovate and adopt early on.

Knowledge can take a variety of different forms in a firm: embodied in artefacts, codified in documents and procedures, but also tacitly incorporated in the skills and capabilities of human beings, or stored as organisational routines. It may be generally available or specific to certain individuals. Due to this dispersion of a its knowledge-base, a firm needs to dispose also of knowledge about its internal capabilities and resources in order to 'match' the internal and external situation with its decisions. The knowledge base of a firm is hence not just regarded as a pool of different forms of knowledge available inside the firm, but also comprises the routines for accessing and exploiting new information sources and for identifying what knowledge is available or missing.

Decision-making and objectives

Decision-making and firm behaviour regarding innovation and diffusion draw on the one hand on the available resources in terms of knowledge and finance and on the other hand on the goals and criteria which serve as orientation for decisions. Profit-orientation and 'survival' are typical goals of private firms, but if public actors are considered, too, then a broader understanding of 'profit' is required. Public interests should e.g. be included in the notion of profit if it is to be applied to public firms and utilities. However, it is sufficient to keep in mind that the goals of public firms can differ from those of private firms, and that they may be imposed by public authorities. The traditional objectives of companies as expressed in profit, survival and growth should also be complemented by long-term, strategic aspects of firm development. They are particularly important for guiding strategic decisions under conditions of uncertainty.

Anticipation and expectations play an important role in decision-making and build on available knowledge. Knowledge is interpreted and extrapolated by routines which are guided by the firm's internal paradigms, thus constraining 'objective' decision-making further. Due to inherent uncertainties, limited capacities to collect and process information, and fast processes of techno-economic change, the standard maximisation hypothesis can not be maintained. It is replaced by satisficing behaviour as the fundamental pattern of decision-making. It means that activities are started only if discontent with the firm's operational results remains under a certain threshold level. Satisficing behaviour can easily be interpreted in self-referential terms because it does not aim at fulfilling subjective preferences with

---

85 Uncertainty in meant here in the sense of Knight, i.e. it refers not only to uncertainties with regard to the extent of diffusion or acceptance of innovations, but also to the complete ignorance of the content of new technologies to emerge in the future.

regard to the environment, but is driven by the internal requirements and organisation of the firm itself.

Although uncertainty is an important constraint on making rational choices, attempts to anticipate future developments are nevertheless important. Qualified knowledge-based anticipations, strategic considerations and resource constraints provide the ingredients for the routinised decision-making process. The strategic aspects represent an element of deliberate decision-making, i.e. firms are not seen as just reacting in the best possible way to environmental constraints, but as actively pursuing 'visions' or 'ideologies'.[86] Strategies could also be seen as a necessary result of uncertainty, because they provide a broad orientation for decision-making.

These self-referential considerations may already help to explain how individual firms' decisions are made and how variety among firms emerges, but further mechanisms are required to understand the adjustment processes among firms and how technological trajectories and paradigms get established. The decisions of other actors are important determinants of the strategy of a firm, its reactions and adjustments. At the technological level, the shared beliefs among engineers of what is regarded to be technically feasible represents a major aligning and reinforcing effect. Selection processes on the market or in any other decision arenas further contribute to the reduction of variety. As part of a dynamic process, they shape the technological development path of the system under study.

Although decision-making processes of actors may follow routinised procedures, it should be taken into account that this is not necessarily a straightforward and rational process, based on well-defined and clear-cut decision criteria. Internal conflicts of interests, reflecting again certain self-referential characteristics, can slow down and bias decisions. Plans may be abandoned and taken up again at a later stage. This reflects a complicated and open process, in which the participants draw on different elements of the firms knowledge base.

Interfaces with the environment - outputs

Two forms of outputs, which relate a firm to its environment, can be distinguished, namely

- Processes of exchange of (predominantly) information and other types of means of interaction (e.g. money);
- Decisions, which can be categorised into innovation decisions, adoption decisions and political decisions; they are in the first instance targeted at the environment and other actors, but can also have an impact on the firm itself and thus represent an element of self-control.

Outputs can comprise several different formal and informal means to affect the social and political environment of an actor. The latter aspect is particularly inter-

86 These visions could be interpreted as political programmes, but also guiding ideas of a future society (e.g. the Information Society) or even the projected roles of large firms in their specific sectors.

esting with regard to political control, because it allows considering the influence of interest groups within the political process.

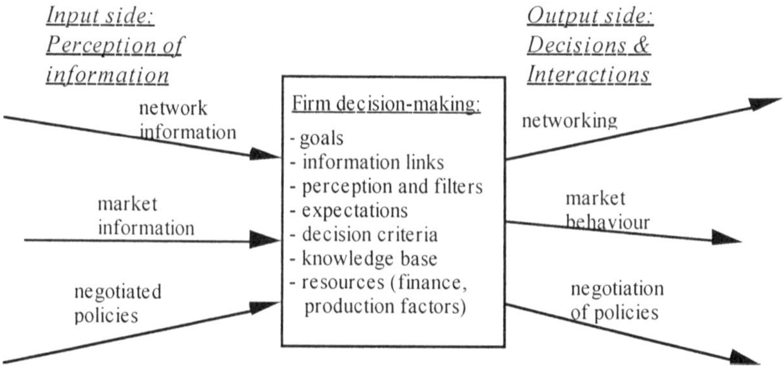

Figure 3.1: A firm's links to its environment

Diversity of actors' behaviour

As a consequence of this conceptualisation of actors, they differ from each other and show heterogeneous perceptions, anticipations and behavioural patterns. The main reasons for this diversity can be summarised as follows:

- Different relations with their environment lead to different pieces of information being received;
- Different internal structures for processing information and transforming it into knowledge are reflected in the search routines of the firms; moreover they build on pre-established knowledge bases.
- Inherent uncertainty of future developments and of new technologies in particular, leading to different individual anticipations and strategies.

If firms differ with regard to the way in which they perceive their environment then it is evident that the knowledge bases as well as the transformation processes differ. This leads finally to a diversity of firm decision-making and behaviour. These differences are reinforced by some of the characteristics technology can have, especially tacitness, cumulativeness and complexity. They further constrain the transferability of knowledge to other actors.

3.5.3 Types of actors considered

The actor model outlined above applies to all the different types of actors who constitute a system. In spite of their diversity, it is possible to cluster them into three main groups which correspond to the three decision realms

- Technology supply;
- Technology demand;
- Public policy-making.

The advantage of the proposed firm model consists of its broad applicability, especially that it can also be used to describe actors who operate within the political system by using insights on the micro-processes of policy-making. This simplifies the structure of the entire research approach as well as the empirical study because the same conceptual basis can be used for the description and interpretation of all the relevant actors' behaviour. However, even if the same model is used for all actors, their motivations, interests and strategies differ and need to be studied for each type of actor individually. In the following, the actors at the technology supply and demand side will only be briefly discussed, because they have been frequently analysed in economic research. The application of the actor model to public policy making will on the contrary be explained in detail in order to exemplify the use of the model. Moreover, Nelson and Winters' actor model has hardly been applied to policy-making, although they had already pointed to the possibility of applying their routine and satisficing approach to other types of organisations than industrial firms (Nelson & Winter, 1982, p. 372-375).

3.5.3.1 Users and suppliers of technology

With regard to technology it is important to distinguish two main groups of actors in the economic realm: technology suppliers and technology users.[87] The importance of the interactions between users and suppliers of technology for both innovation and diffusion has been recognised by several researchers (Hippel, 1988; Lundvall, 1988; Lundvall, 1992a). Rather than studying innovation, adoption and imitation separately as it was done in earlier research work, innovation and diffusion are now usually regarded as interrelated processes.

In more practical terms, it needs to be found out, which pieces of information are finally taken into account for decision-making and which criteria are applied by users and suppliers of technology in their innovation and adoption activities as expressed in different types of investment, organisational decisions etc. Dosi suggested an explanation of the differences in the generation and diffusion of innovations among innovators and technology suppliers, which is based on three categories of determinants (Dosi, 1988a):

- Technological opportunities perceived in an innovation, or more generally in a technological paradigm;
- Appropriability conditions, i.e. the expectations of economic returns of different innovations;
- The expected patterns of market demand for the innovation.

---

[87] User and suppliers can also be part of the same company. In this case the transfer of the technology takes place between different departments.

This formulation is regarded to be applicable to innovating technology suppliers in general, i.e. to radical innovations as well as to incremental improvements and adaptations. A complementary interpretations has been proposed by Kemp et al. for the user side of technology, dealing with the adoption decisions which finally determine diffusion (Kemp & Soete, 1992; Kemp et al., 1991). Three main types of factors are considered:

- Cost-benefit considerations, i.e. the comparative performance in terms of price and quality of different alternative products;
- The ease by which the technology and especially the necessary knowledge can be transferred;
- The general risk and uncertainty to the innovation, including also constraints imposed by the existing capital stock and financing possibilities of the firm.

It is evident that for any specific empirical investigation the actors on both user and supply sides need to be identified. On the technology supply side, for instance, universities, public and private research institutes, technology manufacturers, engineering and industrial associations should be considered, while typical actors on the demand side are companies, private households, government institutions, industrial associations, etc.

3.5.3.2 Political authorities

Delimitation of the range of political actors considered

Public authorities are not a homogeneous body, but are highly differentiated organisations which often pursue diverging interests and strategies with regard to new technologies. Different levels of authorities can be identified, ranging from municipalities and regions to countries and even supranational organisations such as the EU. The question, at which of these levels a study should be focused, depends very much on the research interest (e.g. the impact of EU-policies). With regard to political control, the interaction processes between policy-makers, firm representatives and other stakeholders in the formulation and implementation of policies will be of major interest. Only those political actors who are actually involved in the negotiations are needed as constituent elements of the system. Others can remain exogenous to the system of interactions, and their respective policies be regarded as boundary conditions.

Inputs

Similar to industrial firms, policy-making authorities require expertise and knowledge on technology issues in order to develop adequate political measures. They are embedded in a web of information channels and are confronted with the same types of problems of information processing as any other actor. This reliance on external information sources represents on the one hand an element of dependence, implying the danger of manipulation and dominance by external parties. On the other hand - for example with regard to scientific expertise - it also facilitates the selective use of information by policy-makers to support their own objectives.[88]

Learning takes place by making experiences with the application and implementation of different policy measures. Policy evaluation exercises can support this learning process by systematically assessing the actual effects of a policy as compared to the intended effects. This is predominantly the concern of the political-administrative system which hosts the lion's share of the knowledge base in the political realm. It also represents an element of continuity as contrasted to the legislative cycle of four years.

By focusing mainly on the feedback mechanisms between techno-economic and political realm regarding the creation of technology-related public policies, most other elements of what is usually defined as the 'political system' can remain exogenous to the PET-system perspective. The underlying assumption is that although policies are the outcome of manifold interests and problem perceptions which are transformed in the political system into policies, the groups which are very much affected by a policy are also the ones which intend to play and influential role in its formulation and implementation. However, in practice several interests may not be fully represented among the most powerful interest groups. In other words, the political realm is separated into a part which is located within the PET-system and interacts with the techno-economic actors, and a part outside of the PET-system which is interpreted as part of the environment. This may be unusual, but it is an attempt to integrate at least some of the feedback loops between economic and political realm into an analysis of innovation diffusion.

Policy-making process, interests and criteria

Obviously, one of the performance indicators of government is the public support they have, as expressed e.g. in elections. They impose a kind of selection pressure on governments to pursue policies which are at least perceived as successful by the public. However, there are many other determinants, driving forces of and obstacles to change in policy-making (c.f. Chapter 2). Regarding motivations of actors, at least four different levels can be identified, namely

- Individual interests like self-interest, individual power and prestige, etc.;
- Organisational interests, i.e. the specific modal logic of individual departments and levels of government;
- External forces like public opinion and pressure groups;

---

88 With regard to technology, scientific advise has gained an increasing role to support political decision-making. This is due to the lack of specialised knowledge among policy-makers. It means that much responsibility is delegated to scientists - assuming that they are able to come up with an impartial position. This is obviously in disagreement with empirical findings (Midttun & Baumgartner, 1986) which show that scientific results are often used (and abused) as arguments in the political debates. 'Scientific' should not be equalled to 'true', because there are always several competing opinions on any single theme.

- Ideological motivations and convictions.

Governance systems across countries differ in the ways in which the political realm and the interfaces between economy and politics are organised, e.g. in terms of the degree of co-operation, (de-)centralisation, reliance on market-based or interventionist policy measures, etc. Governance patterns are a structural element of national system and result from path-dependent and historical developments, which can be captured in terms like political cultures, political structures or political regimes.[89]

Outputs

At least two main types of outputs of the political system need to be distinguished. First of all, there are the 'classical' outputs like policies, regulations, financial support, taxes, information, etc. In other words, these are the different means by which government exerts influence on society and economy. Public ownership and procurement can help to reinforce these specific policies. However, public enterprises should in principle be interpreted as economic actors, pursuing particular goals and commitments with regard to the public as defined by political authorities. Secondly, there are also outputs which aim at the internal organisation of the political process. For example, the relationships among policy departments, and between policy departments and the representatives of different interest groups need to be organised. These measures represent elements of political self-control.

3.5.3.3 Other intra-systemic actors

Users, suppliers and policy-makers are regarded as the main groups of actors in the PET-system, but one should not forget other actors which could play a certain role within the system but can not be easily captured under one of the three headlines. Some actors can be assigned to one of the three realms of technology supply, technology use or policy-making, but they do not represent a core element of these areas. In particular, there are institutions which operate at the interfaces between the three realms, for instance

- Industrial and trade associations between government and industry;
- Between the three main groups of actors and the system environment, environmental interest groups and other relevant representations of specific societal interests are active, which influence intra-systemic processes, but are not directly involved in innovation diffusion and political control of the technology under study.

However, in practical terms it will mostly be possible to associate these actors to one of the three major realms in order to simplify the structure of the analysis.

---

89 Similarities to the concept of national systems of innovation are also evident, but NSI do not really focus on policy-making aspects. PET deals rather with particular technological field which is embedded in a system of economic and political interaction.

3.6 Decisions and interactions

The analysis of interactions and decisions requires to study two types of essential links in PET, namely the links between behavioural and structural level, and the interactions between different actors.

3.6.1 Decisions and decision arenas: innovation, adoption, policies

Innovation, adoption and policy decisions are regarded to be the outcome of preceding interactions among the actors, even if they are taken finally by one of them individually. These interactions take place in 'interaction or decision arenas' which are governed by different sets of rules and principles, and by different logics of interaction. In the political realm, these are e.g. the consultation and negotiation processes which precede any important decision. In a company, an important investment is prepared in a process of negotiations with suppliers, experts and researchers, with marketing experts and regulators. Decisions are the final result of these complex preparatory interactions. They also have non-intended side-effects such as the disclosure and leakage of knowledge. Political decisions have a particularly wide-ranging impact in the system, because at least some of the political outputs are "generally binding"[90] and because they define the framework for innovation and adoption. Three main types of decisions are distinguished in the PET-system, and correspondingly also three decision arenas:

- Innovation decisions;
- Adoption decisions;
- Political decisions.

In line with the actor-model outlined above, the exchange of information between the actors is a central element of the interactions. However, this is clearly not the only source of information for them because they also monitor independently the changes taking place in their environment (such as structural changes, prices, technological improvements etc.) as well as the positions and strategies of the different relevant actors. The assessment of other actors' strategies is of crucial importance for the final decision to be taken by an actor.

3.6.2 A typology of interactions

Interactions are regarded to underlie any decision which triggers major changes in the system. These interactions can be classified according to two main dimensions, namely with regard to their content and their organisation.

---

90 According to Easton (Easton, 1965, p. 352-353) one needs to distinguish between authoritative and associated outputs of the political system. Usually, formal outputs of political action are authoritative, and they are generally accepted as binding by the members of the system, while more informal actions like commitments or intentions are associated outputs of a less strict character.

Four types of possible contents or means of interaction can be defined:

- Financial interactions, which are the typical economic means of exchange;
- Information exchange, i.e. information on the technology in question, on its technical, economic, environmental or social characteristics, and on other actors' strategies and positions;
- Power interactions, e.g. 'command and control'-type enforcement of interests, legislative measures, but also negotiations between political authorities and economic agents;
- Exchange of artefacts, i.e. technical hardware which is exchanged as a result of the decisions made in the interaction arenas.

Actors are linked together in organisational structures, which are then activated in interactions. These structures can be classified according to different criteria. A very common classification is based on the following three types of relationships:

- Market relationships;
- Hierarchical relationship;
- Network relationships.

Actors are usually integrated in manifold interactions which correspond to different types of organisational models, i.e. they are linked to markets as well as to networks and hierarchies. For example, large companies are not only connected to their economic environment (where usually market relations dominate) but also to government, and to various sources of technological and scientific knowledge. In addition, the types of content and organisation of interactions seem to be correlated. Financial interactions typically take place in markets, whereas information can be transferred both formally through markets but also informally through networks. Power is often, though not exclusively, exerted through hierarchies, especially in the political and legal sense.[91]

However, as the examples in Table 3.1 show, all combinations of content and organisation of interactions are possible.[92] In principle, the importance of these different types of interactions could be related to other characteristics of a system, e.g. to the structures of a specific industry or to different types of technologies. In other words, this typology opens up the possibility to classify sectors, industries or countries according to characteristic interactions.

---

91 Obviously, in a broader sense, power can also be exerted by market dominance or by the access to knowledge. Power can also be defined as the possibility to mobilise economic, political and any other kind of 'resources'. Here, 'power' is used in a more narrow sense, delimited mainly to legal political power.

92 This typology is clearly related to the forms (market, hierarchy, solidarity) and measures/media of (political) control (law, money, information), but modified to accommodate the needs of PET.

Content / Organisation	Financial	Information	Artefacts	Power
Markets	Standard financial exchanges	Sale of information and information services	Sale of hardware; Maintenance services	Monopolistic large against small firms
Hierarchies	Hierarchical organisations; Principal-agent relations	Standard information flows in hierarchical organisations	Distribution of material; Centralised purchasing	Standard, execution of political power by decree & law
Networks	Corruption of mutual benefit	Informal exchange based on personal relationships	Software exchange; Smuggle	Negotiation of policies; Threat to withdraw trust

Table 3.1: Types and examples of interactions in the PET-system

3.6.3 Institutions, institutional changes and interactions

There are many different definitions of what institutions are. While in political sciences they are often described as formal political institutions, in economic and sociology they have a broader meaning. Carlsson and Stankiewicz define them as "the normative structures which promote stable patterns of social interactions/transactions necessary for the performance of vital societal functions." (Carlsson & Stankiewicz, 1991, p. 109). In other words they are seen as settings, norms and rules guiding interactions. From the PET-perspective, this definition seems to be problematic, because these "stable patterns" become less stable from a perspective of structural systemic change. More appropriate appears to be a conceptualisation which interprets institutions as the guiding but changing principles and patterns of interactions in different organisational forms. They fulfil a stabilising function for the different types of interactions, but are not fixed. Similar to system structures, institutions change, and new institutions can be introduced to the system, be it at a comparatively slow pace. This definition is also consistent with the view that institutions are an expression of historically grown behavioural patterns. This longer-term stability is an aspect which they share with structures: both are the outcome of an evolutionary path-dependent process (Nelson, 1995. p. 81). In addition, they do not only guide interactions, but are also shaped by them in a process of mutual reinforcement (Bijker & Law, 1992a, p. 299). However, this broad behavioural definition of institutions can be linked easily to the more narrow organisational definition. Institutionalised guiding principles of interaction can manifest themselves in a well-defined organisational form (i.e. institutions in the political sense), but they can also remain informal arrangements.

3.7 System dynamics

In the preceding sections the main elements of the PET-system have been introduced. Their relationships are depicted in Figure 3.2. Still, this is a static representation, and in order to transform the PET-system into an approach for explaining patterns and mechanisms of technological change, a dynamic interpretation needs to be provided. In other words, what is missing is a generalised explanation of how changes take place in the PET-system, showing by which forces and mechanisms the changes of the system are driven. This will be done in the following section by using elements of an evolutionary logic to capture the generation, spread and selection of novelties in the PET-system.

Evolutionary economics draws a rough analogy to biological metaphors of evolution.[93] Usually, they can be reduced to three key elements: variety generation, the selection of alternatives, and elements of continuity/stability. Despite the difficulty to apply and operationalise such simple biological analogies, evolutionary reasoning is one of the few useful explanatory models of change. It can be further improved by self-organisational and self-referential elements. In order to develop an explanatory model which is geared towards the needs and the focus of the PET-system, seven elements of an evolutionary explanation are proposed:

- External pressure;
- Variation and selection by innovations;
- Selection by users in their organisational contexts;
- Selection conditions of the regulatory and political framework;
- Stabilisation mechanisms;
- Reinforcing and delaying mechanisms;
- System evolution.

The fact that the generation and spread of new technologies cannot be separated from several other processes of change in the economy, in society and in the political realm represents a major challenge to this model. In fact, innovation, adoption and public policies mutually shape and depend on each other. Therefore, in contrast to 'standard' evolutionary economics where the biological analogy is applied only to technologies and firms as the carriers of technologies, the perspective of PET is much broader by explicitly integrating politics into the evolutionary explanation. Political decisions are regarded as the result of interactions between political authorities and other social actors. The three key decisions to which technology is subjected in the PET-perspective are innovation decisions, adoption decisions and political decisions.

93 This is reflected in the very general definitions of evolutionary economics, cf. section 2.3

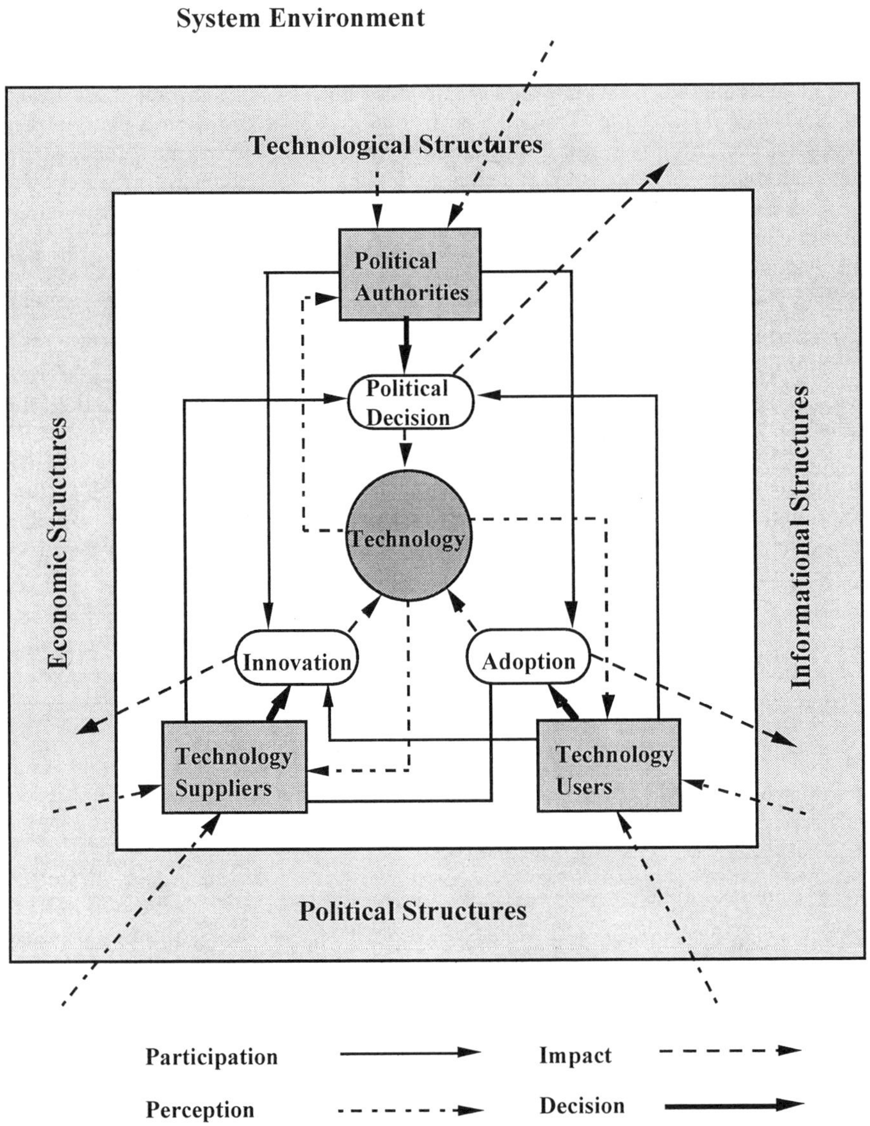

Figure 3.2: The set-up of the PET-system

In line with the focus of this research work, the evolutionary analogy is in the first instance applied to technology. However, when starting from a different research interest, it could also be applied to the individual innovation, adoption and policy decisions. Selection environments and evolutionary processes of change can be formulated for any kind of decision. They become 'relative concepts', depending on the focus of investigation. This is particularly obvious in the case of the formulation and implementation of policies for which an evolutionary argument could be formulated easily. As another example, also adopters have to 'innovate' in order to adjust their organisation, their knowledge base and skills, in other words their routines, to the new technology. Furthermore, the system as a whole - although it can hardly cease to exist - is subjected to selective pressures in terms of performance demands with regard to its function for the embedding social (or global) system.[94] It can decline and shrink, but it can also 'innovate' and adapt. These different evolutionary perspective within the set-up of PET should be kept in mind in the following section. Although the seven elements of explanation are geared towàrds an evolutionary explanation of technological change, it will also be pointed to some of the other possible foci of evolutionary explanation.

External pressure

The PET-system has been defined as an open system, subjected to pressure and impulses from the environment. It needs to adapt as a whole to exogenous changes. Based on autopoietic reasoning, such impulses lead to adaptation processes, which are determined by the internal structure of the PET-system. Different such perturbations can be identified. Technological changes external to PET can induce novelties into the system. Policies other than the ones explicitly associated to the system can trigger adaptation processes as well, like e.g. a sudden rise in environmental awareness. Finally, a variety of typical economic driving forces operate as perturbations on the system, like e.g. international prices fluctuations, volatile exchange rates, unexpected patterns of economic growth and industrial change. In other words, the analysis of external pressures opens up the possibility to consider at least some of the contingencies of real-world processes.

With regard to the technologies under study this means that the PET-system, in which they are embedded, is subjected to adaptive pressures. These pressures induce parameter changes in the system, which can lead to a re-assessment of a technology's characteristics and thereby change the opportunities for its diffusion.

Variation and selection by innovations

Innovation contains elements of both creation of variety and selection. Variety is achieved by a non-random, creative search process in R&D-departments, research labs, etc., often triggered by impulses from the demand-side, from the policy-side or from scientific discovery. The diversity of these actors substantially contributes to variety generation, because innovation conditions and opportunities differ for each firm. Selection takes place at two levels. First, there is the decision to develop a new technology and introduce it on a market, which inevitably implies a decision against an alternative option. At a second level, the actor who carries the innovation must prove its performance within his selection environment (which is in many cases a market).

Creativity and learning are essential cognitive pre-conditions for generating novel solutions. However, cognitive forces also impose constraints on what can be

---

94 The possibility to describe the levels of technological change, of decisions, and of systemic evolution as a whole by means of the evolutionary pattern allows in principle to speak of a multi-level evolutionary system.

imagined as feasible. Expectations of future developments internal and external to a firm play a major role for guiding search processes and represent 'a selective force of anticipated selection mechanisms'. This indicates the importance of concepts such as technological paradigms for guiding search and innovation processes in a self-referential way. Associated to this are technological constraints as experienced by the practitioners of R&D, their understanding of what is technically feasible and fits into a given technological environment. These cognitive, social and economic driving forces of technological change operate within the boundaries defined by natural laws, even if the latter may not be known in detail, either.

Pressures and incentives are required to stimulate the generation of novelty. They can originate from impulses both external to the system and internal to it. Extending the propositions by Kemp at al. (Kemp et al., 1994, p. 71-72) one can identify the following driving forces of innovation:

- The opening up of new but uncertain technological and economic opportunities by new scientific knowledge;
- The emergence of particular technological needs (e.g. "reverse salients", or "presumptive anomalies");
- The discovery or emergence of new markets;
- Uncertainty about what the 'best' technical solution might be;
- Uncertainty about market demand and its evolution;
- Diversity of organisations generating new technology in terms of their capabilities, interests and organisation;
- Political regulations and commitments, which indicate future development opportunities and are, possibly, stimulated by lobbying activities of other social actors;
- Motivational forces at the level of individuals like curiosity, prestige, influence, personal benefits, ideology or altruism.

These forces behind the generation of variety are limited by certain stabilisation mechanisms (c.f. below). They also need to be translated into useful information for the actors. This is achieved in the different 'interaction arenas' where opinions and expectations about innovations and the generation of variety are exchanged. Examples are the interactions of an innovator with his potential clients. They obviously precede the actual investment decision. This combination of creativity, scientific knowledge, mental frameworks, uncertainty, diversity and needs to explain the generation of novelty is not restricted to the realm of technology innovators, but applies in similar ways also to adopters/users and the political realm.

Selection by users in their organisational contexts

The demand side of a technology is clearly its key selection environment, but it is not an independent and straightforward selection process. Users are linked by various channels to changes taking place in other areas of the system. User-producer interactions may already play an important role for defining and refining an innovation in its development phase, and for adjusting the expectations of users, suppliers and policy makers. This process of alignment of uncertain expectations is essential for achieving receptiveness for an innovation among the different actors. In other words, the selection environment of users is actually shaped by the interactions with other actors.

Selection by users is also a process which operates at various levels. Similar to the technology supply side, the cognitive component plays an important role in this process. Routines and the inertia of satisficing behaviour at firm level further channel the opportunities for adoption. Structural and organisational constraints exist at the level of individual users, but also at the structural level of the system. The organisational set-up of an industrial sector can be conducive or detrimental to a new technology. This shows that the concept of 'market success' is too narrow for capturing the complexity of selection processes at the user side of technology.

Selection conditions of the regulatory and political framework

Although the definition of the selection conditions is to a large extent the result of user and supplier interactions, political control represents an important further layer of a technology's selection environment. However, political control also intervenes directly in the innovation process, showing again how the different elements of PET are intertwined.

In practice, regulatory and political framework measures affect the selection conditions of a technology in the following respects:

- Provision of specific advantages in terms of targeted support measures for new technologies, either for the stimulation of innovation or adoption;
- Provision of information on specific technologies;
- Definition of specific regulatory conditions in an economic sector;
- Definition of the general organisational context of an economic sector;
- Moderation of decision processes;
- Definition of the general institutions and rule systems guiding all kinds of socio-economic activities.

Beyond these technology-related activities, the political framework also defines the possibilities of participation in political decision-making processes. These structural conditions of political self-control are important to enable the adaptation of the PET-system to perturbations. Another distinctive feature of the definition of the selection framework by government policies consists of the fact that many political decisions are generally binding, while innovation and adoption are determined by the deliberate behaviour and decisions of actors. These arguments highlight the critical function played by the political realm for the definition of the context conditions of technological change.

However, the definition of these selection conditions is as much an endogenous process as any of the other decisions of relevance to technological innovation and adoption. The involvement of those who are actually subjected to control, i.e. interest groups of stakeholders, citizens, etc., constrains the independence of control. Evolutionary change in the PET-system is 'closed' in the sense that most of the important actors for the area of study are endogenously considered.

Novelty generation and selection are also densely intertwined for the definition of the selection framework. Future expectations whether a novelty might work technically, economically, and structurally, matter in the political realm as well as in any other decision arena. This shows that past performance and future expectations are closely connected to current decisions in a social systems with conscious and forward-looking actors.

Stabilisation mechanisms

Stability and continuity of a system are essential to enable evolutionary change. Without stabilising elements, the interplay of variation and selection would hardly lead to development paths, trajectories and institutionalised settings which are needed for guiding conscious innovation and adoption decisions.

Stabilising mechanisms can be defined at different levels of the PET-system: cognitive, technological, and structural. Guiding paradigms, shaped by communities of practitioners who are linked together in scientific networks have already been mentioned due to their role for decision-making processes. Then there is the local character of search procedures, i.e. the limitations imposed on the possibilities to expand one's searches to areas outside the broad area of established expertise. Self-reference is a further stabilising feature at the cognitive level as well as at the firm level, because the interpretation of newly acquired information is carried out in line with cognitive points of reference. These cognitive considerations obviously apply to all types of actors involved, be it in the techno-economic or the socio-political realm. They jointly generate institutionalised settings and 'rules of the game' in different decision arenas which can be subsumed under the headline of a 'PET-regime'. Such cognitive frameworks need to be established for the new technology to stabilise its formation.

Technological interdependencies, e.g. regarding the existing technology of a firm, constrain and delay change, too. They are due to capital vintage effects and the integration of several technical systems on a single site. Once these interdependencies have been realised for a new technology, it is far more difficult to replace it again.

The interdependence between technological and other types of structures (economic, informational, political) stabilises evolution at the structural level. This issue is particularly important for large technical systems with a complex and pervasive character, inter-connecting economic, technological, informational and political structures. Network externalities can lead to the dominance of a system structure by inhibiting the emergence of alternatives. In order to diffuse successfully, a newly emerging technology needs to be consistent with the different structural conditions and fit into the existing technological system, e.g. in terms of standards and infrastructures. Information networks need to underpin a new technology in order to retrieve and disseminate information about successful demonstration projects and to overcome knowledge gaps. Finally, the economic organisation of a sector and the of political interest structures represent further points of orientation for stabilising a new technology. In essence, the stabilisation of a new technology depends to a significant extent on the possibility to adjust it to the structural context of the established system.

Reinforcing and delaying mechanisms

The establishment of an innovation is often achieved through the effective operation of reinforcing mechanisms. They prevent alternatives from being established and lead to a balance between different technologies. Reinforcements are usually based on positive feedbacks, delaying mechanism on negative ones.

There is an important feedback loop between the micro-level of firms and the macro-level of structures in a system. Firstly, the structural conditions in a system are perceived by the different actors and taken into consideration in their decisions. This corresponds to the traditional structuralist perspective. Secondly, the decisions of the different actors, expressed in terms of innovation/imitation, adoption and political control decisions shape and affect these structural settings. The interactions represent the underlying mechanisms of these decisions, and thus of structural evolution. Actors tend to adapt their own choices to the dominant structural conditions in order to be successful, and thereby close the circular reinforcement process. However, this process does not lead to an equilibrium because the structures are regarded to be in a permanent process of transformation, but at a much slower pace than actor positions and interactions.

Several other self-reinforcing effects contribute also to the establishment and the dominance of a development path. The operate on the basis either of increasing returns to adoption or of critical thresholds (Arthur, 1988a; Arthur, 1988b; Katz & Shapiro, 1994; Katz & Shapiro, 1986; MacKenzie, 1992):

- Learning, knowledge and improvement of technologies, i.e. during its whole lifetime a technology is subjected to improvements, either in terms of the artefacts themselves or due to improved capabilities to use them;
- Network and structural externalities, i.e. the existence of increasing benefits of a technology to the users when an increasing number of other users connect to its infrastructure network;
- Economies of scale, i.e. decreasing cost of production per unit with growing production size;
- Increasing returns to information, i.e. the acceleration of adoption in response to a widespread dissemination of information;
- Technological interrelatedness, i.e. the need to establish further complementary technologies (e.g. infrastructures) jointly with the emergence of a core technology. This makes a switch to alternative systems very difficult. Economies of scope can a have similar effect, i.e. the specific benefits derived from the interdependent production of different products;

- Self-fulfilling prophecy, i.e. the effect that R&D-efforts are concentrated on options which are expected to become dominant, until they achieve in fact a level of superiority which establishes their dominance, even if other options might have been better.

These reinforcing effects trigger 'lock-ins' into certain development paths or trajectories, which can not be abandoned easily, especially in the case of large and complex systems. They explain why large systems are so stable and why it is so difficult to abandon established paths: if a technological system evolves, it needs to remain compatible with the complex arrangements in which it is embedded.

Strategies to circumvent these reinforcements exist in the form of niche strategies, i.e. attempts to stabilise a technology in a limited area with specific selection conditions. Learning processes can then enable the achievement of a critical performance level which allows exploiting one of the aforementioned reinforcement mechanisms.[95] A major difficulty of such strategies consists of the right timing of action. Due to the path-dependent character of such processes of change, specific 'windows of opportunity' exist which are particularly promising for action.

System evolution

Technological change is interpreted as part of the evolution of the broader PET-system. It is interrelated with changes at the cognitive, institutional and structural level and in different realms associated to a technology. Evolution is stabilised by several mechanisms and channelled by reinforcements. This picture appears to be more prone to incremental than to radical changes or even systemic transformations.[96] It suggests that the introduction of major changes in an established system (e.g. by means of public policies) is confronted with major obstacles, especially in the case of large pervasive systems. System evolution and technological change are only possible if the obstacles throughout the entire system are removed. In order to transform a system its main elements need to co-evolve and adapt. Due to the high degree of interdependence it is hard to say whether a single element operates as the 'key driver' of change in a system. There are several sources of major change, which can be driven by reinforcing mechanisms. Some of the typical driving forces of radical change have been summarised by Kemp et al. (Kemp et al., 1994, pp. 67-69), namely

- New scientific insights,
- New technological needs due to bottlenecks or reverse salients in technological systems,

---

95 Even very simple diffusion models such as epidemic logistic diffusion functions are essentially based on some kind of reinforcement mechanism. This is expressed in the hypothesis that once a certain threshold level is achieved, the new species will replace the old one (Nakicenovic, 1986; Nakicenovic, 1991).

96 C.f. the vocabulary used by Rip and Kemp (Rip & Kemp, 1996) and Kemp, Schot and Hoogma (Kemp et al., 1996) who speak of technological regime shifts and the evolution of socio-technical landscapes.

- New demand due to important external events like wars or environmental threads, and
- Limits of existing technological trajectories, e.g. due to increasing marginal costs or - as a cognitive element - the perception of technological limits.

Another option for the stimulation of radical change is government action, especially when it intervenes at structural and organisational level. However, this driving forces of change is subjected to the same types of constraints as the other factors mentioned before. In order to transform a system, well-tuned changes at several levels are required. On the other hand, due to the autopoietic character of the objects of control context control appears to be the only promising policy approach. It is thus a crucial problem to identify the right levels of intervention. A second major difficulty consists of the timing of interventions to meet those windows of opportunity which promise a major impact while using minor perturbing measures only.

3.8 Summary

3.8.1 Basic hypotheses

The structure of the PET-system has been summarised in Figure 3.2 where the main elements and their characteristics are presented. Although the key objective of PET consists of the provision of a guiding pattern of analysis for policy-oriented research, it is nevertheless possible to condense some of the aforementioned considerations into five basic hypotheses on the evolution and emergence of new technologies in their techno-economic and socio-political context.

Structural hypothesis H1: Co-evolution of system structures

Due to the interrelatedness of technological, economic, informational and political structures in the PET-system, processes of technological change are regarded to co-evolve with the other structural elements, which have developed in line with the dominant technology. This is particularly important for pervasive network system technologies because they interconnect many actors with very different perspective on a technology. These interdependent transformation processes are regarded to be based on the co-ordination and adjustment effect which results from the interactions in terms of information, artefacts, power and money among the main actors of a system.

Technical hypothesis H2: Complexity and complementarity of technical systems

Technical systems are often highly differentiated and complex. They need to be characterised by a broad set of dimensions of interest to a large number of actors. Under such circumstances, technology diffusion cannot be described as a simple replacement process. Based on the different characteristics of a technical system and their assessment, a new technology can enter certain application niches within a wider technological structure. It can initially play a complementary role inside the system, before inducing a more pervasive transformation process of the dominant technological regime.

Behavioural hypothesis H3: Diverse perceptions of the environment and actor behaviour

The participation of actors in interactions and decisions is regarded to underlie the structural transformations of the PET-system. The interests, mental frameworks, knowledge bases, and expectations with respect to structural transformations and other actors' strategies within the PET-system shape the behaviour of actors. These factors need to rely on the perception of uncertain information about the actors' respective environments. Consequently, expectations as well as preferences, behavioural patterns and decisions are characterised by a high degree of diversity.

Dynamic hypothesis H4: Structural change and reinforcement mechanisms

It is difficult to introduce and diffuse innovations in a system which is dominated by an incompatible paradigm. Triggering wider transformation at the structural level is even more difficult to achieve because it requires manifold adjustment processes, ranging from the actors' mental frameworks to structural characteristics. Structural change being a path-dependent process, the operation of reinforcement mechanisms and the right timing of perturbations for triggering a change of the dominant trajectory or even paradigm are of crucial importance.

Control hypothesis H5: Inroads for political interaction

Political control can operate at different levels of the PET-system, namely by means of

- Generation of variety, e.g. by means of promotion of R&D- activities in selected areas;
- Changing the dominant beliefs and mental frameworks, e.g. by means of an active information policy;
- Supporting reinforcing effects, e.g. by promoting the emergence of promising niches until they reach sufficient economies of scale, or by providing infrastructures;
- Affecting the incentives (and hence the interests) for technology suppliers and users, e.g. by means of regulations, dedicated taxes or grants, etc.;
- Reducing mismatches between structures and technologies, be they technological, economic, political or informational, possibly by means of reorganisation of industries and research infrastructures, but also by changing the interfaces and organisational structures of government.

In line with these hypotheses, the timing and the selection of a sufficiently broad range of control measures are important for successful control. The measures need to take into account the self-referential character of both the object and the subject of control. This means that the subjects of control also need to understand their own limited role within the system. They need to identify the right window of opportunity, use a broad range of control measures, and take into account that they can only perturb the behaviour of other actors, i.e. objects of control. This implies that detailed and in-depth knowledge of the internal operation of the objects of control improves the possibilities to apply measures successfully. Within the limitations of these constraints, a system can indeed be controlled, but not in the sense of detailed and targeted planning.

3.8.2 The PET-system - a useful starting point

Probably, the PET-approach as presented above does not fulfil all the criteria which are usually required of a 'good' theory, i.e. simplicity, predictive power and reproducibility. However, it is not even meant to be a theory, but rather a guiding approach or model for applied research, which uses available theoretical and empirical findings as widely as possible.

In spite of the attempt to describe it in a very structured way by means of a systems language, PET is a quite complex approach. One of the reasons for this complexity is that the problems dealt with are complex. It would be inadequate to use simple concepts and derive simple hypotheses. The five hypotheses are therefore fairly general and cannot be operationalised easily. They will serve as a starting point for formulating more specific hypotheses on the case of CHP once more detailed information about the technology will have been provided in Part II. Anyway, the importance of the hypotheses for this study should not be overestimated because it does not pursue the objective to test hypotheses but to use them as guiding statements, reflecting the essence of the PET perspective.

In general, the traditional hypotheses-oriented research approach is regarded to be of limited value because the structure of the PET-system and of social systems in general implies that their basic patterns of development come much closer to a biological than to a mechanistic analogy. Even more important, social as well as biological systems are open. In contrast to mechanistic systems it is not possible to define or maintain fixed boundary conditions. Moreover, social systems are composed of conscious and autonomous agents. Simple hypotheses would not be of great value for complex systems in which the number of variables is far beyond what can be captured by means of hypotheses. In contrast to mechanistic models of social phenomena, complex biological analogies such as PET do not allow formulating detailed hypotheses or even predictions (which, anyway, has hardly ever been done successfully in social sciences). This seemingly weak point should however be assessed in the light of the false promises of mechanistic approaches. The recognition that social life is more adequately described by biological than by mechanistic analogies, "reveals the apparent strong predictive power of a simpler theory (...) to be fool's gold" (Nelson, 1995, p. 67).

Reproducibility can hardly be claimed as a typical feature of social theory, and it is note met by PET either. Not even falsifiability as a weaker criterion can be claimed for PET due to the difficulties to formulate well-defined hypotheses.

However, as the approach is synthesising research findings which have already been subjected to rigorous testing, this is regarded as an excusable shortcoming.

Taking these arguments into account, the PET-approach needs to be seen as a guideline for empirical research rather than as a theory in the traditional sense. First of all, it provides a theory-based analytical and interpretative pattern to investigate specific technological cases, and a set of possible mechanisms and interdependencies which can play a major role for explaining processes of change and indicate possibilities of control. This pattern needs to be complemented for each investigation by specific empirical material. Secondly, the PET-system provides a foundation for formulating a general methodology for the study of processes of innovation diffusion and political control. This should be regarded as the major strength and distinctive feature of PET, because it is an attempt to use complex appreciative models as an explicit foundation for research methodology. The next chapter will put emphasis on this methodological dimension of the PET systems approach.

Chapter 4: A General Methodology for Studying Innovation Diffusion and Political Control

4.1 Introduction

To investigate the case of cogeneration technology, it is necessary to define a specific methodology, based on the theoretical perspective formulated before. However, in view of the generic character of the PET-approach it would be useful to generalise the methodology in order to be able to apply it also to other cases in the future. PET being a fairly complex multi-disciplinary approach, this should be seen in the light of the objective to apply social science theory constructively and use it to deal investigate real-world problems. Such a methodology would also represent a condensed and easily manageable translation of the PET system approach into certain guidelines and key elements of applied research.

The notion of 'methodology' must be distinguished from 'research techniques' and 'methods'. While the former represents a general pattern comprising the main elements or steps to be followed in a research study, the two latter comprise the different tools which can be used to collect and process data, e.g. by means of questionnaires or interviews. The issue of research techniques to be used for the empirical investigation of CHP-diffusion will be addressed in Chapter 7.

4.2 The basic steps of the general methodology

The general methodology consists of nine steps (see Table 4.1). It covers the different elements of the PET-system (technology, exogenous elements, structures, actors, interactions, decision-making), their integration into a dynamic analysis of system evolution, and an prospective exploration of system development into the future.

Traditional TA-studies are often restricted to the analysis of the technology and its changes, taking into account different dimensions of interest. Actor analysis in the

PET-methodology goes beyond what is usually called stakeholder analysis in traditional TA because it also includes the political actors involved. Finally, dynamic and especially prospective analysis are rather unconventional elements of technology studies. This last step delivers direct inputs on the control options for the fictive client of the study. It allows deriving conclusions with regard to possible control strategies for one of the different actors in the system, here especially for political decision-makers.

In the following, each of the nine steps will be briefly described and explained. Some indications will be added concerning adequate techniques of data collection.

Step	Description
1	Technology analysis
2	Technological changes
3	Delimitation of the system and exogenous changes
4	Structural analysis and structural evolution
5	Identification of actors
6	Event analysis
7	Analysis of decisions and interactions
8	System dynamics
9	Prospective analysis and scenario building

Table 4.1: Steps of the PET-methodology

4.2.1 Step 1: Technology analysis

It is indispensable to dispose of a good understanding of the technology under investigation. Technology analysis should be understood as an attempt to describe the technology as 'objectively' as possible. It usually represents the major part of a traditional technology and assessment studies. The analysis of a technology's implications and impacts along different dimensions (e.g. environmental, economic, social, etc.) is maintained in the context of PET, even if it is interpreted differently. Usually, it leads to some type of normative recommendations which of the options at stake ought to be regarded as 'better' and should hence be supported by political action. From the perspective of PET, technological information along the different relevant dimensions is needed first of all as a pre-condition for understanding and analysing the different actors' positions and assessments of the technology. In the end, it depends on an actor's priorities which technology best meets his needs. The concept of superiority,[97] which has been used widely in diffusion

[97] The concepts of superiority is very simple and one-dimensional. It is often defined just in technical and/or economic performance terms. Even in this simple form, a number of problems need to be mentioned. For whom is a technology superior? For individual companies or society? And at which time-horizon? Does this take into ac-

research thus needs to be qualified in more detail. It can be applied only from an individual stakeholder's perspective to prioritise options. An aggregation of these individual assessments is confronted with two major problems. Firstly, according to Arrow's impossibility theorem, individual preferences can be inconsistent with each other, even when dealing only with one assessment dimension.[98] Secondly, when taking into consideration a number of different dimensions, it would be necessary to analyse the collective interactions of all actors' preferences and priorities regarding these dimensions. This is hardly feasible in practice. Consequently, a multi-dimensional technology assessment framework needs to be maintained.

Research Issues	**Research Questions**
Options	Which are the different technological options in question (the 'objects of interest')? Which are the main alternatives to these technologies? Which potential future alternatives can be identified?
Characteristics	How can the technologies be characterised in terms of their main dimensions? 1) Technical & innovative issues (trajectories) 2) Economic impact at the firm level (costs) 3) Socio-economic impact 4) Environmental impacts 5) Organisational and power implications 6) Others

Table 4.2: Steps for analysing a technology's dimensions

Moreover, technology analysis can not be restricted to the technology in question (unless one is dealing with something completely new), but needs to include also alternative options. In a fully comprehensive study, these options would have to be subjected to a multi-dimensional analysis along the five main dimensions in the same way as the technology of main interest to the study.

count long-term, external costs? Moreover, a comparison of simple options may allow statements such as 'technology A has lower emissions than B and is hence superior', but when it comes to compare complex systems with far-reaching consequences, such clear statements can hardly be made.

98 Arrow's impossibility theorem says that there is no consistent method of making a fair choice among three or more candidate alternatives. The aggregation of individual preferences is impossible due to inconsistencies among them. With regard to voting procedures, this implies that an election which involves more than two alternatives can not be assured to produce a fair outcome.

4.2.2 Step 2: Technological changes

Going beyond Step 1, the analysis of technological changes deals with the past, present and expected future development of a technology's characteristics along the lines of the different assessment dimensions. Past innovative developments make part of the analysis of technological changes because they allow revealing trajectories and development trends. For the technical dimension, this refers to changes of a technology's performance parameters or its design configuration, for the micro-economic dimension it refers essentially to costs. Similarly environmental impacts have to be dealt with, as well as the external and macro-economic effects which a technology has for society. As an example of political and power impacts, long-term organisational transformations, triggered by the new technology, can be mentioned. The ex-ante identification or even measurement of such changes of parameters is far from being a straightforward task because problems, which might have been perceived initially as crucial, can turn out to be negligible in the course of a technology's implementation. The weight and importance assigned to different impacts is also dependent on the actors' changing subjective priorities. Consequently, although two technologies may be similar in their technical, environmental and economic implications, their acceptance may differ depending on the way they are embedded in a system of social control. For example new information and communication technologies are perceived both as a risk and as an opportunity.

For a prospective application of the PET-approach it is necessary to collect information about the technical developments which might come up in the future. These can be of improvements of existing technologies but the emergence of qualitatively new options.

Past technological changes can be traced by collecting published data on technological and economic performance characteristics. A multitude of methods is also available to assess the environmental implications of technologies, e.g. emission data, Life-Cycle Analysis, etc. Organisational, institutional and political implications are more difficult to analyse and can usually be done only qualitatively. The identification of future developments represents a further challenge. In most cases one has to rely on expert opinion, which, apart from the problem of uncertainty, is subjected to several difficulties.[99] However, experts are a very valuable source of prospective information, but the limitations of the methods need to be kept in mind. They can not produce reliable 'forecasts' of future diffusion paths, but they usually have a quite good notion of upcoming developments in their specific field of expertise. Experts can provide information about actual and promising technological developments, but they tend to be less good when it comes to assess future diffusion paths. Their biased expertise is usually insufficient for assessment the socio-economic processes which underlie diffusion. Their inputs on promising

---

99 Some of the problems, with which expert methods are confronted in general are the following: 1) As experts also determine future priorities by their judgement, their involvement can work as a self-fulfilling prophecy; 2) Experts are usually less inclined to non-standard solutions; 3) The selection of a well-balanced set of experts is very difficult; in practice they are often biased.

technical areas of interest ought to be complemented by information on upcoming developments in the social, economic and political realms.

Moreover, the aggregation of expert opinions from different specialised fields for the construction of future scenarios is a very difficult process.[100] Semi-formalised approaches have been suggested to make this process more transparent, for example "morphological analysis" (Foray & Grübler, 1990) or the quantitative measurement of service and performance characteristics (Saviotti & Metcalfe, 1984). For very general technological developments, diffusion curves represent a means of extrapolation (Marchetti & Nakicenovic, 1979; Nakicenovic, 1991; Nakicenovic & John, 1991), but they do not provide any detailed data of the performance characteristics. Anyway, all these approaches are confronted with the problem of getting meaningful data on the different aspects to be modelled.

Research Issues	**Research Questions**
Evolution	How have the technologies under study developed in the last years in terms of the different dimensions of interest? 1) Technical & innovative issues (trajectories) 2) Economic impact at the firm level (costs) 3) Socio-economic implications at the macro-level 4) Environmental impacts 5) Organisational and political implications (power relations, social impact, prices of final services, democratic control, decentralisation of power) 6) Others
Projection	Which projections can be made for the *potential* of the different technologies in terms of their parameters in the different dimensions?

Table 4.3: Steps for analysing the changes of a technology's dimensions

4.2.3 Step 3: Delimitation of the system

The identification and delimitation of the system to be investigated is an important, though often neglected step. It is necessary to be explicit about the simplifications which are implied by this delimitation, and about the arguments for making them.[101] At the beginning of a research exercise, the delimitation can only be

100 An interesting approach has been proposed by Canarelli (Canarelli, 1996) who aggregates the statements of experts about the future of their field of expertise by means of fuzzy reasoning.

101 Although it is probably correct that everything finally depends on everything else, such a perspective would prevent any useful work. What can be achieved is an as concise and explicit argument which defines what is part of the system (hence including interdependencies) and what is outside of it (exogenous forces).

made in a preliminary way, because the system and its actors are not yet fully understood. It may be necessary to reconsider some of the simplifications at a later research stage. The research interest and the density of relevant feedbacks across the system borders are the key criteria of a useful definition of the system. In other words, the consideration of exogenous factors does not represent a problem in an empirical case study as long as there are not too many important feedbacks from the system to the environment to be considered.

Typical examples of exogenous forces are international development trends, demographic trends, economic growth, or public opinion, but also major events such as a war or an oil price crisis. Similar to structural changes, the exogenous forces need to be traced throughout the period of study analysed in terms of their impact on the behaviour of the actors in the system.

Research Issues	Research Questions
Delimitation of the system	Which is the focus of research interest? Are there important feedbacks from the system to its environment? Does the delimitation need to be reconsidered once the system is known better?
Exogenous changes	Which are important forces impinging on the system? How do they develop over time? Are there major events to be considered?

Table 4.4: Steps for analysing exogenous changes

4.2.4 Step 4: Structural analysis and structural evolution

Step 1 aims at providing an 'objective' assessment of the technology and its implications. By means of structural analysis, the settings inside of which a technology is applied are studied. Four different types of structures have been defined in Chapter 3, and their evolution needs to be traced with regard to the technology in question.

System structures are the result of a historical process. They are subjected to change, but provide a comparatively stable framework interactions. An important part of structural analysis is concerned with the identification of 'matches' between different technological options and the established structures: incompatibilities between technology and structural conditions, structural changes required to implement the technology, opportunities for the introduction of new technologies within the existing structures.

Research Issues	Research Questions
Technological structures	Which are the patterns of diffusion of the technologies considered and their alternatives? Which functions do the technologies fulfil in the respective industries? Which technological interdependencies exist at the level of industries/countries?
Economic structures	How are the main relevant industries structured and organised, both on the user and the supplier side? Which are the main types of economic actors? Have these structures changed and how?
Informational structures	Which are the main institutions/actors involved in the generation and dissemination of information on the different technologies? Which main types of actors can be identified within the information structures?
Political and power structures	Which are the main types of political authorities and other players/actors in the political realm? What do the interfaces between the political decision-makers and the different interest groups look like? Who has an impact on the formulation and implementation of public energy policies?
Structural evolution	How have all of these structures evolved and changed in the past?
Matches and mismatches	Are there incompatibilities between technology and structures? Which structural changes are required? Which structural opportunities do exist?

Table 4.5: Analysis of structural changes

The identification of the main *types* of actors in the system (e.g. different types of power utility companies at the technology demand side of CHP) represents an important part of structural analysis. It is also an important input to actor analysis (Step 5) which focuses in more detail on the individual actors involved in the empirical example.

For certain parts of structural analysis statistical data are available, e.g. on the use and diffusion of different technologies, or on the market shares of major companies. The evolution of informational and political structures can in principle be described by formalised methods such as network analysis. However, in most cases it needs to be complemented if not replaced by a qualitative analysis of formal and informal institutions, regulatory settings, and power relations.

4.2.5 Step 5: Identification of actors

Overlapping partly with structural analysis, the identification of *individual* actors (i.e. not just types of actors) is an essential precondition for the later steps of analysis of decisions and interactions. Based on the definition of three main decision realms (innovation, adoption and policy-making), each individual actors can be associated to at last one of them. As many actors may fulfil several functions in the system or operate at the boundaries different realms, they could be associated to more than one realm. This question will be addressed when dealing with their strategies and decisions in more detail. If necessary, additional or more refined sub-categories of actors can be introduced.[102] In the end, the three basic groups of actors should be interpreted as the starting point for understanding major lines of conflict and co-operation in the system.

Beyond the identification of the individual actors, a very interesting question concerns the actors who can not be identified in the empirical case example, but which could play an important role from a functional point of view. For example, for a new technology it is very helpful if a network of actors is in place which could support it.

Information about individual actors requires to dispose already of a fairly detailed knowledge of the system. This facilitates also the process of data collection, and especially to collect actor-related data in a more targeted way. The identification of actors can be supported by a variety of formalised methods, such as for example network analysis.[103]

Research Issues	Research Questions
Technology supply side	Which major actors have been involved in the supply, research and development of the technologies under study?
Technology user side	Who have been the main direct and indirect users of the technology in question?
Political realm	Which major actors have been participating in political decision-making processes?
Other actors	Which actors have been operating at the interfaces between major clusters of actors or cannot be associated to one of the main clusters?
Missing actors	Has there been a network of actors in place to support the technology in question?

Table 4.6: Elements of actor analysis

---

102 For example, the user-side of electricity supply technologies consists of regional electricity companies, industrial, commercial, private and public service end-users, energy management companies, etc.

103 See e.g. (Marin & Mayntz, 1991b; Schneider, 1989) for some examples of structured identification of actors and networks.

4.2.6 Step 6: Event analysis

After having described the main elements of the system (actors, structures, exogenous environment), the most relevant events to the technology need to be identified for the time period under investigation. While, structural, technological and exogenous changes over time have already been traced before, this step focuses on the detection of individual events and decisions, which have triggered change from within the system. These are essentially

- Political decisions taken;
- Major decisions of actors to commit themselves to innovative activities like R&D or structural reform;
- Relevant decisions with regard to the application of technology, like investment and adoption.

According to the PET perspective, such decisions are regarded as the outcome of interaction processes among the different actors in the system. For example, although a political decision might ultimately be taken by political authorities, it is strongly affected by the participation of different actors in the preceding negotiation processes.

In terms of methods, the event analysis can be based to a large extent on secondary sources, but it should be complemented by direct information from the actors involved in order to be able to assess the actual importance of these events.

Research Issues	**Research Questions**
Political decisions	Which have been major political initiatives and what have the related negotiation processes looked like? Which have been the main issues in the process? Which have been the political alternatives?
Innovation	Have their been major initiatives to improve the technology? Where have the impulses for innovation come from? In which interactions have the technology-suppliers been involved for preparing adoption and policy decisions?
Adoption	Which have been the processes leading to the adoption of the technology? How have they been linked to other decisions of the company (diversification, vertical or horizontal integration, international activities, environmental strategies, etc.) In which ways have the technology-users taken part in negotiating innovation and political decisions?

Table 4.7: Elements of event analysis

4.2.7 Step 7: Analysis of decisions and interactions

Events and important single decisions as identified in Step 6 are prepared in interaction processes. In fact, decisions on innovation, adoption and policies by any of the actors are always the result of preceding interactions, in conjunction with individual decision-maker's knowledge, motivations, strategy and constraints. The perception of exogenous, structural and technological changes completes the decision process. In other words, in order to understand the evolution of the system, it is essential to capture these micro-level decision-making processes. They represent the key node for inter-connecting structural, technological and exogenous changes with the attitudes of the actors.

Research Issues	Research Questions
Interactions, perceptions and expectations	Where have the actors received information on the technological options from? With which other actors have they had close interaction? Which have been major uncertainties perceived and how did the actors deal with them? Which importance has the technology had for the actor? Which future expectations have the actors had with regard to the technology?
Decision-making processes	Which have been the major objectives pursued by the actors in relation to the technology? How have they assessed the different dimensions of the technology? What has been their relative important? Who has been involved in the specific decisions? Which future changes in their respective environment have the actors expected? How have they reacted to them?
Outputs and decisions	Which activities and actions have the actors undertaken in relation to 1) Innovation decisions 2) Adoption 3) Political action 4) Interactions with others Which role have the perceptions, expectations and objectives played for the actors' position in interactions with others, and for the final decisions taken?

Table 4.8: Analysis of decisions

Expectations play an important role for making decisions, and consequently present expectations can already reveal many insights with respect to possible future decisions and developments. This kind of information is very valuable for prospective analysis and should be collected already during the analysis of past decisions and interactions. However, expectations are also an important part of retrospective analysis because they help to assess the quality of information which a decision-maker had when making past decisions. With hindsight, expectations often reveal to be wrong.

As the model of maximising agents has been abandoned in the PET approach, it is necessary to investigate empirically the motivations, perceptions and arguments which have led to major decisions. In-depth case studies are regarded to be an adequate tool for collecting information at a level of detail, which allows to understand the technical, organisational and behavioural issues at stake. This will require time-consuming techniques of data collection such as interviews or questionnaires at actor level.[104] The preparation of interviews or questionnaires requires to be familiar with the different technological, exogenous and structural developments because information needs to be collected on the impact of a broad range of developments of relevance to the actors interviewed.

4.2.8 Step 8: Analysis of system dynamics

The explanation of the system dynamics is achieved by building on evolutionary arguments, though in an appreciative, non-formal way.[105] Technology diffusion being understood as part of a wider changing context, the notion of co-evolution of technology and system has been regarded appropriate. However, emphasis is put on the technology within this co-evolutionary interpretation. This means that in principle other orientations of dynamic analysis are possible, geared for example towards the introduction and selection of policy measures or organisational transformations.

In more practical terms, the dynamic analysis of technological change consists of identifying the empirical data which correspond to the seven key aspects of system dynamics (see Table 4.9). If sufficient data is available about the system under study, it is in principle possible to implement the system dynamics in a simulation model. It should be able can treat both qualitative and quantitative data, e.g. by means of genetic algorithms, neural networks or fuzzy logic (Canarelli, 1995a; Canarelli, 1995b; Canarelli, 1996). However, while such simulation techniques are very useful for capturing the essential elements of system dynamics, several detailed mechanisms remain outside their scope.[106]

---

104 Interviews will presumably be the most appropriate tool for studying motivations, perceptions and their translation into decisions.

105 The three main mechanisms which have to be part of an evolutionary representation are variation/mutation, selection and stability/continuity. Simulation methods have become available which could be used to capture the system dynamics more quantitatively, but they depends on the availability of adequate data.

106 Examples of mechanisms which are difficult to capture by models are the cumulativeness of technological knowledge, or the role of realised technological artefacts as focusing devices for further search.

Research Issues	Research Questions
External pressure	Which have been the main bottlenecks and problems triggering change in the system? Which conclusions can be drawn on the main driving forces of change?
Variation and selection by innovation	Which have been the main technological/scientific driving forces of change? Has the technology matched the interests of the actors at the supply side?
Selection by users	Have there been strong forces and networks of actors supportive or detrimental to the adoption of the technology? What have been the main barriers for diffusion? Has the technology fitted within the set of users' interests? Which have been the users' needs?
Selection conditions of the regulatory and political framework	Has the technology been supported by some major political interests? Has the framework provided detrimental or conducive conditions for the technology? Have there been possibilities to change the framework?
Stabilisation mechanisms	Which role have established mental frameworks and paradigms played for inhibiting innovation among all actors? How open have the actors been to external perturbations, how restricted are they to their routines? Have there been important interdependencies between the technology and the embedding structures and institutions (structural (mis-)matches)?
Reinforcing and delaying mechanisms	Have the parallel developments at the micro- and the structural level reinforced the success of the technology? Have there been possibilities for achieving consensus and co-operation with regard to thc technology? Has learning taken place at a faster pace than for competing technologies? Has it been possible to exploit other reinforcing mechanisms by the technology (economies of scale, network externalities, etc.)?
System evolution	Which has been the joint effect of the operation of the dynamic elements? How resistant or how flexible has the system been with respect to fundamental change?

Table 4.9: Analysis of dynamic mechanisms

4.2.9 Step 9: Prospective analysis and scenario building

The preceding steps have dealt essentially with retrospective analysis in order to understand the system dynamics and mechanisms, which underlie technological change. Based on this knowledge, future development possibilities and actor strategies can be explored.

Prospective analysis and scenario building exercises are often geared towards the needs of a specific 'client' who wants to develop longer-term strategies. From the perspective of the PET approach, this client needs to be interpreted as part of the system. For the purposes of this study with its focus on political control, the main fictive clients are public authorities and government departments. Conclusions for these clients can be drawn on the basis of exploratory scenarios.[107] These scenarios could e.g. explore the impacts of and possible reactions to major exogenous events, or the adjustments of the system to endogenous actor strategies. This has little to do with prediction but with the analysis of robust strategies to dealt with different future development paths.

The major building blocks of prospective analysis reflect the elements of the retrospective PET system analysis, but they are complemented by the following principles for exploration:

- Investigation of major development trends of all elements of PET;
- Identification of potential mismatches;
- Consideration of reinforcing and other dynamic effects;
- Imagination of major unexpected changes;
- Analysis of actors' strategies and their interaction;
- Focus on a client's perspective.

In other words, by checking the different elements of the PET-system with respect to their future evolution, the possibilities of technology diffusion can be explored on the basis of disaggregated information sources. The crucial element is therefore to understand the decision-processes at the micro-level, i.e. the expectations and potential reactions of technology suppliers, technology users and political actors with regard to these future trends and options. Of major importance are also the structural requirements and implications of a wider diffusion of the technology, in particular major the 'matches and mismatches' which this diffusion might bring about.[108]

---

[107] Exploratory scenarios aim to explore different future development paths, starting from the present situation and analysing different future decision options. They can be contrasted to normative scenarios which start from a 'desired' future state and 'backcast' the present requirements in terms of decisions needed to achieve this state.

[108] One major advantage of the study of mismatches is that it allows to exclude several combinations of development trends due to incompatibilities, and thus to reduce the complexity of the scenario development process.

If - as mentioned before - adequate simulation tools and data series are available, then many elements of prospective analysis can be formalised. Others, however, can hardly be formalised and need to be studied by soft methods, e.g. when organisational issues have to be addressed. Several semi-quantitative methods are available which can be used to structure soft, expert-based methods in a more rigid way (Godet, 1993).

Research Issues	**Research Questions**
Development trends	How are the relevant exogenous factors expected to evolve? Which technological changes and new options are expected to emerge? Which changes could emerge at structural level?
Unexpected events	Are there major unexpected events possible (exogenous or endogenous) which might seriously affect the system?
Dynamic mechanisms	Which dynamic reinforcements do these events and developments trigger?
Client analysis	Which are the objectives of the "client" (often political decision-makers)? Which are the policy options? (structural policy, technology policy, diffusion policy) Which are policy strategies?
Analysis of other actors	Which behaviour can be expected from the side of the other actors in the system? What do the incentive structures look like to promote diffusion?
Matches and mismatches	Which matches and mismatches could this bring about between structures, institutions, organisations and actor strategies?
Scenarios	What would be effects of certain political actions? Would they meet resistance? Where are the conflict lines? What are the other actors' counter-strategies?

Table 4.10: Analysing future dynamics

4.3 Conclusions

Building on the scenarios and taking into account the objectives of the 'client' of a study (i.e. here essentially public authorities or government departments), conclusions can be drawn regarding possible strategies to meet the objectives set out.

However, a full-scale analysis along the lines of the PET-methodology remains a difficult task. The crucial elements of investigation are the actors and the processes which lead to their decisions. The simplified model of 'economic man' has been abandoned, and this implies that objectives, expectations and preferences need to be analysed empirically. In practice issues such as profit and growth will remain important, but they need to be complemented by strategic considerations, rules of thumb-type decision criteria, and the inherent limitations of human behaviour. Therefore, much effort needs to be put into the empirical investigation of actors' decisions, perceptions, motivations and interactions with their respective environments. The pieces of information on technological and structural changes are a prerequisite of this micro-level investigation. One needs to be familiar with the socio-technical conditions of a system before the actors' behaviour can be analysed empirically.

While exogenous changes are regarded to be independent of the system's evolution, technology is assigned a certain degree of autonomy. It is embedded and constrained by structures and actors, by institutions and routines, but developing a certain degree of auto-dynamics because it shapes the mental frameworks in research and development. The structural elements are regarded to be the outcome of a historical process based on actors' past interactions. These arguments underline the emphasis given to behavioural aspects in the PET methodology.

The steps defined above are clearly an idealised description of how a research project could proceed. In practice, several of the steps will overlap and merge when applying specific research techniques. For example, the analysis of decisions can not be clearly separated from that of motivations, especially when it has to be derived from the participation in interaction processes. The same interviews will provide data on both aspects. Similarly, the specific content of the technology's crucial dimensions will be better known once the interview-based fieldwork has provided additional information on what the actors regard as important.

In practice, the nine steps need to be further specified for the individual case under study, using as widely as possible already existing information on the case example. Certain elements of the methodology will possibly be unnecessary, e.g. if a technological area is hardly subjected to political interferences. This process of specification of the generalised methodology will be shown in Chapter 7.

The techniques of data collection to be used in the different steps of the methodology cover a broad spectrum, responding to the specific requirements of each aspect of research. Due to the crucial role played by the responses at the micro-level to transformations in the actors' environment, interview-based research techniques can be expected to play an outstanding role for the practical part of the research work.

PART II:

Introduction of the Problem Area and Specification of the Methodology

While the first part of this study was concerned with the general theoretical and methodological research background of processes of technological change and political control, in the following part it will be shown how the concepts developed can be implemented in a comparative empirical study. The technology under study is combined heat and power production (CHP) or cogeneration. Rather than being a technology which is characterised by its hardware, CHP is a principle for electricity and heat generation which differs in many respects from the dominant principles of energy supply.

It may appear exaggerated to dedicate a major part of this study to an introduction of the problem area and a specification of the methodology, but the conceptual and methodological approach developed is a key element of this research. It is therefore necessary to show explicitly how the empirical research work can be implemented on the basis of this approach, and to make the research procedure as transparent and reproducible as possible. Usually, much research work is confined to the presentation of the underlying approach and the presentation of the results, possibly complemented by some general methodological consideration. A critical analysis how the approach is actually transformed into a consistent research procedure is usually missing. Part II addresses exactly this issue and proceeds along the following main steps:

- An introduction will be made to the present problems, challenges and conflict lines faced by the traditional system of producing energy services, and to the characteristics and dimensions of cogeneration technology in this context (Chapter 5). These two elements are needed to understand the empirical analysis in Part III, and especially the reasons behind the different actors' assessment of the technology. It is based on the conviction that a useful analysis of processes of technological change can only be achieved on the basis of a sufficiently detailed knowledge and understanding of the technology in question, going far beyond the usual way of describing it highly generalised conceptual terms. The characterisation of cogeneration technology in terms of the five main dimensions introduced before will therefore be provided.
- A review of findings on the factors and processes affecting CHP-diffusion as revealed in past, mainly socio-economic, research will be given in Chapter 6. This step is crucial from a methodological point of view because it allows

specifying the detailed empirical data and research needs of this study. It addresses the information gap which need to be filled to describe the case of CHP-diffusion in a comprehensive way along the lines of the PET-approach. In other words, this review is needed to specify the research methodology in detail.

- The detailed specification of the research methodology is done in Chapter 7. It consists of two major steps, namely firstly of all an argument for defining the focus of the empirical field work, and secondly an outline of the data requirements and potential data sources for the analysis of CHP-diffusion in the UK and Germany. This outline also follows the pattern of the PET-system approach.

Chapter 5: Issues of Energy Supply and Cogeneration

Electricity Supply is a typical example of a large (socio-) technical system. Its evolution is determined by complex techno-economic and socio-political processes. Similar to other such systems (e.g. water supply, telecommunications, gas supply), a grid structure represents the backbone and physical link between its elements. This physical interconnectedness has implications for the feasibility and introduction of both technical and organisational innovations. The technology user-side is particularly complicated because it is composed of a broad range of service end-users, utility companies and service providers. The example technology chosen, cogeneration, is a grid-based technical system, which integrates electricity and heat supply in one process. Both its diffusion patterns and its far-reaching organisational implications make it an interesting case to investigate.

This chapter discusses the issues which have been of major importance in the electricity sector in recent years and defined the background and boundary conditions for the use of CHP. It can be argued that the historically grown structures and organisational patterns of energy and especially electricity supply do not match the requirements of CHP in many European countries. From the perspective of political control, the recent challenge of liberalisation is of particular interest, because it implies a transformation of these established structures, but it is far from clear whether the result will be helpful for cogeneration or not.

Apart from a historical review of the emergence of electricity and heat supply systems (5.1), and a discussion of the main challenges with which these systems have been confronted in recent years (5.2), the technology of cogeneration will be introduced and assessed along its five main dimensions (5.3). In other words, Chapter 5.3 covers Step 1 of the PET-methodology.

5.1 Historical development of heat and power supply

Heating has traditionally been a decentralised activity in individual households, but with the availability of large amounts of industrial waste heat, some first local heat supply networks were introduced at the end of the last century. Later on, and, in contrast to the case of electricity, the economies of scale of extended heat supply systems could not balance the additional costs for the network any more. The emergence of centralised heat supply was further limited by the heat losses which inevitably occur in extended pipeline systems. In the case of electricity, this problem could by solved by high voltage transmission.

Electricity supply has been introduced at a large scale since the turn of the century. In the majority of industrialised countries a structure for electricity supply has emerged which is characterised by

- Large units which exploit economies of scale,
- The recognition of a specific strategic and political interest in electricity and energy supply in general,
- The emergence of organisations which are able to exploit the economies of scale, i.e. large, interconnected and vertically integrated utilities, operating generally within a monopolistic framework,
- An agreement of the main actors on the general benefits of such a supply system.

The present form of electricity supply has been the result of a historical process of several decades. According to Hughes, small local and regional electricity networks were at the origin of the centralised and interconnected systems which exists nowadays. They were consciously expanded by a few major companies (Hughes, 1983). From a technical point of view, this integration and extension of networks has been enabled by the replacement of direct current systems for alternating current systems, and the emergence of the respective appliances and interfaces for alternating current (David, 1992). The initial opportunities to combine heat and power generation locally turned into a separation of the two supply chains. The major arguments and reasons for the establishment of a centralised power supply system are going to be discussed in the following.

5.1.1 Economies of scale

A couple of rather technical arguments can be mentioned which facilitated and even induced the emergence of a monopolistic structure for power supply. First of all, considerable economies of scale could be achieved by the construction of ever larger power stations. Higher overall efficiencies of large plants and lower specific construction costs drove the maximum plant size up. Furthermore, decisions were made to locate plants in the vicinity of coal mines, rivers or main ports in order to reduce the transport costs for the fuel and facilitate cooling. Electricity transmission in high-voltage grids was cheaper than distributing primary fuels to decentralised generation sites. This favoured the development of large-scale centralised systems, and contributed to making electricity the most universal form of energy.

5.1.2 The political interests in energy supply

With the emergence of an energy- and electricity intensive industry at the end of the last and beginning of this century, governments perceived a strategic interest in the energy sector. Apart from its key importance for the development of a coherent national industrial policy, there were also military interests at stake. In case of a military conflict it was crucial to control the access to energy, and state control was a means to guarantee this. In addition, and especially after the Second World War, energy supply became an issue for social policy, too. Energy was regarded as a public good at that time. Moreover, the profits from state-controlled energy supply could be used to cross-subsidise other public services.

These factors help to explain the emergence of a state-controlled and centralised organisation of energy supply systems. On the other hand, in some European countries, e.g. in Sweden, Denmark and Germany, there are strong local energy companies, which are often closely linked to the local political authorities. They reflect the origins of electricity supply as a local topic.

5.1.3 Emergence of centralised technical and organisational structures

Economies of scale and political interests in energy supply were the key factors behind the establishment of centralised, monopolistic and state-controlled organisations for energy supply. A number of economic arguments supported this development trend further.

Natural monopolies and interconnection

Although in principle the existence of monopolies is regarded as undesirable, in the case of electricity supply the arguments in favour of a monopolistic structure were very strong. Electricity supply was regarded as a natural monopoly, i.e. as a good which can most efficiently be supplied in a monopolistic setting. This conviction was based on the following arguments (Coutard, 1994):

- Electricity can not be stored, and the balance of power supply is easier to maintain in a large integrated and grid-based system;
- Only the construction of one single grid is necessary, which represents a major cost factor;
- In an integrated system it is possible to share reserve plants and therefore to avoid the additional reserve capacity needed for smaller-scale systems;
- The operation of the plant park can be optimised, expensive plants can be operated at peak hours only.

The crucial point in the natural monopoly discussion was that overall a single system might be the most efficient solution, but it could be challenged locally. It was necessary to protect the single system against local competition in order to achieve the social optimum for all customers. Therefore, electricity supply, and in some countries also other grid-based systems like district heating and gas were organised as monopolies and exempted from antitrust law.

Beyond these fundamental arguments in favour of a monopolistic approach to electricity supply, a number of principles were applied by utility companies which reinforced the establishment of a centralised, interconnected system and which acted counter to decentralised generation. They help to explain why it was more efficient to have one single supplier rather than several companies. According to Coutard,[109] these principles were "guidelines for achieving lower costs and greater reliability in electricity supply:

1. Obtaining economies of scale with large generating units, such as steam and water turbines;
2. Massing generation units near load centres or economical sources of energy and near cooling water at giant power plants;
3. Transmitting electricity to load centres through high voltage transmission lines;
4. Cultivating mass consumption by charging low and differential rates and allowing supply to create demand;
5. Interconnecting power plants to optimise their different characteristics;
6. Interconnecting loads to take advantage of diversity and thereby raising load factors;
7. Centralising control of interconnected loads and power plants by establishing dispatching, or system co-ordinating centres;
8. Forecasting load requirements to achieve optimum operations within the interconnected system;
9. Lowering installed and reserve capacity and co-ordinating maintenance shutdowns through the exploitation of power plants interconnections;
10. Accepting government regulation to establish a natural monopoly;
11. Earning a regular and adequate return on investment to obtain capital at a reasonable interest."

System integration

These have been the main arguments in favour of an interconnection between generation plants in a co-ordinated network, but the development towards integration went even further. Generation, distribution and supply services were also vertically integrated in order to optimise and co-ordinate plant operation and scheduling. For this purpose, information about expected regional loads was needed, which in turn required close co-operation between generation and distribution companies. In general, the exchange of information and the control of load and supply is easier in a single organisation where confidentiality is a minor

---

[109] Cited from (Coutard, 1994, p. 168) who refers to (Hughes, 1983).

problem. Furthermore, demand patterns can be controlled by differentiated tariffs for electricity consumption, and thereby be adjusted to the possibilities offered by the available generation plant park. However, this also requires co-operation between generation and distribution which can be achieved more easily in a single company.

Sunk costs and lock-in

The long life time of large power plants as compared to their depreciation, and the sunk costs of past investment in plants (i.e. investment which is definitively fixed and cannot be recovered for other purposes) leads to great cost advantages of a large utility company which operates different plant vintages. A small newcomer to the power business would have to calculate his costs against a plant park which is already partly written off. He would have to invest massively over a long period of time before he could recover his investments. Usually only the utility companies themselves have the financial and economic power to make this type of long-term investment. In other words, once a centralised system has been established it is very difficult to challenge it due to the accumulated costs advantages. This means that in the past decentralised investment was only feasible in countries where the natural monopoly was based on rather small organisational units such as e.g. the German local utilities. However, even in these cases the local companies were often dominated by a second layer of supraregional utilities.

All these factors reinforced the emergence of a centralised power supply system, or to be more precise, the increasing integration of local systems into large supraregional ones. It was a mutually beneficial symbiosis for all actors, matching technological advances on the one hand and organisational patterns on the other.

5.1.4 The typical players in electricity supply

A fierce debate is currently going on about the necessity to protect integrated electricity monopolies, and this issue will be taken up in the next section. However, at the time when the major decisions about monopolistic power supply were made, wide agreement prevailed among those who played a role in energy supply with respect to the advantages of a natural monopoly in this area.

The different actors needed for operating an energy and electricity supply system correspond to the functions which need to be fulfilled in that system (Table 5.1). They comprises the entire energy supply chain from fuels to final energy use, including the customisation of energy services. Moreover, other energy supply chains are associated to the electricity supply system (e.g. gas supply), as well as technology suppliers and regulatory/political authorities.

Functions	Actors
Fuel supply	Gas suppliers Coal mining companies Nuclear fuel providers
Electricity generation at different levels	Large centralised generators Local and regional utilities Auto-producers and Independent Power Producers (IPP) CHP-operators
Electricity transmission at different levels	Large generators/grid operators for high voltage Regional grid operators
Distribution at different levels	Service suppliers/distributors Local and regional utility companies
Facilitation of energy services	Utility companies for heat, gas and electricity Energy management firms, especially for self-generation System packager
Final use of energy services	Final consumers of energy services like households, industry, commerce, public services

Table 5.1: Functional structure of electricity supply and the corresponding actors

5.2 Recent challenges to the traditional organisation of the energy supply industries

The general agreement on the beneficial character of a monopolistic organisation of energy supply has been challenged since the 1970s for several reasons. A new paradigm has emerged for the organisation of the energy supply industries.

5.2.1 The end of the political and strategic role

The oil price crises of the 1970s showed the vulnerability of industrialised countries to interruptions of oil supply. Since then fuel supply has been increasingly diversified. Today, the security of fuel supply is not seen as an important challenge any more which could be used as an argument for state intervention and public monopolies. As the military argument has also lost its significance in Europe, the traditional political arguments have been overridden by economic arguments. At present, the public service function is questioned, socially driven cross-subsidisation is criticised, and the coherent application of market principles favoured by many actors. Furthermore, the active involvement of the public sector has been severely criticised for being inefficient and leading to sub-optimal social outcomes.

5.2.2 The natural monopoly debate and the claim for deregulation

The development towards liberalisation, deregulation and privatisation started in the United States in 1978 with the PURPA-legislation.[110] Later on, similar reforms were introduced in the UK, the Netherlands, Spain and many other countries. Regulations have been relaxed in order to facilitate competition and the emergence of technologies which are regarded to be environmentally beneficial (e.g. CHP and renewables). At European level the debate has recently advanced in response to a Commission White Paper on energy policy (CEC, 1995).

The three key concepts of liberalisation, deregulation and privatisation are closely linked to each other. Although they are used within the same framework of arguments, they need to be clearly distinguished. Deregulation aims to reduce the amount of administrative and legislative rules in order to facilitate investment and to break down unnecessary and costly barriers which distort the operation of market forces. The meaning of the expression becomes obvious when compared to regulation which can be defined as "any state intervention restricting the freedom of trade and contract except those which lay down the general rules of the game of the market economy" (Müller & Vogelsang, 1979, translation by MW).

While deregulation is not directly concerned with the organisational framework of the electricity supply industries (ESI), liberalisation aims to strengthen competition in energy and electricity supply. In other words, it challenges the monopolistic structure of the ESI by allowing new companies to generate and supply electricity, or gas, or heat. Usually, the transmission business is the only part of the supply chain which remains monopolistic for being still considered a natural monopoly.

Privatisation is discussed as an option in countries where the major energy supply companies are publicly owned. The assumptions which underlie the claims for privatisation are that public companies are almost by definition less efficient than private firms, and that different types of cross-subsidisation, which are frequent in the public sector, should be avoided in order to give way to a more transparent and efficient operation of market forces. This would have a positive effect on electricity prices, especially for industry, and thus also on competitiveness and employment.

Especially the liberalisation debate has been fuelled by a number of new arguments which are based on technological changes achieved over the last decades. They have changed the character of electricity supply and can be summarised as follows:

---

110 The Public Utilities Regulatory Policies Act (PURPA) of 1978 was the key decision giving way to the liberalisation of the US electricity supply system.

- The argument of economies of scale in centralised generation does not hold any more due to a number of reasons: size limits have been reached, additional costs for environmental investment need to be considered, and improvements have been made in decentralised technologies. For example, load management systems allow to operate decentralised systems much easier than in the past.
- It is hoped that competitive pressure will lead to lower electricity prices. This argument is based on perceived or assumed inefficiencies of monopolistic state-controlled organisations.
- Vertical integration to optimise operation is not necessary any more due to the possibilities offered by advanced control systems. This development challenges also the public good character of electricity supply because it is now possible to measure power and heat consumption relatively easily.
- A natural monopoly only persists in high-voltage transmission in order to avoid excessive grid construction costs; even the backup function of the grid is increasingly challenged by a higher reliability of plants and by the possibility to provide decentralised back-up capacity which is controlled by advanced information and expert systems.
- The potential advantages of decentralised competitive power production are recognised; higher energy efficiencies can be achieved due to the vicinity of the places of production and consumption of energy.

In practice, liberalisation challenges several of the public utilities' prerogatives. Independent power producers (IPPs) would be able to access the grid and sell to other customers ('Third Party Access (TPA)'), the utilities as the owners of the grid network would have to transfer power on their networks at fair prices ('wheeling'), the pricing of imported and exported electricity would follow market principles based on the calculation of avoided costs, etc. (Coutard, 1994). In other words, the established privileged position of the utilities would come to an end if the existing networks of utility companies, government institutions and major technology suppliers were broken up. At EU-level, the plans for an internal energy market address exactly these issues and are consequently rejected by those who would risk to lose a favourable position.

Although technical arguments have been frequently raised against the feasibility of liberalised electricity supply, the recent technological developments question these points of critique. Decentralisation and system co-ordination are facilitated by load management systems. The very short response times of several small-scale technologies (gas turbines, fuel cells, motors) and even of wind power plants make their integration easier. It has been argued that up to 50% of capacity could be provided by small decentralised plants (Krause, Koomey, & Olivier, 1995). This shows how large the diffusion potential of CHP might be.

5.2.3 The environmental discussion and energy efficiency

Apart from this economic challenge, environmental problem pressure requires to establish a more efficient supply system. It is argued that monopoly utilities, which are vertically integrated and horizontally disintegrated, have an intrinsic interest in increasing the sales of their main product rather than in providing energy services as efficiently as possible. To formulate it in a more general way, the main critique is that the traditional organisational structure does not give the right signals for companies to behave in an environmentally benign way. There are incentives to increase electricity sales, but not for environmental efficiency.

Energy demand has been de-coupled from economic growth since the first oil price crisis. This was achieved because some energy-intensive industries have either been closed down or managed to introduce more energy-efficient process technologies. Electricity production and demand, on the contrary, are still closely tied to economic growth, showing that there is still need for action to increase efficiency at both the demand and the generation side of the power supply chain.

5.2.4 Options and policies for reorganising the ESI

This double shock of environmental concerns and structural changes by liberalisation represents a severe challenge for the ESI. It is reinforced by the introduction of a common European electricity market, but the common market also opens up opportunities for some of the players in the electricity business.

Types of policy strategy	**Example of measures**
Structural policy	Deregulation and introduction of competitive elements like third party access to the grid and an electricity pool Unbundling of services and activities of companies Privatisation of public companies Re-definition of political competencies at different levels Anti-trust control
Technology-supply oriented policies	R&D programmes Subsidies for new applications and demonstrations Information programmes
Technology-demand oriented policies	Specific regulatory measures concerning construction, emissions, safety Guidelines for calculating avoided costs Stimulation of voluntary agreements between utilities and power generators Market-based instruments like taxes on fuels and energy consumption or investment grants Investment control Information programmes

Table 5.2: Energy-related policies, based on (Coutard, 1994; Horst, 1988).

A switch to new technological options is regarded as one possible way out of this dilemma. There are already some coalitions of interest in place which are in favour of change. They bring together such diverse groups as environmentalists and proponents of liberalisation, but there is hardly any constructive agreement on the right measures to be taken to introduce major changes. The question whether a liberalised framework can provide conditions which facilitate a more sustainable energy supply path is still an openly debated issue.

There are many options and strategies available to affect the technological pathways in energy supply. Strategies can focus on the structure and organisation of the sector, on a new taxation and pricing framework, on R&D-support for specific technologies, or on the control of the demand-side for energy technologies (see Table 5.2). The introduction of nuclear power is a good example of the joint operation of these different policies. In view of the rather negative experiences made with targeted policies in the past, structural measures and the deregulation of the framework for energy supply have been the preferred options in the last years.

5.3 CHP technology and its main dimensions (Step 1)

The diffusion of cogeneration needs to be interpreted within the historical and structural context outlined above. It appears to be a very promising option from various perspectives. The crucial question is, in which type of framework it could be developed and applied successfully. It is not sure whether the trend towards liberalisation opens up better opportunities for a technology like CHP than a traditional monopolistic system. However, for both types of frameworks, the question needs to be raised what government could do to stimulate its diffusion in order to move towards a more sustainable energy supply path.

According to the PET-perspective, a change of the political framework is the result of interactions within the system. It affects the positions and attitudes of the actors, but it is also shaped by them. These attitudes can be specified for CHP in terms of its main dimensions of interest. Consequently, it is important to be well informed about the dimensions which characterise CHP-technology and which are relevant to the assessments by the different actors. The actors' positions can only be understood if these pieces of information are available. In the following, cogeneration technology will therefore be analysed in detail. The main dimensions of interest to the different actors in the system will be introduced and discussed as objectively as possible.

The basic features of cogeneration or combined heat and power are very simple. It can be defined as *the technological principle of generating simultaneously useful heat when generating power*. In traditional power plans only electricity is produced, whereas the heat, which, according to thermodynamic laws, is generated simultaneously, is rejected to the air in cooling towers or to rivers nearby. Heat, either as steam for industrial processes or as hot water, is usually generated separately in industrial or private boiler plants.

Despite this simple definition it is very difficult in practice to decide whether a specific plant should be regarded as a CHP-plant or not. Some processes just recover parts of the heat in a later process stage. In others, the cogeneration capacity of a plant is exploited only to a very limited extent.[111] Cogeneration is also a quite complex technical system, composed of several units and interfaces. Its specific systemic characteristics determine the possibility to integrate into a wider energy supply system.

The combined production of heat and power is a technological principle which has been used since the beginning of the 20th century and which has flourished in its first two decades.[112] Industrial applications (e.g. in the chemical industry where much process heat was needed) represented the main share of cogeneration activities. Since the oil price crises of the 1970s there has been a renewed interest in this technology in response to the debates on energy efficiency and liberalisation. This has been accompanied by a number of legislative measures to for promote cogeneration applications and thus to reduce the consumption of fossil fuels. Many of these measures were not designed in the first place for CHP, but they nevertheless had a major impact on its subsequent success. Anyway, the statement certainly holds that "(t)he availability of the technology, combined with its very unequal implementation, shows that CHP is not a technical issue but one of economics and energy policy" (Verbruggen, 1992a, p. 884).

From the PET-perspective, the objectives and expectations of the different actors relevant to CHP and the behavioural options at their disposal are of central interest. In order to analyse the situation in different countries it is assumed that the different assessments of the technology's dimensions by the actors, and their interdependence with the structures of the system, represent the foundation of the technological choices by society. The usefulness of such a multi-dimensional analysis is based on the argument that the actors' positions, preferences and choices with regard to a technology are the result of an evaluation of its different aspects and expected consequences along these dimensions.[113] The interactions and decision-making processes in the system are the means by which these individual preferences are aggregated to a collective choice. National systems differ with regard to structures and actors; this leads also to different assessments of the dimensions of the technology in question, and finally to different collective choices. The main dimensions of interest have already been indicated in Chapter 3. They are used here as a starting point for the analysis of CHP, though adjusted and specified to the needs of the technology:[114]

---

111 This discussion has been crucial for the establishment of a common European statistical data-base on cogeneration, because the national representatives could not agree upon a common definition of CHP.

112 See e.g. (Russell, 1986a; Russell, 1993) for the case of the UK.

113 Such a multi-dimensional view of cogeneration reflects that it is a 'socio-technical system'. CHP is integrated into a set of related technologies, and of interest to several actors and stakeholders.

114 It should be kept in mind that these five dimensions are geared towards the study of CHP. In other cases, it may be necessary to consider further dimensions of interest. For example, the end-user side of services is very important for transport systems.

- Technological dimension: The technological characteristics of CHP need to be understood before the other dimensions can be addressed. In addition to the principles of cogeneration and separate generation of heat and power, the main components and systemic characteristics need to be described. Associated to this are the main application areas of CHP and the respective requirements in terms of knowledge and skills.
- Environmental dimension: The environmental impact of cogeneration is a dimension of increasing importance, including also its characteristics in terms of resource consumption. In principle, this could be translated into economic terms (e.g. long-term social and external costs), but the high uncertainty and unreliability of such calculations requires a separate consideration.
- Micro-economic dimension: This comprises the assessment of CHP from the perspective of individual companies, i.e. technology suppliers, technology users and utility companies. Obviously, this micro-economic perspective can be further differentiated according to the main determinants of decision-making.
- Macro- or socio-economic dimension: The costs and benefits at the macro-economic, often national, level play a major role for the assessment of energy technologies. While environmental and other social costs have become more important in the last years, traditional issues such as security of supply, the overall efficiency of supply, and implications for industrial competitiveness are nevertheless essential parts of a comprehensive assessment.
- Organisational and power dimension: Organisational and power-related considerations concern the medium- to long-term strategic positions of the actors in the system. In this respect, the political and regulatory framework conditions are crucial. The key question is in how far the widespread introduction of CHP would alter the balance of power in the system and the power to define its 'rules of the game'. The interactions and decisions of actors are influenced by these strategic and power considerations, because they affect their medium to long-term positions in the system.

In principle, this could be transformed into a kind of multi-criteria analysis (MCA), but such an approach would be confronted with major conceptual difficulties. First of all, a MCA provides a static assessment of the advantages of a technology and aims to deliver a clear-cut final judgement. This contradicts the evolutionary perspective which emphasises the importance of change, thus implying the need for a 'dynamic MCA'. Secondly, and even more important, the actor-centred perspective of PET assumes that the aggregation of the different main dimensions is done differently by different actors, depending on their subjective interests and positions. Moreover, only the technical and possibly the environmental dimensions are sufficiently objective to be measured properly. Already the assessment and evaluation at the firm level depends on the specific case and the interests of the actor, and this holds even more in the case of priorities for macro-economic and power issues for which structural effects would have to be

---

Dimensions such as convenience, accessibility and comfort would have to be taken into account.

assessed. Thirdly, while technical, environmental and micro-economic aspects can be associated to single plants, the socio-economic and power dimensions refer to the entire plant park set up in a country. Consequently, an aggregation by an MCA would not make much sense, unless it is differentiated for the different actors. Moreover, these dimensions also need to be seen in comparison with the alternatives to CHP, which are also assessed differently by different actors. Some these difficulties could be addressed by available MCA-tools (Davison & al., 1997), but due to the emphasis, which is put here on the dynamics of change, their use would not make much sense in the context of this study.

5.3.1 Dimension 1: Technological issues and applications

Technological characteristics are the foundation for assessment of a technology along the other dimensions. For this reason it is essential to dispose of a good understanding of the technical issues at stake. But the technical dimension has its importance also in its own right, for example when it comes to assess the technical feasibility of a plant, and the compatibility of the technology with a firms' competencies and site conditions.

5.3.1.1 Thermodynamics

In principle, the generation of electricity and heat in fossil plants involves the transformation of chemically bound energy (W_{chem}) into these two products. According to the law of conservation of energy cannot get lost in this transformation process, but it can get lost for actual use when rejected in the form of waste heat.

$$W_{chem}(educts) \dashrightarrow W_{chem}(products) + heat + electric\ work + waste\ heat$$

What is inevitably getting lost in this transformation process is the 'quality' of the energy, because each (irreversible) transformation process involves a loss of high-quality energy ('work') into heat. This is expressed by an increase of entropy. 'High quality' means that certain forms of energy (e.g. electricity and mechanical energy) can be universally and completely transformed into other forms of energy without being limited by the second law of thermodynamics, whereas others (such as heat, chemical energy) can not.[115] Fossil power plants transform chemically bound energy into electrical energy. This is accompanied by the generation of the lowest quality of energy, namely heat.

Chemically bound energy can be entirely transformed into heat, i.e. at an efficiency of 100%. The efficiency of the transformation of heat into work - which is achieved in heat engines - is limited by the thermodynamic cycle which characterises the engine. The ideal Carnot cycle delimits the theoretically achievable efficiency of a single stage process for given inlet and outlet temperatures. In combined-cycle plants (like e.g. in Combined Cycle Gas Turbines, CCGT) a sec-

---

115 In thermodynamics, these two qualities of energy are called exergy (e.g. electricity) and anergy (e.g. heat).

ond thermodynamic cycle for the transformation of the residual heat into electricity is added which can considerably increase the overall electrical efficiency of a plant. Only minor losses during the transformation of work into electricity need to be subtracted from this. In practice, the efficiency of the transformation of chemical into electric energy (W_{el}) is at present limited to about 55%.

$$\textit{Electric efficiency} = W_{el}/(W_{chem}(educts) - W_{chem}(products))$$

The principle of cogeneration consists of making effective use of the residual heat for heating or steam raising. Even in a combined cycle plant, where the residual heat of the first cycle (e.g. a gas turbine cycle) is used to generate steam for a second cycle (usually a steam turbine) to generate electricity, the secondary waste heat still has a sufficiently high temperature to be used in heating systems. The usefulness of the heat which has been produced in a power plant is exhausted when it is completely dissipated to ambient temperature. In a CHP plant, the waste heat is therefore collected and conducted to industrial plants or buildings. Consequently, if efficiency is understood as the amount of useful energy produced from the chemically bound energy of the fuel, a new definition is needed to compare electricity-only and cogeneration plants which considers also the heat term (W_{th}).

$$\textit{Energetic efficiency} = (W_{el} + W_{th})/(W_{chem}(educts) - W_{chem}(products))$$

In the following table several basic cycles are compared with regard to the efficiencies of heat and power generation, heat temperature and the ratio of heat to power production. It is based on thermodynamic considerations only, i.e. static operation conditions, and thus represent just an approximate efficiency indicator.

It is evident that a fair comparison between cogeneration and conventional heat and power production must include the energy balances for separate generation of similar heat flows, i.e. a cogeneration system must be compared to a system which is composed of a power-only *and* a heat-only plant, which together deliver the same amounts of energy outputs (Figure 5.1). Even such a comparison is still problematic, because the different qualities of the products are not taken into consideration. This points to the limitations of a thermodynamic evaluation of cogeneration. Economic or environmental considerations on the values of the two energy products need to be added. Nevertheless, the thermodynamic considerations are essential for the evaluation of different energy supply options with respect to environmental aspects such as CO_2-emissions and primary energy consumption.

In all real-world processes, heat losses occur. They are not adiabatic systems as assumed in the preceding paragraphs. Heat is always transferred to the environment, but within the limits of the thermodynamic cycle used, there are still possibilities to increase the efficiencies of a plant.

Plant type	el. power range [MWel.]	electric efficiency [%]	thermal efficiency [%]	overall efficiency [%]	heat temperature [C]	heat/power ratio
Domestic boiler	< 0.1	0	50-80	50-80	-	-
Industrial boiler	> 0.1	0	80-95	80-95	-	-
Power-only station	10 .. > 100	35 .. 45	0	35 .. 45	-	-
Combined cycle power-only station	10 .. > 100	45 .. 55	0	45 .. 55	-	-
CHP internal combustion engine, gas	0.02 ..1.1	25 .. 40	40 .. 65	80 .. 90	60 .. 110	1.5 .. 2
CHP internal combustion engine, diesel	0.1 ..12	38 .. 44	36 .. 42	75 .. 80	60 .. 110	1.2 .. 1.5
Gas turbine CHP	0.2 .. >100	26 .. 37	28 .. 48	70 .. 80	120 .. 450	1.5 .. 3
Steam turbine back-pressure CHP	0.5 ..60	26 .. 32	44 .. 53	75 .. 85	120 .. 250	1 .. 8
CHP steam turbine extraction/condensing	10 .. >100	32 .. 40	40 .. 50	70 .. 80	120 .. 350	2 .. 5
CHP combined cycle back-pressure	10 .. >100	30 .. 40	40 .. 50	80 .. 90	120 .. 250	1 .. 1.5
CHP-combined cycle extraction/condensing	10 .. >100	38 .. 44	40 .. 50	80 .. 90	120 .. 350	1 .. 1.5

Table 5.3: Comparison of different cogeneration technologies with traditional boilers and power stations, Source: (BMWi, 1994, 1/p. 3; Dufait, Martens, Guilmot, & Verbruggen, 1992; Horlock, 1987, pp. 24-25; Sirchis, 1990, p. 17)

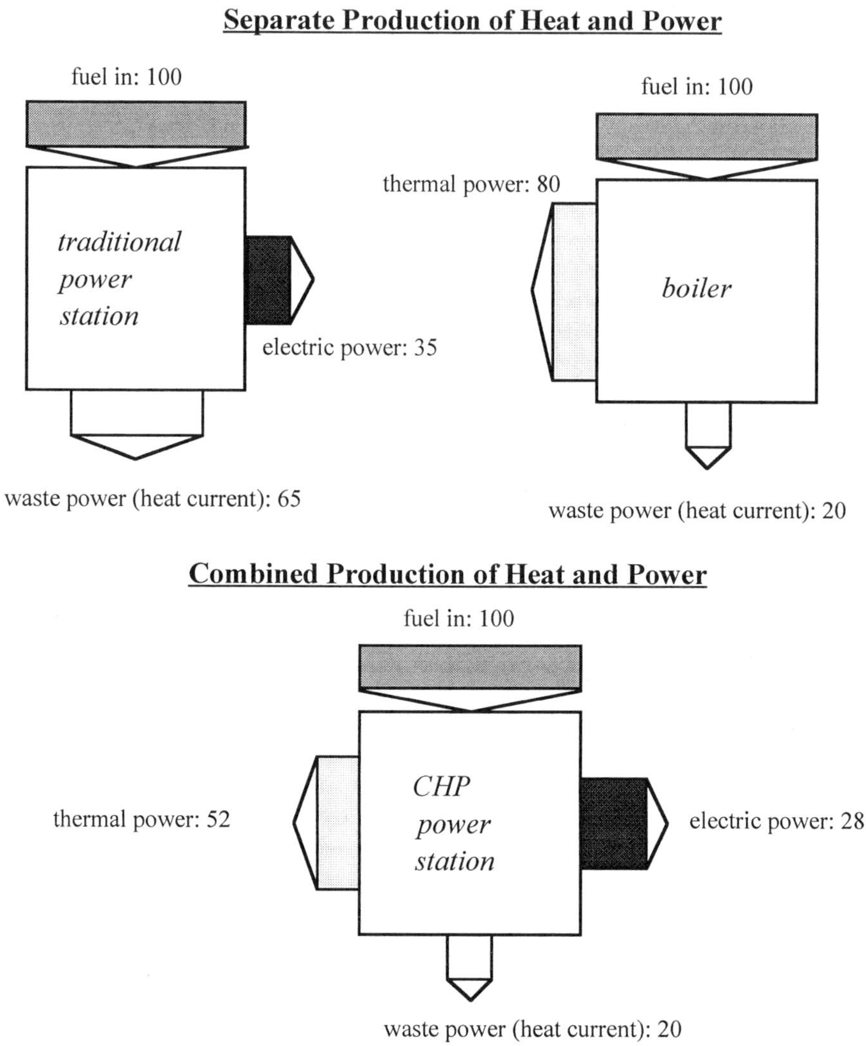

Figure 5.1: Black box model of cogeneration and conventional heat-only and power-only plant, adapted from (Babus'Haq, Probert, & Shilston, 1986, p. 100)

A look at these efficiencies shows that cogeneration seems to be a promising technological alternative to the existing types of electricity- and heat-generating technologies. It can achieve an overall (i.e. heat plus power) efficiency of energy transformation of up to 90%[116] whereas conventional electricity-only generation technologies reach only up to 45% and heat generating boilers up to 95%. It will

---

116 The losses which occur during the conversion of chemically bound energy into heat and electricity are typically about 10%.

be shown later, that these impressive numbers will nevertheless have to be reassessed due to a number of technical and organisational reasons. They can reduce the theoretical efficiencies for both conventional and cogeneration technologies. Furthermore, one should not only consider the single conversion stage from chemically bound energy to heat and electricity, but the complete supply chain (see Fig. 5.2), including transportation and distribution of the fuel.[117]

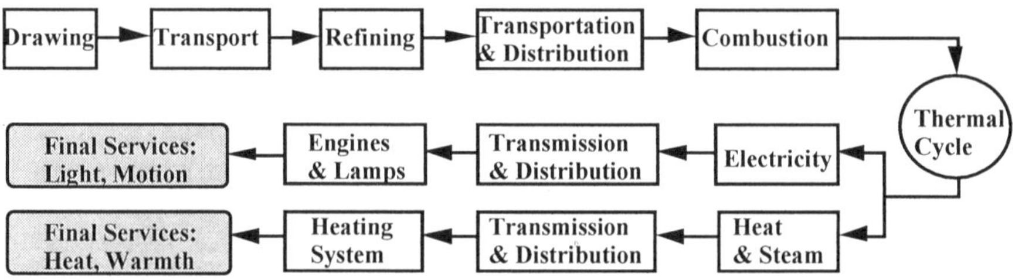

Figure 5.2: The transformation chain from the primary fuel source to final services

5.3.1.2 Structure and components of a CHP-system

Cogeneration plants are technical systems which are embedded in the energy supply and management system for heat and electricity of a site. Beyond the site level, a CHP-system is also part of the wider electricity supply (and possibly also heat supply) system to which it is connected. In the following, the main components of a cogeneration scheme are presented, and it is shown how they compose a technical system (see Figure 5.3).

Prime movers and fuels

The type of prime-mover used in a CHP-plant strongly determines the characteristics of the entire system. Often CHP-systems are even classified by types of prime-movers. Four different types of prime-movers are used in commercially available cogeneration systems: reciprocating internal combustion engines (diesel or gas driven), gas turbines, steam turbines (backpressure or extraction/condensing type) and combined cycles of both gas and steam turbines. Advanced fuel cells are for CHP are a very recent development, but due to the early stage of fuel cell development, they are not yet available on a fully commercial basis. In principle, it is also possible to use the waste heat of nuclear power stations for heating or steam generation.[118]

---

117 The energetic and environmental advantages of natural gas as compared to indigenous coal are reduced by the significant losses which occur during its transportation (~10%). Moreover, the transmission losses of centrally generated power are partly balanced by the necessity to distribute gas, oil or coal to decentralised generation units.

118 For more information on the different cycles and plants, see (AGFW/UNICHAL, 1993) or (Dufait et al., 1992).

Apart from the traditional fuels such as gas, diesel and coal, a variety of other fuels can be used in CHP-systems. Especially all kinds of renewables (e.g. biogas or wood chips) and waste incineration have attracted interest in this respect.

Boilers

The prime mover drives the electric generator and, through a heat exchanger, produces hot steam or water. The operational flexibility of the prime mover is usually limited, both in terms of its heat/power ratio and in terms of its overall output. Additional boilers and a connection to the power grid are needed to adjust the output to variable power and heat loads of a site, to meet peak demands and to dispose of back-up capacity during maintenance and repair periods.

Two main types of boilers for the different CHP-systems can be distinguished. First, in gas turbines and reciprocating engines, the hot exhaust gases are used to generate hot water or steam in a waste heat boiler. Second, in the case of steam turbines, the power source is not directly fired, but the fuel is burnt in a primary boiler to generate high pressure steam, which is then expanded in the steam turbine for power generation. The steam raised can be used for process or heating purposes. The quality of the heat exchangers in the boiler are decisive for the efficiency of these units. However, due to the limited variability of the heat/power ratio of a CHP-unit, there is always a need to complement the system by a peak boiler to cover heat requirements beyond the maximum output of the plant.

Heat storage

Another means to compensate for fluctuations of heat and power loads are heat storage systems. Their application is usually restricted to low-temperature heat as in district heating schemes to balance short-term load variations.

Gas compression and fuel supply

Especially for gas-fired systems, the fuel connection can be an expensive and energy-consuming ancillary equipment. The additional energy demand for gas compression needs to be taken into account in the overall energetic and economic budgets of a plant.

Distribution system

Given the two simultaneously generated products heat and electricity, two distribution networks are needed in a CHP plant. The electric generator is usually connected to the external power grid for back-up and supplementary power, even if it mainly feeds the local network of the site. A short- or medium-distance heat distribution system (pipeline network) supplies the heat consumers. Due to inevitable losses in the distribution system, the range of heat transmission is limited and constrains the choice of location. A cogeneration plant should be situated close to the main heat loads in order to avoid excessive pipeline and insulation costs, as well as heat transmission losses. CHP thus implies the need for decentralised supply structure for both heat and electricity. Consequently, cogeneration schemes are often built in the vicinity of densely populated areas or close to heat-intensive industrial plants.

Exhaust gas cleaning system

Being usually fired by fossil fuels, regulatory rules require CHP-operators to monitor and control emissions. Similar to other fossil power plants, exhaust gas cleaning systems need to be added to the power and heat generating system. Especially for smaller plants it may be sufficient to improve the combustion process itself to fulfil the requirements (e.g. by means of lean-burn engines), but for larger plants the clean-up facilities can represent a major share of investment.

Control system and grid connection interfaces

In order to keep the system in an efficient range of operation and to determine the heat and power output, control systems play an increasingly important role. The more complex and optimised systems are, the more difficult it is to operate them properly. Improvements in this field have considerably broadened the scope of CHP-applications. The flexibility of a CHP-system to respond to changing loads a crucial determinant of its efficiency. Older large cogeneration plants which are usually based on back-pressure steam turbine cycles have a fixed heat/power ratio, whereas modern turbines and engines allow to vary this ratio at least in a limited range. Control systems matter not only for plant efficiency, but also for the interconnection to the electricity grid, because the quality of the electricity supplied (and therefore its price) depends on the reliability of its supply and on the correct phasing of CHP-system and power grid.

Structure of a typical cogeneration system

An example of a typical gas-turbine cogeneration plant is shown in Figure 5.3. It gives a overview of how the main components are interconnected. Natural gas is supplied by a gas compressor to the combustion chamber of the gas turbine where it is burned with compressed air at a temperature of up to 1500 °C. The hot burning gases drive both the turbine and the compressor part of the prime mover which is connected to the electric generator. After the expansion of the gases in the turbine they have a temperature of about 400 °C. This heat is rejected in a waste heat boiler to raise process steam and/or generate low temperature heat. Depending on the requirements on the heat side, an after boiler is available in order to adapt the system to higher heat loads. The exhaust gases can be cooled down to a temperature of about 120°C while high-temperature steam of about 400-500 °C is generated. The first waste heat boiler for the generation of process steam is complemented by a second one for lower temperature heat to be fed into a hot water heating system. Apart from the after burner, a heat accumulator can contribute to use the waste heat as efficiently as possible. Limited amounts of heat are stored in periods of low heat demand and can be used to top up heat supply at peak times.

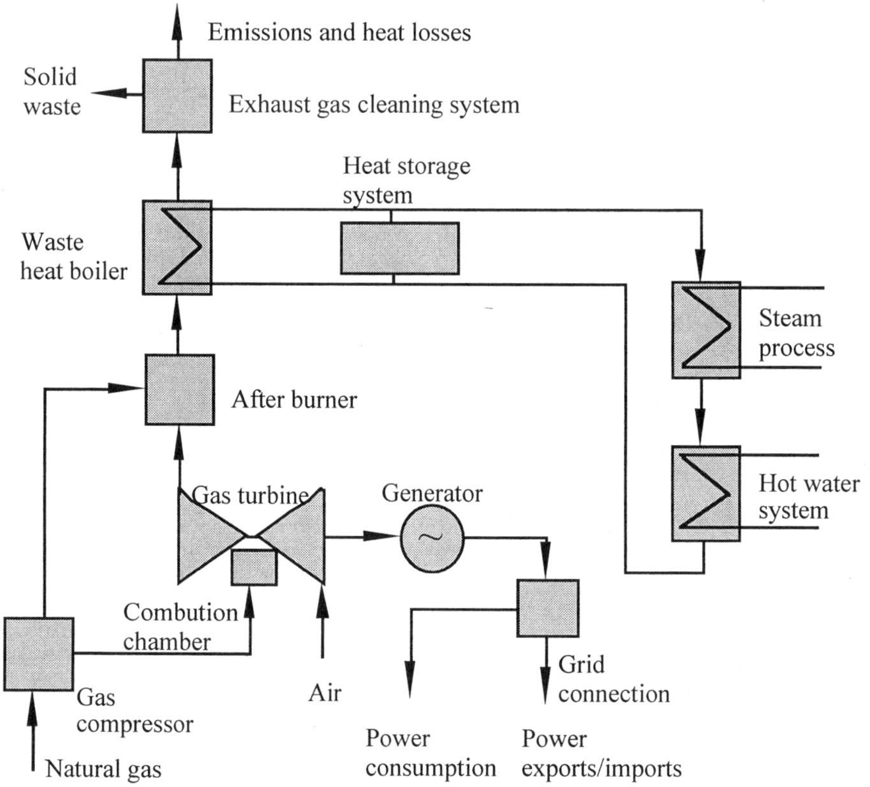

Figure 5.3: Gas-turbine cogeneration plant

In other words, there are three means to adapt to fluctuations of heat and electricity loads: the control system, the after burner and the heat storage unit. In a district heating scheme as presented here, heat is finally rejected at temperatures of about 60°C. The exhaust gas cleaning system is usually an add-on technology. In modern gas-turbines and engines, optimised combustion chambers makes it possible to do without the add-on technology.[119] The more complex and integrated a CHP-system is, the more important is the role of the control system to optimise and adjust the joint operation of the different components.

[119] The competition between and *end-of-pipe technology* (e.g. clean-up systems) and *clean technology* (e.g. advanced combustion chambers) is an important element of the 'greening of industry' discussion. Whereas end-of-pipe technologies were a fast response to legislative pressures, their long-term development potential is limited. Clean technologies often have a greater long-term potential for achieving lower emissions and resource consumption at reasonable costs, but require more sophisticated technologies and development times.

5.3.1.3 *Characteristics of CHP as a dynamic technical system*

While the thermodynamic characteristics of the combined heat and power process represents a static assessment, in practice a couple of further constraints limit the optimal operation of CHP and need to be considered in plant design. This is because a real CHP-system never operates under thermodynamic conditions, but dynamically under changing heat and power loads. Therefore, it is necessary to describe a CHP plant also in terms of its dynamic and systemic characteristics.

Input fuel

Depending on the different types of prime movers, a large variety of fuels can be used in cogeneration systems: all kinds of fossil fuels (coal, oil, gasoline, gas, lignite etc.), biofuels, waste or sewage gas. A few nuclear cogeneration plants are in operation, too. This points to a major long-term advantage of extended CHP systems which connect a variety of sites by a heat network: in order to upgrade such a system, only the core generating unit needs to be replaced, whereas the network stays in place.[120]

Total capacities per unit

In Table 5.3, the power ranges of different types of CHP units are given. Today cogeneration schemes exist from as small as 5 kW up to the 1875 $MW_{el.}$ industrial combined-cycle plant at Teesside (UK) which consists of eight gas- and two steam turbines. In other words, CHP can be applied both for highly distributed generation networks as well as for more centralised ones. Although economies of scale can also be achieved by installing larger units, CHP is particularly suited for stacking several units and operating them in parallel. This increases the flexibility of a plant and its adaptability to fluctuating loads. It is evident that in large systems with several units and different types of heat and power users the optimal operation of the entire system is very difficult to maintain and requires the use of advanced computerised control systems.

System output heat and power: load, distribution and storage

Cogeneration systems supply two products with very different characteristics in terms of distribution, storage, reliability and markets. Some of these characteristics are summarised in Table 5.4.

A further category for cold/cooling could be added to this characterisation of the products, because one way of optimising the operational performance of a CHP-plant consists of complementing the heating load in winter by a cooling load during summer. A number of solutions are available to convert heat into cold, e.g. absorption chilling.

---

120 On this point, see the example of East Germany, where several CHP/DH systems were upgraded after reunification by replacing coal-fired generation unit by modern gas turbines.

Feature	Heat	Power
Storage	Limited by time and temperature levels	Requires conversion, i.e. high investments; Significant losses
Transport	Limited by distance; High investment; Losses proportional to temperature; Pumping power costs	Grid system in place and generally available; Limited losses
Market extension	Local or non-existent	Interconnected regionally, nationally and internationally
Service reliability	High requirements; Inertia in the system limits the impact of outages	Very high requirements; High outage costs and instantaneous effect

Table 5.4: Characterisations of the two products heat and power, Source: (Verbruggen, 1992a)

Depending on the specific site conditions, certain CHP-technologies are more appropriate than others. Industrial applications where high grade steam is needed will usually be operated on gas- or steam-turbine solutions. For heating purposes and small-scale applications engines are more common.

Load flexibility

The operation of a site is optimised usually with regard to economic criteria, not with regard to energy efficiency. For the economic appraisal, the values of the different products play a decisive role. One of the products, heat or power, is the lead product, i.e. the generation of the secondary product depends on the demand for the lead product. For example, the operation of an industrial cogeneration plant is determined by the heat load which is subjected to daily, weekly and seasonal fluctuations (see Figure 5.4). Power generation, either for on-site consumption or for export to the grid, has to follow the heat load fluctuations within the flexibility of the heat/power ratio of a plant. This heat/power ratio can usually be varied by small amounts only or is even fixed. In other words, it is conducive to an efficient operation of a CHP-plant to be connected to relatively constant heat and power loads.

Advanced control systems can optimise operation, but they cannot replace for the missing flexibility. As shown in the Figure 5.4, combined heat, power and cooling is a quite elegant solution to achieve a higher flexibility because heating and cooling loads tend be complementary over the year. Heat storage systems are also used to balance the short-term (daily) fluctuations of loads. Sets of several CHP-modules can also be operated according to load requirements. Modular solutions guarantee also a high security of supply because one module is mostly on standby. In practice, these operational constraints lead to a design of CHP-plants to cover only the base-load of a site. The peak heat loads are then covered by complementary boilers.

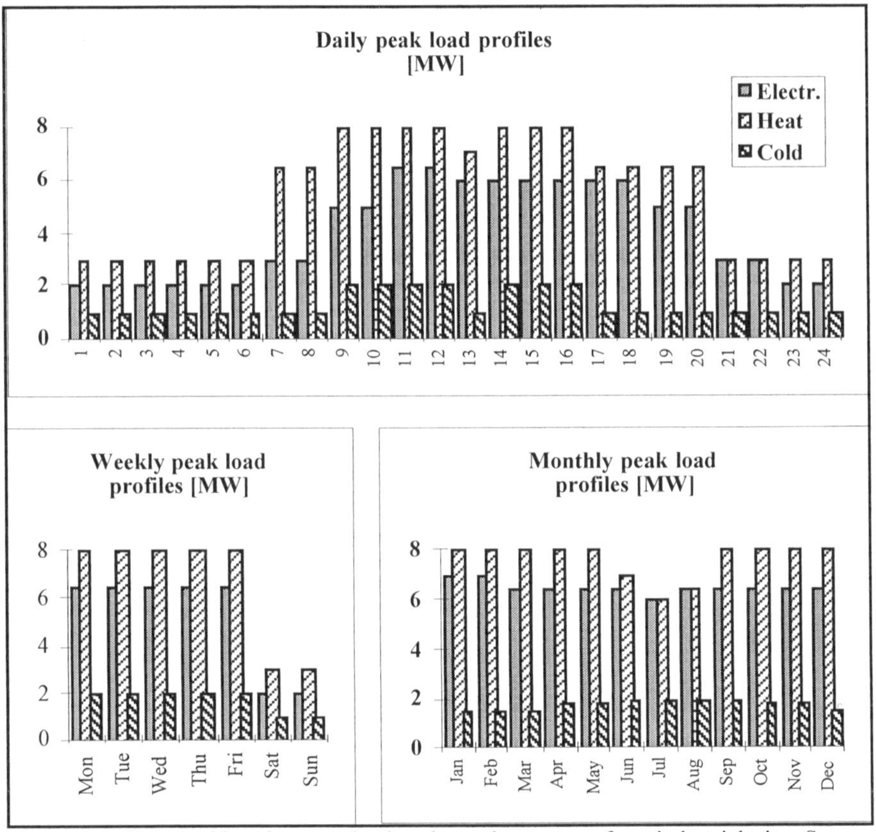

Figure 5.4: Heat, cold and power load and supply curves of an industrial site, Source: (EEO, 1992a; EEO, 1992b; Grohnheit, 1993; Verbruggen, Dufait, & Martens, 1993)

Figure 5.4 shows a typical load example. The sum of heat and cold peak load is about 10 MW, and the power load of the site about 6.5 MW. During working hours at weekdays, a CHP-plant which can deliver 5 MW electricity and 8 MW heat/cold would be a suitable option, to be complemented by limited power imports and a peak boiler. During nights and on weekends, the operation of the plant would not make sense because the load is too low. The CHP-plant would have to be operated during weekdays only. Alternatively, either a smaller unit could be installed to cover the base-load only, or a heat storage system to balance day/night fluctuations. The choice between these options would then depend on their expected economic costs and benefits.

These site-specific fluctuations make it difficult to integrate independent CHP units into an extended electricity network, even if a large number of distributed plants would balance the fluctuations. In other words, it is still necessary for a system operator to dispose of flexible power-only plants to guarantee a permanent power supply in the system.[121]

Quality of output heat

Temperature is the decisive quality characteristic of heat. The higher its temperature, the higher its internal energy, quality and value. High temperature steam can be used for industrial applications, e.g. in the chemical and paper industries, whereas low temperature heat and hot water can only be used for heating purposes in buildings. The interest to generate high-temperature heat has also consequences for the simultaneous generation of electricity: the higher the heat temperature, the lower the electric efficiency of the cycle, because in this case the steam is not fully expanded in the turbine. A second important quality characteristic of heat supply is its reliability, i.e. its permanent availability.

Density of heat load

The distribution system is a source of energetic losses. Therefore, a high density of the heat load is a precondition for operating a CHP plant efficiently. This requirement is the reason for the installation of cogeneration plants in the vicinity of industrial sites, because they constantly require large amounts of process heat. District heating systems are most appropriate in densely populated city areas although in some cases also rural villages are supplied by district heating.[122]

Quality of electricity supplied

Although electricity is often regarded as a homogenous good, its value can differs with regard to availability, voltage and power capacity, stability and reliability of supply. These differences are the reason for differentiated electricity prices, as reflected e.g. in spread tariffs for periods of base-load, medium load and peak load. The predictability of power demand peaks are also an important factor for designing an appropriate capacity of the electricity grid. Grid operators argue that CHP-autoproducers need to pay high capacity charges because of the lower reliability of CHP systems. If a CHP system fails grid capacity must be sufficient to guarantee the provision of back-up power.

A heat-guided cogeneration system which exports power to the grid cannot fully comply with these standard reliability requirements because the amount of ex-

---

121 Simulation models of national or regional electricity supply system allow to calculate and control the plants connected to the grid in order to guarantee permanent supply, but the controllability of a large number of independent plants is one of the crucial topics of debate between CHP-generators and grid operators. However, from a technical point of view about 40% of power generation could be generated decentrally without causing major problems for grid stability.

122 Especially in Denmark, relatively low heat load density have been accepted in district heating schemes. Usually, a load density of about 120 GJ/h km^2 is desired and the load per km of pipeline should not be lower than 20 GJ/h km (Beitz & Küttner, 1986).

ported electricity depends on the heat and power load of the site and on the flexibility of the plant. Consequently, the grid operator has to complement imports by his own plants. These adaptations do not represent a major technical difficulty but an organisational obstacle. Moreover, these arguments have often been used to attack independent power production. They also underline that, in order to understand the constraints of its operation, CHP must be analysed as part of its wider application context.

Availability and reliability of supply

Both products heat and power have to be supplied in a reliable way. While power supply is backed up by national or even international grid connections, heating networks depend on a small number of plants. Single-plant networks are subjected to a certain risk of breakdown. Moreover, the availability of a system is an important criterion for the evaluation of a CHP-project, and it determines the value of the generated electricity. Examples from Denmark show that multi-plant networks can reduce significantly the risk of failure.[123]

Packaged versus individual design

Standardised or packaged CHP-equipment allows to achieve substantial cost reductions, in particular for small-scale applications. Large-scale plants still need to be individually designed, even if several components, including e.g. gas turbines, can be purchased off the shelf. The widespread use of cogeneration systems could therefore induce cost reductions and a higher economic competitivity of CHP. However, there are limitations for packaging because a good plant design needs to take into account the specific site conditions. Small scale heating applications may be characterised by many similarities and are thus more suitable for a packaged solutions. For large-scale system it would be more promising to standardise design procedures rather than the final products.

Technical system, network character and technological complementarities

Cogeneration can not be understood independently of the heat and power networks to which it is linked. It is embedded in the electricity supply structure of a country, and in the heat supply system of a site or a city. In the past, these two supply lines have developed in different directions, but cogeneration requires their interconnection. In other words, the introduction of CHP 'merges' two supply chains and implies a reorientation of the companies involved.[124] The emergence of gas-fired cogeneration has even brought into play a third network technology, namely gas

---

123 This can be interpreted as a network externality, i.e. the interconnection of heating networks to a 'network of networks' implies advantages for the operation of each subnetworks.

124 This need to co-ordinate different activities can also be interpreted in terms of transaction costs. The concept of economies of scope could also be applied. It implies that cogeneration is economically more interesting if the companies running such schemes are involved in service supply for both heat and power.

supply. Gas-based cogeneration offers the opportunity to complement gas supply (for long-distance distribution) with heat network (for short-distance distribution) which are based on small- to medium-scale CHP units (50-1000 kW). In many cases, this could be a good compromise between the two competing technologies of district heating and gas networks.

In other words, a cogeneration system does not only represent a complex technical system in itself, it is furthermore integrated in local (heat) and national or even international (electricity) networks. This network character ties a large number of interests and actors together who are all concerned by changes in the system due to different *network externalities*. CHP generates a number of *indirect* network externalities which differ to some extent from the classical direct effects.[125] A major indirect externality of a CHP-network arises when an growing number of households are connected to it. In this case, the network costs are distributed among a growing number of customers, thus reducing the heat costs for all of them. Other external effects which arise from CHP are the economies of scope for operators of different types of networks, the need to co-ordinate different local planning procedure, and the disadvantages for large power utilities which have to accept decentralised and independently generated power supply to their grids.

5.3.1.4 Applications of CHP

As a consequence of these constraints, CHP is particularly a particularly useful option for three main groups of applications:

- District heating and group heating;
- Industrial applications;
- Small-scale applications.

These types of applications tend to be realised by distinct technical alternatives. Moreover, they are related to certain types of users and/or types of ownership. Due to the universal character of the product power and due to the limitations to transport heat, the discriminating factor for these main types of applications is the heat load on the user side. This topic will be taken up again in the context of the micro-economic assessment of CHP.

Small-scale applications

Small-scale cogeneration is usually the most numerous type of application in the different countries but its contribution to power production is rather modest. Both heat and electricity are used mainly on-site, although it may be profitable to sell

---

125 For more details and examples of network externalities see (Katz & Shapiro, 1994; Katz & Shapiro, 1985; Katz & Shapiro, 1986). Although being a network technology, *direct* network externalities in the classical sense (like for communication services, fax) do not exist in the case of cogeneration. There is no *direct* advantage for an end-user if others are connected to the same network, but there are several *indirect* externalities.

surplus power to the grid operator if good grid connections conditions are granted. Typical examples of small-scale cogeneration plants can be found in leisure centres, hospitals, greenhouse heating, apartment blocks, schools etc. Small-scale systems are mostly based on reciprocating engines, but also small gas turbines have been introduced in recent years. Due to the progress made in terms of standardisation ('packaging'), efficiency and reliability, small-scale applications have been very successful in the last years. They represent the extreme form of decentralised production of power and heat. In the longer term, this type of application might even replace traditional electricity and heat production schemes for buildings, even if a back-up and export connection to the power grid will have to be kept.

'Small-scale' seems to imply a clear definition in terms of plant size, and in some countries such a definition is actually used. For example, in the UK small-scale refers to plants smaller than 1 MW in the UK, while in the Netherlands the upper limit is 2 MW. However, it appears to be more appropriate to use the label 'small-scale' in a different way and reserve it for non-industrial heating applications (i.e. no generation of high temperature steam) which are not integrated into large integrated heating networks but aim at the heat supply of individual buildings or small local networks ('object heating'). Obviously, the borderline between object heating and district heating is fuzzy, but the application context still appears to be a more useful criterion for categorisation than size.

Industrial applications

Industrial sites are a major application area for CHP-technology because high-temperature steam processes represent a stable heat load and provide ideal conditions for cogeneration. It is thus no surprise that CHP was first applied in industrial processes at the beginning of this century. Typical heat-intensive industries are chemicals, paper and oil refinery because the load profiles enable the permanent operation of the CHP-units. Plants are usually of a medium to large scale, i.e. from about 10 MW to a few hundred MW. Traditionally steam turbines were mainly used but in the last years gas turbines and combined cycles have increasingly replaced them.

District and group heating

The first district-heating schemes were a by-product of industrial heat generation. Examples of such co-operation projects between industry and district heating still exist today. In principle, district heating does not necessarily have to be based on a cogeneration plant; heat-only systems, which use highly efficient burners (nowadays achieving thermal efficiency of more than 90%), are a simple alternative. However, there is a clear trend towards cogeneration in district heating, especially if the power can be exported under reasonable conditions. District heating as well as industrial applications are examples of heat-guided processes, i.e. the heat is the primary product. This means that electricity production is largely dependent on the heat load profile. It will be shown later on, that despite these technical reasons for heat-guided operation, the economic conditions may favour a technically and energetically suboptimal power-guided operation, e.g. when CHP plants are used to meet power peak loads and minimise expensive power imports.[126]

Heterogeneity of CHP applications

CHP is far from being a homogeneous and clearly defined technology but cuts across the morphological space of heat and power supply technologies. Furthermore, CHP can be realised in a number of technical ways with different characteristics, opening up opportunities for different types of applications. They are linked by the common principle of the simultaneous generation and use of the two products heat and power. The definition of three main application areas serves the purpose to dispose of a typology of CHP- technologies, which can then be investigated in their respective contexts.

Pervasive character of CHP applications

Energy and electricity technologies are of a pervasive nature, i.e. they have an impact on a large variety of industrial and private activities. Although the use of CHP is subjected to a number of constraints, it can be used in many different contexts. It is not just a niche technology, but a technological principle which challenges key principles of our current electricity supply system. Its widespread diffusion would transform the existing supply structures which have followed a trend towards ever larger centralised power plants and individual heat supply. Due to its technical characteristics, the widespread use of CHP would induce a reversal of these trends: electricity would be produced closer to the final consumer and heat more remotely from him, e.g. in district or group heating plants. Beyond these direct impacts, cogeneration implies major organisational transformations of the energy supply sector, e.g. regarding the role of large utility companies.

5.3.2 Dimension 2: Environmental impact

Due to their long-term impact, the environmental and resource characteristics of a technology are a more important issue for public policy-making than for private enterprises. Firms tend to care about the environment if the political administration imposes environmental regulations, if there is pressure from the customers and the public, or if cost reductions can be achieved by introducing more environmentally friendly technologies. Environmental considerations have also become more important in the public debates about energy supply in the last years.[127]

---

126 In periods of power peaks, export prices are usually very high. This makes it economically very interesting for generators to run their plants during these periods, even if the generated heat cannot be used. Similarly, CHP is run to avoid purchasing expensive peak power or going beyond the threshold level of contracted capacity.

127 The environmental benefits of CHP were already recognised at the end of the seventies in some expert circles, see (Burness, Cummings, & Paik, 1981).

The assessment of a technology's environmental characteristics is difficult. In principle, all the energy and resource flows associated to it should be taken into account and examined over the entire life-cycle. A number of attempts have been made to translate environmental characteristics into monetary terms, but the main problem is to assign an intelligible value to environmental impacts.[128] It can thus be argued that environmental impacts should better be kept as a separate assessment dimension rather than integrating it into a long-term monetarised accounting scheme. In fact, also in the debates about CHP environmental impacts are usually discussed separately from economic considerations.[129]

The following discussion of the environmental impacts of CHP builds on the thermodynamic analysis of comparable technical systems. Depending on the specifications of the systems to be compared, very different results can be obtained. In fact, the decision about these specifications can already predetermine the result.[130] In Chapter 8 and 9, it will be shown how these different interpretations have been used in the CHP-debates in the UK and Germany.

5.3.2.1 *Efficiency, fuel consumption and CO_2: diverging opinions*

The consumption of primary fuel in a cogeneration plant is very low due to the high efficiency of the combined production process. However, the potential thermodynamic efficiency is limited by the process dynamics, e.g. the adjustment of the process to the actual loads. In other words, the complex operational characteristics of CHP limit the theoretically possible efficiency gains. In the most frequent case of a CHP unit covering the base load only, the efficiency of the entire cogeneration plant results from the joint operation of the CHP-unit, the additional peak boiler and the power imports from the grid. This does not curtail the benefits of CHP, but it shows that it can replace only a certain share of the electricity demand of a site.

Calculations by the World Energy Council have shown that up to 80% of the fuel for heating purposes can be saved when switching from coal-based separate heat and power production to gas-fired cogeneration of heat and power (WEC, 1991, p.45). Even the replacement of gas-fired individual heating and power generation by coal-fired cogeneration - a worst case assumption for assessing CHP - would reduce the CO_2-emissions for heating purposes by 50%. More critical assessments

---

128 'Willingness to pay' is one of the methods used, but it essentially delegates the problem of assigning a value to environmental goods to the interviewees.

129 This does not exclude that it may be desirable from a normative point of view to integrate environmental damage in the costs to the end-users. The discussion about the carbon tax deals exactly with this problem.

130 This becomes particularly obvious when looking at the different studies and comparisons carried out in the last years to assess the advantages and disadvantages of CHP vs. large-scale power-plants. In many of these studies the underlying assumptions were decisive for the final results. This discussion was particularly fierce in Germany, see e.g. the studies of (Ardone, Fichtner, Tsai, Wietschel, & Rentz, 1995; Dienhart & Nitsch, 1995; ISP, 1994; Lucas, Rebstock, Roth, & Solfrian, 1992; Nitsch, 1994; Traube & Schulz, 1995).

compare the overall emissions from both power and heat production. For example, an advanced individual heating plus power generation on the one hand is compared with a cogeneration scheme on the other (Conrads, 1992). The conclusion of such a more differentiated assessment is that cogeneration enables the lowest CO_2-emissions in conventional heat and power plants, but points also to the fact that gas-fired individual heating can be environmentally friendlier than coal-based cogeneration.

Combination	**Input [energetic units]**	**Losses** [energetic units]**	**Net efficiency [%]**	**CO_2-emissions [%]**
	Heat/power ratio 100:25			
Individual heating oil + power-only station coal	184	59	68	100*
Small CHP + gas boiler in a heating network	149	25	84	56
	Heat/power ratio 100:50			
Individual heating oil + power-only station coal	225	95	61	100*
Indiv. heating, coal-CHP + peak boiler gas, feed in heat network	198	38	80	80
Condensing boiler & combined cycle gas power only	198	48	76	53
Boiler oil & coal-CHP, both feeding in a heat network	185	29	81	81
Condensing boiler & combined cycle gas CHP, for heat network	176	25	85	47
Indiv. heating, comb.cycle CHP & peak boiler gas in heat netw.	176	26	85	47

* Basis for comparison, ** Including potential network losses

Table 5.5: Comparison of different combinations of heat and power supply for heat power ratios of 100:25 and 100:50, Source: (Enquete-Kommission, 1995, p. 323)

Most of the arguments regarding the usefulness of CHP as a means to reduce CO_2-emissions have been summarised in a report for the Enquete-Commission of the German Bundestag (Enquete-Kommission, 1995). In addition to a comparison of different technological combinations for heat and power generation (Table 5.5), several scenarios for the wider application of CHP are developed in terms of the respective CO_2-emissions.

5.3.2.2 Emissions

For emissions such as SO_2, NO_x and particles it is more difficult to find useful data because in contrast to CO_2 they are not directly coupled to fuel consumption, but depend on the individual technology used. Nevertheless, when comparing systems for similar purposes, considerable reductions have been estimated by the World Energy Council if heat and power are cogenerated for district heating instead of being generated separately in power stations and individual heating systems (see Table 5.6). Especially for coal-fired systems, the comparison is striking, but it needs to be taken into account that emission data are given per MJ heat and not per unit of heat *and* power production. In the latter case, the advantages of CHP would be slightly lower.

Fuel	Emissions	Separate generation	Combined generation
Natural gas	*SO_2, NO_x & particles*	70	45
Oil	*SO_2*	270	130
	NO_x	70	45
	Particles	15	10
Coal	*SO_2*	880	110
	NO_x	110	75
	Particles	270	15

Table 5.6: Emissions for CHP (district heating) and separate generation (individual heating and central power plant) in mg/MJ heat; Source: (WEC, 1991)

District heating schemes need to be built in the vicinity of urban areas in order to be close to the heat load. This may lead to problems of visual intrusion and higher local emissions. On the other hand, the prevailing heating technology - individual fossil heating - is also polluting locally, and usually subjected to less strict emission regulations than larger plants.

Consequently, a 'correct' evaluation of CHP in environmental terms is difficult, and the studies reviewed give evidence of this.[131] A frequently made mistake consists of a comparison of unrealistic alternatives leading to unfair evaluations. It is useless to compare an old power plant with the a new CHP-unit to underline the advantages of CHP (as it has been done by some proponents of CHP), nor does it make sense to compare the latest generation of separate production technologies

---

131 To mention just a number of these studies produced in the last years, see (Burness et al., 1981; Conrads, 1992; ICIE/OPET, 1992; ISP, 1994; Nitsch, 1994; Pfeifer, Fahl, & Voß, 1991; Traube & Schulz, 1995; Verbruggen, Wiggin, Dufait, & Martens, 1992; WEC, 1991).

with rather outdated CHP-systems (as preferred by opponents), possibly even based on different fuels and omitting transportation and other losses.[132]

A fair comparison should confront either different options available for an individual site, or scenarios for the future composition of the plant parks of a country. In other words, what is needed is a comparison of real alternatives. When it comes to replace an old heat and power supply system, the crucial point is to have comparative data on the environmental characteristics of different investment alternatives. In this case, the advantages of CHP-plants reduction of CO_2-emissions tend to be in the order of 15-20%, though depending in detail on site and operation conditions. The other emissions (SO_2, NO_x, particles) depend on the clean-up technology used, but are usually in a similar order of magnitude as for separate production.

A further aspect of environmental importance concerns the path-dependence of technological change. All environmental evaluations discussed before deal with available options. In a long-term perspective, the environmental development potential needs to be considered as well. This is of particular importance from a macro-economic supply perspective. A future supply system based on CHP has different environmental implications than a system based on centralised power plants. However, for the moment there are hardly any reliable data available on the future performance of emerging technologies.

5.3.3 Dimension 3: Micro-economics for CHP-users and suppliers

From the perspective of individual firms, innovation and adoption decisions draw to a large extent on economic arguments, although the other dimensions may play an important role as well. These micro-economic aspects of CHP-investment need to be analysed separately for different types of actors. There are different 'economics of CHP' for users of CHP-equipment (utility companies and service end-users) and for technology suppliers.

The aspects of major interest to the actors are subjected to uncertainty and a lack of information. A clear economic assessment of cogeneration is difficult because the impact of investment needs to be projected into the future. Moreover, it is dependent on the individual site-specific boundary conditions. In spite of these constraints, the major issues at stake can still be discussed in general terms, taking into account their relevance for the different actors. From a supplier's perspective the technology-related, microeconomic characteristics can be summarised in terms of

- Technological opportunity,
- Appropriability conditions, and

---

132 The consideration of these losses (fuel transportation from the source to the combustion site, power transmission losses, heat transmission losses, etc.) should be performed in a kind of fuel chain analysis for the final services delivered (heat, electrical appliances).

- Uncertainty,

and for the CHP-users in terms of

- Price and quality,
- Transferability conditions for knowledge and information, and
- Risk and uncertainty.[133]

These issues are now going to be analysed in more detail with regard to CHP, first for the technology supply side and then for the technology user side.

5.3.3.1 CHP as an opportunity for suppliers

The technology supply side matters for diffusion because of the inter-dependence of innovation and diffusion processes. Technological innovation is subjected to uncertainties which make it difficult to make optimal decisions, and this becomes obvious when looking at the three groups of determinants of innovation decisions for the case of CHP.

Technological opportunities

Cogeneration systems consist of a large number of components, but the cogeneration market is not sufficiently big market to justify the development of specific dedicated components. Consequently, components from other applications are used and adapted for use in CHP-systems, even if in Denmark the importance of district heating has led to the emergence of a number of specialised suppliers, e.g. for piping technology, house connections, heat flow measurement, etc. Furthermore, the potential for technological improvements of cogeneration systems is rather limited because it is a comparatively mature technology.

Energy and especially electricity supply technologies have benefited from considerable economies of scale by plant size. Higher efficiencies and cost-digressions favoured the construction of large units but this trend is now approaching a saturation level. In principle, this effect applies to cogeneration schemes as well. In cases where large heat loads were available, corresponding cogeneration plants have indeed been constructed. On the other hand, the economies of scale for CHP are limited due to the connection to two (or even three) network systems. The local character of the heat network is of outstanding importance in this respect. In other words, economies of scale in generation are limited by diseconomies of scale in distribution.

Economies of scale apply to cogeneration also in a second way. Small-scale units and certain system components have become standardised products which are produced in small series and thus enable considerable cost reductions. Large power plants are usually individually designed, and many core components are produced in very small numbers only. It is evident that a trend towards small- and

---

133 See for example (Kemp & Soete, 1990) and (Dosi, 1988c) whose concepts have been introduced in Chapters 2 & 3.

medium-scale, gas-fired gas turbines and small combustion engines facilitates the standardisation of CHP-systems. Nowadays, small combustion systems as well as medium-size gas turbine units can be purchased as standard products which have benefited from economies of scale in series production. In addition, CHP is only one of the possible application areas for many components. Gas and steam turbines, engines, heat exchangers and control systems are used for a large variety of energy supply systems. Only certain components such as piping systems or house connections are typical for CHP. The main opportunity areas in CHP technology are therefore the design, integration and operation control of these components for specific sites.

Appropriability conditions

Appropriability conditions can be defined as the possibility to exploit the benefits arising from innovative efforts made to introduce or improve a technology. The standard means to achieve this are patents, first-mover advantages and secrecy. Energy technology hardware tends to be strongly based on scientific and engineering knowledge, which is concentrated in a small number of companies, mainly because of the large size of conventional plants. This knowledge is highly specific to the manufacturer and thus relatively easy to protect. However, except for components such as gas turbines or clean-up systems, the issue of appropriability of technological knowledge appears to be of limited importance because most components are relatively mature.

The complexity of plants supports the assumption that the technological knowledge on the planning, design and operation of CHP-systems does not leak out easily. In fact, the tacit knowledge on system integration is critical to the implementation of CHP, not the availability of hardware components. For individual components, the situation can be different because they are often derived from other fields of application.

Knowledge and skills in system integration, design and control are particularly important for small-scale systems. They are usually packaged from established components into standardised basic configurations which need to be adjusted to the specific sites. Similar to other system technologies, the competitive advantage of a packager is not based on the improvements of individual components, but on his tacit capabilities and experiences. System performance and reliability matter more than product innovation in the traditional sense.

Expected market demand

The market demand expected for a new product is an important determinant of innovative activities. Given the limited diffusion of CHP-systems it is evident that the market size does not represent a major driving force for innovative activities. However, the recent revival of cogeneration and especially of small-scale applications in some countries has changed the demand expectations.

Apart from this general trend, the structure of market demand directs innovative efforts in certain directions. Small-scale applications can be regarded as a newly emerging market opportunity in which many technical improvements have been achieved in the last years. Standardisation, new prime-movers and especially sophisticated control systems have enabled the efficient and reliable operation of small independent units.

Consequently, cogeneration should be characterised as a relatively small market of secondary importance for the majority of energy technology supply companies. High uncertainty about the future growth of the cogeneration market and limited technological opportunities inhibit dedicated R&D-efforts for CHP. Some niches of specialised equipment supply have nevertheless emerged in parallel with the wider spread of cogeneration. The very promising developments in the Netherlands and Denmark indicate that the CHP-niche may also become a major energy technology market in other countries in the future.

The rather limited size of the CHP-market, the importance of system design and accordingly the need to integrate elements from other technological areas (engines, turbines, control technology, heat exchangers, electrical equipment etc.) make it necessary for the manufacturers and packagers to dispose of a rather broad technological knowledge base. This applies both to small-scale packaged units, but even more so to larger plants which need to be individually designed.

5.3.3.2 CHP-project assessment at the demand side

In the literature on innovation diffusion, there are three main aspects mentioned as determining adoption decisions: the price and quality of the product, possibilities to transfer the technology in terms of hardware and knowledge, and the economic and technological risks and uncertainties associated to it.

A major difficulty with regard to CHP consists of the complex structure of the demand side which is made up of several different types of actors. They all have a role to play in the realisation of a CHP-plant. Before discussing the three main determinants of adoption, the different types of users are going to be introduced.

Main types of users

First of all, the end-users of energy services need to mentioned as important actors on the user side, e.g. industrial heat users or owners of large buildings. Utility companies operate both as technology users who provide final energy services to end-users and as 'gate-keepers' who inter-connect the networks (see Figure 5.5). Moreover, there are specialised service and contracting companies which share certain functions with utilities because they also manage all kinds of energy-related activities for their customers. In practice, also technology supply companies and packagers provide a range of services to end users which overlap with those by utility and service/contracting companies.

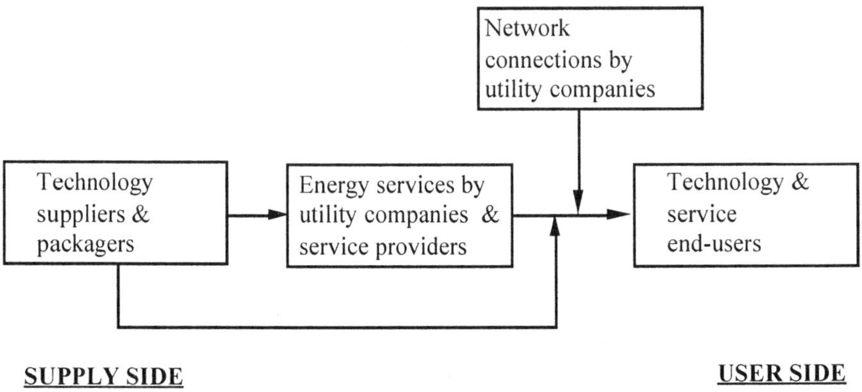

Figure 5.5: The service chains from technology suppliers to service end users

A more fine-grained typology of the user-side, differentiated into utility and non-utility users, is shown in Table 5.7.[134] This typology is based on the objectives and economic criteria of these users. Public utility and service/contracting companies apply different economic criteria than non-utility users. They reflect the long time horizons applied for building up extended networks. In principle, this argument applies similarly to all utility companies for electricity, gas and heat supply. They are all potential users of cogeneration, showing that the technology allows merging energy supply chains which have traditionally been separate.

Non-utility CHP-users	**Utility CHP-users**
Industrial firms	Regional utilities for gas and electricity
Public authorities	Local utilities
Public institutions like schools, universities and hospitals	Independent Power Producers (IPP)
Commercial businesses like hotels, leisure centres and offices	Energy service providers

Table 5.7: Utility and non-utility CHP-users

Industrial as well as the majority of small-scale applications are operated by individual public or private end-users. Local and in some cases also regional utilities are typically involved in district heating schemes, but increasingly also in joint ventures with end-users. A relatively new type of potential CHP-users which fulfils similar functions as the utility companies are independent cogenerators or Independent Power Producers (IPPs). They generate heat and power as their main business, i.e. they sell heat to industry or local distribution companies, and electricity to the grid operator. Traditionally, independent production was limited to

134 The distinction between utility and non-utility users is taken from (Verbruggen et al., 1993).

generation for own consumption of heat and power (IGOPs - Independent Generators of Own Power), but this has changed in recent years.

Price and quality of the technology

For CHP end-users

It is reasonable to assume a profit-orientation of decision-making in private companies. This is underlined by the fact that the expected pay-back time (or a similar measure) of a project is the key criterion for investment. CHP requires higher investment as compared to a simple boiler plant and a grid connection. Waste heat recovery system, a sophisticated control system and complicated prime movers increase the fixed costs. In addition, a higher level of maintenance is necessary. These higher costs can only be justified by savings in the longer term, which accrue from a reduced energy bill as a result of lower fuel costs and the replacement of imported power by cheaper self-generated power. In general terms, CHP is favoured by low fuel and high electricity prices because the CHP-operator carries out the expensive transformation process from fuel to electricity himself. However, the economic optimisation of CHP-operation is a complex issue and depends very much on the grid connection conditions.[135] CHP service users can calculate against the price of purchased power from their (local) utility company and are thus in an economically more promising situation for setting up CHP than the local utility company itself. The latter needs to calculate against electricity purchased from the regional generator whose tariffs are significantly lower than the tariffs which the local utility can offer to its customers. This means that regional and inter-regional utility companies are usually interested only in large CHP-projects which can compete with their own purchase or generation prices.

The additional initial investment in CHP needs to compete with other types of energy-saving investment. Several other investment options promise larger savings than CHP. Typical examples are Demand Side Management (DSM) measures, e.g. the installation of more efficient machinery or passive insulation. However, it has been shown that even in very promising cases of energy efficiency investment, adoption is very slow (c.f. Chapter 6).

For utility and service companies

Public utilities tend to have a double-edged position with regard to CHP because they can be both operators of CHP-systems and distributors of electricity or gas.

---

135 The grid connection conditions are usually expressed in terms of power purchase price, standing capacity charge, back-up power prices and export prices. They have major implications for plant operation. For example, a CHP-plant (like basically any other independent power plant) can be operated to optimise the peak load behaviour of a site. While a CHP-plant is usually operated in a heat-guided mode to cover the base load of a site, it can also be run with the objective to reduce power demand in peak times or to sell as much power as possible at peak times. In these cases, the operation is economically optimised, but energetically suboptimal, because due to the inflexibility of the heat/power ratio, the simultaneously generated heat possibly has to be wasted.

When they are actively involved in CHP-projects, they tend to be fairly supportive partners, but not necessarily when having to accept power imports from independent producers. Anyway, even if utilities carry out CHP-projects on their own, similar conflicts as between CHP-users and utilities may emerge within the utility company itself, because the generation department stands in competition with the distribution department. Still, it can be assumed that these conflicts are easier to settle within a firm than between different firms.

When a utility company participates in a CHP-project as a contracting partner, it becomes an operator and usually sells the final service to end users. This is fully coherent with its business objectives. In the case of a fully independent CHP-project, the reduction of the utility company's power sales is not compensated for by a corresponding increase of its final energy services. In such situations, the terms of economic exchanges between utility company and independent generator are often a highly disputed issues. Due to the need to connect to the grid for back-up power and possibly for power exports to the grid, tariffs need to be agreed upon, but the utility is in a much stronger position. The normal electricity pricing system, differentiated with regard to peak and base-load hours, can also provide disincentives for CHP-projects. A variety of further specific regulations concerning technical requirements for grid connection, wheeling of power, or rights of way represent further sources of conflicts and costs to CHP-operators.

Utility companies need to calculate against the power purchase price of their respective power supplier and/or against their own generation costs. In principle, the same arguments as for CHP end-users hold also for the utilities: investment in cogeneration can only be justified if the additional power generation costs can be balanced by an over-proportional reduction of costs for electricity purchase.[136] Basically, the principle of avoided costs should be applied for remunerating power imports from independent cogenerators as well as for investment in new own capacity, but this is a highly controversial topic. The utilities' avoided costs can be calculated in many different ways and based on different elements. These elements can range from the reduction of their fuel consumption and other variable costs, to the consideration of capacity terms for avoiding new investment and to the contribution of the new plant to an increase in security of supply. Many independent power plants could fulfil the function of a decentralised back-up capacity, especially if a high availability can be guaranteed. Although an individual CHP-unit may have a lower reliability than a centrally controlled plant, the large number of units allows to distribute the risk of breakdown among many plants, and thus guarantees an extremely high level of security.

The situation of private energy service companies is similar to that of the utility companies. The most important difference is that they do not control the economic and technical interfaces, but both coincide in terms of their interest structures. If they set up a service concepts on the basis of CHP, they have an inherent incentive to operate them at highest possible levels of energetic and economic efficiency

---

136 For more details on marginal cost pricing and the calculation of avoided costs see (Verbruggen, 1992b; Verbruggen, 1992c)

because they sell a final service (warmth, light, etc.) rather than an intermediate one (heat, electricity).

Transferability of the technology

Although CHP-technology is fairly mature, it is a complex systems technology which needs to be integrated carefully within each individual site. This holds for all applications, and the degree of complexity rises with the size of a plant. Compared to conventional heat and power supply, the co-ordination of combined production represents a high level of technical complexity and goes often beyond the competencies of a non-specialised company. Utility companies usually dispose of the necessary expertise needed, and also private energy service and management companies. The liberalisation of energy supply has offered the opportunity to management and service companies to provide exactly this type of expertise to energy service users.

Although public electricity utilities usually dispose of the necessary expertise to operate CHP plants, they are often engaged only in the power or the gas business and do not want to offer other energy services. In other words, the transferability of CHP can also be restricted by the willingness to cross established business boundaries rather than by a lack of expertise.

Economic risk and uncertainty

Setting up a cogeneration plant always implies higher investment than in a conventional heat and power supply system. These additional investment costs need to be justified by short payback periods (or any other type of economic investment criterion). For end-users of heat and power services, these criteria are particularly strict if the provision of utility services is not associated to their core business. Even if a project could be economical, there are many uncertainties associated to it which the end-users often try to avoid:

- Electricity and fuel prices are fluctuating significantly and have a substantial impact on the expected comparative benefits of a CHP-system.
- The fact that CHP is an integrated system technology, embedded in the entire energy supply system of a site, implies a high level of uncertainty concerning future changes of loads.
- Investment in energy supply is usually made over a comparatively long time horizon, with lifetimes of installations beyond ten years, in some cases even up to thirty years. It is hardly possible to anticipate heat and power requirements over such long time horizons. Usually 5000-6000 h/year of operation are regarded to be the lower limit for promising CHP-projects.
- The economic risk associated to the technical system can be considerable, depending on the complexity, quality and experiences of the technology supplier. This means that the reliability and maintenance costs of rather new types of equipment carry the risk of failure and of incorrect operation.

- Another source of uncertainty consists of the connection conditions by the regional or local utility company which themselves depend on the wider regulatory framework.
- The regulatory framework also comprises environmental regulations. When the regulations change, adaptations of the plants may be required and the initial economic assessments be reversed.

In other words, there are important economic and regulatory uncertainties which may have a stronger impact than the technological ones. Due to these uncertainties and risks, it is often difficult to find finance for a CHP-project. Specialised service companies can help to reduce the risks associated to CHP, up to the point of taking over the full investment risk, e.g. in arrangements such as contract energy management and discount energy purchase.[137] In practice, several uncertainties remain and are shared between the end-user and the service provider, depending on the detailed contractual arrangements.

Project assessments

The typical assessment procedure of an independent industrial CHP-user is outlined in Figure 5.6 and captures the essential aspects of interest to a CHP-user. This procedure could be easily amended in some details to describe also a co-operative arrangement between a utility company and a CHP-users.

A comparison is usually made between different CHP plant designs and a traditional heat-only solutions with grid connection, but also investment in passive energy efficiency measures can be considered. The procedure is somewhat different for a utility company which intends to set up a CHP-plant, because it needs to take into account the cost structure of its plant park, the possibility to purchase from another utility or an independent generator, and different types of alternative electricity saving measures ('demand-side management').

---

137 Discount Energy Purchase (DEP) is a type of arrangement by which the service end-user pays for the electricity and heat supplied by the service company which operates a plant on the client's site. The service is set at a level which enables the end-user to purchase the scheme after the contractual period and thus to stretch the investment in the CHP-system.

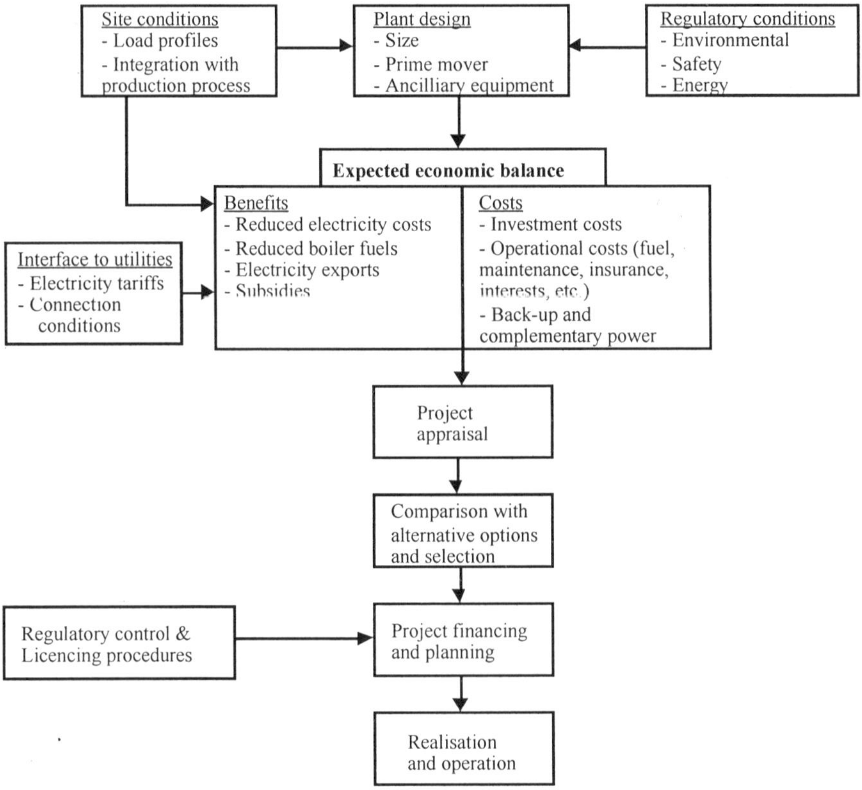

Figure 5.6: Model for the assessment of an independent industrial cogeneration project, Source: adapted from (Verbruggen et al., 1993)

5.3.4 Dimension 4: Macro- and socio-economic issues

There are several socio- or macro-economic implications of CHP which are of particular interest to decision-makers in the political, but also in the industrial realm. Examples are the dependence on fuel imports, the usage of indigenous resources to guarantee secure supply, an efficient organisation of energy supply, or the need to fulfil certain energy-related social and public needs. In general, these considerations deal with longer-term aspects of energy supply, similar to many environmental issues.

5.3.4.1 Security of energy supply and CHP

Security of supply has been the imperative of energy policy for decades. Since the first oil price crisis it has been complemented by arguments concerning the dependence on fuel imports. Another reason why CHP has been become more interesting from a socio-economic perspective was the follow-up of the nuclear debate. As CHP promises a very high energetic efficiency, it has been one of the options of particular interest since then, even if in recent years security of supply has been increasingly superseded by environmental arguments.

It is important to note that the revival of CHP-technology is closely connected to the supply of natural gas because the most successful systems of the last years are all gas-based. Natural gas is regarded a very valuable resource, but at present available at very low prices. The question has thus been raised whether, from a long-term perspective, the extensive use of natural gas for power generation can be justified and whether coal would not be a more appropriate fuel for power generation. However, this is not regarded as a major challenge to the general approach of decentralised CHP, because other fuels and prime movers can be introduced easily once a decentralised system structure is established. Anyway, the current trend towards using natural gas for power generation is expected to continue and will operates as a major driving force in favour of CHP.

The diversification of energy supply chains is another strategy which has been put forward to guarantee a secure energy supply. It means in practice that it should be avoided to favour solely gas-fired CHP, and that other options based on coal or renewables should also be promoted.

5.3.4.2 Efficiency of energy supply and CHP

The supply of energy services at reasonable costs is of vital strategic importance for economic development, especially since the competitiveness of many industrial activities depends on various forms of energy services. The price level at which these services can be provided to industry, taking into account also long-term social and external costs, is thus a key issue for many actors. While energy end-users usually pursue shorter-term objectives, governments should be concerned with longer-term aspects of an energy supply strategy, and also utility companies need to plan for the longer term due to the lifetime of their plants.

Cogeneration is a typical example of a decentralised and horizontally integrating technology. It implies a different model of supply than the currently dominant one, which is composed of centralised, large-scale power-only generation plants and individual heating systems. Due to the long vintage periods of energy supply equipment, and the massive sunk investment in existing plants, it is evident that a transition to a more decentralised energy supply system, based e.g. on CHP, could only be achieved at a slow pace. Otherwise, existing equipment would have to be scrapped or under-utilised. In other words, if there is no actual need to replace or extend existing capacity, the possibilities to introduce a different structure of energy supply are very limited.

The transition to a decentralised supply system is a long-term issue which needs to build on a corresponding long-term strategy. At present, the guiding paradigm and strategy for energy supply focuses on liberalisation and privatisation. The crucial question is therefore whether a market-based and competitive framework can be designed which nevertheless respects and promotes the specificities of CHP.

CHP is also a typical example of a joint production system. This has certain implications for industrial structure, e.g. in terms of concentration and the contestability of the market leader. Generally speaking, entry to an industry with joint production is difficult due to a large number of technical and economic interdependencies and trade-offs. Economies of scope and cross-subsidisation between different products are very common and reinforce industrial structures. Antitrust laws and regulation need to be adjusted correspondingly. The theoretical analysis also shows that in order to make joint production systems work, the 'optimal' structural settings ought to be different than in ordinary single product systems. [138]

5.3.4.3 Excludability and rivalry of CHP-services

Associated to the efficiency question and the potential conflicts with other policy areas is the entire liberalisation debate which aims at disconnecting energy supply from the detrimental influence of other policy areas and interests. It needs to be asked whether the political support for CHP can be justified by the benefits it may have in other policy areas. Energy supply was regarded as a public good and thus as a major social issue in the past. For example, CHP and district heating played a role in the debates about social housing and local energy policies.

Moreover, CHP is not necessarily in line with the existing structure of energy supply, which can only be changed in longer term. This supports further the argument that CHP should be supported by specific measures or a more conducive regulatory framework which fully takes into account its currently under-rated long-term benefits. This stands obviously in contrast to the current trend towards liberalisation in energy supply. However, it seems that the socio-economic characteristics of the generation and supply of heat and power services have changed significantly over the last years. This has consequences for the evaluation of the conflict between a liberal and an interventionist policy approach.

A typology of infrastructure services can be formulated which builds on the two dimensions of excludability/non-excludability and rivalry/non-rivalry of goods or services (Worldbank, 1994). In Figure 5.7, it is outlined how recent development in heat and power supply technology have affected their position in this typology. The closer they are, the easier it is to operate heat and power supply jointly under a common organisational framework.

At the level of heat and power network provision, one can assume a fairly excludable character of these activity, even if in principle it is possible to compete for network-laying. However, once a network is installed, it does not make sense to construct a competing network. This argument applies in particular to the competition between heat and gas networks.

Final power and heat distribution in industrialised countries are fully excludable services, because although there may be an obligation to connect, this connection remains under the full control of the distribution company. Beyond the possibility

[138] For a theoretical introduction to joint production, see (Nadiri, 1987; Schefold, 1987), and for further details (Pasinetti, 1980).

to disconnect an end-user, consumption can even be quantitatively controlled by means of differentiated pricing schemes. This is a comparatively recent development, because smart electricity and heat metering systems were not available at affordable costs in the past. Nowadays, consumption can be monitored in detail over time. The developments in heat metering have been particularly pronounced. In the past heating services in district heating schemes were priced by a flat rate, whereas electricity tariffs already included a consumption term.

Externalities, especially negative externalities of power generation, are at present not fully reflected in the costs for the provision of heat and power services, although this would be technically feasible. In practice, certain non-excludable characteristics are thus maintained. Other, typically positive externalities such as military implications, or security of supply have lost importance, and confirm the move towards excludability. On the other hand, some positive externalities can still not be internalised, e.g. the beneficial effects on economic development.

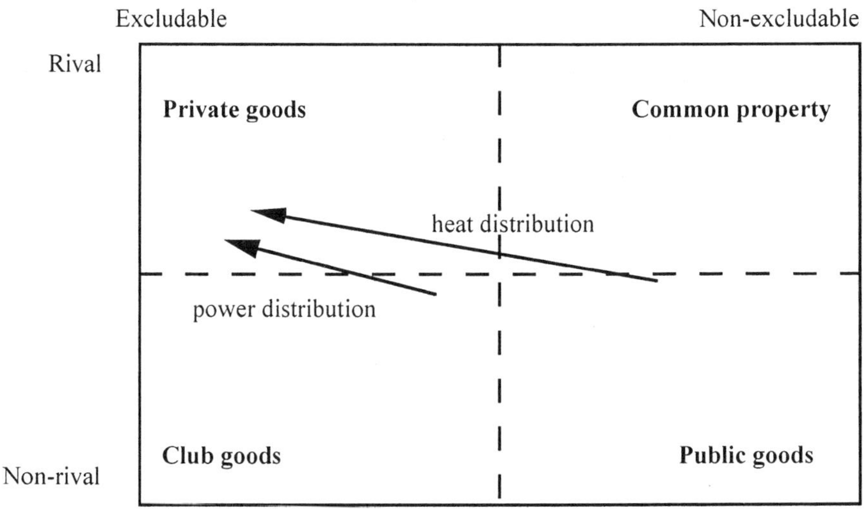

Figure 5.7: Recent development of heat and power supply in terms of rivalry and excludability

At a first glance, heat and power distribution seem to be rather non-rival services. Within the capacity limits of the system, the consumption by one individual does not hamper the consumption possibilities of someone else. Neither does it affect the price of consumption. This situation has not changed in principle. The risk of reaching the capacity limits has remained roughly the same as in the past as well, because the recent improvements of control systems allow to operate the systems much closer to the actual demand profiles. However, when having a closer look at this problem, the argument becomes more complicated because it needs to be taken into account that heat and power are not homogeneous goods, but differ in value depending on the load situation. Peak load plants are more expensive to operate than base-load plants. Customers require secure supply also during peak periods, and utility companies require from their power producers (which can be CHP operators) to guarantee supply also during these periods. While in the past, heat and power were usually priced on the basis of a single unit charge, it has become possible in recent years to reflect the varying qualities of heat and power at different times in differentiated pricing and measurement schemes. This can be interpreted as a trend towards rivalry, even if it is based on indirect mechanisms. Regarding CHP this means that due to the higher reliability of modern plants, their power exports could now be accepted as fully valuable contributions to the guaranteed capacity and be remunerated accordingly.

To summarise this discussion, electricity and especially heat supply are becoming more excludable and at least slightly more rival. Network provision remains in an intermediate position in terms of both excludability and rivalry. These developments are helpful for CHP, because they allow integrating it into the supply system through more precisely defined interfaces in a market framework. From this perspective, a liberalised system which implements the aforementioned differentiation of tariffs would allow operating CHP-plants without specific policy interventions. However, it remains still an open question whether external costs and benefits of electricity and heat supply can be considered properly in such a framework. Additional support measures for CHP can thus still be necessary due to the shortcomings of the current accounting schemes for external costs.

5.3.4.4 The consideration of long-term interests

Long-term interests regarding CHP differ significantly from the short-term interests related to it. It is thus of fundamental importance to include long-term considerations in present decisions regarding energy supply. These long-term impacts of CHP have to discounted to the present in order to be fed into the assessments of the technology by different actors. In practice, however, it can only be done in qualitative terms.

Moreover, the long-term assessment of CHP leads to double-edged results. On the one hand it is the most efficient way to supply heat and power, but on the other it is predominantly based on fossil fuels. Systems based on renewable fuels represent only a minor share. However, CHP can also be seen as a transitory option towards a sustainable energy supply system, which would be based on highly efficient decentralised plants using renewables and on a significant reduction of energy consumption. Given the decentralised character of both CHP and most renewables, a certain compatibility can be assumed between the two options. For examples, district heating schemes require only the replacement of the prime mover to upgrade them to different fuels.[139] A switch to biofuels would thus be possible using major parts of already installed CHP-systems. This flexibility of CHP, together with its energetic advantages, is of importance for any decision-maker who thinks beyond the horizon of the next 10 to 15 years.

---

139 See for example the situation in Eastern Germany where in a relatively short period of time many district heating systems were upgraded by replacing just the primary generation unit.

5.3.5 Dimension 5: Organisational and power issues

While the traditional assumption of profit-orientation may be a useful approximation for several investment decisions, long-term 'strategic' considerations of an actor about CHP can be regarded as an equally important determinant. They are closely connected to the power positions of the actors in the system and their possibilities to influence decisions and interactions. A implications of a technology with regard to the organisational structure of the energy supply system are therefore of major interest. Such long-term structural and strategic impacts of a technology are very difficult to assess; expectations are very vague and based on selected visions of the future state of energy supply.

Cogeneration can be expected to induce a decentralisation of electricity supply and a centralisation of (private) heat supply. Furthermore, it implies the merging of formerly separate services, i.e. heat and power supply would have to be coordinated. While a conventional supply system aims to exploit economies of scale by using larger equipment, a CHP-based system could benefit from fuel economies of scope. These economies of scope could be exploited most easily if heat and power generation and distribution were operated by the same company. Such a situation would also offer strong incentives to generate heat and power at a 'combined optimum' (Verbruggen, 1992b). In any case, co-operation and coordination between the supply chains is needed to achieve the widespread application of CHP. This means that cogeneration would threaten established power positions in countries where horizontally separated services and centralised power supply structures prevail. It can therefore be expected that the established companies would try to prevent the introduction of CHP.

CHP has also certain implications for the political arenas. It implies a delegation of responsibilities to the lower levels of energy policy which would go hand-in-hand with a loss of influence of national government. Similar to a liberalised system, national governments would have to restrict themselves to defining the regulatory framework conditions, and possibly managing certain R&D support measures, but they would definitively lose the direct control of energy supply.

In other words, a technological trajectory towards decentralised CHP implies a switch to a corresponding organisational trajectory. This in turn could endanger the position of some of the established actors, especially of those who are well positioned in the present system. These organisational implications apply not only to the economic organisation of the energy supply sector, but also to the associated political system. The centralised structure of electricity supply are mirrored in the involvement of public authorities. Consequently, the 'political trajectory' would also have to be changed if decentralised CHP-based energy systems diffused more widely. In order to trigger such a transformation process by political action, it might thus even be necessary to change the political structures beforehand.

This extension of the trajectory-concept to economic and political issues means that there is a wide range of powerful actors who can be expected to show resistance against transformations which bear at least the risk of having negative impacts on their power positions. For example, the large centralised power-only utilities which dominate power supply in many countries would be challenged if the position of local authorities and independent generators was strengthened in order to facilitate the uptake of CHP.

These arguments show that apart from strict project- and case-related economic considerations, longer-term organisational implications and "strategic" consequences of cogeneration in terms of economic and political power need to be taken into account to understand the evolution of energy supply and CHP-diffusion.

5.3.6 Technological alternatives for heat and electricity supply

It is not possible to understand the diffusion of a technology without comparing it to alternative options. Such a comparison should ideally be performed at the same level of detail as for CHP, and along the same characteristic dimension. This would certainly go beyond the scope of this study, but at least some background information about fundamental alternatives to CHP will be given to underpin the arguments used in the different decision arenas relevant to CHP.

Alternatives to CHP differ for different actors. Individual heat-users can choose between individual gas-, oil-, coal-fired or electrical heating systems on the one hand and district heating on the other; industrial firms may consider whether to generate heat separately and to buy electricity from the grid, or to generate them simultaneously with CHP; for utility companies the question may be whether to engage in heat generation or not. The comparative advantages of CHP differ correspondingly. In order words, although in principle separate production can be seen in opposition to combined production of heat and power, there are several sets of sub-alternatives once the different actor perspectives are taken into account. Furthermore, unlike a gas-fired combined cycle plant which replaces an old coal-fired power station, cogeneration is not just a technical alternative to a conventional generation unit, but an option that implies changes throughout the entire energy supply system. Such complex transformations are not introduced easily as long as simpler alternatives are available.

Consequently, even if it was possible to compare 'objectively' the different technical options for heat and power supply, one would have to look also at the next higher level of alternatives to heat and power supply. Moreover, due to the importance of current expectations for long-term decisions, these alternatives would have to be assessed with regard to their future developments potential, too.

The detailed comparison of alternatives will thus not be made in this chapter, but it will be a relevant issue in the empirical part of this study. The following paragraphs are restricted to an introduction of the four main levels of competition which are relevant to CHP. They point to implications far beyond the individual decision for or against a specific cogeneration plant. In fact, there is competition between different CHP-options, between separate and combined production, between centralised and decentralised generation, and, as a horizontal issue, competition between all these alternatives with respect to different time horizons.

5.3.6.1 Intra-CHP competition

As mentioned before, CHP is a principle of energy supply rather than a specific technology, and there are a variety of individual options how to realise it. One of its main advantages consists of the possibility to switch comparatively easily to other fuels and prime movers once the piping network and system have been realised. While at present gas-fired turbines and engines are the dominant options, coal-fired steam turbines dominated in the past. This intra-CHP competition is based on national fuel policies, but also on technological advances achieved in turbine technology.

Other options which are discussed at present are first of all the wider use of biofuels as a step towards renewables. Secondly, fuel cells as a new 'generic' energy technology could have a major impact in the future. They would probably stimulate the wider introduction of hydrogen and methanol. Especially for hydrogen, the need to set up a new supply infrastructure is an important obstacle. The future extension of CHP at the low end of the power range will depend on the advancement of fuel cell technology. At present, the way towards small fuel cell systems is paved by the introduction of micro-engine CHP, to be applied by individual households.

5.3.6.2 Combined vs. separate production

Two main groups of alternatives to the combined generation of heat and power for potential CHP-sites can be distinguished:

- Highly efficient power-only plants and centralised heat supply (district heat and steam);
- Highly efficient power-only plants and individual heating system.

For an individual user, both alternatives imply the provision of power from the grid. Heat can be generated on site, either in individual boilers or in a central district heating plant. Furthermore, a variety of intermediate solutions are possible between CHP and separated supply. Especially group heating needs to be mentioned: small decentralised CHP-units for small areas are inter-connected by a gas network. This represents a promising compromise between separate and combined supply systems.

5.3.6.3 The centralisation/decentralisation issue

At the level of entire cities or even countries, the two alternative options are centralised and decentralised supply systems. Both could in theory be based on CHP, but it is more likely to happen for a decentralised system. Other sets of alternatives at system levels are renewables vs. fossil fuels, and nuclear vs. non-nuclear power generation. The choices at this level are particularly important for the major inter-regional utility companies and for national policy-making.

At city level it is also interesting to analyse the competition between gas supply and centralised district heating schemes. The relationship between gas supply and district heating is characterised by both competition and complementarity. While gas supply is needed for the fuel supply of CHP/DH-plants, they compete for direct customers and the coverage of entire city areas by one of the two networks.

5.3.6.4 Competition and improvements over time

The preceding sections have highlighted that the introduction and diffusion of CHP cannot be analysed in isolation, but has to studied comparatively at different levels. Alternatives should not be understood in terms of presently available options only, but also in terms of future expectations, thereby taking into account the different dimensions of interest to the relevant actors. Technologies are improved in parallel, driven by expectations which affect the support lent to them.[140] For example, very positive expectations of the future potential of an alternative option would reduce the incentives to promote CHP. This shows that the future evolution of the supply system as well as the performance of CHP and its alternatives needs to be analysed within the evolution of the entire system. This is not just a technical matter, but a complex and multi-dimensional problem of analysing actors' positions and expectations. The importance and weight assigned to the different dimensions of comparison are the result of interactions and of social preferences. The future performance depends on today's choices, and in turn today's choices are dependent on the expectations of future technological performances.

Consequently, a comparison of CHP with other technologies is always partial and subjective, because it is not possible to consider and weigh objectively all the different aspects of interest. However, in terms of the main dimensions mentioned, the arguments for and against CHP at system level can be made explicit to improve the understanding and the interactions among the main decision-makers and stakeholders. The main arguments concerning CHP are summarised in Table 5.8. Only the two extreme alternatives are shown, but one should keep in mind that the comparison reflects just some rough trends. All the publications on which table is based underline the need for a differentiated analysis.[141]

---

140 A good example is the relationship between fuel cell technology and CHP. If the electrical efficiency of fuel cells reaches a level of 60-70%, there will not be a really good argument for cogeneration any more, at least not in terms of its efficiency. However, in certain cases, fuel-cell-based solutions could also deliver very cheap heat as a by-product and thus be integrated easily in CHP-systems. All these options are still very uncertain. Beliefs rather than knowledge seem to determine whether classical CHP or fuel cell-based CHP is seen as the dominant trajectory in the medium term.

141 In (Pfeifer et al., 1991, p. 575) the authors conclude, that "there are advantages for both the decentralised and the centralised system. An individual analysis of each situation is necessary."

Technology Issue	Decentralised CHP systems	Centralised power generation & individual heating
Organisational & power issues	Difficult to co-ordinate actors; Implies decentralisation; Challenges present power positions;	Separate supply chains require less co-ordination; Compatibility with present organisational structure
Socio-economic issues	Difficult to assess; Promising as shown in many countries; In principle compatible with a liberalised system;	Threshold of economies of scale reached; Long-term externalities unclear;
Microeconomic issues	Benefits depend on actors and organisations; Threat to some major actors; Promises benefits under the right circumstances;	Beneficial for large operators and utilities; Often not the least-cost option for consumers;
Environmental issues	Tends to be seen as globally superior; Locally higher emissions; Higher efficiency;	Possibly lower local emissions, but higher global emissions of CO_2; Others, e.g. NO_x unclear;
Technological issues	Technically mature due to recent innovations; Still further potential for economies of scope & scale;	Mature & fully established; Limits of economies of scale reached;

Table 5.8: Comparison of basic system structures of heat and power supply, Source: based on (Albisu, 1990; Babus'Haq et al., 1986; Grohnheit & Verbruggen, 1990; Pfeifer et al., 1991; Verbruggen et al., 1992)

5.4 Summary

In the first two sections of this chapter, the main embedding development trends in relation to energy supply and cogeneration have been reviewed. The third section has focused on a multi-dimensional characterisation of cogeneration technology. This attempt to characterise the technology in as 'objective' terms as possible is a prerequisite for the analysis of the diffusion of CHP in different countries. It has provided the background knowledge on cogeneration technology and its application context. In other words, it covers the first step of the PET-methodology.

Although it is important for the actors to obtain reliable information on these dimensions, it is not sufficient to explain why a certain technology has been successful or not. First of all, any assessment is subjected to uncertainties, and this prevents to identify 'best' solutions. Secondly, the interests of actors differ. This means that 'best' solutions could in principle be identified from an individual actor's point of view. Therefore, the interactions and the influence of actors on each other are of crucial importance for the overall diffusion process. The decisions on a technology, which stems from the actions of several actors, are the result of a social process which takes place within (mostly national) systems of energy supply. Other main driving forces of change are the different structural transformations of the system by which the application context for CHP changes. From a political perspective, it is important to identify all these different access points for action which could trigger change. In this sense, the proposed five-dimensional static analysis of CHP is only a first step towards a dynamic analysis of diffusion and political control, which illustrates the relevant dimensions and key issues for the different actors. This points to at least two major problems to be addressed in the research work:

- The complexity of CHP-technology and associated decision-making processes;
- The need to reinterpret the concept of superiority of a technology.

5.4.1 CHP technology - a complex research field

CHP is a very heterogeneous technology which not only consists of a variety of different sub-technologies, but - even more important - can be applied in several contexts. It will be explained in Chapter 7 how this affects the specific research methodology which has to deal with these different types of applications in the sample of case-studies. CHP competes also with alternative options, which can be identified at different levels of the energy supply system. They need to be taken into account in the analysis as well.

CHP is also closely integrated with other energy technologies. It is part of a number of embedding network technologies for heat, gas and electricity supply. Beyond these technological interdependencies, a close integration and dependence of CHP on its organisational contexts needs to be noted. Therefore it has a number of implications in terms of economic and political organisation which need to be taken into account for political control.

5.4.2 Superiority, inferiority and the role of social processes

Given the diversity of the dimensions mentioned above, it is evident that no clear-cut decision about a 'best' technology can be made. It depends not only on the conditions of the individual plant or system considered, but also on the emphasis put on some of the assessment dimensions. A technological system can not be characterised by a single value, and any attempt to do this would require a methodology for aggregating its different assessment aspects. As it seems to be highly problematic to aggregate e.g. organisational and environmental impacts, a multi-dimensional characterisation is regarded to be more useful. A single actor may be able to judge which of the different technological options at his disposal is 'superior' from his subjective perspective, but individual perspectives are far too diverse to be aggregated into a quasi-objective evaluation across actors.

Despite these difficulties to make highly generalised objective assessments of CHP, it may still be possible to generalise for different classes of sites. As the main classes of sites in a country are known, different 'diffusion potentials' of CHP can at least in principle be identified. Depending on the constraints considered, a thermodynamic, technical, economic, organisational or even political potential can be distinguished.[142]

Moreover, these considerations need to take into account the simultaneous developments in competing and complementary technologies. This indicates that the actors also need to assess the different options comparatively, using the five dimensions as a proxy for the different issues of interest to the actors. Consequently, the concept of *superiority* which has played an important role in diffusion theory as a driving force of processes of change, needs to be reassessed in a more differentiated way. Given the different assessment dimensions and the multiple interests at stake, one should can only speak of relative superiority with regard to a certain dimension and with regard to certain actor's interests. This means also that superiority applies in essence to attitudes: What does an actor regard to be superior with respect to his set of interests? Attitudes are highly subjective, uncertain, and subjected to manifold influences from other actors. Consequently, what is regarded as superior in the view of an actor is finally also shaped by interactions in terms of information, artefacts, law and money.

Technological choices are thus essentially social processes, which define priorities regarding different aspects of a technology. Diffusion is the result of more or less co-ordinated decision processes along economic, political and informational channels, based on the aims, interests and expectations of the actors involved. This implies that the technology which can achieve the highest degree of consensus among the important actors will diffuse, not necessarily the one with the seemingly 'superior' characteristics. Superiority becomes a subjective, actor-related concept which can be used to represent negotiation processes about technological choices.

---

142 Calculations of such potentials can be found for example in (Fenzl & Pummer, 1993) who distinguish a technical potential and an economic potential, or in (Verbruggen, 1992a) who considers also an institutional potential.

Chapter 6: Determinants of CHP-Diffusion

In the preceding chapter, the main issues and trends in electricity supply have been discussed, and cogeneration technology has been introduced along its main dimensions. This aimed to provide the reader with a better idea of the requirements and consequences which a wider diffusion of CHP would have for the different actors.

In the following, the research findings on the role of different determinants of CHP-diffusion will be reviewed. A separate section is dedicated to the policy recommendations which have been derived in the corresponding research studies. Due to the orientation of the majority of publications, this review gives a rather static picture of the determination of CHP-diffusion. Moreover, the studies tend to focus on single countries and their specific conditions. The synthesis of these findings represents therefore just the foundation and starting point for the comparative empirical work in Part III.

With regard to the methodological approach used, this overview also helps to identify those elements of the PET-system for which well-founded research findings are still missing. Therefore it plays an essential for the definition of the specific research steps and foci of the case-study work, which are derived from a confrontation of available empirical findings with the requirements of the conceptual framework (see Chapter 7).

Obviously, it would have been possible to select countries for the case-studies before reviewing research findings, but three main reasons can be raised in favour of the procedure followed here. Firstly, research on CHP - especially socio-economic research - is scarce, and only a few studies deal explicitly with individual countries. Secondly, a broad overview allows developing a more complete picture of the *potential factors* and *actual findings* which ought to be taken into account in a specific study. Thirdly, the comparative pattern of analysis for this study on CHP can be based on a more solid foundation. This will make it also much easier to apply it to other countries. Finally, the review provides solid arguments to select the UK and Germany as the most interesting cases for comparison.

6.1 An overview of the situation of CHP

The situation of CHP differs among European countries. Its contribution to electricity production ranges from about thirty percent to a nearly complete absence (see Figure 6.1). This diversity is the result of historical processes in each country, of specific resource conditions and of the socio-economic environment in which the process has taken place.

During the Eighties some research on cogeneration systems has been carried out in response to the renewed interest in energy-efficient technologies after the second oil price crisis in 1979/80. The starting point was the identification of different types of general factors and obstacles for the use of cogeneration systems, which should feed into the formulation of targeted support policies.[143] In other words, this first wave of studies had a normative orientation. In addition to several plant case-studies, there are also some comparative studies which try to explain the diversity in Europe. They provide good overview information on the situation in the different countries and figure out some main 'factors' in a static way. There are only a very few in-depth studies on CHP which try to explain the process of the emergence or absence of this technology within its socio-economic context.

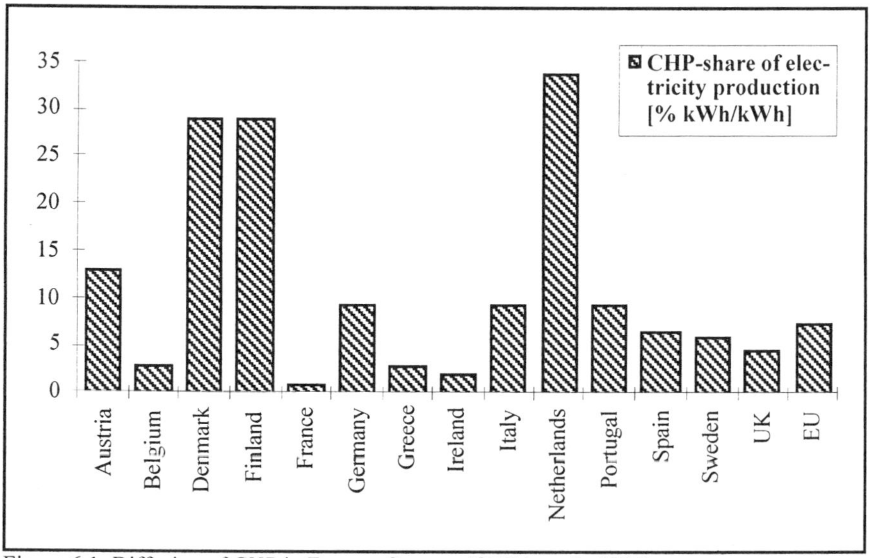

Figure 6.1: Diffusion of CHP in Europe, Source: (CHPA et al., 1995)

143 Several authors speak of different levels of obstacles, so e.g. (Verbruggen, 1992a) who distinguishes between technological, economic and institutional obstacles. In a similar vein see also (CHPA, PWK, & COGEN, 1995). Attempts have also be made to develop a more systematic approach to the analysis of obstacles ("Hemmnisanalyse") (Gruber, Mannsbart, & Walz, 1993).

In spite of its predominantly static character, the results of past research on cogeneration are going to be presented roughly along the lines of the PET-approach, i.e. by distinguishing between exogenous, structural and behavioural aspects. However, due to the structure of available studies, the elements of PET will only be approximated in terms of three levels:[144]

- General exogenous factors;
- Structural level;
    - Technological factors;
    - Economic factors;
    - Institutional and organisational factors;
    - Political factors;
- Behavioural level;
    - Informational barriers;
    - Attitudes and interactions of relevant actors;
    - Case-studies on investment in energy-efficiency and CHP.

6.2 General exogenous factors

According to the studies reviewed, the list of relevant exogenous factors having an influence on the patterns of diffusion of cogeneration systems comprises the following aspects:[145]

- Climate and geography affect especially the heat load patterns with respect to seasonal variations and temperature levels.
- Building, housing and heating patterns affect the possibilities for setting up CHP/DH in various ways. A scattered housing structure is detrimental to CHP while it is facilitated by compact city structures with high-rise blocks. But even in cities, established traditions of individual heating can hamper the acceptance of CHP.
- The political and social cultures affect the citizens' expectations with regard to public services in energy supply, and the acceptance of private companies in this area.

---

144 See also (Reddy, 1991) who develops a systematic view on different possible levels of barriers to improvements in energy efficiency in general. He identifies six different types of actors where a variety of barriers can become relevant (energy consumers, manufacturers and providers of end-use equipment, producers and distributors of energy carriers, actual and potential cogenerators, financial institutions, governments, funding and aid agencies).

145 Based on (Jochem, 1991; Jochem & al., 1986; OECD/IEA, 1983; OECD/IEA, 1994).

- The tax and grant system can favour (or inhibit) investment in technologies such as CHP e.g. if investment grants are offered for the modernisation of houses or if lower taxes are charged for public services.
- The size and structure of industrial activities determine the patterns of demand for industrial heat and steam, both in terms of their location and level.
- Environmental awareness and public opinion on important energy-related themes affect the role which is assigned to CHP as an option for dealing with environmental and resource-related problems.
- Technological developments from outside power engineering can induce technological changes and provide alternatives to CHP, like e.g. low energy houses.

This is a very heterogeneous set of factors; although most of them have not been investigated specifically they are usually taken into account as general boundary conditions in studies on specific countries.

6.3 Structural level

The structural factors of and obstacles to the use of cogeneration systems are differentiated into four groups: technological, economic, institutional/organisational and political factors.[146] The analysis of these factors will be taken up again when discussing political recommendations for establishing 'conducive' structures to promote the spread of cogeneration (Section 6.5).

6.3.1 Technological factors

In contrast to the description and assessment of CHP in Chapter 5, the following considerations focus on the actual research findings dealing with the role of technological issues in decision-making. They can be summarised as follows:[147]

- In most publications it is assumed that CHP is a mature technology, and that the incremental technological innovations achieved do not play a major role for its diffusion. There are no significant technological problems to overcome. Components such as gas turbines, heat exchangers, control and cleaning systems are being improved permanently, and an increasing effort towards the standardisation of equipment is made. However, these findings have never been investigated in depth but are based on rough assessments. Most non-technical studies tend to emphasise non-technical determinants anyway, and

---

146 See (Reddy, 1991) on a similar differentiation with regard to the potentials of cogeneration which are distinguished into a technological, an economic and an institutional one.

147 The main sources for this analysis are (Conrads, 1992; OECD/IEA, 1983; OECD/IEA, 1994; Sirchis, 1990; Verbruggen, 1992a; Verbruggen, 1992b; WEC, 1991).

assume that the technology itself is not evolving at all. At least for the last fifteen years, this does not appear to b correct, because cogeneration technology has in fact achieved comparative advantages with respect to other options as a consequence of technological innovation. Moreover, even if one agreed on the rather incremental character of innovations in CHP, it would still be worthwhile investigating the role of these minor improvements for diffusion.

- In most publications the thermodynamic advantages of cogeneration are underlined, but the practical and operational limitations to these advantages are rarely taken into account. Consequently, the assessment of the environmental benefits of CHP is often too optimistic and neglects major uncertainties.
- The main technological obstacle to cogeneration is the requirement to dispose of sufficiently high and constant heat loads, or to be more precise, the mismatch between temporal and spatial load variations on the one hand and inflexible heat/power ratios of the technology on the other. Although the latest generations of CHP-plants promise a higher operational flexibility, the adaptability to changing loads remains limited.
- In order to avoid excessive heat losses, heat load density is regarded as important, but the case of Denmark shows that relatively low load densities do not represent an insurmountable obstacle.[148]
- As cogeneration is still a small market, there are few incentives to engage in dedicated R&D. Innovation for components is mainly a by-product of other R&D-areas. The example of Denmark shows that an important home market can induce the emergence of a specialised industry for cogeneration equipment[149] which further facilitates the solution of technical (and other) problems associated to CHP.
- Technical conditions for connecting to the electricity grid are often mentioned as a major obstacle to CHP. While this problem seems to be solved from the technical point of view, connection conditions are still used occasionally by public utilities to prevent the implementation of CHP-schemes.

6.3.2 Economic factors

In most of the studies reviewed, economic aspects are dealt with in terms of investment and operation costs for CHP-users and utility companies. In this sense, the economic issues discussed in literature cover a rather narrow range:[150]

---

148 This has been achieved by developing the world's most advanced piping technology. This has also led to the emergence of a specialised industry.

149 See e.g. publications of the Danish Board of District Heating (DBDH, 1982) and of the Danish Energy Agency (DEA, 1993a; DEA, 1993b) on CHP-related R&D-activities.

150 Based mainly on the following references: (Blok, 1993; Jochem, 1991; OECD/IEA, 1994; Rüdig, 1986; Verbruggen, 1992a; Verbruggen et al., 1993; Walz, 1994)

- With regard to district heating, the initial investment in the piping network is regarded as a major problem, mainly due to the heavy construction works required. Certain types of ancillary equipment such as the house connections and heat metering represent additional investment as compared to separate generation systems.
- The difference between fuel prices and electricity prices is one of the fundamental economic determinants of the viability of CHP. Uncertainties associated to the prediction of prices have been recognised as inhibiting factor for CHP. The more volatile prices are, the more difficult it is to make a reliable project appraisal.
- Difficulties to finance projects have been a problem especially for small companies and local authorities due to the massive initial investment which can only be recovered in the long term. The reluctance of banks to provide finance for projects with long pay-back periods adds to this. The World Energy Council concludes that the most favourable conditions for CHP/DH can be found in utilities which have been built up over many years, operate with a long-term time horizon, dispose already of an established energy plant park and benefit of the confidence of financial institutions (WEC, 1991, p. 53).
- Public policies for the direct improvement of the economics of CHP (e.g. investment grants) have been implemented in some countries, and their positive effect on the pay-back periods have been investigated based on simulation models (Blok, 1991). The correlation between these calculations and the success of cogeneration in the Netherlands is striking but the author underlined also that the grant policy could only be effective because a conducive regulatory framework had been introduced beforehand.
- Regarding social costs, several studies have been commissioned in different countries, which deal with the environmental and economic impacts of CHP, but also address many of the economic barriers.[151] It is frequently pointed to the importance of defining conducive conditions for grid connection, both in technical and in economic terms. The tariffs for export, standby and reserve power are crucial in this respect, but also the standard electricity tariffs because they are a key element of the comparative cost-benefit analysis of CHP. While there is agreement that they should be based on the principle of avoided costs, it is unclear which specific elements to consider in the calculations.

6.3.3 Institutional and organisational factors

Technological and economic aspects have been addressed frequently in research work, though mainly from the business perspective of cogeneration user. Institutional and organisational factors have on the contrary been analysed only rarely,

---

[151] Just to mention some of them: (ADEME, 1989; DoE, 1979; DoE, 1984; EEO, 1990; ISP, 1994; Lucas et al., 1992; Traube & Schulz, 1995).

but they often underlie the economic obstacles mentioned before. The main issues in this context are the following:[152]

- The organisation of the power generation industry has frequently been identified as detrimental to the emergence of decentralised and independent generators who are the prime candidates for setting up cogeneration facilities. Plants of regional and supraregional electricity utilities have tended to be located near the energy sources (e.g. in the vicinity of coal mines) and remote from the electricity and heat users. While electricity can be easily transmitted over long-distances, this is not possible for heat. Cogeneration systems thus need to exploit economies of scope.[153] Organisationally, this implies that cogeneration should be organised by horizontally integrated local or regional companies while power-only generation is set up best in large centralised companies. Although these structural and organisational issues are known to be important, their impact on CHP has hardly been studied in detail.
- The separation and competition of different energy and heat supply chains make it difficult to realise CHP systems. Power utilities are usually involved in power supply only, gas utilities in gas supply, heat utilities in heat supply, but cogeneration requires the co-operation between them. The experiences with horizontally integrated utilities have shown that they are more inclined to set up CHP. Consequently, it has been suggested to organise energy supply in a decentralised, liberalised and open supply structure in order to facilitate the merging of services and the emergence of new ones.
- The two preceding issues result from historically grown patterns and structures. The detailed processes which have led to the present status of the institutional and regulatory frameworks of the energy supply industry and for cogeneration have rarely been investigated until now, but they are crucial to understand the path-dependencies on which our current systems depend.[154]
- Decentralised energy utilities can be ideal carrier organisations for CHP, because their interest structure matches the requirements of CHP. In this context, it should be noted that it is possible for decentralised utilities to integrate CHP in wider energy supply concepts and demand-side measures, in least-cost planning and contract energy management.
- Supporting institutions for CHP have been lacking in many countries, but this has changed in the last years. New associations were founded in Spain, France,

---

152 Information has been taken from (Burness et al., 1981; CEC, 1994; CHPA et al., 1995; Dufait et al., 1992; Gruber et al., 1993; Hård, 1991; Hård & Olsson, 1995; Jochem, 1991; Jochem & al., 1986; OECD/IEA, 1994; Rüdig, 1986; vanderDoes, Brown, & Verhagen, 1996; Verbruggen & Marcelis, 1986; Walz, 1994)

153 While power-only generation can benefit primarily from economies of scale and cogeneration from economies of scope, the final distribution business for both heat and electricity is subjected to economies of scale *and* scope by density of demand due to the constraints on heat distribution. This is reflected by the fact that power, gas and heat distribution are usually organised locally.

154 The only comprehensive historical study has been carried out by (Russell, 1986a) on district heating in the United Kingdom.

Germany, the Netherlands and others countries, which try to support cogeneration by providing information and experiences.

- The regulatory framework of the energy sector has been a controversially debated issue in the last years, especially with regard to the advantages and disadvantages of liberalisation. Its monopolistic structure has been identified as an important factor inhibiting the realisation of independent CHP-plants. It has been criticised that there are inherent incentives to expand electricity consumption and production rather than to use and generate it as efficiently as possible. This problem was reinforced by frequent shortcomings of regulatory and investment control. On the other hand, monopolies *at local level* are usually regarded as factor which facilitates the exploitation of the advantages offered by CHP and district heating.
- The right to generate and sell electricity is usually restricted by regulation and thus prevents to set up independent power generation plants. In many cases not even the transfer of power to neighboured sites of the same firm is allowed without severe restrictions.
- Grid connection conditions depend on the organisational structure of the ESI and can be used by grid operators to make independent CHP-systems economically unviable. The grid operators, who are often also power generators and/or distributors, have tended to offer uninteresting tariffs to independent generators. They need to calculate their purchase prices on the basis of their own cost structure, which is determined by the prices of their bulk suppliers and by their own generation costs. Consequently, a grid operator does not have an interest in connecting an independent CHP-operator, especially if heat generation remains outside the scope of the connection agreement. This shows why the horizontal integration of energy supply companies is crucial for many CHP-applications because it allows either the utility company or the independent generator to reap the full the benefits of combined production.
- Stringent environmental regulation leads on the one hand to a promotion of highly efficient cogeneration schemes, but on the other it has been shown that the specific formulation of environmental standards can be detrimental to cogeneration if its specificities are not taken into account.
- The problems of finance of energy projects is regarded as one of the key problems to CHP, not only for industrial, but especially for local utilities. Consequently, a lack of institutions to support and facilitate long-term investment in new energy technologies has been detected.

6.3.4 Political factors

The energy sector has traditionally been the object of particular public interest, and it has been subjected to many political interventions. Public ownership, the definition of the main structural settings and the regulatory framework, but also targeted policies are very common in this area. With regard to CHP, the need to define the 'right' structures for stimulating its diffusion by means of public policy has been highlighted frequently. For example, Verbruggen concludes that "energy sector organisation and energy policy are the major determinants of DH/CHP development" (Verbruggen, 1992b). Many studies derive policy recommendations on the basis of an analysis of the technological, economic or institutional/organisational barriers as described above. However, such normative recommendations do not take those obstacles into account which are inherent to the political process itself; the role of politics on energy and energy technology policy is usually omitted. In other words, the analysis is limited to the obstacles in the energy sector itself, which ought to be addressed by policy-makers, but does not address obstacles in terms of politics and political structures.

With regard to cogeneration, the publications of Marvin and of Russell represent exceptions from this general assessment because they focus on the role of certain political aspects on the possibilities to realise CHP/DH projects in the UK. Marvin investigates the internal cleavages in the Labour Party concerning energy policy and district heating, and points to clear differences between national policy objectives and local party positions (Marvin, 1991a; Marvin, 1991b). In the end, these cleavages inhibited a coherent and widely accepted DH-policy of the Labour Party. Russell (Russell, 1986a; Russell, 1993; Russell, 1994) focuses on the historical evolution of CHP with a particular emphasis on the political interactions between public utility companies, local authorities and government in relation to the failure of this technology in the UK.

Other studies on different European countries,[155] dealing with energy and environmental politics in general, have highlighted further political factors:

- It is very difficult to adjust politically defined institutional and regulatory settings to new technological, economic and social requirements; there is often a structural 'lock-in' which is protected by corporatist arrangements.
- The competencies and power positions of local authorities play an important role for the advancement of district heating projects. The possibility to set up independent local energy policies is one of the key institutional determinants of the diffusion of CHP/DH, which has been clearly identified in literature. Due to the influence of local authorities on local utility companies, this issue is also linked directly to the organisation of the energy sector.
- Political structures and processes determine the substance of policies. This aspect has hardly been investigated with regard to cogeneration, but there are some publications on energy efficiency and energy politics in general. terms They highlight the problem of balanced advice to political decision-makers. Advice is often given by a rather limited range of experts and interest groups, who tend to be inter-connected by vested interests. In addition, political administrations in charge of preparing new initiatives seem to be reluctant to

155 Mainly using material from: (Baumgartner & Midttun, 1987; Boehmer-Christiansen, 1990; Brand & Jochem, 1990; Gruber et al., 1993; Krupp, 1992; Midttun & Baumgartner, 1986; Vogel, 1986). These studies do not focus solely on CHP, but also on energy politics in general.

adopt outsider positions and prefer to follow the mainstream opinion. New strategies and policies are difficult to launch is such an environment.[156]

Although CHP has re-entered the political debate in many countries in the last years (as expressed in the large number of technical, economic and environmental research reports commissioned) there has not yet been any recent investigation of the assimilation and treatment of these new inputs to the political realm. Available studies date mainly from the eighties when many of the new CHP-options such as gas-turbines and small engines were not yet available. Nevertheless, in some follow-up studies, several of these early findings on political factors have been confirmed,[157] especially

- The importance of a strong local level supporting CHP, both in terms of utility companies and in terms of supporting political institutions, and
- The balance of interests at national level and the politics for defining the frameworks and institutions inside of which CHP has to be realised.

6.4 Behavioural level

At the behavioural level, a number of further factors and obstacles to innovation and diffusion processes of CHP have been investigated and assessed. They address the problem of access to and use of specialised information, and the attitudes and interactions of actors. Moreover, the results of different case-studies on investment in energy efficiency provide selected insights in behavioural determinants.

6.4.1 Informational barriers and competencies

A lack of information has been frequently stated by many actors interested in CHP, especially service end-users, planners, utility companies and policy-makers. Experiences with new types of CHP-systems are often not accessible, partly because the companies which operate them would have to disclose information about their production processes, but partly also because there is no institution to organise the dissemination work.

The lack of information and competencies is not sufficient to explain the reluctance with which CHP is generally perceived. In general, the technology is not new, and many of the actors have a clear opinion about it. However, these opinions are often based on pre-established perceptions and negative early experiences with CHP rather than on up-to-date knowledge. In some of the studies reviewed, it is indicated that the recent innovations and improvements have not yet entered current thinking about CHP, and that established preconceptions can not be changed easily. Established engineering rules and practices are regarded to be

---

156 On "issue-led societal construction of technology" see (Krupp, 1992, pp. 25-27).

157 See especially (Gruber et al., 1993; IEA/Novem, 1993; Russell, 1994).

very important in this area. This conservatism seems to hold for many of the actors involved in CHP, from established manufacturers and utility companies (which in principle should have the necessary information), to planners, users and policy-makers. In other words, the mental frameworks lead to a lack of openness and to a general reluctance with regard to new options which would question the established ways of supplying energy services.

6.4.2 Attitudes of actors, actions and interactions

From a behavioural point of view it is important to investigate attitudes and decisions at the micro-level, because they allow to inter-relate different types of macro-transformations. The 'hard' but nevertheless uncertain data on technological factors, and on economic or environmental constraints are only one side of the coin, and it is by way of expectations, preferences, beliefs and decision-making processes that these data enter into the final decisions.

Economic obstacles to cogeneration have been studied for potential operators and users. While public companies might accept longer pay-back periods for their plants, private investment in industry and commerce is subjected to strict pay-back requirements of two to five years. This reflects the situation for other kinds of energy-efficiency investment. Energy supply is usually not a core activity and figures comparatively low on the priority list for investment. Calculations of economic viability usually build on three well-established determinants: the hourly electricity and heat loads, the characteristics of the CHP technology and the electricity tariff conditions (Verbruggen et al., 1993). In practice, however, these calculations can be overridden by the general unwillingness to enter into complex arrangements about power sales and supply across the boundaries of a company.

Due to the higher investment costs and the uncertainties with regard to the tariffs for power exports, the predicted pay-back periods for CHP projects of private companies tend to be relatively long and thus often beyond the acceptable limits. Public companies and utilities pursue also certain 'social' objectives such as guaranteeing safe, stable and clean long-term energy supply and apply longer pay-back periods. In some cases, a CHP plant can even be justified if the investment costs can be recuperated over its life-time. Building on the public mandate of utility companies, it has been proposed that they could play the role of a pacemaker for environmental innovations in the energy sector if their goal system is regulated to incorporate environmental objectives (Zipper, 1988).

In practice, however, public utilities, especially supraregional monopolies, have been suspected to hinder cogeneration because it might erode their long-term strategic position in the energy sector. Growth expectations have already been limited since the end of the 1970s by the stagnation of energy demand, and CHP would further limit their growth prospects. In a study of the attitudes of public utilities towards cogeneration, their activities in support of CHP and the technical and tariff conditions imposed on independent cogenerators have been studied as an indicator of these attitudes (CEC, 1994; Dufait et al., 1992; Dufait & Verbruggen, 1993). It shows that the utility companies dispose of a great variety of means and activities to inhibit or promote CHP.

Apart from studies of the profitability of cogeneration for individual technology users like public utilities and industrial firms, the effects of cogeneration and especially of district heating on final consumers' energy costs have been investigated, showing that the heat cost differences to individual gas heating are only minor and depend on the fluctuations of fuel prices. The unwillingness to give up individual control of their heating systems plays a more important role for explaining the reluctance of some end-users to accept district heating.

While some data are available on the economic side of CHP at the user level, the sociological aspects at the micro-level have hardly been explored. Little systematic work has been done on the different elements which jointly shape decisions on plants and sites. One of the exceptions is a study on a district heating project in a Swedish town, which points to five critical phases for the success of a CHP/DH-project: the decision to build the system, the shaping of the carrying organisation and the acquisition of competence, the enrolment of subscribers, the siting-decision of the main heat plant, and the competition from other forms of heating (Summerton, 1992).

The political side of cogeneration is still an underdeveloped research field; only very few studies can be found. Russell points out that there was a kind of implicit coalition of shared beliefs between Electricity Supply Industry and government departments on the rejection of CHP (Russell, 1986a; Russell, 1986b; Russell, 1993). Similar conclusions can be drawn from some studies on Austria and Scandinavian countries (Hård, 1991; Hård & Olsson, 1995). They all point to the biased influence of some major established actors in the political decision-making processes and the systematic under-representation of proponents of CHP who are, due to the decentralised character of CHP, rather small and less influential players

6.4.3 Case-studies on investment in energy efficiency and CHP

Many individual case studies on energy-efficient investment are available from various European countries, but most of them just describe the plants in question and their technological and economic characteristics.[158] Based on technical and load requirements, economic boundary conditions and regulatory obligations, the question is addressed whether a concrete plant project is promising for a potential investor. These case studies help to understand the technical and cost arguments of an investment decisions, for which regulations and grid connection conditions are just regarded as external constraints and not further questioned.

A number of studies are at least partly based on the study of individual district heating projects and their organisational and institutional conditions (Hård, 1991;

---

158 Just to mention a few of major publications which collect and analyse the situations of individual cases, looking mostly at the site-conditions and economic and technological outcomes of the projects: (Dufait et al., 1992; Gruber, 1992; IEA/Novem, 1993; IVE, 1993; Rydén, Johnsson, & Wene, 1993; WEC, 1991) and many others. Even more extensive case-study material is provided by national associations like the Energy Efficiency Office in the UK, ASUE in Germany, Club Cogénération in France, PWK in the Netherlands, the Danish District Heating Association, etc.

Marvin, 1991a; Marvin, 1991b; Russell, 1986a; Summerton, 1992). The authors try to analyse the political and social processes which underlie the project decisions. Two of them are longer-term historical case studies which trace the fate of CHP/DH over decades. As already mentioned before, they deal with the history of CHP in the UK (Russell, 1986a), and the introduction of district heating in a Swedish town (Summerton, 1992).

Several of the arguments regarding the fairly well-researched field of investment in energy efficiency can also be applied to CHP. It is a similar utility-type of non-core investment which is often neglected even if very short pay-pack periods are promised. From both sociological and economic viewpoints, a number of studies have been carried out which confirm the importance of intra-organisational and mental barriers to energy efficiency investment (DeCanio, 1993; Jochem & Gruber, 1990; Lutzenhiser, 1994). Lack of information and the persistence of established belief systems have led to under-investment in economically very interesting options. In the case of CHP, this has been complemented by prejudices against the technology which can only be partially based on uncertainty and negative experiences. Very important are also 'strategic' arguments which favour the concentration of investment in higher priority production technologies rather than in utility services.

6.5 Policy recommendations

In the available studies, general conclusions are drawn on the possibilities to address the structural and behavioural obstacles to the diffusion of CHP by political means.

6.5.1 Recommendations at structural level

A key question of energy policy is whether and how it would be possible to establish structures, which are conducive to the diffusion of cogeneration. In principle, this applies not only to techno-economic, but also to political structures which frame the politics of cogeneration. Only very few tentative conclusions have been drawn in recent research work on these aspects:[159]

- Decentralisation of energy supply: Power has been generated traditionally in large plants due to economies of scale and higher electrical efficiencies. Heat supply was de-coupled from electricity supply and was based on individualised heating systems in industry and for households. A cogeneration-based supply structure would tend to provoke a more decentralised power generation and a more centralised heat generation. In case of a fast introduction of mini-CHP schemes, which could also be used for households, the decentralised heat supply structures could be maintained whereas the need to generate power in

---

159 Building mainly on (Gruber et al., 1993; Jochem & al., 1986; Verbruggen, 1994; Walz, 1994).

centralised plants would decrease. Heat and power would both be generated as closely as possible to the final place of consumption.

- Decentralisation of political power: District heating projects must be carried out by or in co-ordination with local utility companies. If local authorities do not have the competencies and the political power to carry out their own energy planning, the co-ordination between heat and power supply cannot be realised.
- Vertical integration of energy supply: An all-encompassing integration of energy supply from generation and transmission down to distribution and sale of services would improve the co-ordination of energy supply chains, but it would also lead to the emergence of powerful centralised utility companies with might have a rather critical stance towards decentralised technologies. However, a limited degree of vertical integration is regarded to be rather conducive for CHP, because one of the main problems of decentralised CHP consists of the technical and economic interfaces with the superior supply level (Verbruggen, 1992a, p. 890; Verbruggen, 1994). Vertical integration among these two levels could make the co-operation problems between local CHP-generation and regional power supply easier.
- Horizontal integration of heat and power supply has been proposed frequently as a structural measure conducive to CHP. Centralised horizontal co-ordination of all local energy services is regarded as a measure which facilitates the exploitation of CHP and of integrated energy services (Verbruggen, 1992b).
- Incentive structures and mechanisms of the regulatory framework: They have been mentioned as a lever for reducing the expansion-orientation of utility companies in favour of a stronger service- and efficiency-orientation. Compulsory least-cost-planning of demand- and supply side measures, or the financial incentives for highly efficient generation technologies like CHP are examples of this. Such measures could also help the utility companies to develop a more supportive position towards CHP (Walz, 1994).
- Liberalisation and privatisation of energy supply: Being linked to the preceding point, the liberalisation and privatisation initiatives in several countries are driven by the hope that competition in generation and distribution would lead to stronger incentives for efficiency, both in an economic and in an environmental sense. Nevertheless, liberalisation and privatisation are not uncontested approaches, because the short-term investment and payback horizons of private companies are expected to be detrimental to long-term investment in CHP. Furthermore, the is disagreement on how long-term negative externalities should be internalised in private company investment decisions. Public enterprises are expected to be more inclined to take certain socio-economic criteria into account.

Generalised statements on the 'right' regulatory and political settings for CHP are problematic because of the variety of specific conditions for the different types of CHP-applications. While industrial CHP may find quite good conditions in a liberalised framework, district heating might require more co-ordination and public control. The time horizons for reaping benefits from these two different types of CHP are just too different to summarise them in joint conclusions.

Hardly any statements can be found in research studies on the need to restructure the interface between policy-making and the energy economy, apart from the general conviction of the promoters of liberalisation and privatisation that government involvement should be reduced to a minimum regulatory function to guarantee the operation of market mechanisms.

6.5.2 Recommendations at behavioural level

Apart from these structural measures, some conclusions have also been drawn on how to change the patterns of behaviour of utility companies, manufacturers and users of CHP.[160]

Financial measures to support investment in CHP-plants have been assessed rather sceptically, unless the support is massive and sufficiently lasting to establish CHP widely in the supply system. Financial support also needs to be embedded in a corresponding regulatory framework in order to avoid contradictory incentives. Such as combined approach was pursued e.g. by the Dutch government during the Eighties. Still, the critical question is whether governments should engage in support measures throughout the diffusion phase of technology at all. The experiences from a number of European countries have shown that without support during the early diffusion phase, it is hardly possible to reverse an established centralised and separated heat and power supply system. Support is needed until the technology reaches a critical mass and can fully compete with other options.

Information collection and dissemination of experiences with CHP are generally regarded as useful measures, but they need not necessarily be done by public institutions. Industrial associations can fill the information gap as efficiently. The same holds for raising awareness among users, policy-makers and the public. Commitments campaigns to enrol relevant actors such as utility companies and heat-intensive industries to support energy efficiency measures and CHP go a step further and are regarded as useful if they are combined with an obligation to reach certain efficiency targets

6.6 Missing pieces

The review of recent research findings on the factors influencing the spread and diffusion of CHP has shown that available data are incomplete and focus on certain aspects only. In view of the information needs for analysing political control from the PET-perspective, five main gaps can be identified in available research work. These are the absence of a process view on diffusion, the consideration of

---

[160] Mainly based on (ADEME, 1994; CHPA et al., 1995; Foster, 1994; Gruber, 1992; Hård, 1991; Jochem & Gruber, 1990).

an incomplete set of actors relevant for decision-making on CHP, the nearly total neglect of the political dimension, an insufficient treatment of the role of recent innovations for the diffusion of CHP, and in general an incomplete analysis of the forces operating at the behavioural level.

Lack of a process view on diffusion

Nearly all the studies surveyed follow a static approach to the analysis of cogeneration. Apart from a few historical studies, it has not yet been attempted to capture the transition *process* of the introduction and diffusion of CHP in a research study. Analyses of 'obstacles' have been carried out frequently, and conclusions have been drawn on conducive regulatory conditions, but the questions of timing of measures and of dynamic government strategies to change structural settings have been not been addressed so far.

Incomplete set of actors

Associated to a process issue is the necessity to look at interactions between the main actors in the system who determine the innovation and diffusion process of CHP. If ever a study is based on a micro-level perspective, it is only applied to a very selected range of actors, mainly utility companies and technology users. The inter-relationships with manufacturers and with policy-makers are only mentioned at the margin. While the neglect of manufacturers might be justified by assuming (erroneously) that technological innovations do not play a major role for CHP-diffusion,[161] the political realm is just regarded as an external factor without calling this assumption even in question. Consequently, the crucial role of associations and interest groups for shaping CHP-related policies remains also out of the scope of most research work on CHP.

Neglect of the political dimension

When investigating different policy-aspects in relation to CHP, the political decision-making structures and processes as co-determinants of diffusion have hardly been questioned - despite the highly political character of energy supply technologies. These aspects are particularly important in the case of CHP-diffusion because it is related to fundamental issues such as the local-national policy dichotomy, the centralisation-decentralisation debate and the problem of horizontal integration.

Neglect of the role of innovation for diffusion

Although CHP may not be a highly innovative technology, the improvements made have radically changed its competitive position in relation to other options. It has become a competitive threat to established types of energy systems throughout a wide size range and especially at the low end of the power range

---

161 CHP may have been improved incrementally only, but the improvements were decisive for its breakthrough as a commercial technology.

below 10 MW. However, except for explicitly technical studies, technical aspects are usually regarded to be of minor importance for diffusion. Moreover, the organisational challenges which are implied by the technological innovations have not yet been explored either.

Insufficient micro-foundation

To carry out an analysis of processes and strategies for policy-making, it is necessary to have a close look at the decision processes at the level of individual actors. This has been done for certain types of actors only (see before), but in these few cases the interactions with other actors' strategies are not considered. This interplay needs to be founded on the individual actors' perceptions of their environment, both in terms of structural conditions and in terms of other actors' behaviour. Present work has concentrates too much on one partial aspect at the micro level only, namely costs and benefits from a business perspective.

6.7 Summary

The diversity of diffusion patterns across Europe points to the complexity of the issue, and this diversity cannot be explained on the basis of some general contextual factors only. Neither can a technology-push argument be supported, because the technological possibilities of CHP in Europe do not differ significantly among countries. A fairly broad set of structural and behavioural factors has been investigated in the past in order to provide more differentiated explanations of the diversity of the diffusion patterns of CHP. Available research work has until now focused predominantly on technology analysis and business economics of different *single* actors. At national level, some major structural obstacles to the diffusion of CHP have been analysed which were used to derive normative suggestions for policy-making, e.g. regarding regulation or direct support measures of CHP. These conclusions, however, are based on a static vision of 'how things should be' and do not take into account the evolutionary character of the transformation processes. In other words, available research results should be seen just as a first step towards the type of process analysis outlined in Chapter 4.[162]

The importance of institutional and organisational structures as obstacles to the diffusion of CHP has been indicated frequently, but hardly been studied in detail. Moreover, their transformation depends strongly on the willingness and power of the policy-makers to initiate structural changes. While the crucial role of policy-making is often recognised, empirical information is scarce, as it is generally the case for all aspects which go beyond technical and microeconomic analysis. The few exceptions from this rule are some historical country and case studies. They

[162] Even a approach which took organisational and structural 'obstacles' into account in a static fashion, it would be unclear how the new structures and organisational settings could be achieved. Therefore, an analysis of processes and pathways of transformation is crucial.

underline the importance of long-term structural changes for the success of CHP and point to the necessity to take the co-ordination of and interactions between the different relevant actors into account.[163]

At a first glance, the brief account given on the state of research on CHP may appear to be unnecessary for the purpose of this study, but it fulfils a crucial function for the implementation of the research methodology outlined in Chapter 4. Only now, after this review, it is possible to identify the actual empirical research gaps which need to filled in order to complete the pattern of analysis of the PET-approach. The five 'missing pieces' pointed out above indicate, on which elements of the PET-system the empirical work on CHP should focus in order to complement available data.

---

163 See especially (Hård, 1991) who attempts to combine the concept organisational cultures of firms with the notion of social shaping of technologies to explain the spread of cogeneration systems.

Chapter 7: Methodological Transfer - From the General Approach to the Empirical Investigation

The major objective of Part II is to show how the theoretical approach and the basic methodology can be implemented in practice. As a first step, an overview has been given of major developments in energy supply and CHP technology. The main dimensions of the technology and their relevance for the analysis of actor decisions have been highlighted. Secondly, the main findings on the innovation, diffusion and political control of cogeneration technology as identified in past research work have been reviewed, in order to identify the empirical research needs and missing pieces of information for a PET-based analysis. Although this has delivered only 'static' statements on the importance of certain factors of innovation diffusion, it provides a useful foundation for the further specification of the detailed empirical research methodology. In the end, the intention is to draw a more dynamic picture. This specification is going to be done in this chapter, following three main steps:

- Specification of the general hypothesis of Chapter 3 for the case of CHP, based on the empirical information presented before;
- Delimitation of the research focus, and justification of a comparative methodology, both based on the initially formulated research interests, the information on cogeneration and the propositions of the PET-approach. It will be argued for the selection of the two countries UK and Germany, for the selection of the types of cases and actors to be studied in the two countries, and for the choice of an interview-based research method.
- Discussion of the possibilities to operationalise the concepts on which the PET-approach is built for an empirical investigation, and presentation of the potential data sources to be used for the case of CHP. This will be complemented by some remarks on the implementation of the empirical field work by semi-structured interviews, and the problems and limitations associated to this approach.

7.1 Specification of the basic hypotheses

Five general hypotheses have been proposed in Chapter 3 as guidance for the empirical research part. These hypotheses can now be reformulated more precisely by taking into account the available empirical information on CHP and energy supply. The original five hypotheses are the following:

- A structural hypothesis which addresses the co-evolution of technology and system structures;
- A technical hypothesis on the complementarities in technical systems;
- A behavioural hypothesis on the perception, expectations and decision-making processes of actors;
- A dynamic hypothesis on the mechanisms of change;
- A control hypothesis on the possibilities to control processes of technological change by political means.

However, the role of these hypotheses should not be misinterpreted. It is evident that they do not represent a comprehensive set of all possible hypotheses but a selection of relevant statement with respect to the research interest of this study. Furthermore, as they are complex, interdependent and sometimes vague, they are not meant to be 'tested' in this study. They rather synthesise the theoretical framework into some condensed guiding statements about processes of technological change.

7.1.1 Co-evolution of structures and technology (H1)

<u>H1.1</u> Due to the pervasive characteristics and implications of cogeneration technology within a wider system, both the exogenous and the endogenous conditions must be conducive in order to enable its diffusion. It can be expected that in countries where cogeneration is successful (i.e. where a corresponding change of the technological structure of the energy supply sector occurs) this is paralleled by the emergence (or already the existence) of corresponding technological, economic, informational and political structures, and by conducive exogenous changes (e.g. concerning primary energy supply or public opinion). It is not sufficient for CHP to be 'superior' to other options in a techno-economic sense, but it needs to fit into the entire context of technological, economic, political, informational structures.

- <u>H1.1.1</u> Technological structures favouring the diffusion of cogeneration technology are characterised by decentralisation, by horizontal integration or co-ordination of supply chains and by an easy interconnection between network and generators. The latter is facilitated by standardised agreements on technical connection conditions.

- H1.1.2 Conducive economic structures for CHP are characterised by a decentralised organisation of power generation, locally centralised heat supply and the existence of horizontally integrated heat and power supply companies which can exploit the economies of scope of CHP. It is crucial that the interests which are conducive for CHP can be incorporated in the economic organisation of energy supply, i.e. there should be organisations or actors which can fully reap and internalise the benefits of CHP, e.g. by providing integrated energy services. In addition, the regulatory conditions should provide incentives for an efficient generation of heat and electricity.

- H.1.1.3 At the supply side of technology, companies need to recognise technological and economic opportunities to benefit from the provision of CHP equipment and/or associated services in order to carry out targeted R&D-work.

- H1.1.4 The economic structures of the energy economy being strongly determined by political decisions, a transformation process towards CHP-based system structures depends on the existence of sufficiently flexible and open negotiation structures at the political level. Otherwise, the necessary decisions to trigger change can not be taken, because the political decisions can be blocked by established interest groups, representing the prevailing techno-economic structures. Cogeneration being a 'local' technology and centralised power generation being a 'national' one, financial and planning competencies at local level represent an important element of conducive political structure.

- H1.1.5 Informed decision-makers are needed to trigger an adjustment process between cogeneration technology and the system structures. In order for the experiences with and the benefits of CHP to be recognised, information has to be made available and accessible. Well-established information networks and structures with respect to CHP need to be in place to achieve this.

H1.2 The existing system is the result of a specific match or mismatch between the four types of structures and the dominant technology. A 'lock-in' into a non-CHP economy can therefore be traced back to structural factors.

- H1.2.1 Existing technological structures for heat and power supply have been shaped by economic structures, which were established in response to past technological arguments in favour of centralisation and monopolies. Within such a centralised energy supply industry only those technologies can diffuse which 'match' the economic structures.

- H1.2.2 Centralised political structures tend to reinforce centralised economic structures, and represent a major obstacle to triggering change towards a decentralised technology like CHP.

- H1.2.3 Informational structures to support a transformation process towards a wider use of CHP emerge among actors whose visions and interests are compatible with CHP. If there are no established actors who could adopt a CHP-friendly position and thus develop a information network, pressure from outsiders is needed to give positive examples of the benefits of CHP.

H1.3 A 'lock-out' or transition from a non-CHP to a CHP-based system can be triggered only if *all* the structural interdependencies are taken into account and broken up where necessary.

- H1.3.1 Structures are the result of a historical process. A transition from a structural lock-in can be either smooth or fast. A smooth transition requires the transformation of existing structures and structural interdependencies, a fast transition their break-up.
- H1.3.2 Consequently, stimulating diffusion is a complicated task, which requires a multitude of long-term and pervasive efforts. The improvement of the economic conditions for CHP can remove certain obstacles but is not sufficient to overcome institutional and political ones. The existence (or creation) of organisational interests and alliances in favour of CHP is thus of crucial importance.
- H1.3.3 The pressure for major structural changes can be induced by exogenous developments like e.g. emerging environmental problems or rising energy prices. Problem pressure can increase the acceptance of politically initiated changes.
- H1.3.4 It is necessary for the emergence of cogeneration that no 'bottlenecks' remains, i.e. all the relevant actors must co-operate with regard to CHP. This is due to the close integration of CHP into different network technologies. Without the co-operation of end-users, energy utilities, technology suppliers, industry and policy-makers cogeneration cannot be realised. As some of the actors involved would not necessarily benefit from CHP,[164] the enforcement or stimulation of co-operation can be an important political tool.
- H1.3.5 Even incremental technological innovations can be sufficient to trigger the adoption of CHP if they are complemented by organisational innovations which enable the integration of CHP into the sets of interests of the major actors and therefore into the structural context.

7.1.2 Technical complementarity (H2)

H2.1 The diffusion of energy technologies is a long-term and time-consuming process. The more pervasive the impact of a new technology on the entire large technical system of energy supply is, the higher the barriers to its diffusion. Tech-

---

164 This means that decision situations regarding CHP are not necessarily of the 'prisoner's dilemma' type where one option is of some benefit to all actors involved. The co-operation case may still have negative effects for some of them,. Only the threat of political sanctions increases the probability of co-operation.

nical barriers can be very high in energy supply systems due to technological interdependencies. Cogeneration is even inter-connecting the two energy supply systems of heat and power. Consequently, if heat and power supply chains are technically separated, diffusion will be extremely difficult.

H2.2 Cogeneration is a medium-level element in the hierarchical (technical) organisation of the supply systems for heat and power. The possibilities for its diffusion therefore depend on the technical organisation of the higher level of the electricity supply system as well as on the lower level of energy concepts of firms.

H2.3 Diffusion as a result of adoption decisions does not mean the complete replacement of a technology A by another technology B (as e.g. assumed in the concept of 'superiority'). Due to the specifics of each individual projects, one technological solution may fit better the local conditions in one case, while in other cases a different solution is implemented. In other words, the diversity of actors and site-specific conditions for CHP lead to a diversity of technological choices and diffusion patterns.

H2.4 Improvements of the technology can lead to a better match with the requirements of the different actors, but these improvements of the objects' dimensions do not automatically lead to its diffusion because it can still blocked by technical constraints and interdependencies.

7.1.3 Behavioural hypothesis (H3)

H3.1 The availability of information on the different characteristics and dimensions of cogeneration (and on other competing options), on interactions and on structural and exogenous changes is a precondition for informed technological choices. By reducing uncertainty it would favour the diffusion of CHP. In other words, information flows represent the key interface between structural and behavioural level as well as among actors.

H3.2 Different actors are interested in different aspects or dimensions of CHP: technology suppliers need information about alternative technological application opportunities and their application areas, technology users and energy end-users about different supply options, and political authorities about macro-economic and environmental impacts of cogeneration.

H3.3 Strategic and power-related implications of CHP are an important determinant of decision-making. First of all, by raising issues of long-term consequences of CHP, they may put into question preferences which have been formed on the basis of standard assessment. Uncertainty about long-term and strategic impacts is detrimental to the introduction of new options such as CHP due to the tendency to make worst-case assumptions and behave in a risk-averse way. Secondly, CHP is expected to be strategically problematic for a number of the key actors in energy supply, especially for centralised utility companies.

H3.4 Uncertainty about future public policies and regulations on CHP is particularly problematic for technology users and suppliers, because it affects the possibility to make reasonable planning. Enterprises require reliable foundations for planning and long-term investment in energy supply. This highlights the need for clearly defined long-term political goals to increase the receptiveness to the introduction of structural changes.

H3.5 The quality of available information and the limitations to information processing by actors prevent that fully 'rational' decisions can be made. Expectations play an important role for long- and medium-term decision criteria, which feed into a kind of multi-criteria analysis by each individual actor.

H3.6 The mental frameworks and paradigms of decision-makers strongly constrain the possibilities for CHP to be adopted. Actors' expectations on CHP are shaped by these dominant frameworks, leading in many cases to a mental lock-in which complements similar mechanisms at structural level.

H3.7 The control of information interfaces and expectations is very important for efficient public policies. Supporting the supply of more reliable information on the different dimensions of cogeneration can reduce the uncertainties for users and suppliers. Political decision-makers need to be informed also about the internal interests and decision-making structures of suppliers and users (i.e. their autopoiesis) in order to use the information interface effectively for political control.

H3.8 In principle there is no difference between the strategic behaviour of political and other decision-makers who are all trying to operate and control their social environment, based on similar operational principles but different interests and internal structures.

7.1.4 Dynamic hypothesis (H4)

H4.1 The mutual adjustment of structural and behavioural level, the path-dependent character of technological change and the possibility of lock-in situations show that comparatively small events can trigger substantial transformations in critical moments, even if the complexity and interdependence of systems tend to resist such transformations. If the right technological improvements are available at the right moment in time, a development path towards CHP can be triggered. In order to achieve self-reinforcement it is necessary that a threshold level of confidence into CHP is reached. This can be facilitated by disseminating information on successful applications.

H4.2 Any attempt to break up a lock-in situation requires not only that the new option is 'superior' for different dimensions and actors, but also that the right timing of supportive action is met. In this respect, time-dependent aspects such as the long vintage times of energy supply equipment need to be taken into account.

H4.3 Due to both the structural and behavioural interdependencies, there are only small windows of opportunity for a technology like CHP to trigger phases of fast and radical change, and to transform these structural interdependencies.

H4.4 Successful examples of CHP-plants can trigger reinforcing effects by leading to more conducive attitudes towards CHP and by incrementally changing the structural context for CHP. This can induce a transition process towards a supply system which is based on a substantial share of CHP.

7.1.5 Control hypothesis (H5)

H5.1 Due to the historically grown technological regimes of energy supply in different countries, there is no unique and fixed structure which would be most conducive to CHP for all countries and which just needs to be introduced to make it diffuse. The dynamics of transition towards CHP and the transformation process of the structural context depend on the elements of the established regime.

H5.2 Political control can operate in different areas and at different levels, including political auto-control, structural policies and targeted support policies.

H5.3 Information policies can play a particularly important role for CHP, because they shape the mental frameworks and reduce the uncertainties associated to the long-term character of investment in the technology.

H5.4 Specific policies for promoting CHP are regarded to be most successful in periods when a new structural frame is being established ('windows of opportunity'). The stimulation of reinforcing effects to trigger a transition path is particularly effective in these periods.

H5.5 It is important to keep the control framework flexible and open to change in order to be able to adapt to changing longer-term objectives and conditions. On the other hand, it should also be sufficiently stable to enable the co-ordination and exploitation of economies of scope of CHP .

H5.6 It is not of decisive importance whether CHP is applied in a liberalised or a monopolistic framework, but rather whether incentive mechanisms exist within the framework, which enable the respective actors to internalise the advantages and benefits of CHP. Transition phases to a new (e.g. liberalised) framework represent opportunities to induce a switch towards a CHP-based supply system because the structural settings are going to be changed anyway.

H5.7 In more stable phases, a strategy of niche development can be helpful to establish a first limited space for CHP which is adapted to the basic structural settings of the dominant supply system. The niche provides the foundation for triggering a wider regime shift in technological, economic, informational and political terms.

7.2 Focus of interest and comparative methodology

In order to implement the underlying theoretical and methodological approach of this study, it needs to be specified how this can be achieved in detail. The refinement of the five hypotheses has been a first step in this direction, but the general methodological steps (see Chapter 4) need to be made explicit for the case of CHP. It still needs to be defined which cases to study and why, and how the missing pieces of information will be collected.

7.2.1 On the usefulness of a comparative methodology

In order to derive some generalised conclusions based on a fairly complex theoretical framework like PET, a comparative methodology appears to be most appropriate. A comparison allows gaining insights into the relative importance of individual determinants and mechanisms. More specifically, a comparative approach fulfils two main purposes:

- It allows assessing the usefulness of the theoretical and methodological approach for the explanation of diffusion phenomena by applying it to different cases. Therefore, a standardised pattern of analysis is needed to carry out a comparison. Such a pattern, based on the PET-perspective, will inevitably be quite broad and difficult to handle. In order to keep the work manageable it is thus necessary to specify explicitly the research approach for the example of cogeneration in more detail, and to narrow down the scope of the empirical fieldwork. Fortunately, several of the elements required by the PET-approach have already been tackled in other research projects on CHP. They can be used but need to be complemented by further empirical work on those aspects which are still unclear or which are of particular interest. These empirically novel aspects need to be clearly distinguished from those which are already available.

- In order to study the role of public policies and of different types of structures for the diffusion of technical systems like cogeneration, a nationally comparative methodology appears to be reasonable. It can be argued that energy supply structures and policies are areas which are strongly determined by national regulations and characteristics. In addition, there is still a strong influence of national power engineering and manufacturing companies on their home markets - at least in Europe. This may change in the future with the introduction of the Internal Energy Market and with a rising trend towards internationalisation of the energy supply and engineering business. For this study, a comparison of different national PET-systems for cogeneration, reflecting different control philosophies, appears still to be appropriate.

7.2.2 Target areas

As a result of the research review in Chapter 6, five main areas have been identified which should be addressed by additional empirical fieldwork in order to complement the analytical pattern of PET:

- The transformation of the object of interest, i.e. of cogeneration technology;
- The different types of structural transformations;
- The processes which underlie the formulation and implementation of public policies;
- The processes which lead to innovative activities;
- The processes which underlie the wider adoption of cogeneration.

These five areas can be further prioritised, depending on their relevance for the case of CHP. It is evident that the first area cannot be neglected even if CHP does not appear to be a particularly innovative technology. Similarly, the fourth area will not represent the main research focus, either. These conclusions can already be drawn on the basis of available material. The two areas which seem to merit most attention are the third and the fifth, the latter being more economically and the former more politically oriented. In line with the PET-approach, it will be indispensable to study the second aspect (i.e. the co-evolution of national structures) before addressing the two other areas at the behavioural level because the structural transformations are key determinants of decision-making. Moreover, the third area and the interdependencies among all five represent topics which have hardly been addressed in preceding research. Therefore, emphasis will be put on the adoption decisions of potential and actual cogeneration users. However, the other aspects will be taken into account as well in order to maintain the broad perspective on the diffusion phenomenon as proposed by PET.

7.2.3 First specification: Selection of countries

It will not be possible to follow an 'ideal' comparative approach (such as the "most-similar" or the "most-different" models[165]) for this study because both structural and behavioural determinants are considered in the PET-system. How-

165 Przeworski and Teune distinguish "most similar" from "most different" approaches to comparative study (Przeworski & Teune, 1970). A most similar approach is characterised by as many similarities as possible among the cases to be compared in order to be able to filter out effects of the remaining variables of interest. However, this is difficult to control if the number of remaining variables is very large. A most similar approach is most appropriate to investigate the impact of a limited number of systemic factors. Most different approaches are useful if the expected determinants are to be found at the micro level, so that systemic factors can differ widely without affecting the results.

ever, due to the focus on possibilities and limitations of political control, the selection of cases with rather different political and regulatory framework seems to be most appropriate. Anyway, the restriction to EU-countries provides a common ground in the sense that these are all industrialised countries facing similar problems in terms of energy supply and environment.

There are several dimensions by which countries, their political and economic structures can be characterised. Examples are the centralisation/decentralisation patterns of heat and power supply, both in technological and economic terms, dominance of private or public ownership of the different elements of the supply chain, monopoly vs. competition, centralisation/decentralisation of political power and planning competencies, co-operative vs. conflictive patterns of political negotiation, or liberal policy approach vs. pro-active energy policy.

Exogenous factors can influence the usefulness of a comparison because they can affect the possibility to draw conclusions on the impact of individual factors. For example, climate plays an important role for district heating and small-scale applications because they determine the heat load of a site. Consequently, only countries with relatively similar climatic conditions should be chosen.

An overview of factors and barriers of diffusion of CHP has been given in Chapter 6, though without differentiating explicitly by countries. By building on the results of some of the studies reviewed[166] the selection of the two cases of the UK and Germany can now be justified. The situation of cogeneration in the EU member states is very heterogeneous (Figure 6.1). In the Netherlands and Denmark cogeneration contributes over 30% to power generation capacity, while in many others its share is negligible. Different factors can explain this diversity. In Norway, the availability of cheap hydropower makes CHP an uninteresting option. In Belgium and France, the regulatory framework which is characterised by strong national monopolies with vertical integration but horizontal separation of supply chains, is detrimental to a wider use of CHP. Although this has changed in recent years, and utilities as well as governments are taking a more conducive stance towards CHP, the situation is not yet sufficiently clear to allow any detailed analysis of these changes. In addition, nuclear power is still the favourite source of electricity supply in these two countries. In Spain specific government targets for CHP have accompanied a moderate liberalisation policy, and substantial support has been given to industrial CHP. However, being a Mediterranean country with a warm climate, a comparison with Nordic countries would be very difficult, especially with regard to district heating. Also in Italy first steps towards liberalisation have been made, but the situation has not yet stabilised. The most interesting examples of success stories seem to be the Netherlands and Denmark. In Denmark, the success of CHP can be traced back to the mid-Seventies when a very active policy in

166 The most important of these overview studies are (CEC, 1994; Chadburn, 1994; CHPA et al., 1995; Dufait et al., 1992; IEA/Novem, 1993; OECD/IEA, 1983; OECD/IEA, 1994; WEC, 1991).

favour of energy efficiency was implemented as a reaction to the first oil price crisis. In the Netherlands, the liberalisation of 1989/90 provided very conducive conditions for CHP which have boosted its share in only a few years. A grant scheme for CHP-plants had also a major impact. The German case should be regarded as a fairly positive example, too, although the CHP-share is well below that of the Netherlands or Denmark. From a structural point of view, it is also very interesting because of the decentralised, but monopolistic, utility structure. This stands in sharp contrast to the situation in the UK where the former national monopoly structure has been broken up entirely since liberalisation in 1990, leading to a boost in industrial and small-scale CHP-capacity. The UK is also particularly interesting due to the model character it has for the transition processes to privatised and liberalised energy supply systems in other countries.

The UK and Germany also share a number of very basic conditions like climate, resources, housing patterns, etc. In both countries CHP has undergone a transition period during the Eighties and early Nineties, but within very different regulatory and political frameworks. This makes is particularly promising to study these two cases. In the UK, there was a revolutionary change of the framework from a monopolistic to a liberalised one, whereas in Germany the monopolistic settings were maintained, but slowly improved in favour of independent and decentralised power production. The comparison of the UK and Germany since the early 1980s can thus be characterised by the following main features:

- In the UK the share of CHP is among the lowest in Europe, and CHP is concentrated in the industrial sector. Since the late eighties, this share has been rising sharply. Germany, on the contrary, disposes of the highest CHP-capacity in Europe, corresponding to a moderate share of electricity production. Industrial, district heating and small-scale applications are all widely used.
- Both countries have a similar energy supply history in terms of the role of the coal industry and industrial production patterns. One main difference lies in the emergence of the British petroleum industry at the end of the Seventies, but due to the stable supply situation for fossil fuels, concerns about security of supply have declined in importance in both countries over the last decade.
- In Germany, energy supply is characterised by an influential role of local utilities and authorities whereas in the UK the local level is very weak with regard to both energy supply and political competencies.
- Liberalisation and privatisation of the British energy supply industries have overturned the system since the end of the Eighties. This has not only altered the structure of the industry (e.g. in terms of horizontal and vertical integration) but also changed the specific conditions for setting up CHP-projects.

The case of the UK is particularly interesting because it is often used as a model for liberalisation. However, it is vividly disputed whether the developments of the last years have really been beneficial. In the past, i.e. before 1990, government and electricity supply industry were closely intertwined. The transition process since then has been a major experiment which is expected to prove the feasibility and the advantages of a liberalised framework within an industry which has traditionally been monopolistic and under state control. Germany, on the other hand, represents a model of a supply system which is still based on a monopolistic set-up, even if it is not an as extreme example of centralisation as France. There are multiple links between political and industrial decision-making at different levels. The powerful role of the local level not only guarantees a higher degree of diversity of energy supply practices than in many other countries, but also quite good opportunities for CHP and district heating. In other words, the comparison of the UK and Germany confronts the new liberalised test-case with a fairly CHP-friendly monopolistic system. However, in the case of Germany an additional difficulty needs to be considered, namely the situation in Eastern Germany since reunification. This is obviously a very special environment for CHP which would merit an investigation in its own right. It will be addressed separately in the investigation of the German case.

In principle, it would be possible to characterise the two different national systems more precisely in terms of centralisation, privatisation, industry structure, openness, degree of co-operation and strength of the national political systems, etc. but such a multidimensional space and the resulting typologies are regarded to go beyond the scope of this study.[167]

7.2.4 Second specification: Selection of actors and cases

Although particular emphasis is put on the user side of cogeneration, the decisions of other actors considered in the PET-approach will also have to be studied. This means that not only several examples of CHP-plants have to be selected, but also some of the actors who participate in the other decision-arenas associated to CHP: technology supply companies, energy utility companies, energy service companies and government institutions. The analysis of their positions and decisions in relation to CHP needs to complement the investigation of the user-side.

Coverage of CHP-applications

The selection of CHP-cases in the two countries should cover different *types* of applications, different *dates* of the investment decisions during the diffusion path of CHP, and (if possible) both *successful* projects and *failures*. These different aspects should be equally reflected in the two samples of cases. The different types of applications (district heating, industrial applications, small-scale) are

---

[167] Such an approach implies to develop a first comparative pattern along the structural dimensions of the system, i.e. based one the main characteristics of technological, economic, informational and political structures. For examples see (Kitschelt, 1983; Vogel, 1986)

roughly correlated with the size of application (small, medium, large scale) and also with the type of ownership (private, public, mixed).

It is also desirable to choose similar and comparable cases in the two countries. The selection should provide material on the main types of application contexts. This could be difficult, but for both the United Kingdom and Germany data-bases or reference lists of cogeneration plants are available from which the cases can be chosen.[168] Given the different types of plants and the restrictions of time, it seems reasonable to study about fifteen to twenty single cases in each country.[169]

Consequently, the following types of projects will be included in the study:

- Public district heating project of more than 5 MW;
- Industrial small- to medium scale CHP applications of less than 30 MW;
- Small-scale commercial applications of cogeneration;
- Small- or medium-scale projects in the public sector (local utilities, hospitals, universities etc.).

Failed projects are difficult to identify because information on failures is usually not very widely published. However, in practice the selection procedure allowed to identify a number of less successful examples. For this purpose the data-bases revealed to be very useful tools even if they obviously contain only projects which in the end were realised. By choosing randomly from the data-bases cases to be studied in more detail, also projects with bad economic and technical performance could be included in the sample.

Given the fact that large scale applications are very rare, especially in the UK, and that very large plants are extremely site-specific, the upper size limit of CHP-applications to be considered is about 30 MW. According to the data-bases, there are only very few examples of CHP beyond that limit in the UK, but several DH-

---

168 The data-base on the CHP in the UK is run by the Office of Electricity Regulation (OFFER) and is continuously updated. In 1995, when the fieldwork in the UK was carried out, it contained data of 800 plants, out of about 1000 operating overall. In 1997 contained data about 1400 plants. Although in principle a wide range of technical and economic parameters is available in the database, it is very incomplete and could be used only as a reference for contacting plant operators. In Germany, a first database was set up in 1992 by the VDEW and ASUE, mainly for small-scale engine-based CHP. It contains about 1500 units, also with rather incomplete information. The 1996 update (which is much more complete with 2500 plants) could not be considered any more for selecting case-studies.

169 Other empirical research options have been considered but finally rejected. A written survey could have covered more examples but would have provided more superficial and descriptive data only, similar to that which is available in existing case-studies. In-depth research on a few cases only would not have been representative for the situation in a country and would have made a comparison difficult. Consequently, the approach adopted was regarded as the best compromise.

projects in Germany. In the case of the UK one can hardly speak of a 'diffusion' of district heating anyway, because only a handful of DH-schemes are actually in operation. Nevertheless, several DH project plans are currently reconsidered. The inclusion of large DH-schemes is justified by the argument that they give clear evidence of the political character of technological choices, whereas the diffusion of smaller and industrial applications can be expected to follow predominantly an economic logic. Furthermore, district heating highlights the important role of institutional framework conditions which in the case of Germany favoured the construction of DH-plants by local energy supply companies.

Period of investigation

The selection of cases should cover the time period from the early Eighties, when new types of CHP-technology became available, up to the mid-Nineties. The study periods thus comprises the major reforms in the UK and the adaptation processes in Germany. The cases are clustered according to two major dates which are particular interesting for both Germany and the UK, namely the beginning of the Eighties and the early Nineties. The first date provides evidence on cases which were initiated before any major change in the regulatory framework took place. It also reflects the early diffusion phase of small-scale CHP and gas-turbine CHP. The second date addresses examples after the major changes in the regulatory systems, i.e. after the stabilisation of the technology and the main application conditions.

Therefore, sixteen cogeneration cases are investigated per country, covering four types of applications and two dates of investment decisions. In order to dispose of a broader base of evidence, two cases in each of the eight categories are chosen.

7.2.5 Third specification: Interview-based methodology

In principle it would be possible to carry out a questionnaire survey to collect data on the different actors and their decisions on CHP-related issues, but this idea was abandoned in favour of a more selective data collection technique, namely semi-structured interviews. Here, a well-defined set of issues or questions is used as guideline for the collection of the pieces of information needed to complement a pattern of analysis which is based on the PET-systems approach. Answers to the questions are given in free form, leaving the interviewee the possibilities to provide additional details on his/her motivations and perceptions at the time of decision-making. Semi-structured interviews provide on the one hand a well-structured set of results, but are on the other hand sufficiently open to be adjusted to the specific situation of the interviewee. The application of such an interview-based methodology can be justified by two reasons:

- The high *diversity of actors* addressed in the study. It would not be possible to set-up standardised questionnaires, not even for the CHP-users only, because the circumstances of each case are very different. Other questions would have to be asked to technology users than to representatives of government institutions. For the semi-structured interviews, this problem can be circumvented by using standardised subsets of questions, which allow to compose each interview guideline according to the type of actor addressed. Obviously, only the questions on the perception of structural changes can be used for all actors, while questions on their individual situation need to be adjusted to each type of interviewee. The sets of questions on these latter aspects are thus specified differently for users, suppliers, utilities/service providers, government institutions and interest groups. Although it is very work-intensive, this approach allows to carry out a very flexible investigation.

- Interviews allow to go much more into depth and to maintain an *exploratory element* in the inquiry, whereas a written questionnaire requires to know already quite precisely in advance what types of answers can be expected. This has been a decisive argument in favour of interviews, because the main interest of the inquiry is not to find out *what* the different actors are doing, but mainly *why* they are doing it and *how* certain decisions were made. A written questionnaire hardly allows to address these issues; it is more useful for gathering descriptive data than for investigating motivations, perceptions and decision-making processes.

7.3 From theory to comparative data collection

The general methodology described in Chapter 4 represents the first stage of a translation process from the PET systems approach into an empirical equivalent. The first step of this methodology was refined in Chapter 5 by specifying the different dimensions of cogeneration technology. In the following, the pieces of information to be collected, and the methods and data sources for the remaining steps will be described. Equally important, the stepwise description aims to make clear how the meaning and the content of the theoretical concepts underlying the PET framework can be transformed into a manageable pattern for the collection of empirical evidence on the cases under study. This 'operationalisation of PET' is geared towards the case of CHP, but it shows also which forms of data collection could be applied in general. It is specific to the sector and the technology under study but independent of the countries to be compared. Such a common empirical pattern is essential when carrying out a comparative study across countries.

In the description of the different steps of the PET-methodology, the main variables relevant for the study of CHP will be listed first ('identification'). Then suggestions will be made of potential sources of evidence and of techniques of data collection. This means, that each paragraph explores the full range of possibilities to provide as comprehensive empirical information as possible on the different elements of the theoretical approach. With regard to the actual empirical work to be done, the possibilities to simplify the empirical work on CHP in the UK and Germany will be discussed. Finally, this will allow to indicate the specific types of data to be collected and the appropriated methods to be used.

As mentioned before, particular emphasis is given to the user side of CHP-technology. However, also the other actors and aspects of the PET-approach need to be investigated by means of interviews and secondary sources. Before carrying out the interviews at the micro-level of actors, the researcher needs to be familiar with structural changes, technological improvements and political actions, because the actors' behaviour is a reaction to and an anticipation of such structural transformations. Similarly, exogenous conditions and changes play an important role, too, and need to be studied early on by means of available literature. This implies that two major stages can be identified in the course of data collection, namely first the study of structural and exogenous changes using mainly secondary sources, and secondly semi-structured interviews on the decision-making processes and behaviour of different actors.

7.3.1 Technological changes and innovations for CHP (Step 2)

In Chapter 5, the main dimensions of CHP-technology were introduced. Based on this information input, the appropriate methods for identifying technological changes can be discussed. The changing technical characteristics of CHP are first of all a constitutive part of the description of the evolution of the PET-system. Secondly, they need to be analysed because the perception and the assessment of these changing technical characteristics by the different actors are crucial for understanding decision-making processes related to CHP.

The question could be raised whether it makes sense to study innovations of CHP-technology separately for different countries, but in fact these national technological differences still matter for a number of reasons:

- In some countries, there are close links between technology manufacturing companies and the often nationalised energy supply industry. For example, the national nuclear power industries tend to rely to a considerable extent on their respective home markets. Due to the lack of competition with foreign companies, national differences in technological performance remain.
- CHP is not 'big business', it is a local technology system which requires support from the local manufacturers and service providers. In principle, this implies that one could speak of local differences of technology, even if one might argue that several local conditions depend on national frameworks.
- Technology suppliers which are entirely or partly owned by foreign companies, or buy key components from abroad, have usually a national base organisation. Moreover, the national organisations know best about the needs of their

customers through their networks of user-supplier-relationships and adjust the technology to the user needs.

- The more technology suppliers are involved in providing services rather than in selling hardware only, the more they have to be familiar with the local and national regulatory conditions. This is due to the integration of CHP systems into the local or regional supply context which requires the technology to be tailored to the technical and regulatory requirements of the different countries.

In order to trace the improvements of the technology one needs to distinguish between component characteristics and service characteristics (c.f. Chapter 5). In principle, it is possible to use data from the technology supply companies, and for some of the components like gas turbines and engines such data is indeed available. Cogeneration technology is expected to be characterised by rather incremental technological changes, because it is mainly composed of mature components. Therefore, *technological* uncertainty in the narrow sense is expected to be a problem of minor importance, but system integration and the improvement of service characteristics of the system remain crucial.

Consequently, it is not the component side which represents the main technical bottleneck for the spread of cogeneration but system integration on the site. Useful data on system performance are hardly available and difficult to measure. If such data exists then it is usually not provided by the manufacturers for reasons of confidentiality, or it is provided in a format which makes a sensible comparison with other plants impossible. For components, technical data are already quite difficult to get, especially for small-scale equipment. The situation for cost data and price digression over time is even more difficult. In some cases, monitoring exercises within government funding schemes have made interesting data accessible. Parameters of particular interest are reliability, failures, maintenance costs and performance improvements. However, they are usually highly site-specific and difficult to compare.

For similar reasons, patents as an innovation output indicator are not useful in the case of CHP, either. As for most of the power engineering industry, appropriability is mainly guaranteed by firm-specific, tacit knowledge on system integration and design. Statistical data sources are not sufficiently disaggregated to derive information on research activities and results on cogeneration, either. Furthermore, improvements of cogeneration technology are often based on R&D for other purposes, i.e. improved components and technologies from other applications are just integrated into cogeneration systems. Direct R&D-spendings on CHP are thus meaningless in comparison to these spin-offs.

Similar arguments hold with regard to environmental performance characteristics. Information is available about the compliance of equipment with environmental regulation, but its costs are hidden in the price calculations of the manufacturers.

Neither do emission characteristics of prime movers say very much about the actual system performance under variable load conditions.[170]

One can thus conclude that the main problem consists of the lack of comparative and quantitative data on the technological changes of CHP. Consequently, important complementary data sources are the experiences and expectations of the interviewees concerning major technological changes, including also the expectations they have for the future. The interview information has the advantage of reflecting not only the 'objective', but also the subjective characteristics, which are particularly important for the explanation of decision-making.

Identification/Indicator	**Data Source**
Technological performance of components	Technical data from suppliers Assessment of users Published performance data
Economic performance	Cost digression curves Assessment of users
Environmental performance	Information from suppliers
Economic and political dimension	Interviews and existing studies
Future improvements	Information derived from interviews

Table 7.1: Identification of technological changes

7.3.2 Delimitation of the system and exogenous changes (Step 3)

The delimitation of the PET-system requires to interpret a number of determinants of diffusion processes as exogenous factors. This delimitation is not definitive from the beginning of a study but may have to be modified once more information on the case under study becomes available. Exogenous factors of CHP-diffusion are the events or developments which trigger changes in the system without being substantially affected by the system itself. With regard to CHP, these are changes of the attitudes and cultural orientations of society (e.g. emergence of environmental values), long-term energy price fluctuations, climatic and geographic factors, or international political and economic developments. Given the focus on national systems, European energy policy can also be regarded as an exogenous

170 The other dimensions of CHP-technology (national economic, political, organisation) are only indirectly associated to technological changes, e.g. when new technologies widen the promise for a widespread and economic use of CHP. This is then reflected in the studies (e.g. on the economic implications of wide CHP-use) and interactions in the system rather than in any measurable parameters.

factor. However, in principle it is subjected to the same negotiation game of political authorities and interest groups as national policies. For tracing these exogenous changes, available information sources can be consulted.

Identification/Indicator	Data Source
World energy prices	Statistical data
Environmental awareness and interest groups	Key events, publications
Dependence on energy imports/existence of indigenous resources	Statistical data
International policies	Key events, secondary literature
Climate and geography	Secondary literature

Table 7.2: Identification of exogenous changes

7.3.3 Structures and structural changes (Step 4)

Structures have been defined as the relatively stable but nevertheless changing characteristics of the system under consideration. Technological, economic, political and informational structures have been distinguished (c.f. Chapter 3). It is possible to introduce typologies of these different structures. For example, technological structures can be characterised in terms of centralisation/decentralisation or the degree of horizontal integration. Similar categories can be formulated for political structures (e.g. open/closed), economic structures (e.g. centralised/decentralised, monopolistic/liberalised), and informational structures (e.g. parallel/sequential). Although this would be an interesting analytical exercise, it would only be useful if a larger number of case-studies was available for comparison.

7.3.3.1 Technological structures

Technological structures can be characterised by the number and size of the existing CHP-systems, but also in terms of the constitution of a country's power and heat generation plant park in general. In principle, statistical data is available on the use of different types of energy technologies and fuels in the UK and Germany, as well as on their centralisation/decentralisation patterns, even if the data are not always fully comparable. With regard to cogeneration systems, the situation is more complicated. Data have not been collected at comparable standards; they are incomplete and difficult to interpret. In the UK, a CHP-data base has been set up, based on regular CHP-surveys by the Department of Trade and Industry since 1991. In Germany, the best data are available from the industrial associations in the field of electricity supply. EUROSTAT is about to establish a common framework for data collection in Europe, but it is not fully operational yet. The fully comparative presentation of evolutionary patterns of change in CHP diffusion is thus quite problematic.[171]

The technological interdependencies between CHP and the rest of the energy supply system are also regarded as a structural element and can be analysed by using qualitative technical information from a variety of secondary sources.

Identification/Indicator	**Data Source**
Profile of energy technologies used	Statistical data
Centralisation/decentralisation patterns	Statistical data
Diffusion patterns of cogeneration	Statistical data as detailed as possible

Table 7.3: Identification of technological structures

7.3.3.2 Economic structures

Economic structures comprise both technology supply side and technological user side. Issues included are the organisational and regulatory set-up, price and investment control, and patterns of horizontal and vertical integration of industry, etc. Statistical data are available on some of these structural issues but due to the fact that CHP is used across a variety of economic sectors it only makes sense to look at the economic structures of the energy supply and power engineering industries. Due to their diversity, it is hardly possible to derive any useful information on the different CHP-using industries.

The structure of the energy supply industries is closely related to the regulatory framework as described in literature and legal documents. Its organisational structure can be sketched also by using some quantitative data like e.g. the number, size and distribution of utility companies. Also the specific regulations which are applied to the energy supply industries can be taken from secondary sources. Ownership patterns are more difficult to identify due to their confidential character and the difficulties to evaluate the quality of cross-ownership and share-holding.

The supply side of CHP is a less complicated issue, because it is a limited market for which information can be obtained from industrial associations and other sec-

[171] It became obvious in the course of the research, that it is very difficult to get comparable statistical data on the case of cogeneration. International statistics (OECD, WEC, Eurostat) do not even differentiate their data with regard to specific technologies like cogeneration. National statistical data sources are better in some countries, but highly diverse and difficult to compare due to different assumptions and specifications of the data. Industrial associations revealed to be the best source of detailed statistical material.

torial organisations. Issues of horizontal and vertical integration of energy-related industries is usually well documented and require complementary information from interviews.

Identification/Indicator	Data Source
(De)centralisation of energy supply	Statistical material
Industry structure of manufacturers/users	Existing material, cross-ownership
Control and vertical integration of generation, transmission and supply	For all these aspects statistical data if available and descriptive literature
Privatisation & liberalisation, market access	Available publications, also government material
Energy and environmental regulation	Available material, interviews
Grid connection conditions	Material from utilities, interviews
Horizontal/vertical integration of energy services	Mainly interviews

Table 7.4: Identification of economic structures

7.3.3.3 Structures for the provision of technological information

Due to the fact that cogeneration is not a radically innovative technology, it can be assumed that the structures for the provision of technological information are not particularly important for the technology supply side. However, information on cogeneration technology may be important for the actual and potential users, for example to identify new opportunities for the use of CHP systems, to optimise and integrate CHP-plants, or to exchange experiences on operational problems.

Identification/Indicator	Data Source
Informal contacts and networks	Interviews
Research institutions for energy efficiency and CHP	Publications Interviews
Users' information sources	Interview information
Suppliers' sources of innovation	Interview information
Information sources of political authorities	Interviews, formal consulting services from policy consultants
Network promoters/carriers	Existence of promoting associations

Table 7.5: Identification of information structures

Institutions for the collection, distribution and promotion of information play an important role as crystallisation points for activities associated to CHP. In many countries one can find examples of industrial and public associations which play this role. With regard to political decisions, technological and scientific information support has become more and more important in the last years, not only with regard to environmental and energy issues. Institutions operating at this interface to the policy-making process are an important element for the understanding of political control.

In principle, the analysis of informational structures could be further extended to the realm of R&D, but for cogeneration this is regarded to be less relevant. In the interviews, information can be gathered on the involvement and role of research institutions for technology development and use.

7.3.3.4 Political structures and structure of power relations

All information on political structures is of a rather qualitative nature. Obviously, a wide range of publications and official documents exists on 'polity', i.e. on the formal arrangements for policy-making but they usually do not reflect the reality of decision-making. Consequently, much information needs to be derived from interviews and other studies on the decision-making process in the energy field.

Identification/Indicator	**Data Source**
Integration of utilities and industry in the policy-making process	Formal arrangements & network analysis by interviews
Importance of other interest groups/actors	Interviews
Informal influence on policy-making	Networks of interest groups, Interviews
Co-operative/conflictive character of policy-making, openness of the political system	Secondary literature
Vertical distribution of political/planning power for energy for implementing policies	Secondary literature Interviews

Table 7.6: Indicators and measurement of political structures

The distribution of competencies and power among the different levels of government can be discussed on the basis of available literature, but, as before, interviews represent a complementary source of information.

7.3.4 Identification of actors (Step 5)

The identification of the main actors is an important precondition for the analysis of interactions and decisions, but it is also linked to structural analysis. While the identification of the main *types* of actors has been addressed already when studying the technology, the objective of Step 5 consists of identifying the individual players in the two countries considered. According to the theoretical approach, three main types have been proposed for classifying the different actors, namely technology users, technology suppliers and political authorities. In the energy sector, a broad range of firms, associations, institutions and individuals need to be classified into these three types of actors. The final selection of 'relevant' actors is based on information from the interviews. The interviewees are asked which of the actors they regard to play an important role for CHP.

7.3.4.1 The cluster of actors on the user-side

A typology of CHP-systems has been suggested in Chapter 5 which reflects the main application contexts of the technology: district heating, industrial and small-scale applications. This typology can be used as a first element to structure the investigation of different countries. A distinction has also been made between different sub-types of CHP-users, namely service end-users and utility/service companies. Examples of the latter group are grid operators, generation companies, energy management firms, electricity and gas distribution companies, etc. This does not exclude the possibility that organisations such as e.g. local utilities can be both CHP end-users and fulfil functions of intermediate utilities or service providers. In other words, the tasks of utility companies, service companies and end-users can overlap and complement each other in different ways. This is also the reason why both groups have been clustered jointly on the user side of CHP.

Identification/Indicator	Data Source
Supraregional and regional electricity utilities	For all of them: • Interviews • Company and associations publications • Other publications and energy statistics
Local utilities for heat, gas and power	
Industrial power & heat generators	
Independent Power Producers	
Small-scale Producers	
Industrial heat and power users	
Industrial and other associations	
Heat, power and gas grid operators	
Final consumers of heat and power services	

Table 7.7: Identification of technology users

An important type of actors are users' associations. Apart from facilitating exchange of information among their members, they concentrate and represent the interests of their members in various interactions, e.g. in negotiation with government. Although in principle they are meant to be the 'devices' by which different users aggregate and transmit their interests, in practice these institutions pursue their own internal interests and thus need to regarded as independent entities.

7.3.4.2 The cluster of actors on the supply-side

Regarding the technology supply side, several different firms are involved in the construction and development of power plants or their components. Usually, a plant project which has been commissioned by e.g. a public utility is carried out by a main contractor who co-ordinates several sub-contractors and their respective contributions to the construction works.

In general, this study does not aim to investigate the detailed processes and determinants of innovative R&D-projects, but to study the contribution of the supply side to technology diffusion. This contribution consists first of all of providing better, improved and innovative equipment for power and heat production. However, there are also other inputs in the form of new types of services, which blur the boundaries to energy service supply companies. All actors who are carrying out R&D associated to cogeneration are thus of potential relevance. They range from industrial firms to universities and research institutions. Moreover, technology suppliers also provide the expert knowledge needed in the different negotiation processes, e.g. as consultants to policy-makers.

Identification/Indicator	**Data Source**
Component suppliers	CHP-database and reference lists
R&D institutions, universities	Interviews, R&D-project databases as far as available
Supplier associations	Official publications, interviews

Table 7.8: Identification of technology suppliers

Interest groups of manufacturers can be expected to lobby in the fields of technology- and energy policy making. Technical and engineering associations also play an important role as facilitators of information exchange among their members and as support organisations for standardising technological knowledge.

7.3.4.3 Political authorities

'Political' are all those actors who participate, either formally or informally, in the processes leading to binding decisions such as regulations and legislation. It is evident that in the process of political bargaining which precede these decisions, interest groups as representatives of the supply- and user-side of CHP play a major role. However, in the terminology used here, the political authorities at different levels are treated separately from industrial interest groups which are grouped within the clusters of actors on the user and supply side of CHP.

Identification/Indicator	Data Source
National and regional government	Publications
Local authorities	CHP-data-bases, publications
Planning departments	Publications, interviews
Interest groups & promotional organisations	Publications, interviews

Table 7.9: Identification of policy-makers

7.3.5 Event analysis (Step 6)

While the analysis of decisions and interactions refers to the level of individual actors, 'events' are interpreted as having a major a signal function throughout the system. Several of these events are easy to detect because they are usually widely discussed in the media of the energy sector. Regulatory reforms in the fields of energy and environment, support programmes, major new plant projects, adoption of CHP by important users, major technological novelties or the foundation of new actors like e.g. support associations for CHP are always discussed in the public. Events can also be hidden, e.g. decisions not to adopt CHP and to promote other options, or in general failures to achieve agreement about CHP-related issues.

Information about most major events can be retrieved from available publications, but complementary interview data is needed to interpret their importance.

Identification/Indicator	Data Source
Regulatory reform	Publications
Initiatives for support measures	Publications
Foundation of new actors/associations	Publications
Agreements and commitments by actors	Publications, interviews
Major R&D programmes and new systems	Publications
Major investment programmes	Interviews

Table 7.10: Identification of major events

7.3.6 Analysis of decisions and interactions (Step 7)

As the PET-approach covers both the macro-level of structures and the micro-level of actors, decision-making processes are a crucial of the entire analysis. While it is relatively easy to describe the structural evolution in the energy sector, the causal links between these developments and their micro-level foundation can only be clarified by investigating the behavioural patterns and motivations of the main actors in the system. The next step of the methodology thus consists of investigating these decision-making processes, i.e. the identification of decisions, the study of interactions, and the analysis of actors' motivations and perceptions. Interviews are the main research technique used for this purpose.

Interactions have been classified into four categories: the exchange of technological information, the provision of hardware, financial interactions, and execution of power. They represent the *content* of interactions as contrasted to their *organisational patterns* in form of hierarchies, networks and markets. Moreover, they correspond to the four aspects of the structural context (technological, economic, political, and informational)[172]. Institutionalised 'rules of the game' further shape interactions, either as tacit institutions (e.g. trust among actors) or as formal organisations for the control of the interactions (e.g. monopoly control).

7.3.6.1 Decisions

Decisions are the results of interactions, even if they are finally made by individual users, suppliers or policy-makers. As a first step, they need to be identified, i.e. the meaning and content of important decisions regarding innovation, adoption or political action needs to be carved out.[173]

Identification/Indicator	**Data Source**
Innovation and R&D-priorities	R&D-expenditures, (patents), interviews
Adoption of plant-projects	Statistical data, publications, interviews
Political decisions and their implementation	Official policy publications on new policies and major regulations, press releases

Table 7.11: Identification of decisions

[172] Technological structures differ from the other structures because they are not explicitly of a 'social' nature, but reflect the ensemble of technological objects.

[173] Walz (Walz, 1994, p. 65) distinguishes between five main categories of political decisions of particular relevance for CHP: price control, investment control, internalisation of external costs, regulations/standards ("Auflagen"), financial support, information supply. These measures can operate through of price-oriented effects, demand-oriented effects and supply-oriented effects.

7.3.6.2 *Interactions*

The types of interactions introduced can be identified and studied in various ways. While some information on project partners and their objectives can be taken from publications and PR-material, interviews are essential to gain insights into the quality and effectiveness of these interactions. If possible, the different perspectives of *all* the participants should be taken into account, but this may be difficult in practice. Especially for political interactions, several aspects are also formalised in participation procedures which can be extracted from existing publications on policy-making.

Identification/Indicator	Data Source
Financial interactions (financial arrangements, co-operations, joint-ventures for projects)	Possibly published material Interviews
Political and power interactions (planning procedures, formal and informal influence/participation in energy politics)	Publications Interviews
Exchange of technological artefacts	Mainly interview case studies & literature
Information exchange (information links between companies and R&D-institutions, importance of secrecy, information consulting to government)	Interviews, if accessible minutes of Parliament or government committee meetings

Table 7.12: Identification of interactions

In principle, the combination of different types of content and of organisational patterns of interactions could be used to set up a multi-dimensional matrix for the comparison of the interaction constellations in different countries.[174]

---

[174] This matrix would consist of nine main dimensions.
- Dimensions 1-3: The three types of content of interaction, each to be characterised by an intensity measure (How strong are the interactions?);
- Dimensions 4-6: The three types of content of interaction, each to be characterised by an organisational indicator (hierarchical, network-type, market-type), or by the degree of informality of the relationships;
- Dimension 7-9: The three types of results or outputs of interaction, each to be characterised by an intensity measure (How efficient are the interactions?)
Finally each country could be located inside this fictive nine-dimensional matrix, with each dimension having three levels. This could allow to identify the main types or clusters of PET-systems of innovation diffusion. As a further stage of elaboration of this matrix, technological characteristics could be added.

7.3.6.3 Motivations, perceptions and decision processes

The processes leading to a decision are influenced by interactions, but finally based on the motivations, perceptions and interpretations of the decision-making actor. Information on how decisions are made are preferably collected by means of interviews, despite the problems imposed by the subjective filtering of information by the interviewee. While motivations are a fairly straightforward concept to investigate by interviews, perceptions (and associated to them also expectations) are more difficult to deal with. Actors perceive on the one hand structural and technological changes of the PET-system, but also other actors' decisions and strategies. In order to adjust their own strategies and decisions, they try to anticipate future structural changes and strategies and build up expectations. When it comes to study past decisions as for the CHP-plant case studies, this implies that the interviews have to reveal past perceptions and expectations, and how they influenced decisions.

Political decisions are usually well documented, but the underlying decision-processes represent a complex set of interactions of government officials, interest groups and the public. Often the official interests and motivations are documented in position papers, but they need to be complemented by information on how the actors perceive the other actors' positions and strategies in the political game.

In order to complement the study of technological innovations and improvements, the motivations and impulses which have led to innovative activities can also be studied by interviews. Interviews with some important technology supply companies will allow to collect data on their attitudes, expectations and strategies regarding CHP and the changes in their environment. The role of national (and European) R&D-policies for energy technologies needs to be addressed, too.

Identification/Indicator	**Data Source**
Technology users' motivations, perceptions and expectations	For all of them: Interviews, complemented by published material on official positions and strategies
Technology suppliers' motivations, perceptions and expectations	
Political actors' motivations, perceptions and expectations	

Table 7.13: Identification of decision-making processes

Adoption decisions are clearly the core aspect of the study. First of all, data on the positions of potential and actual adopters are required in order to explain diffusion patterns. The main sub-issues to be addressed deal with the initial interests and motivations regarding CHP, the assessment of the technological options, the information and knowledge-related problems, the decision-making processes, the relationships with the other actors (especially utilities and manufacturers), financial and contractual arrangements, regulatory issues, and the impact of as well as the feedback to the political arenas.

7.3.7 System dynamics (Step 8)

The next step consists of integrating the data collected into a dynamic interpretation of the processes of change in the two countries. Seven basic elements are proposed to give a rough structure to this interpretation, namely:

- The constraining force of exogenous events as a first element which can trigger changes in the national systems, like e.g. a sharp rise in price for fossil fuels.
- Innovation opportunities for technology suppliers (but in principle also for other actors) to generate variety in the system. These are for example advances of fuel cell technology, major cost-reducing innovations for heat transmission, or higher heat/power flexibility of CHP-plants.
- Decisions of technology users to select technologies, based on the aggregation of the different arguments for or against CHP in the main decision arenas.
- The regulatory and socio-economic selection context as defined largely by political decisions. The entire rule system applied to CHP in the energy supply industry is an example of this, but also environmental regulations.
- Stabilisation mechanisms due to e.g. mental frameworks/paradigms and structural (including technological) interdependencies. For CHP, this comprises e.g. established principles of electrical engineering, or the difficulties to control and operate decentralised plants in a centralised supply system.
- Reinforcing and delaying mechanisms, i.e. the different mechanisms in the system which strengthen or weaken the advancement of CHP. The model function of successful CHP-plants and the competitive lock-in into gas-supply instead of heat supply are examples of this.
- System evolution as the integrated picture of change, which comprises technological change as one of its elements.

These seven elements as described here are not yet very specific for the case of CHP, but further details depend on the empirical data to be collected. An important aspect of dynamic analysis is the comparison of two countries. It will not only include the different elements of the system and their individual transformation, but also a comparison of the evolutionary mechanisms at work. Reinforcements and delays as well as structural and technological interdependencies in the British and German supply systems are essential to understand the differences in diffusion patterns.

7.3.8 Prospective analysis (Step 9)

Prospective analysis is closely related to dynamic analysis because it aims at exploring the mechanisms and development trends of the dynamic interpretation of the PET-system into the future. Dynamic analysis integrates the structural, technological and behavioural elements. Consequently prospective analysis needs to build on future projections of structural patterns and behavioural (re-)actions. Beyond the exploration of trends, major unexpected events or changes should be taken into account ('wildcards'). In the end, prospective analysis should provide a limited number of scenarios to capture the range of possible future development paths in technological, structural and behavioural terms, giving particular emphasis to the operation of reinforcing mechanisms. For the moment, very little can be said about how prospective analysis will look like for the case of CHP because it depends on the results from the preceding research steps.

There is a wide range of methods of prospective analysis, which can not be discussed here in detail.[175] In general, the well-structured and systemic character of the PET-approach opens the way for both qualitative and quantitative prospective analysis. Depending on the availability of appropriate data, even advanced simulation techniques, which require a well-defined system structure, could be applied.[176] Although this option was considered when designing the research approach, its implementation would have gone beyond the scope of this project.

7.4 Issues and problems of research practice

7.4.1 Limitations of data sources

One of the main challenges of the above sketched procedure for data collection consists of capturing the interdependence between the behavioural and the structural level in a dynamic way. Interviews with decision-makers, covering a period of about fifteen years, are the main tool for gathering the corresponding information at the micro-level. Although structures are relatively stable longer-term set-

---

175 Different prospective methods and approaches are introduced in (Dror, 1988; Godet, 1993; Hourcade, 1993; Schwartz, 1991).

176 The most widespread prospective methods use expert knowledge in different ways, mainly because mathematical modelling on the basis of extrapolation or system dynamics approaches can hardly deal with qualitative changes in the system. This is different in the case of methods which try to integrate expert knowledge in evolutionary simulation models. An example of this can be found in (Canarelli, 1996) who uses the principles of fuzzy logic to aggregate expert knowledge (which is reliable only in the narrow field of the expert's specialisation) from a variety of areas in a model for system evolution.

tings of the system, they are nevertheless not static but also change continuously as a result of the different types of interactions. The objects of interest, i.e. the technologies considered, are similarly subjected to transformations, innovations and improvements. The changes of these elements, together with the actor-specific objectives and motivations, induce the actors' reactions and decisions. They are expressed by their participation in different interaction arenas. The final decisions contribute then to the shaping of both the system structures and the trajectory of CHP-technology., The aforementioned interdependencies appear obvious n such abstract terms, but they are difficult to illustrate by empirical information. This is partly due to a lack of adequate data, but partly also due to the fact that the range of data requirements is enormous. The selection of interviews inevitably needs to be restricted to some representative cases. The empirical investigation is therefore not a full-scale survey, but a well-balanced sample of case-studies which aims at filling the most important information gaps at behavioural level. They are complemented by further interviews with the other major actors in the system. Their selection obviously depends on the situation in the energy supply systems of the two countries under study.

In Appendix 1, the outline for the semi-structured interviews is provided. The interviews aim to collect qualitative rather than of quantitative data. Three reasons support this choice:

- The theoretical approach is still too immature to be operationalised for quantitative analysis. It does not yet provide sufficiently clear-cut hypotheses which could be translated into similarly clear-cut research questions. From the perspective of PET, quantification would not even be particularly important because it is overridden by the complexity and inherent uncertainty of system evolution. The PET-approach aims rather at a 'procedural rationality', at guidelines for how to structure sensibly an empirical study of a complex technological system of.

- Qualitative information from semi-structured interviews is regarded appropriate to describe patterns of behaviour, motivations and decisions. The reasons is that the specificity of each individual decision would make a quantitative aggregation rather questionable. Another reason consists of the exploratory character of the research work. Quantitative analysis is only feasible if the *types* of answers are already known and just need to be refined by hard data.

- The type of data which interviewees are willing to provide is usually of a rather qualitative kind, especially when it comes to analyse motivations and perceptions. But even for seemingly quantitative data on cost estimations, the interviewees often indicate that these data have in fact been interpreted as rough qualitative indications only.

In order to deal with the problem of having to carry out very different types of interviews (with technology users, manufacturers, utilities, service providers, government institutions), the structure of questions and topics to be addressed has been organised in a number of research blocks. While questions on the perception and assessment of structural changes have been asked to all interview partners, the other issues have been addressed only to those interviewees who dispose of the adequate expertise. The selection of the right interviewees has been made on the basis of their expertise and the position in their organisation. In the case of technology-users, the energy or engineering managers of the firms in question were usually selected. They dispose of the technical and economic, but also of strategic knowledge about the decision processes in relation to CHP.

7.4.2 Constraints on the general validity of the results

The formulated arguments for the delimitation and structure of the empirical research work may sound like an ideal and unrealistic vision. In fact, a number of practical difficulties prevent the final results from being as reliable and valid as one might wish them to be:

- It is difficult to distinguish clearly between companies on the user and on the supply side of CHP. It may be possible to identify the core activity of a company, but diversification of services has become an widespread strategy of many firms in the energy sector. Many utility companies which were initially active in distribution services only, have recently started activities in generation and contract energy management. This rapidly changing picture makes it difficult to make any clear statements about the main types of actors.
- When investigating a specific plant project it is difficult to identify the right company to talk to. This is partly due to the fact that several constellations are possible to carry out a project, ranging from companies which finance, build and operate their own plants (e.g. large chemical firms) to different types of contract energy management and third-party-financing schemes. In fact, the boundaries between owners, operators and heat/power service users are blurred. In addition, due to the limited knowledge of a single interviewee it would be good to carry out more than one interview for a specific case, but in practice it is hardly possible to interview all the actors required for drawing an all-encompassing picture of a case example.
- The questionnaire guidelines can be completed comprehensively only in very few cases. An interview situation using semi-structured guidelines is always difficult to manage. It is a mixture between an predefined set of questions and a rather open discussion of topics of major relevance to an individual case. Both interviewer and interviewee are involved as individuals in the research situation, and this adds inevitably a subjective element to the results.
- This has obviously consequences for the type of data which can be obtained from the interviews. The results are not as quantitative as the ones of a written

questionnaire,[177] but go into more depth with regard to motivations, perceptions and interests. Although in principle, it would be possible to generate quantitative results out of the different interviews, this was not regarded to be adequate for this project. It would pretend a level of precision which obscures rather than enlightens the complex processes underlying technological change. Consequently, the results can be criticised easily for not being representative, but it should be kept in mind that the aim is not to provide fully representative results, but rather to complement more representative but superficial overview work by the richness of an interview-based study which nevertheless follows a well-structured and theory-based format.

7.5 Summary

At the beginning of this chapter, a set of hypotheses on the diffusion and political control of CHP has been presented as an attempt to refine the theoretical part of the analysis. Then, the research focus and the further methodological approach have been outlined and justified. The United Kingdom and Germany have been selected as the two country case-studies to be compared. As the user-side of CHP has been recognised as one of the main research bottlenecks, a comparable set of single plant cases was defined which represents the backbone for a series of interviews on the plant decisions and the role of structures and public policies for these investment decisions. They will be complemented by interviews with the other main actors in the energy sectors of the two countries on their respective CHP-strategies, i.e. manufacturers, government representatives and industrial associations.

In the remainder of the chapter, the possibilities for data collection on the different aspects related to CHP-diffusion according to the nine steps of the PET-methodology have been explored and delimited for the fieldwork on the energy sector and cogeneration. The conceptual elements of PET have been specified and linked to indicators and data sources, based on the set-up of the PET-system, the dimensions and research findings on CHP, and some practical considerations on

---

177 It should be kept in mind that for *both* a written questionnaire and an interview-based procedure, subjectivity is a crucial problem, especially when it comes to translate qualitative information into quantitative indicators - as often done in questionnaires. In an interview-based procedure, this translation of qualitative answers into numbers is done by the researcher who thereby maintains full (though biased) control of the translation process. In a written questionnaire this is done by the addressee himself, e.g. when deciding about the type of quantitative answer he gives, using a scale defined by the researcher. In other words, the difference consists of the control of this translation process which is in the one case kept by the researcher and in the other by the addressee of the questionnaire.

empirical research work. This description has been kept fairly general so that it can be used also for other case studies than the two selected countries.

The nine steps represent the analytical element of the research work, but it is evident that it needs to be followed by conclusions on the strategies and measures at the disposal of the study's fictive client - which in this case has been defined as political decision-makers at national level. In view of these conclusions, the importance of the comparative approach becomes evident, because the two country studies represent two basic policy strategies for controlling the energy sector and more specifically the diffusion of CHP. Only on the basis of an analysis of the different mechanisms, incentives and pitfalls operating within a liberalised and a monopolistic framework it is possible to develop strategies for the promotion of cogeneration technology.

PART III:

The Empirical Account: Cogeneration in the UK and Germany from the Early 80s to the Middle of the 90s

In the two preceding parts, first a theoretical approach and methodology for the study of innovation diffusion and political control has been presented which should be regarded as the background of the whole CHP-study. Then, the different elements of the approach have been introduced, starting with general introductions to innovation diffusion in the energy supply sector and to CHP-technology. The main characteristics relevant for an assessment by the main actors were discussed, as well as the technological requirements imposed by CHP. This has mainly served as a static specification of the 'object of diffusion'. After that, a review of research findings on the determinants of diffusion of CHP has been given, as far as identified till now in the relevant literature. The discussion of these findings has finally led to a refined specification of the research methodology to be applied in the empirical part of the study. It addressed the arguments for selecting the UK and Germany as country cases, the choice of semi-structured interviews as the main research method, the types of cases to be studied, and the specification of the data to be collected.

Now the case-studies on Germany and the United Kingdom are going to be presented along the line of the PET system approach. Most of the results are presented in a qualitative way, though complemented by quantitative data where available. In addition to the limited number of cases and the diversity of interview partners (users, suppliers and policy-makers), the main reason for this should be seen in the qualitative character of most of the replies given in the interviews. Although in principle the transformation into a quantitative representation would have been feasible, the emphasis on assessments, motivations and processes has favoured a qualitative illustration of the argument.

Neither should the case-studies on the UK and Germany be seen as all-encompassing historical studies. They aim at providing insights into the complex relationships and interdependencies between structural and behavioural levels of change. Theory has been used as guidance for identifying potential patterns and mechanisms in the complex world of innovation, diffusion and political control. The two cases are meant to analyse processes of change, and what can be achieved by their comparison is an evaluation of the relative importance of different mechanisms of change in different techno-economic and socio-political contexts.

Chapter 8: Liberalisation of Energy Supply in the UK

8.1 The history of cogeneration in the UK

Cogeneration has had a very mixed history in the UK and has never been established to a significant extent. Building on the work by Russell (Russell, 1986a; Russell, 1993; Russell, 1994), its history can be separated into six distinct phases.

Phase 1 (till 1940)

The early history up to 1940 is characterised by a relatively high interest among engineers in the efficiency advantages of CHP, and it led to the construction of a small number of plants. Especially the fuel shortage after the first world war raised interest in CHP among government officials. However, the take-up was limited, mainly due to a preference for large scale, power-only generation to exploit economies of scale.

Phase 2 (1940 - 1960)

The second phase started in the last war years and ended in the late 1950s. The planning approach which had been adopted during the war continued to be applied later on. The interest to improve living and housing conditions, and the destruction of some city centres led to proposals for a number of district heating projects. These plans were substantially reduced in the course of the time, when costs escalated and when nationalisation in 1948 brought about a new constellation of actors. Local authorities then needed to negotiate with regional electricity boards and did not dispose of the planning powers and competencies to carry on with projects on their own. Support by government also declined and was inconsistent so that in the early 1950s, most of the schemes were shelved. Projects envisaged by the regional electricity boards themselves did not advance either. In the early 1950s, government became also more reluctant to interfere with the British Electricity Authority's (BEA) plans for power supply, especially after 1951 when the Conservative Party won the elections. In general, the task of BEA to optimise power generation and transmission did not leave space for heat supply activities in its organisational culture, and thus helps explain its reluctance to support CHP/DH (Hård & Olsson, 1995). The few surviving project plans were assessed on a project by project basis, but despite the advantages e.g. in terms of fuel policy, they were rejected by BEA on the grounds of the disadvantages for its own strategic position. As CHP was increasingly seen as a marginal activity, it could go ahead only if it did not interfere with the main business.

Phase 3 (1960-1973)

In the Sixties and early Seventies, CHP saw a slight revival in the UK, though neither for concerns over energy conservation nor for social reasons. New types of housing patterns (high rise blocks) and an increasing interest of the fuel industries in CHP enabled the progress of a number of district heating schemes. It nevertheless remained a sideline within the energy and electricity supply system. At least a number of important innovations could be made at that time, especially concerning the costs, the efficiency and the reliability of the piping networks. This period is also characterised by an expansion of industrial CHP capacity, in parallel with industrial and economic growth in the UK.

Phase 4 (1973-1983)

The oil price crisis in 1973/74 marked a major change in British CHP-policy. A number of potential cases were re-examined, but little progress was made in practice due to the reluctance of Conservative government (which was in power from 1979 onwards) to interfere in economic activities. The Marshall Report (DoE, 1979) which intended to identify the potential for CHP in the UK led to the introduction of the so-called 'lead-city scheme' to start up a number of model projects for district heating. The other main proposal of the report, namely to set up a National Heat Board, which ought to co-ordinate activities and operate as a carrying organisation for CHP, was rejected. There are many reasons why most of these lead city projects failed in the end. One of them is certainly the lack of support by the ESI for local energy supply activities. However, this phase of consciousness of the advantages of CHP finally led not only to the implementation of some modest CHP/DH projects,[178] but also to the establishment of organisations of actors interested in CHP, especially of the metropolitan local authorities and of industry. Although industrial CHP may have been less prominent in the political debates than district heating, its cost and efficiency advantages favoured the construction of most British CHP-capacity in the Sixties and Seventies. It declined again with industrial recession, lower fuel prices and a policy of the regional utilities to discourage CHP schemes in industry.

Phase 5 (1983-1990)

It is difficult to say whether the Eighties should still be regarded as part of the fourth phase or not. While the Energy Act 1983 was a first step towards the trans-

178 Surprisingly, the most successful schemes were not those supported by government but the ones which were characterised by a high degree of local commitment.

formation of the energy supply industries in the UK, very little changed in practice. It was not before the decision to liberalise and privatise the Electricity Supply Industry in 1990 that a significant diffusion of CHP took place. However, during the early Eighties most new technological options for CHP became available and important preparations were made which proved to facilitate the take-up of CHP later on.

Phase 6 (since 1990)

The liberalisation and privatisation of the Electricity Supply Industry was the decisive event which marked the beginning of a new phase and overturned the structures and actor positions regarding power supply. It required a strong ideological backing by Conservative government in favour of liberalisation, and a high level of discontent with the performance of energy supply, to start to think about alternative solutions for organising the sector. This interpretation reflects the 'satisficing behaviour' of the main actors who needed a strong push to start moving ahead in a more innovative way.

This study on the diffusion processes of CHP in the UK focuses on the phases 5 and 6 because it was not before the early Eighties that new impulses facilitated the take-up of this technological principle. The situation at the beginning of the 1980s was the result of a long historical process which had led to the establishment of a highly stabilised energy supply structure. In the early Eighties, a number of new developments started, which foreshadowed further transformations and broke up existing lock-ins. Numerous examples of this multiple lock-in can be noted, not only in the technical sense of the existing plant park, but also with regard to the main organisations which had the power to decide about energy supply strategies and policies. Even at the political level, where a break-up of this lock-in could have been initiated, the situation seemed to be stuck due to the influence of the nuclear and coal lobbies on the one hand and the Central Electricity Generating Board (CEGB) on the other. Until the late Eighties, there was not enough momentum to induce a process of change in the ESI. In spite of the importance of liberalisation and privatisation in 1990 for removing several economic and organisational obstacles to the uptake of CHP, the preceding phase was already decisive for the preparation of the fast diffusion process of the Nineties. This is underlined by the fact that CHP capacity started to grow already before the major reforms took place in the energy sector of the UK, namely in the mid-Eighties.

8.2 The context conditions

Energy sources and supply context

The UK is a country rich of indigenous fossil resources (Figure 8.1). The exploitation of the North Sea fields since the late 1970s has made the UK a major exporter of oil and gas, and the coal reserves - in spite of numerous pit closures - represent a security for the future. For this reason, the problem of security of supply is less important for the UK than for other European countries. Only in the longer term, there is some fear that the present expansion of the use of natural gas ('dash for gas') might lead to an early exhaustion of this valuable resource.

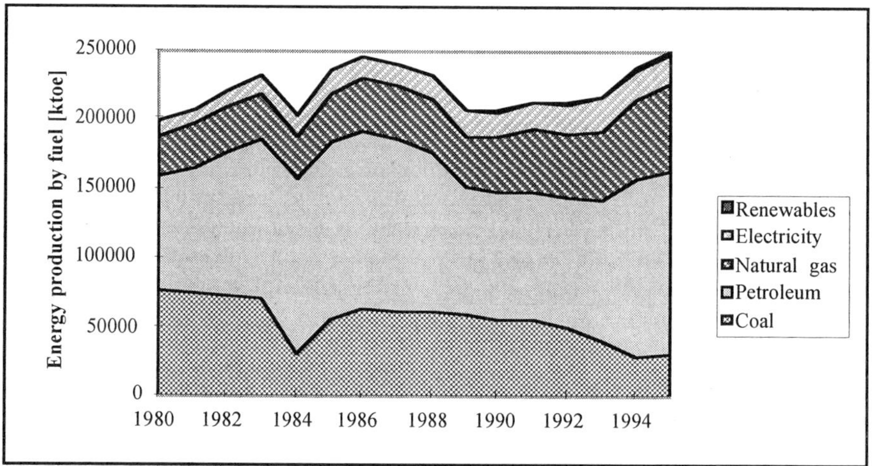

Figure 8.1: Energy production in the UK by fuels, Source: (DTI, 1996; ENERDATA, 1996)

In the Eighties, the nuclear debate figured prominently on the political agenda, but has come to standstill recently. The critical decisions about the Sizewell B power station and the Windscale reprocessing plant have finally led to a moratorium on nuclear power and to a search for alternatives. Higher energy efficiency, until then hardly a topic in the UK, is one of them. While coal was the major fuel for cogeneration plants in the past, this has been replaced recently by natural gas - at least for most new plants. In this sense, the coal debate may have played a role for CHP in the past, but it has not mattered very much significantly in recent years.

The price differential between primary fuel (predominantly gas) and electricity underlies the cost estimations of CHP and hence determines the viability of individual projects. The relatively low fuel prices in the last years have not been conducive for a switch to highly fuel-efficient technologies like CHP. Probably the most important change on the fuel side was the possibility to use natural gas for power generation after British Gas was obliged by a European Court decision to remove a premium on gas when used for generation purposes. In addition, privatisation of British Gas and liberalisation of gas supply in 1986 led to a drop in gas prices. It had a double-edged impact on CHP because on the one hand the price differential to electricity rose, but on the other hand the absolute fuel price level decreased as well - which is rather detrimental for energy-efficient investment.

The role of the environmental debate

Environmental issues have always played a rather modest role on the British political agenda as well as in the public debates. Consequently, the environmental arguments in favour of cogeneration could hardly ever influence the discussion on energy policy and technology options. Economic arguments and - in the context of district and community heating - social arguments were much more important, at least until the mid- and late Eighties. International and European commitments have been the main drivers which obliged British government to take environmental considerations more seriously into account and impose them also on the ESI. Harmonisation of European environmental legislation and the Rio-conference triggered not only new policy strategies,[179] but also to the implementation of environmental regulations under the control of Her Majesty's Inspectorate of Pollution (HMIP).

Governance philosophy

The entire debate on public industries like energy, telecommunications or transport cannot be separated from the ideological discussion on the need to 'bring the market to the public sector'. Until the late 1970s, British government exerted a very strong influence on many key industrial sectors, in most cases by means of public ownership. The privatisation and liberalisation philosophy of Conservative government was one of the main driving forces behind the reforms for the ESI in the Eighties, both those for electricity (1983 and 1989/90) and gas (1986). But also the recognition that the British energy supply system was not performing well in international comparison urged for a change. A second key element of the British governance approach must be seen in the role played by local and regional authorities. Compared to many other European countries, they have very limited competencies, and their position was further weakened during the Eighties.

Economic and industrial development patterns

The fate of especially industrial CHP needs to be seen within the context of economic recession and restructuring in the UK since the mid-Seventies (Figure 8.2). The possibilities for investment were limited in the private as well as in the public sector. In addition, local authorities which play a crucial role for district and group heating, have been confronted with the problem of limited planning competencies. In industry, a reorientation has been made away from highly energy-intensive processes - a trend shared with most other European countries. On the one hand the demand for high-grade heat (and thus for cogeneration) declined, but on the other hand the parallel trend towards higher energetic efficiency of the remaining processes boosted the scope for CHP. In addition, the increase in power consumption called for solutions with a higher electrical efficiency.

---

179 See for example (DoEnv, 1990) where the principles of British environmental policy for the next decade are described.

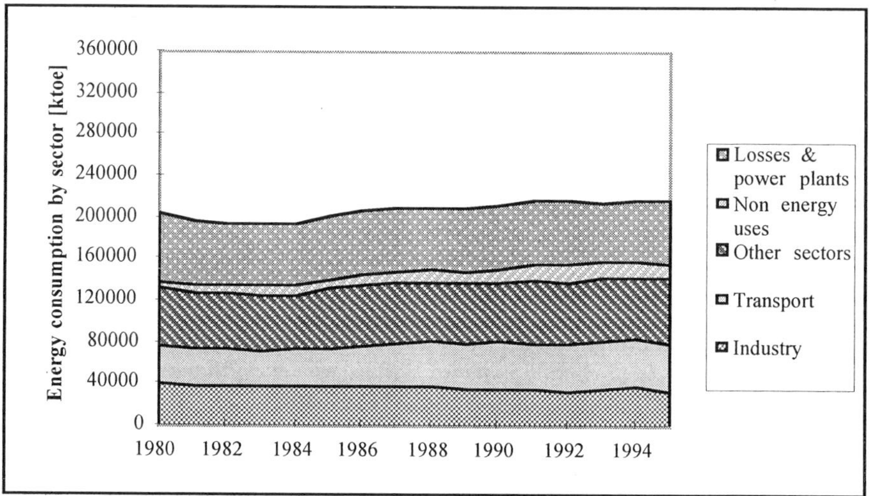

Figure 8.2: Energy consumption in the UK by sectors, Source: (DTI, 1996; ENERDATA, 1996)

European policy issues

The plans for an internal energy market in Europe should be mentioned as a further external force affecting the British energy system. The UK being often seen as a test case for liberalisation, its industries are already prepared for future changes of the European framework. Especially the big players prepare themselves to operate Europe-wide, while the others are prepared to face competition from abroad.

8.3 Innovation patterns of CHP-technology in the UK

8.3.1 Technical improvements

In general, CHP is regarded as a mature technology. The principle has first been applied at the beginning of this century in Manchester, but it has never been used widely in the UK. Research and development on CHP has been limited, but a couple of improvements have been made in the last decade which have been crucial for its wider diffusion.

Although, the components for energy supply systems are traded on an open international market, this does not necessarily hold for smaller systems like most CHP-plants. Several foreign companies have entered the British CHP-market in the last years, but they had to adapt to the specific national context, its regulation and requirements in terms of design, technical specifications and interactions.

The improvements which have been regarded by the interviewees as of major importance in the UK are summarised below.

Improvements of engine technology

The introduction of new generations of engines was essential to make small-scale CHP acceptable. The new engines require less and less costly maintenance and show a considerably improved reliability and availability. This was achieved mainly by introducing computerised, remotely monitored control systems and preventive maintenance of critical components. Slightly higher electrical efficiencies further improved the scope for small-scale CHP. Especially the development of gas engines was pushed forward in order to be able to exploit the advantages of natural gas for power generation. Finally, the low end of the power range for CHP has decreased significantly. Fully operational units are now available below 30 kW, even if their operation may be economical only for special sites.

Availability of new gas turbines

Higher electrical efficiencies of gas-turbines imply a higher share of the high-value product electricity. This development is favoured by rising power loads and stagnant heat loads of many (industrial) sites. Turbine technology is still in a process of permanent improvement in technical terms, whereas most other components mainly benefit from cost-reducing incremental innovations (e.g. heat exchangers). In 1994 about 20% of CHP-capacity was already based on gas-turbines, and a similar share was operated in combined cycle plants, i.e. in combinations of gas- and steam-turbines.

Interconnections with the grid system and control systems

Improved computerised interfaces facilitate the connection of decentralised plants to the electricity grid. In principle this applies also to district heating house connections and individually controlled DH-radiators, which have improved the acceptance of CHP/DH. Although for a long time mechanical interrupters were required to protect the grid from failures of the power supply by CHP, these can now be fully replaced by electronic switches. In other words, the advancement of control technology has enabled various critical steps forwards to improve the operation and interconnection of CHP-systems.

Packaging

Packaging of small and medium scale CHP equipment has allowed to pre-assemble units and install them on site as 'black boxes' for the customers. Initially this was only done for small-scale system below 1 MW, but now also gas-turbine system up to about 5 MW are provided as packages.

Environmental performance

The environmental performance of CHP-systems has been improved in parallel with all other kinds of combustion technologies. Catalytic converters, lean-burn technology, etc. have been developed to cope with more stringent environmental regulation. For gas-driven units, one of the major problems consisted of dealing with varying gas qualities, whereas for diesel engines the development of adequate catalysts represented the key problem. In the UK, the obligation to be able to run on both gas and oil was one of the major challenges in this respect.

Technical issue	Improvements	Approximate introduction
Gas engines	Use of varying gas qualities Dedicated design Less maintenance Higher reliability On-line control system	late Eighties/early Nineties, especially for small engines
Gas turbines	Higher electric efficiency Higher reliability	from mid-Eighties onwards
Grid connection and control systems	Introduction of electronic switches On-line monitoring Preventive maintenance	early Nineties late Eighties, now standard
Packaging & standardisation	Pre-assembling of plants into standardised packages, to be installed as black box on site	mid-Eighties
Environmental performance	Catalytic converters	early Nineties
System integration	Integration of CHP into the entire energy system of a site	slowly introduced in the early Nineties
Metering equipment	Cheap and smart metering equipment for power and especially heat	mid-Nineties

Table 8.1: Improvements of technical characteristics of CHP and their approximate timing in the UK

System integration

A better integration of CHP in the entire supply systems for all energy-related services (heating, electricity, cooling, heat storage, steam-raising, etc.) of a site was repeatedly mentioned as one of the key issues for CHP. This was less a problem of adequate hardware but rather of an accurate design, which in turn required comprehensive data-sets on the site's energy load profiles.

Metering equipment

The development of equipment for heat and electricity metering was needed first to facilitate an accurate design and integration of CHP-systems. But even more important, liberalisation induced an urgent need for cheap and accurate metering of energy services in order to be able to quantify and monetarise their supply ('smart metering'). Especially for small-scale equipment this represented a major cost factor.

8.3.2 Expectations for the future

A first major important novelty expected for the next years are new types of prime-movers at the low end of the power range. This will initially be mini- and micro-engines, but in the longer term Stirling engines and especially fuel cells are expected to replace combustion engines due to their low emissions and high efficiency. At present, most of these options do not yet play any role for investment and it will still take several years until they will fulfil the reliability standards and the economic criteria required for energy supply equipment.

Technology	Characteristics	Importance	Expected introduction
Cooling	Combination of heating and cooling to achieve smoother load curves	medium	mid-Nineties
Ancillary components	Adaptation of various system components to the requirements of CHP	medium	permanent
Stirling engines	Efficient, multi-fuel combustion with low emissions	low	2000
Gas turbines	Further efficiency, reliability and emission improvements	medium	permanent
Fuel cells	Highly efficient power source with low emissions, fuelled by Hydrogen or Methanol, flexible operation by stacking	high	2000
Micro engines	Reliable, low maintenance and low investment engines for highly decentralised heat/power supply	low/medium	late Nineties

Table 8.2: Needs for and expectations of future technical developments of CHP in the UK

A second field of improvements is seen in the different kinds of ancillary equipment such as heat storage, heat pumps, integrated system control, etc. They will facilitate the co-ordination, optimisation and operation of the entire energy system of a site. Integrated design, operation and monitoring of energy supply is expected to be the most important step towards higher efficiency. System control goes hand in hand with improved metering equipment which has already become of crucial importance in the context of liberalisation of energy services.

These statements are based on all interviews carried out, but a particularly interesting issue can be discovered when looking at the differences in opinion among actors. This issue will be discussed in more detail in the chapters on the decision-making processes.

8.3.3 Non-technical implications

Obviously, the technical improvements made have had a clear impact on the overall costs of CHP plants. Investment costs have been reduced by packaging, by the onset of series production and by improvements of the production processes. This holds especially for gas turbines and for reciprocating engines in the 100 to 500 kW range.[180] Higher reliability, less downtimes, lower maintenance needs and optimised system integration had a major impact on operation costs. These developments have extended the range of economically viable CHP-plants down to smaller unit sizes. In the UK, the usual limit is seen at about 30-50 kW, depending on the specific site conditions.

These cost reductions for the end-users of CHP represent a new challenge from a macro-economic point of view, because the new situation brought about by technical improvements makes CHP more interesting for a large range of potential power producers. If these possibilities are taken seriously, this would be a real threat for the centralised power generators in the UK, having serious implications of CHP for the national organisation of the energy supply system and the power constellations in the sector.

8.4 Structural changes

8.4.1 Technological structures

8.4.1.1 Electricity Supply Structure in the UK

In order to be able to assess the situation of CHP in the UK, it is important to be familiar with the electricity supply and demand context and its evolution.

---

180 Unfortunately, most of the cost-related data are confidential and were therefore not provided by the interview partners. It is also very difficult to single out different cost components in an integrated energy supply package or even in a Contract Energy Management or Discount Energy Purchase scheme. However, for the years between 1991 and 1994 agreement prevailed that investment costs for medium scale CHP-engines in the range between 100-50 kW decreased by 20-30%.

Electricity production and consumption

Similar to other European countries, energy demand has stagnated in the UK since the late 1970s, while electricity demand has been slowly but constantly growing. The bulk of electricity is generated in coal-fired and in nuclear power stations. Since the beginning of the 1990s, a switch of fuel sources has taken place: the 'dash for gas'. From 1990 to 1995, the share of natural gas used for power generation rose from 1% to 16%. A number of new gas-fired Combined Cycle Gas Turbine (CCGT) power stations have started to change the structure of fuel supply for power generation.[181] This trend was triggered by the fact that natural gas became available as a fuel for electricity generation. The new plants have led to a massive over-capacity in power generation in the UK, but a number of older units will be decommissioned in the next few years.

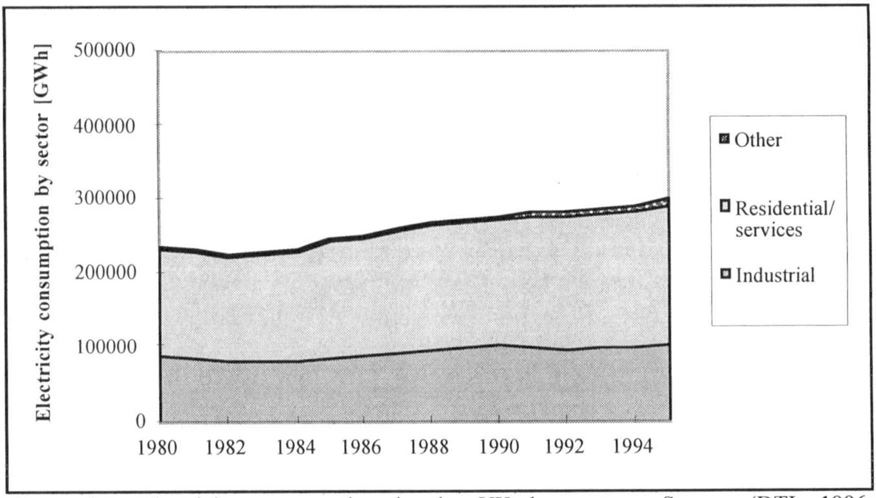

Figure 8.3: Electricity consumption in the UK by sectors, Source: (DTI, 1996; ENERDATA, 1996)

[181] This has led e.g. to fears that gas - a very valuable fuel - would be wasted in power generation over the next few decades. In fact, if 20% of power generation was based on natural gas, the British reserves would be depleted by 2030.

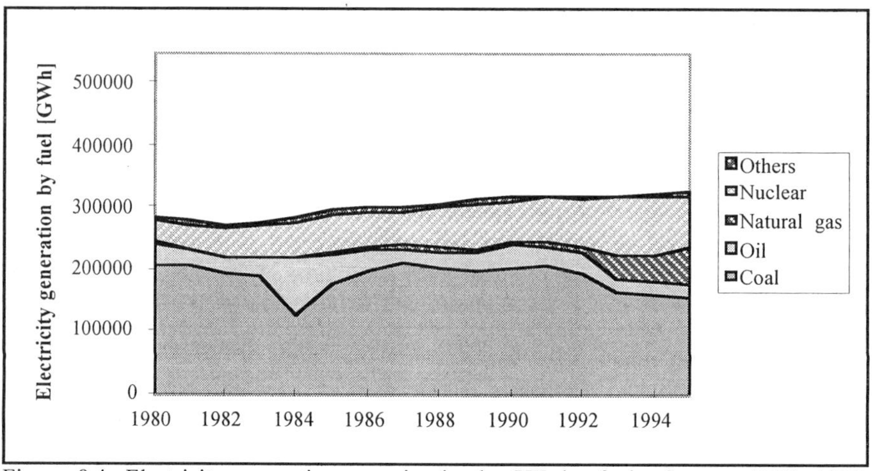

Figure 8.4: Electricity generation capacity in the UK by fuels, Source: (DTI, 1996; ENERDATA, 1996)

Power generation is still highly centralised in the UK, but the old coal-fired power plants, which will presumably be taken off stream in the next ten years, open up an opportunity to introduce more decentralised technologies. The extent of decentralised power generation is of particular interest because it can serve as an indicator for the possibilities to set-up CHP. A situation of highly centralised power supply in combination with a highly decentralised heat supply structure is rather detrimental to most CHP-applications.

Independent and auto-production of electricity

Usually, independent power production is an indicator of the level of liberalisation in a supply system. The figures for the UK show that since the early Eighties power production of companies which are not committed to electricity supply as their main business has remained fairly stable, namely in the order of 6-7%. Privatisation and liberalisation have until now changed very little in this respect.

8.4.1.2 Diffusion of CHP in the UK

The analysis of recent diffusion patterns of CHP is based on available statistical information sources, the CHP data-base of the Office of Electricity Generation (OFFER) and some international publications which consider the case of the UK. Cogeneration had been on a constant decline in the UK until 1987/88 when the lowest power generation capacity was reached of about 1800 $MW_{el.}$,[182] corre-

[182] This decline can be found for independent power production in general, not only for CHP. It implies a concentration trend in the power generation market.

sponding to less than 3% of overall capacity in the UK.[183] There are no other statistical data available on CHP than for the years given in Figure 8.5. Since 1988, i.e. already since before privatisation, capacity has nearly doubled, though starting from a very low level. The latest available statistical data for 1996 indicate a total CHP-capacity of 3560 $MW_{el.}$ (5% of overall capacity). This capacity is distributed over about 1350 CHP-sites, having thus only a minor decentralisation effect. Even if the data on units below 100 $kW_{el.}$ may be incomplete, the contribution of small-scale CHP to power generation is still very low but it represents the vast majority of sites (more than 1100). Expectations for the next years indicate a continuation of the present trend, i.e. the installation of around hundred new units per year. The years 1995 and 1996 were not particularly good years for CHP, but in 1996 a number of larger industrial projects were still under way so that in 1997 capacity is going to rise further.

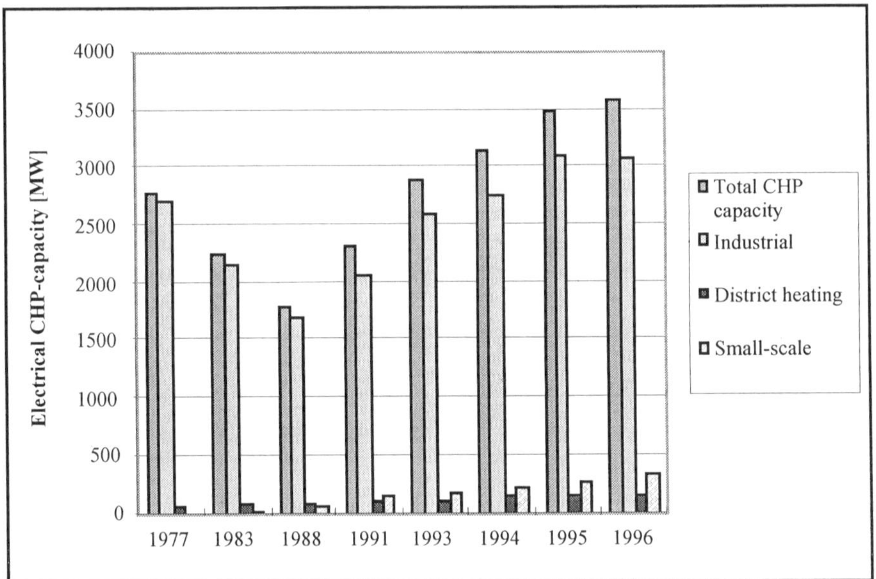

Figure 8.5: The diffusion of cogeneration in the UK (in $MW_{el.}$), Source: (DTI, 1997), data prior to 1991 is not fully comparable to the later survey years due to different standards for data collection.

In order to give an idea of the distribution of applications, the capacities for different sectors are given in Appendix 3. Chemical industry still represents a major share of capacity, and also refineries. The high share of other than the specified types of industrial applications from 1993 onwards is due to the inclusion of the

---

183 Capacity is used here as an indicator for diffusion, but it should be taken into account that an enormous overcapacity has been built up in the UK since privatisation, corresponding to nearly 50% of maximum power load.

privatised utility companies' capacity. The data also show that there are hardly any large-scale district heating hardly in the UK. However, a limited number of projects is currently being reconsidered. In general, the re-introduction of district heating is difficult because there are already extensive gas networks in British cities. On the other hand, there is a mushrooming of decentralised small-scale power plants. Object heating and group heating schemes have been installed since reliable small-scale technology became available. They help to avoid the construction of complex heat piping networks, while having all the advantages of CHP and representing a complementary rather than a competing system to gas supply networks. The main applications can be found in hotels, in the health and education services, on sewage, water treatment and landfill sites, in private or public leisure centres, in residential buildings and large single estates like public buildings. The lion's share (75%) of the nearly 950 buildings with CHP is installed in hotels, in the health system, and in the leisure sector (DTI, 1996).

The statistics thus show a clear growth of both small-scale and industrial CHP-plants, but no take-up of traditional district heating. The role and the mechanisms behind these diffusion patterns are going to be studied in the next sections.

8.4.1.3 Future potential

Any discussion of the future potential for CHP needs to be differentiated with regard to its technical, economic and institutional level. Technically, the limits are in the area of what has been achieved in some other countries like the Netherlands and Denmark, where more than 30% of electric capacity is based on CHP. For most other European countries (including the UK) this is at present an irrelevant order of magnitude because their CHP-capacities are far below this level.

In earlier studies of the future CHP-potential in the UK (DoE, 1979; DoE, 1984; DoE, 1988; DoE, 1990), the economic potential for the different application areas was assessed rather pessimistically. For example, still in 1988 an additional capacity increase of only about 200-500 $MW_{el.}$ was expected over the next years. These expectations changed with liberalisation, and this is highlighted by the CHP-target of 5000 $MW_{el.}$ for the year 2000 which is regarded to be quite realistic. Beyond that horizon, the perspectives are unclear, but a share of 20-25 % of power generation is regarded to be achievable in the course of the next thirty years. The expectations for the nearer future within the liberalised framework are more modest, pointing to a slow capacity increase only. The longer-term perspectives are unclear and highly dependent on regulatory and organisational conditions.

8.4.2 Economic structures and regulatory framework

8.4.2.1 Regulation of the British ESI - The transformation of a monopolistic structure

The year 1979 represents a good starting point for this investigation because it is not only marked by the beginning of Conservative government, but also by the publication of the 'Marshall-Report' on the future possibilities for CHP and DH in the UK. Since then, a revolutionary transformation of the British energy supply system has taken place, breaking up the monopolistic structure of the old Central Electricity Generating Boards (CEGB), of the Regional Boards and of British Gas.

A first attempt to introduce more competition in electricity supply was made with the Energy Act 1983. Although privatisation of the Electricity Supply Industry (ESI) in 1990 may have been the most influential single decision in the last decades, the Energy Act had already opened up first possibilities for independent power production, because it obliged the - then - electricity boards to accept power sales from independent producers. Before that time, it was nearly impossible to set up independent generation at all. After 1983, the terms of exchange between utility company and generator needed to be negotiated individually, and some first attempts were made to establish new CHP-plants which not only generated power for a site's own needs but also exported power to the grid. However, although the boards were obliged to accept independently generated power imports, they maintained their role of 'gatekeepers'. It enabled them to prevent independent generation, because they still controlled the technical and economic grid conditions: there was no clear specification of what was meant by and 'adequate' remuneration of exported power as stated in the Energy Act.

The privatisation of British Gas in 1986 was the next important event for the energy supply sector, because gas supply plays a major role for CHP. While British Gas had restricted itself to gas supply until then, it has developed into an important player with respect to cogeneration. It is interested in CHP because it represents a potential market for gas sales.

After all, the experiences with the Energy Act 1983 were rather disappointing because the Boards maintained the possibility to obstruct the new regulations by setting up a variety of technical and economic barriers. In 1989 the decision was finally made to reorganise and liberalise the entire Electricity Supply Industry.

The Electricity Act was accepted in 1989 and became effective on 1 March 1990. Its major elements can be briefly summarised as follows:

- The Central Electricity Generating Board (CEGB) in England and Wales was broken up and its main generating assets were transferred to the new private companies National Power and Power Gen. In Scotland, the situation remained roughly as it had been before, i.e. the two major electricity companies Scottish Power and Scottish Hydro continued to operate as vertically integrated utility companies.

- Privatisation of the major companies in the ESI, with the exception of the nuclear industry.
- Separation of generation, transmission and distribution. In contrast to the two main generating companies, grid operation remained under the control of the public National Grid Company (NGC). The regional boards were privatised and become the Regional Electricity Companies (RECs).
- The Electricity Pool was set up as an artificial market for bulk generation. Based on half-hourly bids for the next day, the pool prices for these intervals are determined and thereby also the operation times of the different plants. For power generation outside the pool prices are negotiated individually between generator and supply company. The RECs power prices for franchise customers continue to be subjected to price control by OFFER.
- The Office of Electricity Generation (OFFER) was founded as central regulatory body under the responsibility of the Dept. of Energy, and later of the Dept. of Trade and Industry. The competencies of the regulator are far-reaching, and he is formally independent of government. His main functions are to guarantee competition in the ESI and to protect the consumers. In practice, he plays a very important role for all questions of licensing and regulatory control. Although he has not intervened very often, his decisions had a major impact on the ESI. Especially the price-cap, introduced in 1994, needs to be mentioned and also the obligation of National Power and Power Gen to sell about 10% of their capacity (6 $GW_{el.}$) to independent generators in order to reduce their market power.
- Liberalisation in generation and supply is being introduced in three major steps. From 1990 onwards customers beyond 1 MW electrical capacity demand can choose their electricity supplier, and from April 1994 also customers between 1 MW and 100 kW (competitive market). Below that limit, customers are bound to buy from their REC until 1998 (franchise market) when all customers will be able to choose their supplier. In addition to this competition in supply, there is also competition in generation through the pool system and through direct contracts between power generators and large users.
- The only specific consideration of CHP in the entire regulatory set-up of liberalisation consists of the regulator's duty to monitor the diffusion of CHP-technology in the UK.

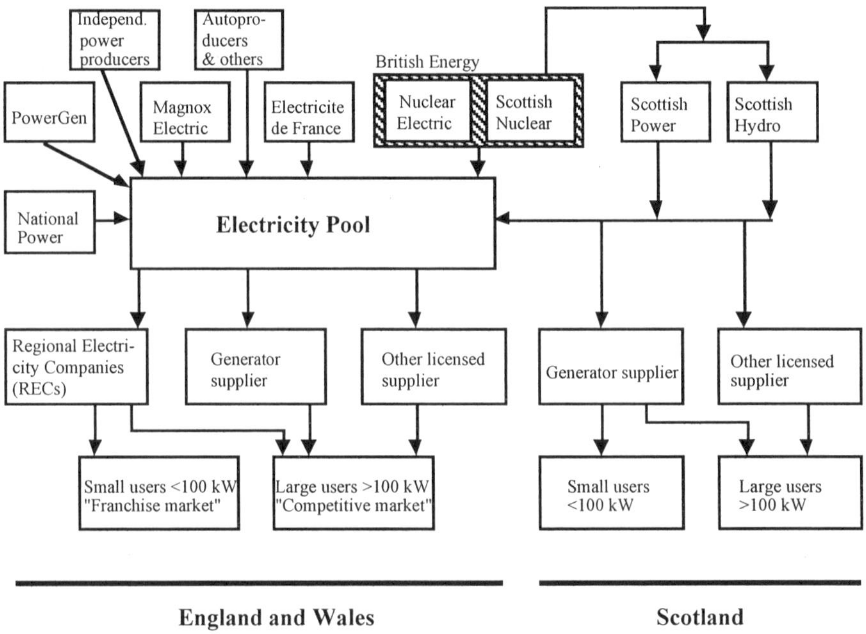

Figure 8.6: The organisation of the ESI in the UK, adapted from (DTI, 1996)

8.4.2.2 The reality of privatisation

In practice, this regulatory set-up has led to the establishment of a set of actors in the ESI which can be grouped as follows:

- Major generating companies: Power Gen and National Power in England and Wales, and Scottish Hydro and Scottish Power in Scotland;
- Grid operator: National Grid Company (NGC);
- Regional Electricity Companies: 12 Regional Electricity Companies (RECs);
- Independent Power Producers, often in partnerships with RECs;
- Service providers like contract energy management (CEM) companies and - to a certain extent - CHP-manufacturer/packager;
- Regulators: Office of Electricity Regulation (OFFER), Office of Gas Supply (OFGAS), Her Majesty's Inspectorate of Pollution (HMIP);
- Government departments, especially Dept. of Trade and Industry and Dept. of Environment;
- Energy service end-users.

Apart from the large generators and Regional Electricity Companies, a number of private companies are now participating in the pool system. They can directly buy electricity from the pool and sell their excess power to it. However, the majority of smaller generators continues to deal with their local REC, although they are not obliged to do this any more.

Several competitive elements in energy-related services have been introduced. Direct contracts between customers and generators can be concluded, supply licences for different areas can be granted to a company, and a variety of generation and supply services can be offered by any private company in the sector. As a consequence, new types of companies have emerged: independent power producers and different kinds of energy service companies.

Although privatisation and liberalisation was initially planned as a system with vertically separated generation, transmission and supply, the situation has changed in the meantime because generators and suppliers have moved into each others' businesses. Moreover, also a trend towards horizontal integration can be observed. Most RECs and generators have set up gas divisions or subsidiaries, and plan to move into other utility activities such as water supply or telecommunications. While on the one hand the widespread availability of gas networks prevents the construction of competing district heating networks, it offers the opportunity to combine gas and electricity supply by CHP on industrial sites or in large buildings.

A major shortcoming of the British energy supply system with regard to the requirements of CHP consists of the absence of strong local energy utility companies. They would be best placed to exploit the benefits of many non-industrial CHP-applications. This deficit is clearly related to the weak competencies of the local political level in general. In theory, privatisation facilitated the set-up of utility companies by local authorities, but first of all they would have to compete with any other supply company, and secondly they would have to dispose of substantial financial resources during the start-up phase. In practice, there are only some sporadic attempts of local authorities to become more active in energy supply, mainly by those local authorities which already played the role of pacemakers in the DH-debates of the Seventies and early Eighties.

For CHP, the most important element of this restructuring process of the ESI consists of the emergence of a new type of companies: Integrated Energy Services Companies (IESC). They could bridge the gap between final service users on the one hand and equipment suppliers and power generators on the other. Privatisation put an end to the strict organisational separation of energy supply chains, offering the opportunity to integrate generation and supply of different energy services in one hand and to exploit economies of scope by these new companies. Before that time, the aim of the different supply chains consisted of optimising their individual production, mainly by exploiting economies of scale. There were hardly any incentives to inter-connect parallel organisations and exploit economies of scope. In other words, there was competition and not co-operation between the supply chains. Finally, all major companies in the sector have started to move in the direction of horizontal integration: RECs, large generators and private contract energy management companies. This trend towards service integration has met - as will be shown later - with many industrial energy users' attempts to outsource their utility services, especially for those services which have become increasingly sophisticated and thus require a higher level of specialised knowledge.

8.4.2.3 *Electricity and gas prices*

Liberalisation also had a major impact with regard to the determination of power prices. The bidding approach of the pool system led to very unstable generation prices with very high fluctuations during the day. Generators who do not sell to the pool but to a REC are confronted with the problem that the RECs are no longer obliged to purchase independently produced power - in contrast to the situation under the Energy Act 1983. These agreements are now fully determined through bilateral negotiations. Distribution prices in the franchise segment of the market are still subjected to price controls by OFFER, and since 1994 even to a price cap.[184] In the liberalised market segment beyond 100 kW prices are negotiated individually between customers and supplier, subject only to monopoly control, but not to price control. In general, more differentiated pricing regimes seem to become common practice, but confidentiality about these individual negotiations makes it difficult to draw any specific conclusions. With regard to CHP, this flexibilisation of tariffs is reflected in a large variety of arrangements for back-up tariffs, export tariffs, connection charges, etc., and in the sophisticated rules applied to adjust these prices to changing economic conditions. However, in most of the cases studied, the tariffs are coupled in some way or another to the pool price, and differentiated at least for peak and night times.

With respect to electricity and gas prices (Figure 8.7) which are key determinants of the economic viability of CHP, it is important to note that a reduction of electricity prices was one of the main aims and expectations of liberalisation, and it has been enforced in several cases by the regulator. While the figures shows that industrial gas prices have dropped on average, it hides that this trend does not hold for interruptible gas (which is the most common type of gas used for CHP).[185] Overall, the price differential between electricity and gas has decreased and had a negative impact on the economic viability of several CHP-projects.

Within the liberalised regulatory framework, it is evident that investments and economic operations of the firms can only be monitored to a limited extent by the regulator; the ESI is not a publicly owned industry any more. Investment control is exercised by the regulator only if he regards competition in the electricity markets to be in danger. Similarly, monopoly power is monitored, and in case of abuse of

---

184 This price cap is based on the so-called 'RPI-X formula'. It means that the maximum average permitted supply charge may not increase more than the rate of inflation (RPI) less a fixed amount X. The formula has been complemented later on by adding a term Y to cover certain business costs of the RECs which are out of their own control, like e.g. costs for certain levies and the administration of the pool.

185 The two main types of gas supply in the UK are interruptible and non-interruptible gas. Non-interruptible gas has become cheaper since liberalisation, but less so interruptible gas. The price differential between the two has thus decreased. In addition, while in the past interruptible gas was hardly ever interrupted in practice, the increase of gas demand has led to interruption problems during peak periods in winter. In other words, both the price development and the quality development of interruptible gas has been detrimental for CHP.

market power, the companies accused can be referred to the Monopoly and Mergers Commission (MMC).

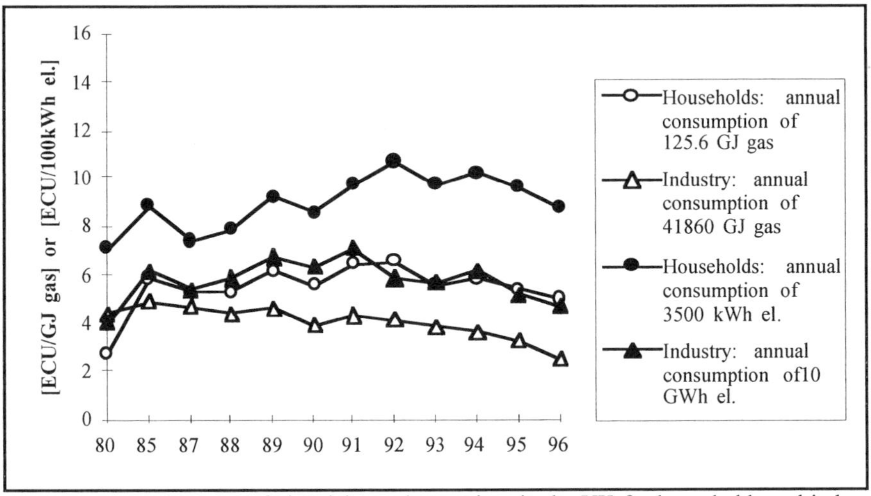

Figure 8.7: Development of electricity and gas prices in the UK for households and industry, Source: (EUROSTAT, 1997)

Some of the intentions of liberalisation have been blocked by the specific regulatory conditions in individual sectors. For example, the financing regimes for local authorities and the health services make it difficult to realise economically interesting projects for which a major initial investment is required.

Not only the electricity side of the CHP-business has undergone liberalisation, also the gas side. Privatisation of British Gas in 1986 and liberalisation of the gas market have led to the emergence of several new gas suppliers, many of which are now controlled by electricity supply companies.

Overall, the new framework has opened up new opportunities for CHP which were blocked before, simply for the reason that independent generation became possible after all and was no longer hindered by formal and informal regulatory barriers. However, new obstacles have emerged due to the complexity and internal logic of the framework itself. These issues will be dealt with in more detail from the users' perspective (Section 8.5.3).

8.4.2.4 Environmental regulation

Nine basic principles can be identified on which the implementation of environmental regulation is based in the UK. These are (Foster, Cooper, & McCulloch, 1994):

- Prevention is better than cure;
- The polluter-pays principle;

- The precautionary principle;
- Integrated pollution control of different polluted media;
- Consideration of critical loads for specific environments;
- Preference for economic instruments and market mechanisms
- Recognition of the right of the public to know about environmental agreements;
- Orientation towards sustainable development;
- Application of Best Available Technologies Not Entailing Excessive Costs (BATNEEC).

While providing a rough orientation, they tend to be interpreted flexibly in practice. In general, the regulatory authorities do not attempt to dictate or prescribe specific regulations and limits, but rather try to achieve the use of the best solution which can be achieved at reasonable cost under the specific conditions of a site.

The BATNEEC-principle as applied by HMIP has been of major practical importance for cogeneration projects. It has the advantage of being formulated in an open and flexible way, thus allowing to take measures which are well adapted to a site's conditions and problems. On the other hand, it does not provide any reliable guideline for planners and investors who can be suddenly confronted with new regulatory requirements. Very much depends on the success of the negotiations with HMIP and the local environmental authorities. This implies also that there is no inherent incentive built into the system which would intrinsically stimulate environmental innovation, and in fact most environmental upgrades of CHP have come from abroad. In other words, BATNEEC turned out to be a very flexible method of lagging behind the standards set elsewhere.

The formal responsibilities for environmental regulation of combustion plants are split between local authorities for small-scale applications and HMIP for larger scale plants above 50 MW of heat output. As a result of the flexibility of the regulatory rules, their application by the local authorities is rather generous, depending on local industrial policies and the availability of staff for regulatory control.

8.4.2.5 Technology supply side

In the early Eighties, the technology supply side of energy technologies was - as in many other countries - a nationally dominated industry, built on close links between the energy supply industry and the manufacturers.

The re-orientation of the energy utility and service companies after liberalisation has been paralleled by an extension of the activities of traditional energy system and component manufacturers. They extended strongly the range of their service, maintenance and financing options. A number of new companies were founded which now cover a broader range of services beyond technology supply, especially in the field of packaged small-scale CHP-systems. Some of these ‘first-movers' had to pay the price for gaining first experiences with CHP and were taken over after a few years by one of the utility companies. In any case, the close symbiotic relationship between energy equipment and energy supply industries has undergone massive changes, driven by restructuring trends on both sides.

In terms of CHP, there was hardly any specialised CHP-supplier in the UK in the early Eighties. The few large industrial CHP schemes were all individually designed, mostly by the industrial operator himself or by an engineering company on his behalf, who integrated the different components on site. In the mid-Eighties the first specialised CHP-suppliers entered the market, at first in the small-scale segment. These small-scale packagers came up with several new innovative ideas on how to provide CHP-based service solutions.

While the large-scale area of equipment manufacturers remained rather stable in terms of industry structure (mainly due to the fact that large-scale plants could not be packaged) the small-scale industry was characterised by a mushrooming of companies before and after liberalisation. Many of the newcomers, but also several of the first movers from the mid-Eighties went out of business later on in the stabilisation phase of the early Nineties. At the general level, this development can be sketched by an industry life-cycle curve which was triggered by the service innovations made in the late Eighties (see Figure 8.8).

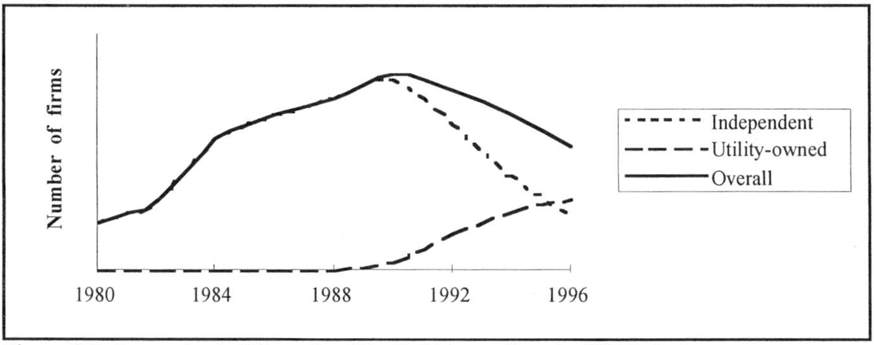

Figure 8.8: Approximate life-cycle of the British small-scale CHP industry

After privatisation and liberalisation, the interest of the established players of the ESI in CHP rose sharply. Due to the fierce competition in the small-scale CHP market, and the need for capital, many small-scale packagers were taken over by RECs.[186] The large generators National Power and Power Gen built up their own CHP business activities, mainly in the larger-scale area beyond 5 MW.

186 In 1996, the five major small-scale packagers were subsidiaries of different utility companies. Combined Power Systems was partly owned by NORWEB, Northern Electricity and London Electricity, Nedalo UK by Eastern Electricity, Lincoln Green Energy (former Applied Energy Systems) by East Midlands Electricity, Cogen Systems by Midlands Electricity and Petbow Cogeneration by British Gas.

8.4.3 Structures for the provision of technological information

Cogeneration has long suffered from a bad reputation, especially due to the poor performance of the few early district heating schemes. Low reliability and bad economics were the main prejudices against it. Often this was just due to mistakes made during plant design, which is far more complex than the design of heat- or power-only systems.[187] The supply of information on new and improved equipment as well as on the environmental and efficiency advantages of CHP was crucial for making potential investors aware of this option and its benefits.

The structures for the generation and supply of information on cogeneration in the UK are dominated by five types of actors:

- Government agencies: Energy Efficiency Office (EEO), Energy Technology Support Unit (ETSU), Building Research Establishment (BRECSU);
- Technical associations, such as the Watts' Committee on Energy, the Institute of Mechanical Engineers, and others;
- Industrial associations, especially the Combined Heat and Power Association (CHPA), the Association of Industrial Electricity Producers (AIEP), and the Major Energy Users' Council (MEUC);
- Research institutions: British Gas and power utilities' research facilities, manufacturers, universities;
- Networks and co-operations between manufacturers and users of CHP.

Until the middle of the 1980s, only very sporadic activities in support of CHP can be identified. Then, the Combined Heat and Power Association (CHPA) and especially the Energy Efficiency Office (EEO, now part of Dept. of Environment) started to become more active in promoting CHP by supplying information on successful applications and by giving some limited financial support to innovative applications. The EEO represented a kind of key node in the network of information supply, mainly because it could draw on the competence and independence of the Energy Technology Support Unit (ETSU)[188] and the Building Research Establishment (BRECSU). The Combined Heat and Power Association, which is the successor of the former District Heating Association, strongly reinforced its promotional activities, supported by a growing number of members from the CHP-industries, and since recently also from utility companies and users.

In addition, the trend towards energy service integration changed the information links between service users and technology/service suppliers. The installation and

---

[187] One of the typical mistakes made in the design of the first generation of small CHP-plants was to size them according to the peak load of a site, i.e. according to a procedure which is common for boilers. This mistake was not only environmentally but also economically disastrous because the fluctuating loads made it impossible to run the plants efficiently.

[188] ETSU is part of the Atomic Energy Agency, and its conducive stance with regard to energy efficiency and CHP came as a surprise for many observers, but it also gave them a high level of credibility.

operation of a CHP-system requires detailed information about heat and power loads of a site, but this type of information is only provided if a considerable level of trust exists between the partners. Hence, especially in the late Eighties when several new types of applications were customised, very close relationships had to be established in order to gain and build up the necessary knowledge about CHP-systems, their design requirements and operation. The information networks which emerged in the late Eighties enabled a wide range of learning processes to take place among their members, involving government institutions, technology suppliers, service providers, and service users.

Neither research institutes nor the technical associations seem to have played a major role in the information networks related to CHP, even if some of the innovative CHP-packagers were spin-offs from university research work. The limited importance of basic technical research for the success of CHP is probably major reason for this, but also the conservatism of the technical associations. It was repeatedly pointed to the problem that engineers which were trained at a time when CHP was technically still quite unreliable could hardly be convinced of its benefits when it became a really good alternative. In the very early days of cogeneration, i.e. in the first decades of this century, the situation was different in this respect, because heat engineering associations were at the forefront of the promotional activities for district heating.

8.4.4 Political and regulatory structures

8.4.4.1 CHP and the political decision structures

Beyond the question which actors actually had a decisive influence on the political decision-making processes in relation to CHP, the mechanisms and procedures through which this influence is exerted are of utmost importance for understanding the role of political structures. Before privatisation, the Dept. of Energy and the Electricity Supply Industry were closely linked to each other, enabling the ESI to shape most relevant political decisions due to its expertise on energy-related issues. The Dept. of Energy was mainly occupied with 'running the ESI', and the symbiotic relationship between the two prevented any structural development.

A major study on the possibilities of city-wide CHP/DH schemes was commissioned in the late Seventies in response to the first oil price crisis,[189] but when it was finished energy policy in the UK was already concerned with other issues: the nuclear debate, North Sea Oil, and some first proposals for liberalisation. CHP became just a 'non-subject'. None of the major players in British energy policy at the time was a powerful force in support of CHP. The Central Electricity Generating Board and its members, the regional power supply boards, did not have any interest in decentralised power generation, because it would have weakened their influential position. Due to the horizontal separation of energy supply chains, they could not even potentially benefit from the economies of scope offered by CHP.

189 These investigations ended with the so-called 'Marshall Report' (DoE, 1979).

Heat and gas supply were not their businesses just as power supply was not the business of British Gas. The strong ties between the established energy supply industries and government excluded other potentially interested actors, such as e.g. local authorities, universities, hospitals or industry, from the major decision-making processes. However, in the end industrial concerns about an inefficient power supply systems helped to build up support for liberalisation in the Eighties.

Given the strong influence of the ESI, the decisions of 1983 and especially of 1989 represented unexpected ruptures. The Conservative government managed to carry on with its privatisation and liberalisation plans against the resistance of established power constellations. The strong ideological background as well as the lamentable state and efficiency of British power generation are complementary factors which help to explain these developments.[190]

The interface between policy making and energy supply system was restructured by privatisation, and the focus of interest moved to OFFER as the new regulatory body. It is in charge of most decisions regarding the regulatory framework for CHP and the privatised electricity supply industry in general. Its decisions in the last years have had a considerable impact on the uncertainties and hence on the investment decisions for power generation. In contrast to this, government has been very reluctant to interfere in the energy markets; it has in fact given up an active energy policy except for the nuclear and oil industries.[191] The mutual symbiosis between government and the ESI ceased to exist.

As part of its attempts to raise the importance of energy efficiency on the political agenda, cogeneration has in the last years been supported by the Dept. of Environment. It is now responsible for most government activities in support of CHP, mainly by means of information provision and finance. However, its influence is limited due to the responsibility of the Dept. of Trade and Industry for the regulation of the ESI.

Obviously, there are also a number of other decision arenas and structures which play an important role for energy policy-making. First of all the Parliament's Select Committees need to be mentioned. Especially during the privatisation phase, the Select Committee on Energy commented government decisions very critically, but with the dissolution of the Dept. of Energy in 1991, energy issues became part of the Trade and Industry Committee. The hearings of the Select Committees are also one of the main platforms for the major interests groups to introduce their positions into the political process. Their position statements are collected and fed into requests to the different government departments. External advice also plays an important role in this context. When looking at the participants of the major CHP-related sessions of the Select Committees, it becomes apparent that industrial

---

190 In the mid-eighties, the performance of British power stations was among the poorest in Europe, both with regard to system failures and with regard to fuel efficiency.

191 The official position of government is that since 1990 there is no need for any energy or electricity policy. It is regarded to be sufficient to regulate a market-based system to protect the consumers and guarantee the efficient operation of the market. Although the regulator is formally independent, informal agreements between the regulator and government are common before any major decisions are made.

interests are strongly represented, whereas interests in favour of energy efficiency and the environment are not very prominently considered (HoC, 1991; HoC, 1992; HoC, 1993) This, however, reflects the comparatively low importance of environmental issues on the British political agenda.

In spite of the possibilities of formal and - perhaps even more important - of informal lobbying in Parliament and Departments, British energy policy was very hard to access from outside in the past. Due to the close links with the experts from the ESI until privatisation, no need was seen for major external advice. From a structural point of view, it appears to be surprising that privatisation could actually be realised despite the 'lock-in' between Dept. of Energy and the ESI and their isolation from other policy arenas.

It has already been recognised in earlier studies that the competencies of the local political level are of major importance for the success of decentralised energy supply technologies and especially of CHP (Gruber et al., 1993; Jochem & al., 1986). The situation in this respect has not changed very much in the UK, rather on the contrary. The financial situation of the local authorities is very critical, and the objective of government to stimulate more private-public partnerships is confronted with major barriers.[192] As long as local authorities are not given more financial and planning autonomy, decentralised energy supply systems with public involvement can be realised only in exceptional cases.

8.4.4.2 The control of policy-making structures by policy-makers

Government and regulator are reluctant to interfere in specific decisions in the ESI, which means that structural policies become particularly important for political control. Privatisation and liberalisation are the best examples of this, but also the way in which the interactions between government and energy supply sector are organised. In other words, structural policies are also applied to the political process itself as a means of political auto-control.

Privatisation and liberalisation had a strong impact on these interfaces between politics and economy. In the UK there has been a transition from an interventionist governance approach to a more passive one. The interesting point in this context is that it was indeed possible to break up the existing structures, to put an end to a highly problematic involvement of government in a key industrial sector, and to overcome the lock-in between political and economic realm. At the same time, it was decided to manage policy-making structures in a way which allows for more open outcomes of the policy-making process. The entire system of political input and output structures changed, making it more difficult for single partial interests to dominate energy policy.

---

192 These barriers are often of an administrative nature. Till very recently it was very difficult for local authorities to stretch investment over longer time periods than a single budget year, but this problem was addressed in the initiatives of UK government to stimulate private-public partnerships.

8.5 Behavioural aspects: Decisions, interactions and motivations

This section investigates the processes and factors which affect the decisions at actor level. The discussions of the main types of decisions (innovation, adoption and public policies) are divided into a section on the decisions themselves, and a section on the underlying mechanisms, interactions, motivation and perceptions of actors. Obviously, this division involves the risk of duplication because the motivations of actors are closely related to the decisions made but on the other hand it makes the interdependence between these two aspects clearer .

8.5.1 Political realm

8.5.1.1 Decisions

The decisions made at the political level can be categorised in two different ways, namely either according to the target area (political self-control, technology policy, energy policy, environmental policy, etc.), or according to the type of policy tool used (regulation, market-based instruments, financial incentives etc.). In the following discussion the political decisions and policies are structured along different target areas.

Measures of political self-control

Apart from its impacts on the ESI, the liberalisation and privatisation decision of 1990 fulfilled an important function of political auto-control. It reduced the density of interactions between industry and interest groups on the one hand and government on the other. Also the competencies of government to interfere in the energy sector were drastically reduced, thereby re-directing the activities of interest groups to other actors, especially the regulators.

Although it had been recognised that a decentralisation of competencies would be a helpful measure to facilitate the wider use of CHP at local level, rather the contrary was done in the UK, namely a reduction of the financial and planning power of local authorities.

Energy policy regulation

The key decision of British government was the liberalisation and privatisation of the ESI (see above). In addition to the structural elements of energy liberalisation, several more specific policy and regulatory measures were implemented in parallel which have had an impact on CHP. Many of them were controversially discussed in the British policy debates.

In order to complement the new regulatory framework, a number of specific regulations and programmes have been introduced in the last years. In 1990, British government defined a target of 4000 $MW_{el.}$ of CHP-capacity for the year 2000

(DoEnv, 1990). Later on, in 1993, as an outcome of the Rio-Conference and the fast progress towards reaching the target, it was raised to 5000 $Mw_{el.}$. Moreover, CHP started to figure prominently as an element of the UK's CO_2-strategy. This represented a major shift compared with the few concessions which the promoters of CHP had been able to achieve during the negotiation phase before liberalisation. At that time, the regulator's only duty in relation to CHP was to monitor its diffusion by setting up a plant data-base.

In 1992, the Non-Fossil Fuel Obligation (NFFO) was introduced as a support for non-fossil generation technologies. A levy of initially 11 % was raised on electricity bills of franchise customers.[193] While 95% of the funds were used for financing the decommissioning of nuclear power plants, 5% went into renewables projects. Some of them were based on CHP and represented a very specific niche. In the fourth round of the NFFO in 1997, also waste-to-energy schemes were included, so that some more CHP-plants could benefit from the levy.

While investment and financial control of the ESI was based on clearly defined criteria in the nationalised industry (with returns on investment of initially 4%, later of 8%), the privatised companies operate under a market regime which gives less influence to the regulator. However, in the first years of regulation a number of measures were taken to reduce especially the influence of the two major generating companies National Power and PowerGen. The rising electricity prices had led to complaints about their manipulating the pool prices by abusing their market power (the two companies still dominated 60% of the market). In 1993 they were referred to the Monopolies and Mergers Commission (MMC), which decided in 1994 that PowerGen and National Power had to sell or decommission 6 GW of their generation capacity. In addition, a price cap was introduced on pool power prices.

In the last years, several regulations have been modified in order to facilitate autoproduction and cogeneration. For example, the lower limit beyond which generation licences are required was raised from 50 MW to 100 MW, pre-existing licences have been extended regularly, connection regulations for generating plants have been relaxed, self-supply of generators in the pool does not need to be traded through the pool any more, etc. All these measures reduced administrative and technical burdens on all kinds of self-generation.

Environmental policy

The basic principles of environmental regulations have been discussed above. Their specifications were implemented during the early 1990s (Foster et al., 1994), based on the following main policy actions:

- The White Paper "This common inheritance" 1990 and the yearly follow-up reports describe the specific objectives of British environmental policy (DoEnv, 1990; DoEnv, 1991). Since then, the importance of energy efficiency

193 Later on the levy was reduced to 10%, and in the course of the reorganisation of the nuclear industry to about 4%.

for achieving CO_2-targets and for reducing energy consumption has been recognised in British government policy.

- The Environmental Protection Act 1990 which specifies the regulations for water, air, and waste on land.
- The Clean Air Act 1993, which is geared towards industrial production and describes limits for various emissions. It also specifies the BATNEEC-principle for power generating plants. For examples, an authorisation by HMIP is prescribed for plants above 50 MW and by the local authorities for smaller plants.
- Internalisation of external costs: Value added tax (VAT) on fuel was introduced in 1995, but this decision was not really driven by environmental concerns. Despite the emphasis given to market forces, the key market instrument of environmental taxes has not yet been introduced in the UK.

Obviously, British environmental policy complements measures at European level where most of the guidelines for national legislation are defined. However, regarding emissions, the EC has also defined specific standards in some areas.

Diffusion policy

In terms of support for later phases of technological change, British government has - in contrast to other European countries - refrained from substantial support for CHP. The information requirements of users were regarded as a priority area and addressed by the activities of the Energy Efficiency Office. Written material, case-studies about experiences with CHP, guidelines, workshops etc. were the main lines of activities. They were associated with the different innovation programmes (see below) from which the main results were extracted. This case-study material complemented the general monitoring function of OFFER's CHP-database.

Financial support was limited to very specific areas. In the Eighties, CHP could still benefit in some cases from grants which were intended for the use of indigenous coal, but this did not play a major role once gas took its leading position for CHP. The financial support of the Energy Savings Trust, (EST) was mainly building on contributions from the Office of Gas Regulation (OFGAS).[194] It has been a minor programme of about 2 Mio £ (CHPA et al., 1995) to co-fund about 100 residential schemes since 1992 (51 in 1993). The Energy Saving Trust is also involved in a number of other small programmes to promote energy efficiency.

As mentioned before, the NFFO provided some funding for renewable energy schemes from which CHP-plants could benefit. As the support was granted on the basis of the power capacity of a plant, it was biased in favour of power-only

---

194 The financial basis of this programme was in danger when a new Director General took over responsibilities for OFGAS in 1994. Whereas her predecessor accepted to co-fund the trust, she did not see any justification for charging the gas customers higher prices in order to fund a support programme.

plants. The recent fourth round has been re-formulated in a more CHP-friendly way, and has also included waste-to-energy schemes.

The Energy Management Assistance Scheme (EMAS) is a programme which promotes energy efficiency monitoring by small firms, and it also provides funds for efficiency projects. The costs for the initial monitoring by a consulting firm is meant to raise awareness among small firms for the potential of energy efficiency. Also the so-called Making a Corporate Commitment campaign aims at raising awareness and commitment, here at the board level of companies. Till late 1994, about 1800 organisations subscribed to the campaign (Foster, 1994). Although both programmes are not targeted towards CHP, they should be seen as elements of an energy efficiency diffusion policy, which also includes measures on the generation side.

Innovation policy

Two major innovation programmes have been set up in the UK in recent years to support CHP. The Energy Efficiency Best Practice Programme (since 1989, replacing the Energy Efficiency Demonstration Scheme) aims predominantly at monitoring demonstration projects and at diffusing the experiences made. Financial support is usually limited to the additional costs for monitoring. Since 1994 it has been complemented by an Environmental Technology Best Practice Programme which aims specifically at improving environmental performance, and CHP-schemes can be considered within this programme. The New Practice Programme focuses on new technologies and provides a substantial share of finance. This as well as preceding programmes have been limited in terms of overall amount of money available (Foster, 1994).

Other policy measures

According to (CHPA et al., 1995), there are no other measures to support actively the innovation and diffusion of CHP in the UK. There are no special tariff arrangements nor any kinds of taxation advantages for CHP foreseen in the regulatory and policy framework. An interesting recent initiative concerns the financing possibilities of local authorities. The Private Finance Initiative (PFI) facilitates the set-up of private-public partnerships for energy efficiency investment like Community Heating by providing extra funds to local authorities (DoEnv, 1996).

Programme	Description	Time period
Energy Efficiency Demonstration Scheme	Demonstration of the successful application of modern reliable technologies like gas turbines and spark ignition engines for CHP	before 1989
Energy Efficiency Best Practice	In addition to demonstration and R&D-projects ("New Practice" and "Future Practice"), dissemination activities by case studies, monitoring and information guides ("Good Practice")	1989 - present
Non-Fossil Fuel Obligation (NFFO)	Levy on the electricity bills to finance the decommissioning of nuclear power plants and the set-up of renewables energy projects	1992 - present
Greenhouse Programme	Run by the CHPA on behalf of the EST, provides up to 25% of funding	1992 - present
Making a corporate commitment campaign	Encouragement of organisations to make an effort towards energy efficiency as part of positive environmental management	1993 - present
Energy Management Assistance Scheme (EMAS)	Assistance towards consulting costs by providing grants of 30-70% of costs for identifying efficiency opportunities, efficiency projects and project management.	1994 - present
Environmental Best Practice	Improving environmental performance of UK industry by monitoring exemplary projects for emission reductions	1994 - present
Private Financing Initiative (PFI)	Facilitating public-private partnerships for setting up CHP projects by providing additional budgets to local authorities	1996 - present

Table 8.3: Overview of policy programmes to promote CHP in the UK during the Eighties and Nineties

8.5.1.2 Decision arenas and interactions in policy-making

The link between the ESI and the Department of Energy

Till 1989 the close interaction between Dept. of Energy and the ESI determined energy policy in the UK. The possibilities for interest groups to exert influence on these decision-making processes were very limited. However, the complaints about the operation and performance of energy supply could not be ignored, and neither the critique of the electricity boards' attempts to undermine the intentions of the 1983 Energy Act. Although the Energy Act 1983 made it *in principle* easier to set up self-generation of electricity, technical and economic connection conditions were still defined in a very detrimental way by the boards. The demands for change came thus predominantly from industry and met with the general privatisation and liberalisation philosophy of British government.

Although CHP had been applied in industry for a long time, it is interesting to look at the way in which CHP was perceived, and how it became an issue on the agenda in the course of the time. Since the Marshall-Report in 1979 which had raised interest in some new city-wide DH-schemes ('lead cities') after the first oil price crisis (DoE, 1979), the topic disappeared again from the agenda, at least in terms of active government support. Some of the lead city schemes were continued at a reduced scale by the cities themselves. The first fully independent generation schemes were triggered by the Energy Act 1983, but CHP did not start to raise substantial interest before the mid-Eighties when new generations of gas turbines and small-scale CHP-units were introduced. These new technological options became available in many countries simultaneously, i.e. their emergence can not be interpreted as an outcome of the political decisions of 1983. Since then, CHP has attracted increasing attention in the UK policy-debates, especially since privatisation, when more actors became interested in it. The success of CHP in other European countries, especially in the Netherlands, further improved its reputation.

Government position and ideological motivation

It is hardly possible to understand the reasons and the decision-making processes which led in the end to privatisation without taking into account the ideological driving forces behind government policy. The crucial point is that strong political motivations and power were needed to go ahead with privatisation against the resistance of the established industrial organisations. Although the detailed implementation of privatisation was negotiated between ESI and DoE, the major preparatory work was done independently by DoE alone. This strategy of staying independent of those to be privatised and re-organised was crucial the success of the entire operation. Other examples of privatisation, e.g. the gas industry, show that it is not sufficient just to privatise a large company without actually transforming the structure and the culture of the main actors.

Apart from the ideological element, there were also deep concerns about electricity costs for British industry and its competitiveness, but opinions differ whether the drastic cure of privatisation and liberalisation was really successful in this respect. At the time, there were internal conflicts in government about the desirable extent of privatisation and liberalisation. The structural problems of the British ESI made some kind of reform inevitable, and this was also recognised by the Labour opposition.[195] With hindsight, arguments have been raised again that some degree of long-term energy policy beyond applying market principles may be beneficial and necessary. In other words, although liberalisation is sustained as the

---

195 These structural problems were both of a technological and of an economic nature. For example, it was known that the performance of British ESI as compared to other European countries was rather poor, and the external environmental costs very high - even if they were never quantified.

basic principle for running the ESI, the opinion has started to swing back in favour of a slightly higher degree of intervention.

The departure from an active energy policy and the role of OFFER

According to the official position of British government, energy policy in the UK consists of the reliance on market principles in a private industry. The issue of security of supply which has been on the political agenda for decades is not regarded as a major problem nowadays. An active involvement is deemed necessary only to maintain competition, to protect the consumers and to address negative externalities, especially environmental ones. These aspects are dealt with by regulatory measures, either by electricity regulation for competition and consumer issues, or by environmental regulation to guide technological choices in a more sustainable direction. However, till now environmental regulation in the UK has not been perceived as a constraint on investment decisions, especially for small-scale CHP-applications.[196] Both standards and technological improvements are driven by developments on the continent, where stricter levels apply, which are then introduced sooner or later also in the UK. In other words, the UK lags structurally behind other countries in terms of its environmental policy and regulation.[197] The remaining responsibilities have been shifted to OFFER and HMIP who are now the central foci of the energy and environmental debates in the UK, and thus also the main targets for lobbying.

The role of the Parliament Select Committees

The preceding analysis has already indicated that the possibilities for different interest groups to participate in the policy-making process were rather limited. Before privatisation they could hardly enter the ESI-DoE network, and afterwards the scope for lobbying was rather limited. Interests in favour of CHP could therefore not enter the industrial and political decision arenas before the late Eighties. This lack of platforms for debate gave the Select Committees of Parliament an important function for the articulation of different positions on energy policy issues. Their reports and hearings played a major role for expressing dissent, both of Parliament itself and of interest groups and experts who were invited to make statements on specific subjects.

The lists of participants of the Energy Committees (and later of the Trade and Industry Committees) show the balance of represented interests. Although industrial interest groups dominate, since the late Eighties also environmental and con-

---

196 The most important distinction made by regulation is between plants smaller or larger than 50 MW thermal capacity. Below 50 MW the emission regulations do not represent a real constraint. Anyway, as most CHP-units which were built in the last years are based on gas-fired system, the only critical emission is NOx.

197 An example of this is the BATNEEC-principle (Best Available Technology Not Entailing Excessive Costs) applied by Her Majesty's Inspectorate of Pollution, because what 'best available technology' means is defined by what companies in continental Europe manufacture. British users have to conform to the (very flexible and uncertain) interpretation of the BATNEEC-guidelines by the regulator.

sumer interest groups have entered the debates. Particularly relevant for CHP has also been the consideration of the associations of local authorities as one of the few comparatively powerful proponents of CHP (although also for them CHP was a rather minor topic). However, it needs to be admitted that the effective political influence of the Select Committees was limited. They were able to raise issues and provide material and arguments in the CHP-debate, but beyond questioning Ministers and Secretaries, their impact on shaping the framework for CHP was limited.

Reorganisation of interest groups and the limits to lobbying

Privatisation and liberalisation also had a major impact on the interest groups themselves. The fact that new actors had entered the industry, emerging from the old public companies, altered the traditional conflict lines. An example of this change is the interest of the large generators and the RECs to enter the Association of Independent Electricity Producer (AIEP). This means that the former opponents of the independent generators (e.g. in negotiations about tariffs) then tried to enter their association, because they had also become private companies.

However, industrial electricity producers as potential proponents of CHP are still weakly represented in the political process. Only some large companies interact directly in the political process. The key associations such as the Confederation of British Industry (CBI) voice industrial interests in the political process which are of a more general nature. Although their influence should not be overestimated, the imbalance of interests in favour of the large and powerful industrial organisations certainly affects government positions. The most important industrial associations for CHP-related issues are the Association of Independent Electricity Producers (AIEP) and the Combined Heat and Power Association (CHPA). The AIEP was founded in 1987, i.e. at the same time when the CHPA (which evolved out of the District Heating Association DHA) became more active in the CHP-debate. In other words, also from the side of interest groups, there was not much activity in favour of CHP before privatisation. In the meantime, especially the CHPA has acquired an influential position; it is well integrated in the energy policy debates and co-operates successfully on some issues with the Energy Efficiency Office.[198]

With regard to small-scale CHP the situation is even more difficult because in fact neither CHPA nor AIEP nor the Major Energy Users Council (MEUC) are fully appropriate platforms for them. There is hardly any possibility to articulate their positions of small-scale users in the political process. Neither do technical associations play an important role. Although in some of the Select Committee sessions, scientific experts were invited, the technical dimension of CHP was overshadowed by the more fundamental debates about independent power production and the advantages and disadvantages of liberalisation. Anyway, except among those who actually dealt with it, CHP had a rather bad reputation in the technical communities, though often due to a lack of up-to-date knowledge and information.

---

198 For example, since 1992 the CHPA manages the Greenhouse Programme for the EEO, and they also co-operate informally on other issues.

A major change with regard to lobbying took place in 1991/92, i.e. after the initial phase of privatisation had been finalised. The Department of Energy was dissolved and its remains were integrated into the Dept. of Trade and Industry. Only the Energy Efficiency Office (EEO) became part of the Dept. of Environment. Consequently, one of the major focal points for lobbying disappeared.

Anyway, the impact of lobbying on the diffusion of CHP in the UK was rather limited, at least during the first years after privatisation. Its success was mainly an unintended by-product of wider changes - in contrast e.g. to the situation in other European countries like Denmark and the Netherlands, where governments have been actively promoting cogeneration. In the last few years, lobbying activities have led to a reconsideration of a number of regulations because they clearly put CHP at a disadvantage. While until the late Eighties, CHP was a hardly known alternative in the political debates, it is now at least recognised as a result of successful lobbying, but also due to the prominence and the success of CHP in a number of other European countries.

Addressing controversial topics since 1990

Since privatisation the debates about energy policy and CHP have been restricted to minor adjustments of the regulatory framework. Especially the limited consideration which was given to CHP in the new regulatory framework has been the target of lobbying by CHPA and AIEP, but also of other organisations. They mainly aimed at correcting unjustified disadvantages for CHP and the lack of recognition of its environmental benefits within the regulatory framework at the time. For example, initially all major generating companies needed to get a generation licence from OFFER which entailed a long and costly procedure. In 1995, this limit was raised from 10 to 50 MW and later on to 100 MW, making self-generation much easier. The bias of the NFFO in favour of power-only generation was removed in 1997. Also a number of administrative obstacles concerning connection and financing have been changed.[199] Environmental regulation of CHP-plants was often unclear, both in terms of responsibilities (local authorities or HMIP) and content. For instance, it was unclear whether the emission limits for power or for combustion plants had to be applied to CHP. Due to the experiences gained in the last years, most of these problems have been settled in the meantime, even if there are still differences among local authorities.

8.5.2 Suppliers' decisions on innovation

8.5.2.1 Decisions: Innovation and supply of technology

The Energy Act 1983 opened up first opportunities to develop technological solutions for self-generation and CHP, but the potential users, especially small ones, were hardly willing to operate a fairly complex technology on their own. Often

---

199 For more details on these specific regulatory changes, see (CHPA et al., 1995; DoEnv, 1996).

they did not dispose of the specialised knowledge and competencies to operate CHP properly. Beyond the technical improvements of CHP, additional innovations were thus needed to make it successful. The decisive step forward to overcome this reluctance was an extension of the technology supply companies' services to turn-key and full maintenance solutions. The first moves in this direction were made in the mid-Eighties, but in the meantime many of the companies involved in the manufacturing of CHP-system have started to propose at least full service solutions to energy users. Due to financial constraints, especially of small users, their offers include also financing schemes. These schemes were marketed under names such as Contract Energy Purchase (CEP) or Discount Energy Purchase (DEP). They comprise the finance, operation and maintenance of CHP-schemes on the costumer's site. The customers pays for the electricity purchased, but at a price below the tariffs of his REC. Moreover, he usually gets the by-product heat free of charge. In some of these arrangements, the plant becomes the property of the customer after some years. The customer's potential savings are obviously limited by the costs for the expertise, the external services and the investment risk which is born largely by the service provider. DEP has become very popular since privatisation, even if there is still some reluctance to embark on long-term energy service contracts over ten or even fifteen years because it is hardly possible to calculate the financial risk of such arrangements. Some of the service companies now conclude the vast majority of their contracts on a DEP-basis. While this kind of standardised service solutions was initially limited to small-scale packaged CHP, it has become in the meantime also more common for medium-scale industrial applications (Figure 8.9).

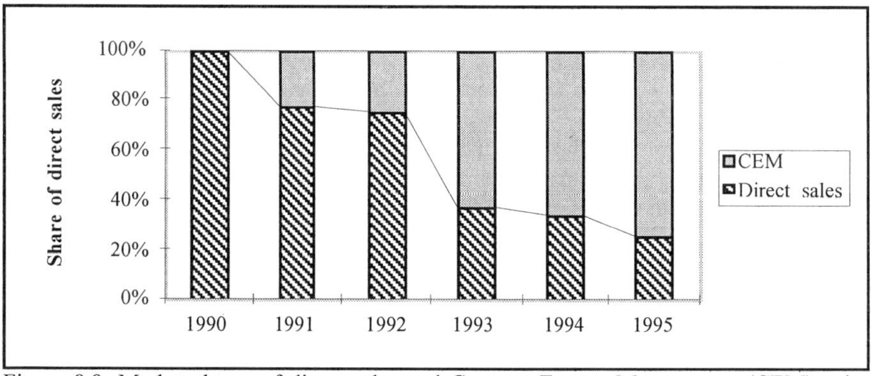

Figure 8.9: Market shares of direct sales and Contract Energy Management (CEM) solutions in the British gas turbine market since liberalisation

Another major change on the technology supply side must be seen in the reduction of R&D-activities by utility companies. Before privatisation, much power engineering research in the UK was done or at least influenced by the utility companies. The CEGB had a major research centre of its own, and it fulfilled a co-ordinating function for the research done by the manufacturing industry. They defined certain standards which had to be met in order to be part of the collaboration network in energy research. Similarly, British Gas had and still has a major research facility but focuses now on technology development for the world market rather than for the domestic energy sector. The large generators and the RECs have substantially reduced their own involvement in research. They argue that in a privatised industry research should be done by the manufacturers, not by the supply companies. Consequently, the innovative impulses coming from the utility companies have become less important since liberalisation (Moncada Paterno Castello, 1996).

8.5.2.2 Decision arenas and interactions on innovation

After the first step towards liberalisation of power generation in 1983, several companies identified the market opportunities of CHP. Especially the field of small-scale reciprocating engines developed very fast, but also gas turbines. The first generation of CHP-engines were just derivatives of available standard engines to which a heat recovery system was added for the exhaust gases and for the jacket cooling water. The deception about the poor performance and reliability of these units which, in addition, were often badly designed for the requirements of the installation sites, confirmed initially the relatively negative reputation of CHP among engineers.[200] This first generation should be regarded as an intensive learning phase for the manufacturers. First of all, there were technical issues to be addressed, because CHP-technology in the early Eighties was not sufficiently stabilised and reliable to represent an acceptable option for most customers. Secondly, apart from these technical aspects, CHP-suppliers had to respond to the non-technical needs of their customers, who were often not only interested in maintenance contracts but also in financing and full energy services. As mentioned before, this has led to a great variety of contractual arrangements for the planning, financing, installation, operation and maintenance of plants on site. The customers only purchases the final electricity and heat services.[201]

A crucial problem for the manufacturers in the Eighties was to understand the needs of their potential customers which could not be easily satisfied by just selling hardware to them. Their new customers did not dispose of the knowledge and skills to operate CHP-systems properly on their own, and there were hardly any operational experiences available with small-scale CHP. Data on load curves of industrial or building sites were generally missing, but essential for a proper design of a scheme. The EEO's Demonstration and Best Practice programmes pro-

---

200 A typical example was the Fiat TOTEM-system, one of the first systems on the market, which was just a converted car engine, and hence not designed to run permanently. Due to a complete lack of service support, failures were frequent and reliability low, creating a bad reputation for small-scale CHP in general.

201 The 'first mover' in this respect was presumably a company called Combined Power Systems (CPS) which was a spin-off of university research in Manchester. Having a strong background in computer control systems, they managed to improve the reliability and reduce the maintenance costs of small-scale CHP by introducing preventive maintenance schemes based on intelligent data analysis. This advanced technology was combined with a full service and discount energy purchase package. In the late eighties and early nineties CPS clearly dominated the British CHP market.

moted a number of smaller projects, not only to improve available technology, but also to monitor and disseminate operational data of CHP-plants and to improve their reputation. As a consequence of these learning processes, manufacturers and planners could not afford to oversize plants any more - something they had been accused of doing in order to sell larger units than actually needed. Anyway, the optimisation of schemes was in the interest of manufacturing and service companies once they started to offer full energy service solutions .

Cases of close co-operation between large users of small-scale CHP-equipment (e.g. hotel chains) and manufacturers led to designs of CHP-systems which were well adapted to the customers' needs and advanced CHP-technology decisively. As a second step, the experiences gained through these model projects fed back into other similar applications. In the end, economies of scale through packaging and standardisation of equipment could be achieved, contributing further to improvements in terms of reliability and costs. Co-operation was thus crucial during the stabilisation and standardisation phases of the technology, i.e. when the manufacturers still required much information from their clients in order to develop the design principles for providing optimised solutions. For reciprocating engines, stabilisation was achieved in the early Nineties by packaged schemes with on-board control systems and on-line monitoring. Until the middle of the decade, stabilisation was achieved at a higher power level, namely for gas turbines up to about 10 MW. User-groups of specific types of turbines were founded for some of the more widespread units which provided a sufficiently broad basis for collecting and exchanging experiences. The manufacturers were interested in getting more information about operational experiences with their units beyond the usual maintenance issues. They needed to learn more about the conditions and possibilities of providing better service solutions to their future customers in a competitive market. Such integrated turn-key and service solutions were exactly what users had often been missing before.[202]

In principle the aforementioned arguments hold for both engine and gas turbine applications. Gas turbine plants are usually larger and more site-specific in character, and they cannot be packaged easily. Reliability and maintenance were less problematic issues due to the accumulated experience with gas turbines in other fields of application. Anyway, the rotating masses of gas turbines make them less sensitive to wear than reciprocating engines. While the technical transfer from aero- and off-shore applications of gas turbines to stationary CHP was thus quite successful, considerable changes had to be introduced with regard to the user-producer relationships. For example, low maintenance requirements were imperative for CHP while regular and often costly maintenance work was not a real concern for aircrafts or off-shore sites.

The impact of liberalisation on the technology suppliers can be best understood when looking at the major changes with which they have been confronted in the

---

202 For example, there is a user group for EGT Typhoon turbine operators. A sufficient number of CHP-units are operated in the UK to make it worth for the manufacturer to co-operate with the user group. Initially, this user group was also a means for exerting pressure on the manufacturer if necessary.

last years. In their perception, privatisation has substantially expanded the potential for which they can provide technological solutions, and made it also possible form them to extend the range of their services to areas which had been monopolistic and closed before. Now, they have much better possibilities to internalise the benefits of CHP. On the other hand, they have started to recognise the pressures of a more competitive environment. However, with hindsight the decision of 1983 may have been of greater importance for manufacturers than liberalisation in 1990 because it gave the first signals and incentives for the development of CHP-systems. The technologies they had developed in the Eighties were then sufficiently advanced in 1990 to be fully exploited immediately.

8.5.3 Users and their adoption decisions

While the major decisions on the technology user side are going to be presented jointly for all types of users, the analysis of their motivations and interactions is separated into two sections; one dealing with the service end-users and one on the utility and service companies.

8.5.3.1 Decisions: Technology use

The diffusion patterns of CHP and the structural changes at the user side have been described in Section 8.4.1, but they have only captured the aggregate changes. These aggregate patterns are based on decisions taken by users, which are shaped by several detailed mechanisms, e.g. concerning the willingness to adopt new practices, to learn and to co-operate with other actors.

Regarding the utility companies, one can first of all identify a change of their general position towards CHP, which has become far less hostile, but also an extension of the range of services they offer. For example, the activities of the former public electricity utilities overlap now with those provided by some of the technology suppliers and by energy service companies. In fact, they have all broadened the spectrum of their energy related-services, which now also comprise CHP-systems. Another major change concerns the horizontal and vertical integration of utility and service companies. Many of them have started to get involved in other than their established energy supply chains (gas, electricity) or have at least set up complementary activities. This is expressed by their financial partnerships with CHP manufacturers and packagers, and by take-overs or foundations of gas supply divisions, generation divisions or end-supply divisions. In fact, all former utility companies and the new Integrated Energy Service Companies (IESC) have converged towards full-energy service providers. At the operational level, these organisational changes of the energy utilities have been complemented by the adoption of demand side planning and energy service contracts, reflecting an increasing interest in efficient energy supply.

In order to realise new CHP-based service projects, the acceptance by service end-users was an important uncertainty. After some initial reluctance, they have indeed increasingly adopted the new service approaches. In addition, the notion of 'adoption' does not only apply to the actual use of a CHP-system, but also to the *facilitation* of the final use of CHP. This is important with regard to the 'intermediate' types of CHP-users such as the RECs and energy service companies. The changes which have taken place in this respect are probably more drastic than on the end-user side. The RECs have become first of all more interested in CHP applications which they operate themselves, mainly in the form of service supply contracts. Secondly, they have also become more open to independent generation of power. They have understood that it can be a comparatively cheap source of power for their supply business. Perhaps more important, they also had to face the risk of losing important clients in their supply areas if they did not accept their power sales at reasonable tariffs.

8.5.3.2 Decision arenas and interactions I: Energy service end-users

Motivations

The utilisation of CHP must always be seen as part of a wider energy supply strategy of a firm. The main alternative to CHP for heat and electricity service end users is the separate generation of heat on site, combined with electricity imports from the grid. While the latter option requires the lowest initial investment, over a longer period of time CHP can be and actually is in many cases the economically better solution. The interest of companies, for which energy supply is a decisive part of their production process, in CHP is thus obvious, and they are willing to invest major financial and human resources into it. When heat and power supply do not represent core activities of a company, but just a utility service, then the ease of implementation and operation of the technology becomes an important argument for adoption. In this respect, CHP has made much progress in the last years.

In general, the main interest in CHP is based on the cost savings it promises. Moreover, it also promises more independence from the volatile electricity prices in the power pool. On the other hand, energy utility services being a non-core activity for most companies, CHP must promise very good payback periods in order to get approved, especially in times of scarce financial resources.

Another important characteristic of utility services is that most companies do not want to worry about them. They cannot afford to build up additional competencies to operate sophisticated utility service schemes. CHP being a more complicated, less standardised and less proven technology, they are often reluctant to adopt it even if the economics are very promising. This effect is partly compensated for by the general trend towards outsourcing of secondary and utility activities. It met perfectly with the service orientation of energy supply companies which were able to provide and operate well-designed cogeneration systems on site.

Environmental arguments play a limited role when deciding in favour of CHP, and especially in the private sector this is often done to build a 'green' reputation and show environmental commitment. They are just of secondary importance as compared to the dominant economic arguments. In the public sector, environmental concerns matter more due to political commitments. However, in most of the cases under study the persistent support by a strong promoter seemed to be far more important for carrying a CHP-project forward during years of resistance than any kind of formal commitment.

Decision-making and assessments

Although the great uncertainties associated to long-term investment in CHP are fully recognised, payback calculations or similar methods (e.g. discounted cash flow) are widely used assess the economic viability of additional investment in CHP instead of going for a simpler alternative. However, the results are usually regarded as orienting data only which need to be joined with strategic considerations.

Due to the limitations of available capital for investment in energy efficiency, at first the best opportunities with the shortest payback periods are exploited. In spite of the different barriers to energy-efficient investment, several minor and passive measures had priority over the comparatively costly and long-term investment in CHP. In the UK, the re-emergence of CHP was only possible after several other types of efficiency investment had already been exploited. This was the case from the late Eighties onwards. In earlier years, CHP was considered only if the main power generation system had to be replaced anyway. Obviously, the availability of energy supply services also facilitated the adoption of CHP in the early Nineties, because it was a low risk opportunity for the energy users to reduce their costs.[203]

Public and private users apply different criteria to energy supply investment. In industry, payback requirements of two to three years are typical for CHP-systems, whereas in the public sector longer periods of five to seven years are common, reflecting also the dichotomy between core- and non-core investment. Liberalisation and the associated uncertainties have led to ever shorter payback requirements for investment in CHP, because hardly any company is willing to accept longer-term investment risks.

Public CHP-investors also need to calculate differently than private ones. Usually they need to export power to the regional electricity company, because the heat loads in their buildings are less stable than on industrial sites. In order to achieve nevertheless sufficiently long running hours of their CHP-units, they need to find different operational modes. Consequently, their calculations are based on a comparatively small amount of autogenerated power to replace expensive power purchases from the REC, and a quite large share of power exports for which they used to get a quite low remuneration only. In some cases, agreements between RECs and public CHP-users were concluded which were even based on the export of all the power generated. Industry, on the contrary, usually assumes that the full capacity of the CHP-system can be used to replace power imports.

Rough expectations of fuel and electricity prices add to this mix of arguments at the level of individual users. The investment decisions may be supported by quan-

---

[203] Obviously this depends on the individual contractual arrangements. Often the risk of failure was borne by the energy service company, i.e. the energy user was guaranteed a certain price for electricity, and any additional benefits were shared between service provider and end-user.

titative estimations, but it is widely acknowledged that they are not very reliable. On the other hand, the difficulties involved in the prediction of highly volatile electricity prices in the liberalised system can also have a conducive impact on CHP and self-generation in general, because several companies became interested in being more independent from the pool price. This argument is particularly important in long-term contracts (up to 10 or 15 years) with energy management and supply companies. In these cases, the suppliers/energy management companies often take over the risk of investment in CHP, while keeping a good margin of the long-term savings in exchange for their expertise in carrying out the project and for the risk they bear.[204] The fact that the needs of the end-users can be matched with the extended range of services provided by the suppliers was helpful for CHP, but on the other hand it is still unclear whether these long-term contracts will be economically beneficial in the end or not.

Although during the Eighties and early Nineties awareness about environmental issues rose, it must be clearly stated that it has played until now only a secondary role. There is at best a discussion on the 'double dividend' or the 'win-win promise behind CHP-investment, i.e. the possibility of making profits while also doing something good for the environment. Obviously, this is not always the case. Especially public authorities have given stronger emphasis to environmental issues, trying to achieve best possible solutions under given economic constraints. However, economics is a 'conditio sine qua non' whereas the environmental arguments are an add-on which may tip the balance in favour of one of two similarly economical options.

Perceived problems and difficulties

Given the aforementioned difficulties to achieve a reliable assessment of an adoption decisions to be made, it is obvious that very much depends on the individual perception of the users, the quality of their information sources and the different constraints which they regard to be of major relevance.

In general, in the early Eighties there was hardly any reliable data available about CHP in the UK, especially not about smaller-scale engine and gas-turbine applications. It was widely recognised that a major help was given later on by the dissemination activities of the EEO, especially its publications and seminars. As soon as a number of plants were actually installed, site visits became a second source of information, based on formal or informal contacts with the operators. Surprisingly, according to many interviewees the role of the CHPA for information supply was rather limited, mainly because it was regarded to be biased by the influence of manufacturers. Information from the manufacturers and packagers themselves was generally received with much suspicion, too, despite some examples of very good and close co-operations. However, it needs to be underlined that none of the actors had the experience and knowledge to design plants properly, and they were lack-

---

204 Again, it is unclear whether the additional costs for the contractor are outweighed by the cost-reducing impact of his specialised knowledge. If this is the case, then the co-operation would be a non-zero-sum game with benefits for both sides.

ing essential information about site conditions to implement a project in a reliable way. CHP was still in a phase of stabilisation until the late Eighties/early Nineties.

Sources	Importance	Characterisation & timing
Industrial associations (CHPA, AIEP, MEUC)	high	Technical and contracting support especially from CHPA since late Eighties
Utilities and their associations	low	Little input until early Eighties, now changing
Engineering associations	low	Hardly mentioned as relevant for CHP
Engineering consultant offices	medium	Support for planning and designing plants
Government (EEO)	high	Technical information and best practices, establishment of contacts to other users
Other users	high	Informal contacts with users of similar applications
Manufacturers and packagers	high	Close interaction with some manufacturers for planning, design and operation
Own expertise	medium	Intensive learning process due to cooperation with others, high energy expertise of some industrial users

Table 8.4: Main information sources of CHP-users in the UK

In this context, it is also interesting to study the differences between technology users and technology suppliers in assessing CHP-technology. In the Eighties, the main issues for the suppliers were a higher efficiency and automated operation. For the users, low maintenance costs and high availability ranked highest. In many cases the high efficiency and sophistication of new engines and especially of new gas turbines were sometimes even perceived as disadvantages by the users because the output heat was lower grade and the turbines less reliable. Neither could they intervene themselves easily in the operation of plants, obviously depending on the operators' level of technical competence. In the course of time the differences in assessments and requirements of CHP levelled out due to closer interactions between users and suppliers.

A major issue for CHP is the high dependence on specific site conditions. It is difficult to transfer experiences and designs from one site to another, because the more sophisticated a CHP-system is, the better it is integrated into the entire energy supply system and/or production process of a site. In such highly integrated cases, it is also hardly possible to separate the specific investment for CHP from that for other production purposes.[205] In order to design a well-integrated system, detailed data on demand profiles are needed, a type of information which was often not available in the past. This had consequences both for the users (who did not get a plant which matched their needs) and the suppliers (who could not build up a knowledge-base).

The complexity of the technology itself represented another problem for the users, up to the point that they preferred less advanced, but more reliable, solutions. Together with the general trend towards outsourcing, this facilitated the emergence of energy service companies. The complexity and the sheer volume of the regulatory procedures was conducive to this development, too. Although administrative and regulatory processes were not perceived as a real problem in the UK, they required specialised knowledge about the new rules, especially those of the electricity pool. The administrative costs for participating in the pool have thus deterred several smaller generators. Finally, the financial situation contributed to the emergence of IESCs, because the lack of finance and the economic recession in the late Eighties favoured the concentration on core-investment; utility services like energy supply were of secondary importance. The promise that no additional financial outlay was needed for most CEM-solutions was exactly what many industrial firms were looking for. However, the sharing of benefits between service user and service provider was not always justified by the quality of the specialised competencies of the latter. Moreover, while in some cases CEM was the only way to realise CHP, in others it made projects unviable because of the additional costs for the external expertise. The finance problem was particularly pressing in the public sector where funds had been substantially cut. Even private-public partnerships, one of the solutions preferred by government, were confronted with major obstacles such as alien business cultures, unclear economics and administrative delays.

These arguments show that learning from initial failures has a double-edged function for diffusion. Failures (which are inevitable in the early phases of diffusion) can entirely stop the diffusion process, but when learning through failure is fast enough, the improvements override the negative experiences and define the reputation of the new technology.

The relationship with the RECs

The RECs and before privatisation the electricity boards have always been regarded as one of the main obstacles to a wider diffusion of CHP. Although the case studies from the early to the late Eighties seem to confirm this judgement, there are also exceptions which provided the first test-fields for the technology. The transformation of the interest structures of the RECs after privatisation has led to a far more conducive stance towards CHP, and this is also reflected in the users'

---

205 This argument applies in principle to all integrated clean technologies (as contrasted with end-of-pipe technologies). The tight integration in the entire production process is one of the key characteristics of clean technologies, making it impossible to single out additional investment costs and benefits of the integrated plant. Consequently, clean technologies can only be assessed on a case by case basis.

experiences. This transformation process is not yet finished; it will probably take several years to change the corporate cultures. Several recent examples show also that there are major differences among the twelve RECs in terms of their positions towards CHP. The technical and administrative obstacles to prevent CHP in the REC's home areas have been widely removed, but the tariff system remains a problematic issue. Export tariffs are still too low for many auto-producers, and the introduction of highly differentiated tariffs is perceived as detrimental to CHP. The traditional separation into four tariff zones (summer/winter, day/night) has been replaced by more complex arrangements which are difficult to predict economically. Industrial schemes are less affected by these export tariffs because their calculations are made in comparison to the costs of purchased electricity. In the public sectors, where heat is the main energy service, a large share of power needs to be exported due to the strong heat load variations and the need for long operation times of CHP-plants per year. In this respect, the RECs have still many possibilities to discriminate against CHP, especially with respect to those public sector applications which are below the franchise limit of the liberalised market (100 kW until 1998). This means that the operators cannot choose their supply company. Above the franchise limit, the defensive position of the RECs against auto-producers has been given up in favour of a strategy of co-operation with these clients, e.g. by setting up joint CEM-solutions. The main objective of the RECs in the competitive power market is to now protect themselves against competing supply companies.

In spite of the claims which are still raised sometimes for a more conducive position of the RECs, this problem seems to have been solved in the meantime. Most of the formal and informal means to prevent CHP have been abandoned by the RECs, and the difficult question how cogenerated power should be remunerated falls rather in the political realm where environmental considerations might become more important in the future. This does not mean that the RECs are now particularly CHP-friendly, but they have accepted it as a complementary, and sometimes competing, element of the energy supply landscape.

Regulatory framework and public policies

The 1983 framework obliged the then Regional Boards to purchase independently generated power, but the prices which were actually paid for it did in many cases not even cover the fuel costs of the CHP operators. Since 1990, this obligation to purchase power is not in force any more, but in general the price level for exports has risen, be it in a highly differentiated way. At the same time, the decrease of electricity prices, enforced by the regulator, reduced the general incentives for investing in CHP. The differential between electricity and gas prices as one of the key economic determinants has become smaller. As the environmental advantages of CHP are of minor importance for decision-making, CHP-plants have become more difficult to justify due to the falling and highly volatile power prices.

From the users' point of view, electricity liberalisation and privatisation had the main effect of improving their negotiation position with the electricity supply companies. Equally important for them was the liberalisation of gas supply (and especially the availability of gas for power generation through an extended network). Technical options became feasible due to the access to the gas infrastructure which were excluded before, thus opening up the potential for gas-based cogeneration. Environmental regulation, although being fiercer, is not yet perceived as a real constraint on CHP. It is generally regarded as an environmentally friendly option by the regulator, and the flexibility of the regulatory framework makes it possible to negotiate the specific conditions with HMIP or the local authorities.

Expectations in past and present

A decision for or against the adoption of CHP is not made once and for all. Especially industrial companies reconsider permanently their energy supply options, looking up to ten or even fifteen years ahead in terms of site development plans. The present uncertainties of the energy supply system make any reliable planning and predictions of future conditions difficult and have a detrimental effect on additional investment like in CHP. Extended projects, involving the co-ordination of several sites are hardly feasible due to such uncertainties.

Although prejudices against CHP and a lack of knowledge played a role among the older generation of decision-makers, especially at the board level, the generation change since the early Eighties has brought people to the forefront who are more open to non-conventional solutions. This holds for all actors involved in CHP-related decisions. The successful experiences with CHP have also changed the minds of many users who were initially critical of CHP. Similarly, the structural conditions for CHP are generally seen as more positive now than in the past, mainly due to the fact that the utility companies cannot obstruct CHP-plants any more. On the other hand, CHP is not expected to become a widely used option because of the uncertainties of the liberalised market. Reluctance to invest in CHP has always been particularly strong when major changes were about to take place, namely in 1990 when the new framework was established, and in 1994 when the franchise limit was reduced to 100 kW. A similar attitude is expected for 1998 when the franchise limit will be entirely abolished.

The expectations of all the actors having a stake in the CHP business have thus converged in a conducive way. Nowadays, extreme positions are hard to find, showing again that the technology has stabilised. Future technological expectations of users do not go far beyond incremental improvements of existing engines and gas turbine systems. Radically new systems like fuel cells are seen as being still too remote from economic and technical viability to think about their widespread application. As long as they have not proven to be as reliable as conventional CHP-technology, hardly any end-user will show interest in them.

8.5.3.3 Decision arenas and interactions II: Utility and service companies

The Regional Boards were not very responsive to the Energy Act 1983. On the contrary, by setting up technical or economic barriers they tried to circumvent its objective of facilitating self-generation. Although they were obliged to purchase power from independent generators at adequate costs, there was no clear guidance what this meant in practice. Technical grid connection conditions imposed by the

Regional Boards, and very unfavourable tariffs for export and back-up power, made many projects uninteresting. Administrative procedures were delayed, showing that the boards still maintained a gate-keeping function in power generation. In the end, the expected expansion of self-generation did not happen, but at least a few new plants went ahead. This was sufficient to enable first learning effects and to establish CHP at a slow pace. In addition, some of the Regional Boards became interested in power generation, thereby breaking the unwritten law of division of tasks between CEGB (generation) and Regional Boards (distribution). CHP was also problematic for them for another reason, namely it required to take an active role in heat supply which would have created problems with British Gas.

Privatisation and liberalisation were the forces behind the change of attitudes of the utility companies. Due to the regulatory restrictions of the profits which they were allowed to make from their supply business, they became interested in expanding to other services (power generation, telecommunications, etc.), including heat supply. In practice, they transformed themselves into energy management firms, though maintaining their specific supply monopoly for franchise customers until 1998. RECs as well as large generators became horizontally integrated providers of energy services under the roof of holding companies. Most gas and electricity utility companies have now cogeneration divisions or subsidiary companies. While the large generators also focused on larger scale applications of CHP, the RECs were also interested in small-scale CHP and general energy services. Even British Gas started to cover a wide spectrum of gas-related activities in the power generation area. In other words, diversification became a key strategy to adapt to the changing systemic conditions and opportunities after privatisation and liberalisation.

In addition to these positive developments, competition had further beneficial effects on CHP. First of all, the possibility to compete with the other RECs in their respective areas paved the way for the RECs to provide competitive energy services country-wide. They may still benefit from their monopolistic home base in the franchise market, and from established business co-operations, but there is a clear incentive to develop CHP in order to be able to invade other areas. Secondly, liberalisation forced the RECs to be more open to self-generation if they did not want to lose clients. Hence, whereas they were very reluctant with regard to cogeneration till privatisation, they then tried to set up their own CHP-activities, usually in co-operation with users in their home areas. As long as the RECs had not build up major generation capacities of their own, they even had an interest in the comparatively cheap imports from independent and auto-generators.

However, in their regions, the RECs still hold a strategic position because they control the conditions for grid connections for those CHP-operators who do not deal directly with the pool. In 1995, grid connection has been liberalised and opened to regulated competition. This meant for the RECs the loss of their monopoly control over this important issue. The possibilities to obstruct projects which are not desirable for them have thus more and more disappeared. Another critical issue concerns the power tariffs. It is still a matter of dispute whether autoproducers should be remunerated for their contribution to reducing the need for new generation and local grid capacity. These two impacts would have to be considered as long-term avoided costs to the RECs, but the parties concerned have not yet come to any agreement. Neither does the regulator seem to be very interested in interfering in this question of avoided costs.

However, in spite of these positive developments for CHP on the side of utility and service companies, it should be kept in mind that it is still a rather marginal activity for the large utility companies. Their position with regard to major regulatory issues is dominated by their core activities, i.e. by power supply for the RECs, by power generation for the large generators, and by gas supply for British Gas.[206]

8.6 The evolutionary dynamics of CHP-diffusion in the UK

In order to interpret the preceding analysis in a dynamic way, the pattern of evolutionary elements as described in Chapter 3 is going to be applied. The different development trends need to be related to each other along these lines in order to capture the dynamic mechanisms which underlie the diffusion of CHP in the UK.

External pressure

Innovation and variation in different domains of the PET-system can be driven either by external developments and/or internal ones. In the UK, neither a strong environmental pressure nor particular difficulties with regard to fuel supply in the UK required an adaptation of the system. The *release of natural gas* as a fuel for power generation represented an external event which opened up opportunities for new power supply options. The *global and European environmental decision* context affected the situation in the UK, e.g. by triggering the introduction of a number of environmental regulations. There was no particular need for new power capacity, because growth of overall demand was slow, following roughly the economic growth of the British economy. Several *technological innovations* of importance to CHP in the UK can also be interpreted as external driving forces of change, because they were based on experiences made abroad.

Variation and selection by innovation

Technology manufacturers introduced several technical innovations but especially also non-technical novelties. They were on the one hand a response to changing regulatory framework conditions in the UK which from the mid-Eighties onwards made CHP possible, and on the other based on the perception of new technological opportunities. By combining *improved hardware* (engines, turbines and packaging) with advanced control systems and a new *service-oriented approach* to

---

206 In 1997, British Gas was finally broken up into separate companies for supply and distribution. It had been privatised in 1986 but remained intact as an organisation. Frequent complaints about abuses of its market power have finally led to the separation.

energy supply, they made a qualitatively leap forward in CHP-technology. These innovations were not randomly driven, but building on expectations of developments in their external and internal selection environments, as well as on the interest structures and mental frameworks of the main actors. This is particularly obvious in the case of the utility companies which in the post-privatisation years moved into the manufacturing/packaging and energy services business.

The *liberalisation* decision had opened up these opportunities for the technology suppliers; they could set up activities which had been controlled by public utility companies before. Breaking up the monopolistic economic structure enabled the development of a variety of technological and organisational solutions which had been regarded as not feasible in the Eighties, because they would have conflicted with the interests of the utility companies. After liberalisation in 1990, the changing power positions in the energy markets exerted more pressure on the utilities.

At the technical level, there were several adaptation problems of CHP to be solved, but basically the technological knowledge was available. The innovative step was rather to use the existing elements in a novel way and improve them to match the requirements of the customers. However, specific *links between users and suppliers* needed to be established in order to achieve the necessary improvements. The integration of new technologies, such as microelectronics, into traditional engine design, and spin-offs from other applications, provided a hardware which fulfilled the users' requirements, even if this process of adaptation and learning took some years. New technological opportunities, e.g. fuel cells and Stirling engines, are now explored, based on the expectation that there will be a significant potential for their application. The *catalysing effect of some government activities*, especially those run by the EEO, proved to be helpful for spreading information and creating networks of exchange. Similarly, the promotional activities of the CHPA, which is mainly supported by the equipment industry and to an increasing extent by CHP-users, helped to strengthen the information network and provided a solid structure to facilitate exchange between users and suppliers.

Although the technology supply side is basically a source of variation and innovation, it also represents a first selection step. Based on their expectations on technological opportunities and constraints of CHP, technology suppliers decide in the mid-Eighties already to pursue the pathway of a new generation of CHP-systems.

Selection by users in their organisational contexts

The key selection step for a technology must be seen in the decisions made by its potential users. They can be more or less successful in *choosing a technology or energy service* option which matches their requirements and thus improves their own performance. Their choices and experiences are fed back to manufacturers and/or other users. Especially during the early years of application, these individual choices and experiences were decisive for *building up the reputation of CHP*. Small events or a series of failures can fully discredit a promising technology. In the case of the UK, the positive experiences made from the mid-Eighties onwards contributed to improving the reputation of CHP. This was facilitated to a large extent by information dissemination and monitoring programmes like the ones run by the EEO.

This selection metaphor also applies to the facilitators, utility and energy management companies, which moved from a detrimental to a fairly conducive position towards CHP. However, a large power generation over-capacity has been built up in the last years due to the expected opportunities in gas-fired power supply. They limited the *incentive for the utility companies to support small-scale power supply*. Liberalisation and hence the possibility for small- and medium-sized power users to 'escape' from their regional supply companies (or at least threaten them to do so) forced the utilities to be more open towards options like decentralised CHP and offer better connection conditions to independent CHP-operators. Embedded in their strategy to extend the range of services, they even started to perceive limited opportunities for their own commitment to CHP. In other words, their interest structure and the scope of their activities changed radically as a response to the new liberalised framework, even if until now there is only a slow transition from a philosophy of selling electricity to one of supplying services.

The large generators, who are operating the major share of the centralised technical infrastructure, have started to look at decentralised options as well, but are still constrained by their capital stock. Their interest structure favours traditional large-scale and centralised generation of power only. The technological system they are carrying has been built up according to a completely different logic than the one underlying the operation of CHP. While their system is based on economies of scale in terms of size, so that they cannot exploit the economies of scope and scale from series production, packaging and standardisation which are typical of CHP. The two corresponding system architectures are based on different reinforcement mechanisms, both in terms of technological and organisational structures which mutually stabilise each other. Hence, although *technically* there have been possibilities to combine the two different system worlds by means of new computerised grid connections, it would have been necessary that the structural settings facilitate the integration of CHP-related interests into the individual organisations. This has clearly not been the case. In the UK, the diffusion of cogeneration has long suffered from this *mismatch between organisation of the energy sector and the requirements of CHP*. The higher degree of flexibility of the liberalised framework has at least enabled the incorporation of CHP-related interests into the strategies of the large power generators.

Similar arguments hold for gas suppliers. The *change of the organisational structures of both electricity and gas supply* overturned the established areas of activity and opened up the boundaries between them. The importance of matching technology and organisation becomes obvious in this respect: As long as companies work in gas or electricity supply only, they have hardly an interest in a technology which requires to combine both worlds, and thus questions the traditional business territories.

Hence, we can identify a '*gateway function*'[207] *performed by service integration* which bridges between heat and power supply, but also between centralised and decentralised organisation. As this integration is easier to achieve at the local level, it is of crucial importance that there are local actors who can incorporate interests which are compatible with CHP. Despite the emergence of IESC, the situation of CHP is still problematic in the UK, because local authorities and utilities are not sufficiently powerful to fulfil such a gateway function in organisational terms. Technically, the gateway function is fulfilled by modern microelectronics which allows to control complex and decentralised systems on-line. Economically, it is driven by the trend towards Integrated Energy Services Companies which encompass interests in both service branches. Politically, it has been enabled by liberalisation. In other words, it has been crucial to establish a 'match' between technological characteristics of CHP, organisational patterns in the energy supply industries and a political environment which facilitates the establishment of a conducive context for these matches to emerge.

Selection conditions of the regulatory and political framework

The decisions which triggered the changes in the UK (i.e. especially liberalisation and privatisation in 1983 and 1990) have been strongly disputed, because they *threatened the established technological and economic structures of the system.* The Energy Act 1983 disregarded that several possibilities still remained for the utilities to undermine by informal means its intentions. Consequently, the transformation of the established organisational structure of the British energy supply system, which beforehand did not match with the envisaged objectives of government, was only achieved in 1990.

However, it is important to take the intentions of privatisation into account which had little to do with any specific objective in the field of energy policy. It was rather *ideologically motivated* by the aim of putting an end to an active involvement of government in energy supply and of abandoning the approach of providing clear guidelines and visions of how energy supply should look like in the future. The established interfaces and political structures between Dept. of Energy and the ESI could only be overcome and 'locked out' due to this ideological background. Both at the level of government and at the system level, this can be regarded as a *process of systemic auto-control*, because, in fact the, structures for political decision-making were also re-negotiated. The definition of the detailed regulatory selection framework and innovation context in the ESI was determined by policies which were decided upon in interaction among several actors. Before liberalisation the Dept. of Energy was the focal point of these interactions, but now the responsibilities for shaping the political framework conditions are shared by the regulators, the Dept. of Environment and the Dept. of Trade and Industry. Their decisions on regulations and technology support measures are perceived by technology developers as factors shaping decisively the opportunity structures for CHP.

---

207 For the concept of gateway technologies c.f. (David, 1992) who investigates the example of inter-connecting direct current and alternating current systems.

The reality of liberalisation clearly shows that *CHP was not an issue when defining the regulatory set-up* (which is no surprise). It was driven by a number of fundamental convictions about the appropriateness of liberal governance structures in the energy sector. The secondary role of CHP is reflected also in the inherent shorter-term logic governing the operation of the ESI. It makes the operation of CHP as part of a long-term strategy difficult, if not impossible, in many cases - despite the advantages of liberalisation for CHP in other respects. This implies that the situation for CHP in the UK is still problematic for two major reasons: the *short-term logic and high level of uncertainty* induced by the pool system, and the instability of the regulatory rules. Once the set-up is fully established it may be possible to re-introduce a longer-term perspective.

Stabilisation mechanisms

The past resistance against CHP was a matter of *mental frameworks and engineering paradigms*, but both in the utility companies and in the engineering associations the generation of sceptics has in the meantime been replaced by younger staff having fewer prejudices against CHP and against the microelectronics equipment needed to make it work properly. District heating is still very weak in the UK, and this is not only due to the extensive availability of gas networks. It has long suffered from the bad reputation which it had acquired in the past; it was often said to be unreliable and difficult to control properly by the heat users. This implies that in the case of district heating the service end-users (i.e. individuals and households) play a quite important role. However, although it takes a long time to change prejudices, the reputation of district and - of greater importance in the UK - group heating has also changed due to *information campaigns and the success of plants constructed during the past decade.* Furthermore, the examples of other countries (and also the technological transfers from abroad) point to the importance of information supply. It is difficult to assess how important these information and equipment networks have been for the success of CHP, but they have at least accelerated diffusion in the UK by transforming the reputation of CHP.

The supply structures were clearly stabilised in the past by the *multiple interdependencies or lock-ins* between the centralised and horizontally separated generation and supply structures in the UK, between the organisational structures to carry the supply system, and between the political decision structures which reflected the patterns of centralisation (e.g. weak local authorities) and separation (e.g. in the Dept. of Energy). Such long-term lock-in effects can in principle also operate in favour of CHP, but the current capacity is still too small to speak of strong interdependencies which make the establishment of CHP irreversible. The technological and organisational structures are still in a phase of stabilisation.

Reinforcing and delaying mechanisms

In order to achieve this stabilisation, reinforcing and delaying mechanisms play a crucial role. Their operation can also be identified in the UK with regard to CHP. A first example are *technical improvements which are perceived as new opportunities and attract further investment in innovation.* Also the pressure exerted to improve the framework conditions in favour of these opportunities initiates such a positive feedback loop. If successful, it *broadens the economic scope for CHP and triggers further support for CHP and pressure to improve the framework conditions and policy measures*. As a secondary effect, efforts are induced to improve the technology and to drive investment costs down by *economies of scale and scope*. In parallel to this, the positive experiences with new CHP-applications have *changed the mental frameworks* of most actors; this effect has facilitated a more favourable attitude towards specific projects as well as towards regulatory changes. One of the mechanisms which had a detrimental influence on CHP consists of the *success of CCGT power stations*. which seems to confirm the large-scale, power-only trajectory of electricity supply. Its economic success in combination with its environmental benefits - though comparatively limited - challenge the arguments in favour of CHP and a more CHP-friendly regulatory framework. For these mechanisms to operate, information networks played a crucial role. The establishment of CHP as a real economic and political option - a technology which had never attracted much public and political attention - was enabled by the *provision and spread of information* about its possibilities, advantages and experiences during the last decade.

Associated to the successful operation of such reinforcements and delays is the question of timing, i.e. the achievement of the right temporal matches or '*windows of opportunitie*s' for CHP. It is a long-term investment and once the opportunity has been missed, a lock-in into other technologies might occur like it has already happened to some extent for CCGTs. The transitory phase immediately after privatisation provided a unique opportunity for government to define incentives for a fundamental re-direction of its energy supply trajectory. The 'iron links' between economic, technological and political structures had been broken up, and even if the vintage effect of large power plants delays change, there was a clear opportunity to set new signals. Although this opportunity to establish CHP more widely during this unstable phase was not seized by government, some beneficial mechanisms actually worked in time. The technology was available and fully developed into a variety of applications. Moreover, support networks were in place when liberalisation was introduced, so that the potential, which had been blocked for years, could have been opened up. However, the more active government support for energy efficiency and cogeneration since about 1992/93 may have come too late to exploit fully the reinforcing mechanisms which operate in favour of CHP. A considerable amount of investment has already gone into other options, which usually have a lifetime of at least fifteen to twenty years.

System evolution

When trying to synthesise these evolutionary elements, it becomes obvious that all three types of decisions and interactions defined in PET for the innovation and diffusion of CHP have mattered. The development path over the last fifteen years has been determined by the integration of these driving forces in interdependent processes, which were stabilised by elements of continuity, and directed by the balance between positive and negative reinforcements.[208] The major changes were triggered by liberalisation and privatisation in 1990, but prepared already by changing conditions and opportunities for CHP during the Eighties. The evolution of CHP has thus been part of the evolution of the PET-system as a whole.

Applied to the situation in the UK, this means that in spite of the long-standing paradigmatic refusal of decentralised cogeneration, *the reorganisation of the ESI triggered a transformation of the relevant actors' sets of interests which led to a better 'match' with the principles of combined heat and power production*. It was driven by ideological motivations and a poor performance of the ESI. The diffusion of CHP was then the result of a *triple inducement: technical innovations, a new regulatory framework, and networks to enable learning processes*.

On the other hand, it needs to be taken into account that the new framework introduced also several detrimental effects on cogeneration. The most important one is probably the switch to a *shorter-term investment logic* (and associated uncertainties) which is expressed in very tight payback requirements for energy services investment. The present procedures and regulations for assessing the viability of a plant project, which do not fully consider the environmental, long-term dimension of energy supply, are certainly not helpful for a wider diffusion of CHP.

It is hardly possible to assess what the net contributions of these different forces have been. One can identify a sequence of events and requirements as partly random and partly intended reactions to the basic impulse of liberalisation/privatisation: Changes of interest structures and of organisational structures, new links and coalitions. Although they opened up a number of new opportunities for CHP as well as for other more sustainable energy technologies, the overall effect is rather contradictory and double-edged due to the dominance of short-term commercial interests, and the high degree of uncertainty. There was certainly the *opportunity in the early 'turbulent' post-privatisation phase to switch to a decentralised trajectory*, but this would have required a more active policy approach - an apparent contradiction with to the liberal ethos of Conservative government. Nearly ten years after privatisation, efficiency improvements have been achieved, but to a limited extent only. Hence, the interesting conclusion is that the *initial driving force of the very promising process of change with respect to CHP - liberalisation - also contained the reasons for its partial failure*. A well-balanced solution for a liberalised framework for energy supply which is economically efficient and sustainable has not yet been found in the UK.

8.7 Outlook

Building on the preceding analysis, three possible scenarios for the longer-term development of CHP in the UK over a twenty-years time horizon are now going to be sketched. They can be characterised by the following three headlines:

---

208 For a similar interpretation of evolutionary processes of change, c.f. (Scharpf & Mayntz, 1995).

- Adjustment of the liberal framework to the requirements of CHP;
- Full application of liberal principles;
- Political enforcement and support for CHP.

All three scenarios exclude major external events such as e.g. a new oil price crisis or climatic disasters. Such 'wildcards' could obviously change the perspectives. The differentiating dimensions of the scenarios is the orientation of government policy. A change of government in the UK is expected to have an impact mainly on the probabilities of the three scenarios, but hardly on their content. Finally, it is assumed that the liberal approach to energy policy will be maintained, though with decisive differences in detail. Four aspects will be addressed in the following exploration: the basic principles of each scenario, their main elements, the implications for CHP and the positions of major actors.

8.7.1 Scenario 1: Moderate CHP-friendly liberalisation

Already in the last few years, British government has underlined the importance of CHP for environmental and efficiency reasons, expressed in the targets set for CHP-capacity. A number of adjustments of the initial regulatory and policy framework have already been made to facilitate the uptake of CHP. This trend is expected to continue in this scenario. While maintaining the regulatory framework in general, a number of further issues would be addressed by regulatory control and policy, and lead to the following developments:

- A fair remuneration for exported power based on clear guidelines for the calculation of avoided costs;
- Strict and reliable environmental regulations;
- Further simplification of administrative rules, especially for the electricity pool;
- Introduction of CO_2-taxes to stimulate energy efficiency by market-based instruments;
- Measures to stabilise electricity prices in the pool system in order to improve the long-term planning basis for CHP-projects and reduce uncertainties;
- Definition of long-term objectives or targets for energy supply to provide reliable planning conditions for the sector;
- More financial autonomy and planning competencies of local authorities in order to facilitate the introduction of district and community heating schemes;
- Limited research funding to improve critical technologies for CHP;
- Promotion measures to disseminate information about experiences with CHP.

These measures would not entail massive interventions by the regulator or by policy-makers in the energy market, but aim to adjust and improve the conditions for CHP within the current framework. This scenario excludes two important possibilities, namely a stronger co-ordinating role of authorities to optimise the integration of different energy supply chains (gas, heat, power), and substantial financial support for CHP-projects to accelerate its diffusion. The improved conditions for CHP would stimulate some limited R&D work on CHP, which will reinforce its wider establishment. With regard to diffusion, a moderate growth of capacity would be probable, but it would be limited to a fairly large market niche. District heating is not expected to be taken up widely because of the lock-in into direct gas supply and the detrimental organisational conditions. Only smaller community-heating solutions promise to be introduced more widely.

In terms of the positions of the main actors, this scenario appears to be quite acceptable to many of them. It would not represent a major departure from the present situation. Obviously, in the longer term it would imply a stronger decentralisation of power supply and therefore weaken the position of the large generating companies, but basically they would maintain a key role.

8.7.2 Scenario 2: Full application of liberal principles

Scenario 2 reflects the uncompromising implementation of the principles of liberalisation. Government and regulator's interventions would be restricted to the minimum needed to maintain an efficient operation of energy markets. Environmental commitments would be addressed by regulation and supported by the efficiency gains expected to emerge in a market environment. Overall, the scenario can be roughly characterised by the following features:

- Fierce competition between energy service companies at all levels;
- Concentration trends towards an oligopolistic market structure based on a limited number of horizontally and vertically integrated energy companies;
- No decentralisation of planning and financial competencies to the local level;
- High uncertainty about electricity prices;
- Frequent adjustment of regulatory rules concerning the environment;
- No additional measures like taxes or regulations to promote energy-efficient technologies like CHP ;
- Little specific consideration of the characteristics of CHP in terms of co-ordination of energy supply chains;
- Short-term time horizons applied to all operations in the energy sector.

Despite the lack of any specific support for CHP, the growing service orientation in the energy sector would lead to a moderate expansion of industrial and small-scale CHP, if the conditions are promising and reliable. However, alternatives which promise shorter term benefits (e.g. CCGT power-only plants) would continue to represent the main share of new power generation capacity. Due to the lack of horizontal co-ordination and decentralisation of responsibilities for energy planning, district heating would remain negligible. Small community heating systems would be realised in large individual buildings or new residential areas. Overall, the growth of CHP would be limited, because even in the most promising area - industrial CHP - uncertainty and co-ordination problems would make it a risky business for both users and suppliers of the technology.

Although from the perspective of several actors this scenario may look less desirable, e.g. for local authorities or small industrial energy users, they do not represent particularly powerful and influential actors. The scenario also reflects a consequent continuation of the policy of the last decade, which has become widely accepted by many of the powerful actors in energy supply. Resistance to the developments outlined in this scenario will therefore be rather limited, despite its disadvantages in terms of long-term strategic considerations.

8.7.3 Scenario 3: Political enforcement and support for CHP

Scenario 3 is characterised by stronger political intervention than the first scenario. Beyond the measures of scenario 1, there would be a more active policy in favour of CHP, but within the liberal framework. Especially the following measures would be added to those of scenario 1:

- Financial incentives and grants for CHP along the lines of the NFFO;
- Obligation to co-ordinate supply chains horizontally, especially at local level;
- Strong emphasis on energy efficiency and a reduction of CO_2-emissions, including the possibility to prohibit non-CHP power plants.

Scenario 3 would be the framework which would make a take-up of new district heating projects probable as a result of the obligation to co-ordinate heat and power supply locally. In those cities where gas supply has not yet diffused widely, the exploration of district heating would become a promising option. Few differences to scenario 1 are expected with regard to industrial applications, but the localisation of energy supply would lead overall to a substantial expansion of CHP-capacity. In addition, the financial support given to CHP would directly stimulate its diffusion and establish it as a dominant trajectory. It would also provide incentives for R&D work to improve the technology further.

It is difficult to regard this scenario really as a realistic one, because it would have a major impact on the actor constellations which have emerged in the last years. The decentralisation of competencies and planning responsibilities might be acceptable to many energy service companies, but the utility companies would lose parts of their power to new local actors. The only viable solution would be a widespread co-operation between local heat and power system operators and the utility and service companies. In addition, it is hard to see why central government would re-distribute power to the local level. In terms of government - industry relationships, this scenario would also imply major changes, namely on the one hand a reorientation of lobbying to central government, and on the other a greater attention to local political authorities. However, as the past liberalisation history shows, a radical reorientation of energy policy can not be excluded.

Chapter 9: Transforming the Monopoly - CHP in Germany

9.1 The history of cogeneration and electricity supply in Germany

The development and application of CHP in Germany has been much more successful than in the UK, especially since the second world war. But already before that time some capacity had been built up, with some first applications of CHP at the end of the last century. Overall the history of CHP in Germany can be split up into five main phases.

Phase 1 (till 1945)

The development before the second world war is characterised by the construction of both industrial and district heating CHP plants. Often the two applications were combined. Examples from Berlin and the Ruhr-area show that at that time the co-operation between industrial heat users and producers on the one hand and local utility companies on the other facilitated the construction of integrated solutions for co-ordinated heat and power supply. District heating was nevertheless often based on heat-only generation, whereas industrial applications tended to combine heat and power production.

Phase 2 (1945-1973)

After the second world war, the destruction of entire cities represented the starting point for building up extended heat networks. Although most of them were initially coal-fired and heat-only, they provided the foundation for later upgrades of the systems to CHP, especially during the Seventies. In industry, cogeneration was also applied frequently, because the technology was well adapted to the needs of the growing heat-intensive industries in Germany. At that time, the adoption of CHP was driven neither by substantial cost advantages nor by any kind of efficiency considerations, but were a useful by-product of the production process.

Phase 3 (1973-1980)

The situation changed with the first oil price crisis when security of supply became an important issue, and the interest in more efficient energy technologies was rising. During the period from the mid-Seventies to the early Eighties financial support was offered by government for CHP-based district heating, which led to a steady increase of capacity. Industrial cogeneration capacity stagnated at the same time for several reasons. Of major importance were the changing structure of industrial production towards less energy-intensive industries (induced by high energy prices) and the attempts of electric utilities to exploit their capacities as efficiently as possible (implying favourable tariffs for electricity sales to large customers and therefore less incentives for independent production). The remaining heat demand was increasingly satisfied by CHP as part of a strategy to increase the energy efficiency of production processes.

Phase 4 (1980-1990)

The decline of fossil fuel prices in the early Eighties changed the situation for CHP again. The relatively small price differences between power and fossil fuel prices made several projects unviable. In addition, two new and associated topics had emerged and started to play a major role in the energy policy debates: environmental considerations and the nuclear question. CHP, together with a number of other promising technologies such as heat-pumps, solar collectors, passive energy efficiency measures etc., started to be considered as desirable options which should be applied more widely - in spite of concerns about their cost implications. The main delimiting event of this fourth phase should be seen in the first agreement between the association of industrial power producers ("Verband der Industriellen Energie und Kraftwirtschaft - VIK") and the public electricity utilities' association ("Vereinigung deutscher Elektrizitätswerke - VDEW") concerning the remuneration and tariffs to be applied to CHP and independent power production in general. This Associations' Agreement of 1979 laid the ground for the systematic exploitation of the CHP-potential in Germany. New technologies were developed and explored during the 1980s which broadened the scope for the application of CHP within the new framework. In addition, public funds were made available for supporting the widespread expansion of CHP, both of traditional district heating schemes and new types of applications.

Phase 5 (since 1990)

Reunification in 1990 marks the beginning of the present phase for cogeneration in Germany because it shifted attention to the situation in East Germany where there was an enormous potential for upgrading old heating networks. The debate about a liberalisation of energy markets in Europe started to have an impact on the context for cogeneration, too. Finally, the fear of a global greenhouse effect due the increase of CO_2-emissions started to play a decisive role in the environmental and energy policy debates.

As in the preceding chapter on the UK, the focus of this empirical study on the developments in Germany deals with the time period since about 1980, i.e. the phases 4 and 5. Although the preceding history of CHP had already led to a diffusion of CHP in some application areas, the early Eighties mark a new quality of change, characterised by the emergence of new technological solutions based on engines and gas-turbines. In addition, the Associations' Agreement was introduced, natural gas became widely available for power generation and the new priorities of environmental policy changed the conditions for CHP significantly. In other words, although, in contrast to the UK, no major reform of the regulatory regime of the energy markets was introduced, the period under study is characterised by important modifications of the context in which CHP has diffused, too. These changes will be analysed in more detail in the following chapter. Although for reasons of consistency of analysis and availability of data, the focus of analysis is on West Germany, the developments in East Germany will also be considered; a special section will be dedicated to the situation in the New Länder.

9.2 The context conditions

Energy sources and supply context

Germany does not dispose of noteworthy indigenous oil reserves, but of major resources of both brown and hard coal. While security of supply had been an important issue after the first oil price crisis, when the dependence on oil imports was very high, it has played a minor role in recent years. The diversification of fuel sources and the establishment of a much more efficient energy supply system have been very successful. Dependence on foreign fuel sources may still be a concern, but it is not a priority topic.

Associated to the issue of security of supply are also the coal subsidies in Germany and the obligation to use a certain amount of indigenous coal for power generation. This clearly had a retarding impact on the introduction of gas-based instead of coal-based CHP, and hence on the use of the new gas-fired, small-scale technologies. On the other hand, the comparatively high electricity prices which resulted from the subsidies improved the economics of CHP and the incentives for investment in energy efficiency.

The nuclear debate in Germany played a major role in energy policy during the Seventies and Eighties. CHP as well as other solutions have benefited from the de facto moratorium on the nuclear option since then, especially since the CO_2-debate has transformed improvements of energy efficiency into a key option.

Finally, the reduction of oil prices in the mid-Eighties interrupted the growth of CHP in general and of district heating in particular. It could not maintain its cost advantages compared with direct natural gas supply, which increasingly penetrated German cities. On the other hand, the availability of gas for power generation since the mid-Eighties has facilitated gas-based CHP and enabled complementarities between gas supply and CHP.

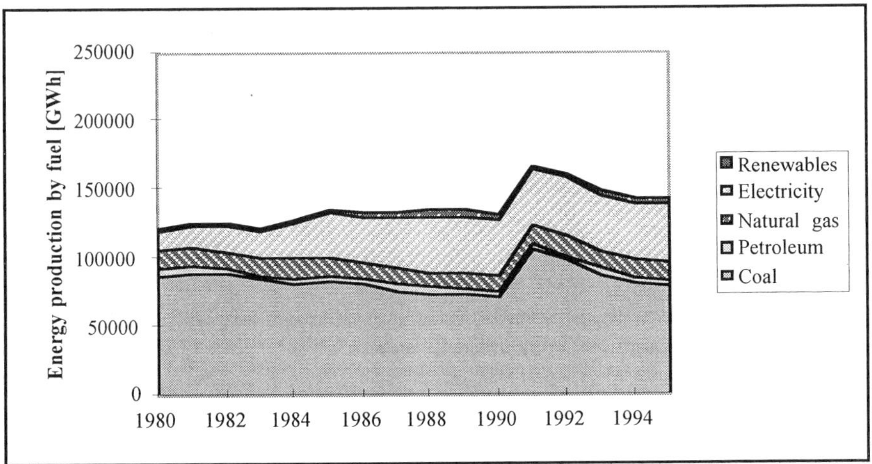

Figure 9.1: Energy production in Germany by fuels, East Germany included since 1991, Source: (BMWi, 1996; ENERDATA, 1996)

The environmental debate and public opinion on energy supply

The environmental debate has played a major role on the German political agenda in the Eighties. Jointly with emissions, energy efficiency was one of the major topics in that debate. The Clean Air Code ("Technische Anleitung (TA) Luft") (Jarass, 1995) was the main legislative act in response air pollution. Since then, Germany has on several occasions played the role of a pacemaker in the international environmental debate. Both with regard to European legislation and the commitments made at the Rio-conference, it went beyond the proposals of most other countries. However, the challenging targets should be seen in the context of the situation in East Germany where many plants had to be upgraded anyway. These commitments have led to a number of further initiatives concerning CHP.

Governance approach

The German governance approach is usually regarded as consensus-oriented. It relies on many corporatist elements such as the delegation of responsibilities and competencies to the top organisations in industry and the economy ('the social partners'). While this approach leads often to constructive outcomes, it has without doubt also disclosed an early consideration of radically innovative proposals and limited the possibilities of 'outsiders' to exert an influence on the energy policy debate. In spite of a growing emphasis on market-based solutions during the Eighties, the motivations for 'bringing the market to the public sector' were less ideological than in the UK. This is reflected in the recognition of the social benefits of public utilities, especially at local level, and in a rather technical orientation of the policy debates by drawing on the expertise of many external experts. This results often in pragmatic policy initiatives: evolution rather than revolution.

Economic and industrial development patterns

CHP was always an attractive option in many of the energy-intensive industries in Germany. This helps to explains its widespread use for industrial applications. In certain areas like the Ruhr there was even much scope for integrated supply, e.g. along the heat chain Ruhr ("Fernwärmeschiene Ruhr"), which inter-connected industrial steam supply and local utility district heating at a large scale. The growth phase of these industries until the Seventies was thus paralleled by a similar extension of district heating. After the first oil price crisis, this growth process slowed down, mainly for reasons of a stronger orientation towards less energy-intensive products and the energetic optimisation of production processes. Restructuring continued during the Eighties, and it was further reinforced by the use of technologies which responded to the imperatives of environmental policy. In other words, even if industrial CHP-capacity has declined in absolute terms, it now covers a higher share of industrial heat demand because industrial production processes have changed simultaneously and are less energy-intensive nowadays.

Reunification brought about the decline of much of East Germany's industry and its CHP-schemes, which was highly integrated with district heating. Simultaneously, district heating was upgraded by substantial public support for CHP in the New Länder, at least until 1994 when funding for the modernisation of plants was reduced and finally cut.

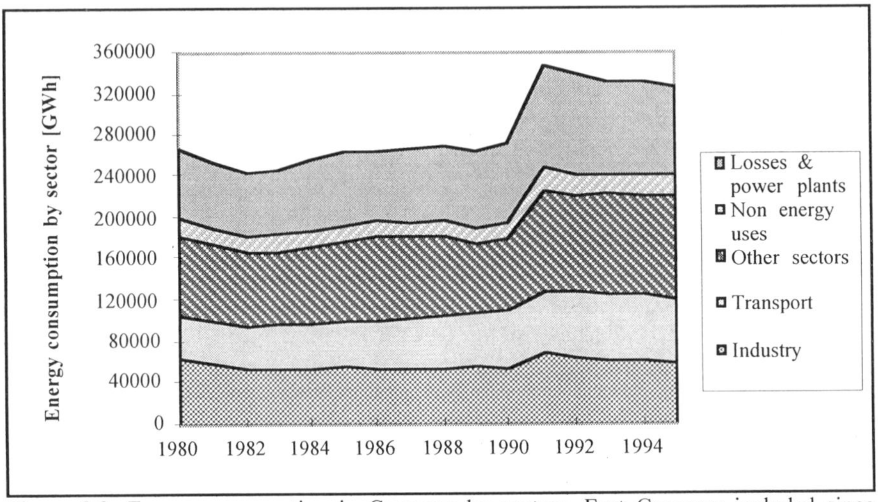

Figure 9.2: Energy consumption in Germany by sectors, East Germany included since 1991, Source: (BMWi, 1996; ENERDATA, 1996)

European policy context

Several European policy areas need to be mentioned as driving forces behind measures at the national level. They initiated changes both of the competition rules for energy supply and of environmental regulation. The German liberalisation debate has been strongly driven by the discussion on a framework for an Internal

Energy Market in Europe, but also by concerns of German industry about disadvantages in comparison to European competitors. The European environmental debate was less decisive for the situation in Germany due to the fact that the TA-Luft had already been introduced earlier.

9.3 Innovation patterns of CHP-technology in Germany

9.3.1 Technical improvements

Generally, most technologies for CHP are mature and widely tested in Germany. The long-standing experiences with coal-based and oil-based CHP for industrial as well as for district heating applications represent a rich knowledge-base for the design of good plants. In many respects, the technical developments in Germany may be similar to those in the UK, but there are also a number of significant differences, especially in terms of timing of these innovations and the emphasis given to certain areas.

Since the late Eighties the newly developed CHP-technologies based on gas-turbines and engines can be regarded as established. They had become reliable, their design was standardised and incremental improvements were made continuously. However, these seemingly minor innovations played a major role for the subsequent success of these new technologies.

Despite the use of certain components - even key components - from abroad, CHP-technology supply is dominated mainly by national companies. Small- and medium-size gas turbines are imported from abroad, but the entire packaging, design and implementation business is dominated by German firms. The specific national context of energy supply requires to be very well familiar with the prevailing regulatory and organisational conditions, but also to dispose of good networks of informal contacts.

Improvements of gas engines and micro-CHP

Until the early Eighties, diesel engines were most common for all types of stationary engine applications, but stricter environmental regulations and the availability of gas for power generation required new solutions based on gas engines. Initially, diesel engines were fitted and adapted for gas and dual-fuel applications, but the experiences with the first generation of adapted diesel engines were disappointing. In the course of the second half of the Eighties new generations of dedicated gas engines were developed which performed better and were much more reliable.

In general, availability, reliability and maintenance were major areas of development work (Bauer, 1994). This was due to the fact that maintenance costs of engines are comparatively high (compared e.g. to gas turbines), as well as emergency costs in case of a engine failure. Sophisticated maintenance regimes, initially based on the operators' experience, later on in the Nineties by using on-line control systems for preventive maintenance, helped to reduce these risks and costs to a level which was acceptable to end-users. The new generations of engines played an important role for this because they are composed of fewer moving parts. Many lessons for engine design were also drawn from experiences in the automotive industry.

Major efforts have also been made to develop micro-CHP units below 20 kW. Especially after the second oil price crisis they raised high expectations. In order to develop reliable micro-CHP systems, some manufacturers built on previous experiences in related fields such as engine-driven heat pumps but it was not before 1997 that the different technical and economic problems could be overcome, and a fully reliable product be marketed.[209]

An issue of permanent interest has been the improvement of the electrical efficiency of reciprocating engines in order to boost the output of the high-value product electricity. The improvements made over the last decade have led to diesel engines with electrical efficiencies of over 50%, which, in combination with heat recovery from both the exhaust and the jacket cooling water, allow to reach overall efficiencies of more than 90%. However, in order to keep hot water temperatures sufficiently high for heating purposes, further increases in electrical efficiency may be counterproductive.

Emergence of gas turbines

As in the UK, gas turbines were increasingly introduced from the mid Eighties onwards, with now more than 250 gas turbines being installed in Germany. They range in size from less than 1 MW up to more than 200 MW (Meckel, 1996). For large-scale district heating schemes, gas turbines are used as well, in recent years mainly in combination with steam turbines.

Although higher electrical efficiency is also an issue for gas turbines, their predominant use for raising high-temperature steam implies that a high power production is usually of secondary importance compared to the quality of the heat output. Reliability and availability have been more important preoccupations of the users. In the case of aero-derivative gas turbines, high maintenance costs were, and still are, a critical issue. Consequently, the challenge for gas turbine developers was to provide reliable and low-cost units rather than highest levels of efficiency. This objective was met only in the early Nineties when new stationary gas turbines were tailored for the requirements of CHP.[210] The development of this new generation of turbines only started once a sufficient market potential had been perceived by the manufacturers.

---

209 Fichtel und Sachs developed a 5.5 $kW_{el.}$ packaged unit on the basis of earlier work on heat pumps. It was technically very promising, but economically viable only if a series production could be achieved to exploit economies of scale. Otherwise the units would have been too expensive. Full commercialisation started in 1997.

210 The Tornado turbine by Ruston-EGT was one of the first examples which specifically targeted the CHP market, but also most other gas turbine manufacturers followed the same strategy.

The development of smaller gas turbines was another major R&D area. Now, very small gas turbines are available down to 200-500 kW (Fulton, 1994) and even below, but they are not very sophisticated and reliable yet. Despite the principal advantage of rotating masses (implying less friction and wear), engines are still the preferred option in this size range.

Improvements of control systems, grid connection and system control

Reliability and maintenance costs are key problems especially for reciprocating engines, but also for gas turbines. The introduction of computerised control systems, remote monitoring and preventive maintenance was of major, if not decisive, importance for the success of small-scale CHP. Due to the typical operational profile of gas turbines, i.e. permanent operation without major interruptions, they benefited less from these developments. A second crucial problem concerned the varying gas qualities in Germany. In order to avoid mechanical damage to the engines by changing gas qualities, control systems with very short response times had to be developed to monitor and control each individual combustion process in the cylinders. However, the widespread introduction of sophisticated control systems started later than in the UK, namely only in the early Nineties.

Other areas in which the advances in computer control technology was of major importance are grid connection and safety switches. The introduction of electronic switches and synchronisation for grid connection, which replaced expensive mechanical devices, was a major cost-reducing step. In a transitory phase in the late Eighties the exclusive use of electronic devices was not yet accepted and delayed the widespread diffusion of these new elements.

Packaging, standardisation and norms

Although packaging should be seen as a first step towards providing standardised solutions for different customers, it is still limited to small applications. The larger the applications, the more site-specific are the requirements to be fulfilled. Even small-scale equipment still needs to be fitted to the context of each individual site.

The conditions of each site being different, it was thus considered of major importance to standardise the procedures and norms for design and planning in order to accelerate and improve the implementation of CHP. Technically, this was achieved by defining guidelines for standard specifications, in particular of 50 kW units. In terms of norms this was achieved by the work of the technical associations in Germany, which usually provide the guidelines for planning and designing plants. This represented a major step forwards towards the full recognition of CHP by the communities of engineers in charge of the design and planning of energy systems. Since the introduction of these norms and design principles in the early Nineties,[211] packaged 50 kW units have become the standard small-scale CHP-option in Germany, and they are now provided by several packagers.

211 Guidelines and semi-official standards have been issued by the Association of German Engineers ("Verein Deutscher Ingenieure - VDI") and the Association of German Machine and Plant Manufacturers ("Verband der Deutschen Maschinen- und An-

Environmental performance

Emission regulations in Germany are among the strictest in Europe, with the TA-Luft being formally applied for plants with more than 1 MW of heat output (Jarass, 1995). This means that lion's share of CHP-capacity in Germany is directly affected by it, but not necessarily the majority of plants which are of a smaller size. However, in practice the same or even stricter criteria are informally applied to small units below 1 MW. In the tenders of public investors, for instance, it is often required that the emissions of these small units have to be below half of the values indicated in the TA-Luft. The improvements of technology have kept pace with these requirements. Catalytic converters, lean burn engines and particle filters are now widely used in Germany to comply with regulation. Substantial R&D efforts went into the improvement of the environmental performance of CHP-systems, building in some cases on spin-offs from automotive research.

System integration

A typical learning effect can be identified in relation to system integration. A CHP-unit is part of a larger energy supply system of a site or a city, and it needs to be fitted optimally into this context. The experiences made in the last years allow to optimise CHP for specific conditions, if detailed information about the site itself (e.g. its daily, weekly and annual load profiles and the locations of major loads) can be provided. In order to achieve this optimisation, CHP is increasingly integrated with ancillary equipment such as heat storage, absorption chilling, heat pumps, waste incineration, etc. Such systems are far more complex to design and operate than the standard combination of a CHP-unit with a peak boiler. The optimised operation of such a complex energy plant requires an intelligent control system. In other words, the different innovations mentioned before have to complement each other.

It should be underlined that system integration was mentioned as one of, if not *the* major innovation for CHP in Germany. Building up experiences and learning about how to design and integrate different components into a complex energy supply system allowed to exploit in practice the benefits of CHP which were only known at an abstract and thermodynamic level before.

Ancillary equipment

In order to achieve a good system integration, development efforts were needed to improve the different types of ancillary equipment for CHP-systems. Heat storage systems are a good example in this respect: the reduction of heat losses, operation control and lower investment costs were major R&D-objectives in order to be able integrate them effectively into a site. With regard to district heating, major steps forward have been made, and are still under way, to improve pipeline networks, not only in terms of investment and construction costs, but also to reduce friction and heat losses (Neuffer & Witterhold, 1995; Witterhold, 1995). Small-scale CHP

---

lagenbauer - VDMA"). Both are very influential organisations of engineers and planners.

is now also integrated in gas networks. This means that the competitive gap between gas supply and group/district heating could be bridged, at least technically. As will be described later, such combined heat/gas supply solutions can have adverse economic effects when they are set-up within larger DH-networks.

Technical issue	**Improvement**	**Approximate introduction**
Gas engines	Development of dedicated gas-engines Reliability Use of varying gas qualities Micro engines	late Eighties/early Nineties late Eighties late Eighties mid-Nineties
Gas turbines	Smaller units Efficiency Reliability	from mid-Eighties onwards
Control systems	On-line monitoring Sophisticated operation control Preventive maintenance	early Nineties, now standard
Packaging	Packaged small-scale systems	late Eighties
Standardisation of design	Small-scale standard specifications and design norms	early Nineties
Environmental performance	Catalytic converters Lean-burn engines	early Nineties
System integration	Integrated design of energy supply systems	early Nineties
Piping technology	Cheaper pipe-laying technology	mid-Nineties
Ancillary equipment	Integration of various components	early Nineties

Table 9.1: Improvements of technical characteristics of CHP and their approximate timing in Germany

9.3.2 Expectations for the future

At present, CHP-technology is regarded to have stabilised. The next major impulse is expected to come from the introduction of fuel cells as a radically new type of power source, but several problems regarding reliability and costs need still to be solved. The diffusion process will take a long time due to vintage effects and the reluctance to adopt a new technology for applications that essentially require high reliability. One of the reasons why fuel cells are expected to play a major role in the future is their generic character. They can be widely used for many types and sizes of energy production purposes. In spite of a number of ongoing field trials with fuel cells and first modules being sold to customers,[212] it is expected that it will still take a number of years until fuel cells will become available on a fully commercial basis. In the longer term, one of the great promises of fuel cells consists of the possibility to generate high-temperature steam.

Technology	**Characteristics**	**Importance**	**Expected introduction**
Cooling	Combination of heating and cooling to achieve smoother load curves	high	mid-Nineties
Niche systems	Development of systems dedicated for specific application niches	medium	late Nineties
Fuel cells	Highly efficient power source with low emissions, fuelled by Hydrogen or Methanol, flexible operation by stacking	high	about 2000
Stirling engines	Efficient, multi-fuel combustion with low emissions	low	2000-2005
Micro engines	Reliable, low maintenance and low investment engines for highly decentralised heat/power supply	medium	mid-/late Nineties

Table 9.2: Needs for and expectations of future technical developments of CHP in Germany

New very small units are expected to become available in the next years, although the prospects for engine-driven solutions are rather uncertain for reasons of investment and maintenance costs. Apart from the positive assessment of fuel cells, also Stirling motors are currently considered. Although some first systems have been commercialised since 1996,[213] there are still serious doubts about the reliability of Stirling engines. For the types of systems presently available on the market, little scope remains for further improvement of their core components. Higher temperatures and efficiencies of gas turbines, more reliable engines and control systems, and especially a further price digression for all components are regarded to be the major trends (Nitsch, 1994). Moreover, it is expected that current CHP-technology will be adapted to new niche applications. Several areas appear to be promising in this respect, e.g. the incineration of waste and produc-

---

212 For example, at Ruhrgas in Essen, at HEAG in Darmstadt and at Thyssengas in Essen, 200 kW fuel cell modules have been successfully tested (Lezuo, Knappstein, Langnickel, & Nymoen, 1995). For further information on recent developments of fuel cell technology for power generation, see (Seidel, 1994).

213 See (anon., 1995) and (Menzler, 1996).

tion residues on industrial sites. The main problem to be solved is the integration of CHP into the main production process of such plants. Heating applications of CHP are generally confronted with the problem of insufficient heat load in summer. The combination of heat, power and cold generation is one of the options for addressing this problem. In spite of the moderate climate conditions in Germany, chilling and cooling devices have been introduced as new elements of CHP-systems, and they are expected to find more widespread application in the future. Absorption chilling is one of the key technologies in this context.

9.3.3 Non-technical implications

Together with the onset of small series production, technical standardisation and simplification are still driven down the capital costs of CHP, especially for the most attractive sizes of engine in the range of 100-500 kW and of gas turbine in the 2 to 10 MW area. The expectations concerning micro-CHP (based presumably on fuel cells rather than maintenance-intensive engines) need to be treated with care, but if new units became available at competitive prices for individual households, possibly even supplied by natural gas, they would be a major challenge to the traditional power supply system, questioning also established organisational principles. However, at the moment the perspectives for micro-CHP are still regarded to be highly uncertain. Fuel cells are seen as the major future option for CHP in Germany and represent also an opportunity for making a first step towards a hydrogen-based energy supply system. Notwithstanding the uncertainties associated to hydrogen, such CHP-applications are thus an important strategic niche for companies which plan to get involved in hydrogen technology in the longer term.

9.4 Structural changes

9.4.1 Technological structures

9.4.1.1 Electricity supply structures in Germany

In general, electricity supply patterns in Germany are quite similar to those in the UK, especially in terms of fuel use, centralisation and demand patterns. Heat demand patterns are very complicated to single out, but can be roughly approximated by the differences between non-transport energy demand and electricity demand.

Electricity production and consumption

The general patterns of energy and electricity consumption in Germany reflect the developments in other OECD countries: stagnation of energy consumption, but in parallel with economic growth still an increase of electricity consumption. The composition of primary energy sources for power generation is characterised by a decreasing role of oil, an increase of natural gas, and - in recent years - a stagnation of nuclear power. The changes in Germany have however been rather gradual, aiming at a balanced spread of energy sources.

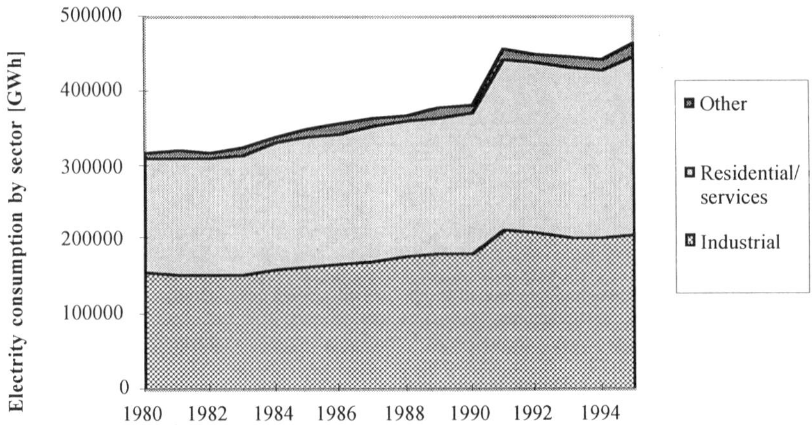

Figure 9.3: Electricity consumption in Germany by sectors, East Germany included since 1991, Source: (BMWi, 1996; ENERDATA, 1996)

Power supply is highly centralised, in spite of a substantial contribution of independent power producers (about 14 % of generation capacity). In terms of size, 92% of power generation capacity is represented by power plants above 150 MW, of which about 8% is provided by industrial plants (VIK, 1994a).

Independent and auto-production of electricity

Although a quite large capacity of new CHP-plants was built in West German industry, industrial and auto-production declined from 19% of power production in 1980 to about 14% in 1995, implying a clear dominance of public power supply companies in the generation business. This trend could only be reversed in the mid-Nineties when the industrial decline in East Germany was stabilised. Apart from industrial and other private companies, the Deutsche Bahn AG operates a number of power plants for auto-supply.

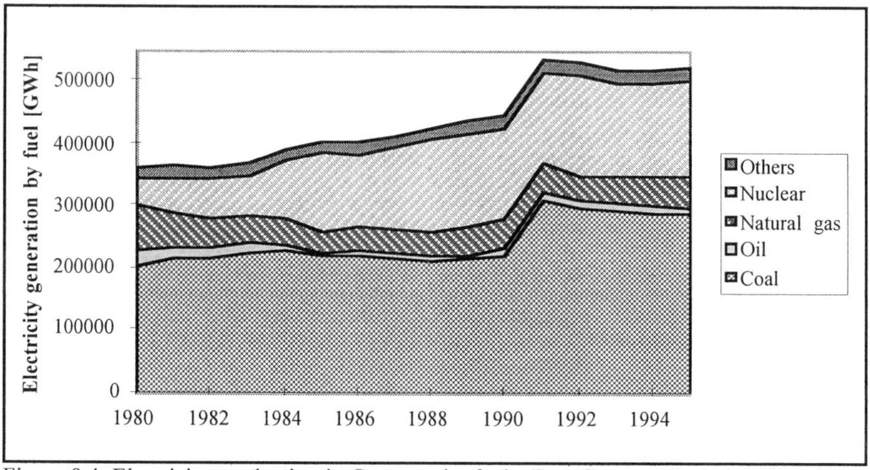

Figure 9.4: Electricity production in Germany by fuels, East Germany included since 1991, Source: (BMWi, 1996; ENERDATA, 1996)

9.4.1.2 Diffusion of CHP in Germany

Several different sources of information had to be consulted to collect diffusion data of CHP in Germany. Despite these difficulties concerning the reliability, comparability and completeness of data, the overall patterns of diffusion can be identified clearly. There is a balance between the two major applications of district heating and industrial use with about 10000 $MW_{el.}$ each (1996). Of this capacity, East Germany contributes about 1700 $MW_{el.}$ to district heating and 1800 $MW_{el.}$ to industrial applications. The capacity of small-scale CHP is comparatively small, but it is the fastest growing application area. Overall, CHP-capacity in Germany reached about 22 $GW_{el.}$ in 1996. However, in terms of capacity share, CHP has changed very little, namely from 18.5 % of installed power generation capacity in 1980 (West Germany only) to 18 % in 1995 (unified Germany). In terms of number of plants, there are now about 3000 small-scale CHP-plants in Germany (ASUE, 1997; HMU, 1996) and about 530 CHP/DH plants in 1996 (AGFW, 1997). Data for industrial applications are not unequivocal, but can be estimated between 500 and 800 plants above 1 MW.

In the last two decades there has been a slow but constant growth of district heating capacity. Industrial CHP declined until about 1990, but has recovered since then. Figure 9.5 reflects this only insufficiently, because from 1991 onwards East German capacity, which has declined due to industrial restructuring, is included. For West Germany only, industrial capacity has grown from 8200 $MW_{el.}$ in 1990 to 8900 $MW_{el.}$ in 1995. Part of this renewed increase can be attributed to the use of gas-turbine CHP systems. The fastest diffusion took place in the field of small-scale CHP which since the early Eighties multiplied its capacity and represented in 1996 about 5.5% (= 1600 $MW_{el.}$) of the entire CHP-capacity in Germany.

Despite the increase of gas-turbines and small-scale reciprocating engines for CHP-applications, there is still a large share of steam turbine systems in operation. According to data from VIK, they represent 60 % of industrial CHP-capacity, but in terms of new plants, gas turbines and engines clearly dominate (VIK, 1996). Single steam turbine cycles are hardly built any more, they are usually combined with a gas turbine in Combined Cycle Gas Turbines plants (CCGT).[214]

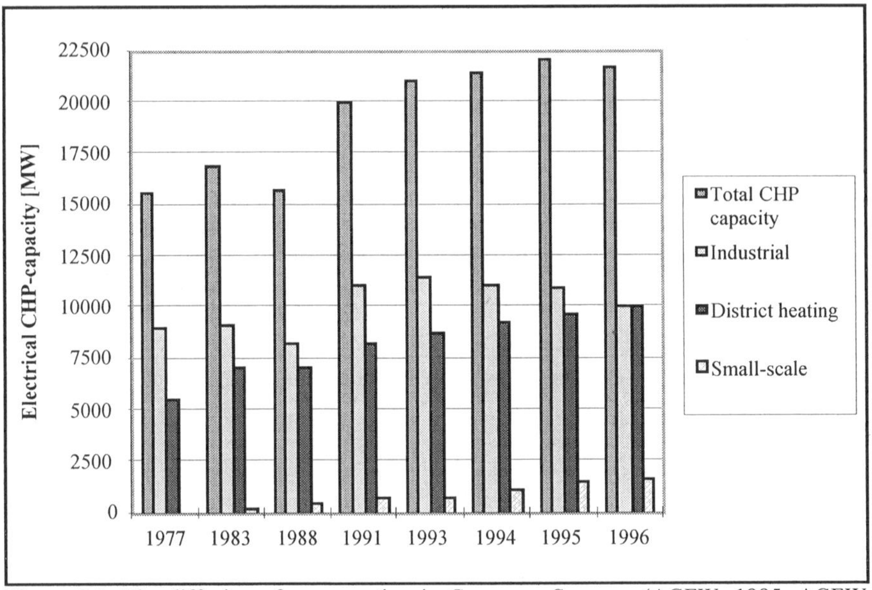

Figure 9.5: The diffusion of cogeneration in Germany, Sources: (AGFW, 1995; AGFW, 1997; AGFW/UNICHAL, 1993; ASUE, 1996; ASUE, 1997; BMWi, 1996; VIK, 1991; VIK, 1997), Data for DH[215] and industrial CHP include East Germany from 1991 onwards, for small-scale CHP from 1983 onwards.

A more refined breakdown of the distribution of CHP by sectors is provided in Appendix 3. It shows that despite the growth of chemical production in Germany, its CHP-capacity has remained fairly constant. The same holds for petrochemicals/refineries, paper and board, and textile industries. Only in the field of mining the capacity has increased substantially, but these changes are partly the result of modifications of statistical accounting principles. The detailed distribution of small-scale CHP can be retrieved from a recent database (HMU, 1996) and its analysis (ASUE, 1996). It is shown that about 35% of small-scale capacity is operated in industry and commerce,[216] another 30% in public buildings such as

214 The German expression is "Gas- und Dampfturbinen (GuD) Kraftwerke".

215 Data for district heating are estimated to be about 10% higher, because the AGFW survey does not cover all DH-plants. A comparison with official statistics leads to estimations of 11200 MW_{el} of district heating for 1995 but even in these figures it is unclear in how far small-scale CHP is included.

216 Data on small-scale CHP below 1 MW_{el} is not included in the industrial statistics.

hospitals, swimming pools and administrative buildings, and 30% in landfill and sewage plants. Gas turbine CHP-plants have been studied separately and their capacity has grown at a similar pace as that of engine-CHP (Meckel, 1996).

These data show that the overall diffusion of CHP is slow but stable in Germany. Small-scale and gas turbine applications grow comparatively fast and contribute to a decentralisation of generation capacity. Apart from East Germany, where many plants are currently being modernised, district heating is stagnating. The picture for industrial applications is more complex. The typical large-scale industries seem to have reached capacity limits, whereas in those industries where small gas-turbine applications have become viable, a continuous expansion can be observed.

9.4.1.3 Future potential

A number of studies have been carried out to assess the future potential of CHP in Germany. Some of them distinguish between technical potential and economic potential. The former, however, is a rather irrelevant number, because the situation in Germany is far below the technical potential. Under the specific conditions assumed by Fenzl et al. (Fenzl & Pummer, 1993), the technical potential for small-scale CHP is estimated between 57 and 113 $GW_{el.}$.[217] For comparison, the power capacity of the entire German power plant park amounts to about 120 $GW_{el.}$. Once economic constraints are considered, the authors arrive at a potential of 5.7 $GW_{el.}$ in Germany.

Other sources come to different results, depending on the assumptions made. According to COGEN Europe, the European association for CHP (CHPA et al., 1995), using data from the German gas industry, capacity will further expand based on gas turbines and gas engines by 1600 $MW_{el.}$ over the next ten years. Dufait et al. come to similar results by estimating between 800 and 2000 $MW_{el.}$ (Dufait et al., 1992). For district heating, a slow growth is expected, depending on the general regulatory framework conditions for the energy supply industries. The development could be faster if the possibility to combine gas supply with group heating schemes was widely exploited. This technically and economically promising alternative combines the advantages of gas supply (no heat losses in networks) with the advantages of CHP (high efficiency) in a kind of decentralised CHP/DH-system.[218] However, the current organisational settings are rather detrimental to its diffusion.

---

217 The higher number is based on the use of 14 kW_{th} CHP-units, and the lower one for the use of 55kW_{th} units. To compare the calculations with other data on electrical capacity potential, it is assumed that the smaller of the two units has an electrical capacity of 8 kW_{el} and the larger one of 25 kW_{el}.

218 In the report of the Enquete-Commission of the German Bundestag, a number of estimates, scenarios for CHP potential in urban areas at different time horizons (2005 and 2020) are provided, both for East and West Germany. They are not reported here in detail, because they use different units of comparison (basically CO_2-emissions) and would require a quite complex analysis.

Reunification and its impact on the East German energy supply system clearly affect the perspectives for CHP, because the situation in East Germany is quite different from that of the Western part. District heating plays a more prominent role due to the prominence of high-rise blocks and the availability of a well-developed heat network infrastructure. In terms of capacity, a stabilisation is expected, combined with the replacement of old coal-fired DH-plants by modern CHP/DH-unit, which will be either gas- or coal-fired. Industrial CHP in East Germany has strongly declined in absolute terms as a result of the breakdown of industrial activity. Any assessment of the future potential depends therefore on the perspectives for industrial recovery in the New Länder.

There is no explicit target for CHP-development in Germany, apart from the general commitment of government to extend its capacity in response to environmental and resource-related concerns. However, in a study carried out for the German Ministry of Economic Affairs, only a small increase of district heating is expected for the next 25 years (BMWi, 1995).

9.4.2 Economic structures and regulatory framework

9.4.2.1 The evolution of the regulatory framework in Germany

In contrast to many other European countries such as the UK, the Netherlands, Spain or Italy, no major reform of the electricity supply industry (ESI) in Germany has been attempted during the Eighties. The Energy Supply Act ("Energiewirtschaftsgesetz") from 1935 was still in force, but a debate on the need for a reform has been going on for some years already. The EC White Paper on Energy Policy (CEC, 1995) has accelerated that process, and at the end of 1996, a first proposal has been made to liberalise the German ESI.[219] Below, the main features of the system until then are summarised.

The regulatory set-up of the ESI in Germany[220]

- The German Energy Supply Act is based on the principle of closed supply areas for gas and electricity companies. They have the right to negotiate exclusive demarcation contracts among each other and exclusive concession contracts with local authorities.[221] This right is linked to the obligation to guaran-

---

219 When finishing this book, i.e. in 1997/98, the new framework was about to be implemented, but it was still too early to assess the impacts of the changes in practice.

220 Based mainly on (Walz, 1994), (CHPA et al., 1995), (Gröner & Smeets, 1988), (Berkner, Herrmann, & Schmitz, 1990; Topp, 1993) and various recent volumes of the journals *Elektrizitätswirtschaft*, *VIK-Mitteilungen* and *Energiewirtschaftliche Tagesfragen*.

221 These concession contracts are concluded over periods of up to 20 years, after which they can be re-negotiated. In a number of cases in the last years, local authorities did not want to extend the contracts with the regional supplier and took over electricity supply under their own responsibility by setting up a local utility company.

tee connection and supply, and consequently to an obligation to invest in power supply. Similar demarcation agreements are also concluded among supraregional electricity supply companies ("Verbundunternehmen") to determine their supply areas. In addition, all electricity supply companies need a supply permission from the regional State authorities. This framework is based on the assumption that the entire supply chain shows the characteristics of a natural monopoly.

- In order to avoid an abuse of the monopolistic supply position, regulatory control is carried out by different authorities in Germany. First of all, electricity pricing and tariffs are regulated for low voltage customers only, i.e. usually small customers such as households and small commerce (price control for tariff customers). Tariff increases need to be justified and approved by regional authorities. Large customers can negotiate their tariffs individually with the regional or supraregional supply company, but the lack of competition gives the suppliers a much stronger position in these negotiations. The pricing principles are outlined since 1990 in a new framework regulation for electricity tariffs ("Bundestarifordnung Elektrizität"), but they apply to tariff customers only. They are based on the requirement of secure and economic supply, but indicate also the need to take the rational and efficient use of electricity, the protection of the environment and the protection of resources into account, though without specifying them in detail.
- A second element of regulatory control is investment control ("Fachaufsicht"). In practice, supply companies have to inform the regulatory body about major investment projects. They can be accepted or rejected, depending on whether they fulfil the principles of economic and secure supply.
- As a third element, cartel or monopolies control ("Kartell- oder Mißbrauchsaufsicht") is carried out by the Monopolies Commission, especially for those customers who are not subjected to price control, i.e. large industrial customers with individually negotiated contracts ("Sondervertragskunden").
- Technical grid connection conditions are not subjected to any regulation, but the norms as defined by technical associations and by the association of electricity utilities (especially VDEW) play the role of a flexible, but very important orientation for all public utility companies.
- Basically, two regulatory levels need to be distinguished for the three regulatory tasks of investment control, price/tariff control and monopoly control. The two former are supervised by the State authorities at regional level ("Bundesländer"), and the latter is jointly controlled by State and Federal cartel commissions. However, legislative responsibility for all energy supply affairs is with the Federal level.
- In contrast to electricity supply, gas supply is subjected to competition among different supply companies at regional and supraregional level, but not at local level. The situation at local level still differs from that of electricity supply because there is no obligation for gas utilities to connect customers. Correspondingly, customers can choose the form of heat supply they prefer, i.e. there is competition between coal, gas, district heat and electricity.

- The possibilities for competition in electricity supply are still limited in Germany, despite some first attempts to facilitate contract energy management and independent power generation. CHP is not specifically considered in the German regulatory framework. The new principles of the framework regulation for electricity tariffs have changed very little, either.

9.4.2.2 The conditions for CHP in the ESI

In contrast to the UK, the basic structure of the German energy supply system has remained intact over the last decades. It is based on energy utility companies ("Energieversorgungsunternehmen - EVU") which operate at three levels (see Figure 9.6). The types of major actors playing a significant role for the economic operation of energy supply are the following:

- Nine supraregional generation companies; these are RWE, Bayernwerk, VEW, EVS, Badenwerk, HEW, BEWAG, PreussenElektra, and VEAG which was founded after reunification to supply East Germany. It is owned by the West German supraregional generators, mainly RWE, PreussenElektra and Bayernwerk. The supraregional generators also in charge of bulk transmission grids and high-voltage distribution.
- About 70 regional supply companies, which operate a certain amount of generation capacity and are responsible for medium voltage transmission and distribution, especially in rural areas.
- About 850 local utility companies, which are often owned by local authorities, even if in many cases a regional or supraregional company holds a major share. Many of them are horizontally integrated electricity, gas, heat and water suppliers.
- Industrial and other independent power producers, including e.g. also the Deutsche Bahn AG.
- State and Federal regulatory bodies for environmental and energy regulation.
- Government departments, at Federal, State and local level, dealing with economic, environmental, technology and construction/building policies.
- Energy service end-users, including local authorities, public institutions like hospitals and universities, commerce and industry, etc.

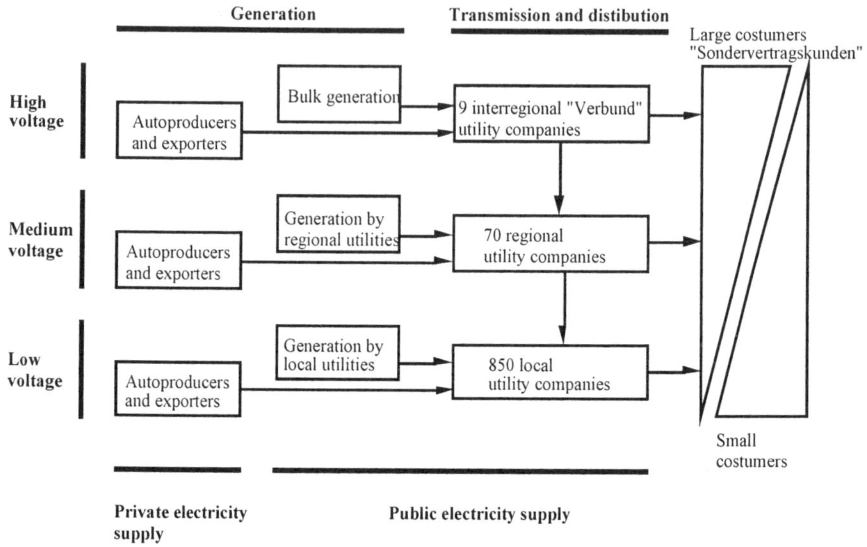

Figure 9.6: The organisation of the ESI in Germany

There is a complex network of formal and informal links between the different supply companies (Gröner & Smeets, 1988; Ruddy, 1994; Walz, 1994). Especially the supraregional companies hold shares in a variety of regional and local utility companies. This gives them a central role in the German system.

While electricity supply is based essentially on closed supply areas ("geschlossene Versorgungsgebiete") to guarantee the monopoly position of the supply company, gas supply is a more competitive business. Ruhrgas and Wintershall are the main supraregional competitors, while at local and regional level many electricity utilities also run gas supply divisions.

The possibilities for entering the electricity market, either as a generator or a supplier, are strictly limited. Due to the monopolistic set-up, costumers are bound to be supplied by their regional or local utility. Self-generation and supply are subjected to a number of barriers. Different licences need to be obtained, and due to the rights of way given to local authorities, they can impose charges for connecting sites across public roads. Third Party Access, i.e. the supply of electricity from outside the demarcated supply area by another company is not possible in Germany, but considered in the currently ongoing reform plans.

Generation licences were difficult to obtain in the past. According to §12 of the 3rd Power Generation Act ("3. Verstromungsgesetz"), units larger than 10 MW required a generation licence. The complexity of the administrative procedures made the realisation of gas-fired CHP in the range between 10 and 30-50 MW nearly impossible.[222] In practice, however, these rules are hardly applied any

222 This legislative act was based on the argument that coal should be preferred to oil or gas as fuel for power generation, but coal-fired CHP-plants are economically viable

more. For units beyond 1 MW an operator is now only obliged to indicate its construction plans to the regulatory body which can intervene if deemed necessary. Plants larger than 10 MW need to be checked by investment control. If third parties are supplied (e.g. the local utility company), a supply licence is already needed beyond 1 MW (ASUE, 1992).

Most local and regional EVUs are horizontally integrated, i.e. they supply both gas and electricity, in many cases also heat. Local utilities are often involved in other services as well, e.g. water supply, swimming pools, or public transport. In other words, at the local level horizontal integration allows internalising the benefits from cogeneration and from the co-ordination of heat and power generation. It is further facilitated by the possibility to exclude competition between network construction for gas and district heating.[223] Horizontal integration is also common at regional level, but heat and power supply tend to be separated in different divisions ("Spartenunternehmen"). This leads to an optimisation of the individual supply chains rather than to their co-ordination.

Although in principle the different levels of energy supply appear to be separate, this is not the case in practice, because of the shares held by regional and supraregional utilities in companies at the lower levels. In addition, the major companies are mutually interwoven by cross-ownership. Consequently, the power generation policies of local utilities are not fully independent and have to obey in many cases the interests of the supraregional owners regarding the distribution of their centrally generated power.

However, the comparatively strong position of local utilities and the difficulties encountered by independent generators in the German system have another consequence: in practice, most energy-related services are under the control of public utility companies, including also most of the technical interfaces. Due to this key role, energy management or contracting solutions tend to be operated by jointly owned companies of utilities and end-users, a model which has been implemented frequently in East Germany. Some local utility companies have even explicitly subscribed to an approach of energy service supply as opposed to energy product supply, applying least-cost-planning principles and demand side management. Consequently, private contract energy management (CEM) and similar service activities are not very widespread in Germany. They are viable solutions for a limited number of cases only, mostly in industry. The situation has changed slightly in the last years with the emergence of companies which concentrate specifically on service contracting. Also some of the small-scale CHP-packagers have

---

only beyond approximately 30-50 MW.

223 The importance of the local utilities and their horizontal integration for the relative success of DH in Germany has already been highlighted in (Jochem & al., 1986) and (Gruber et al., 1993). In contrast to the complementarity of gas and electricity supply for CHP, there is competition between gas supply and district heating. Usually, the two pipeline-bound supply systems for heat services compete for the same clients, and both require the construction of costly networks. In horizontally integrated local utilities, this competition can be excluded by defining areas to be supplied by heat or by gas. Even if gas is supplied by a different company than heat, informal agreements are often made to assign supply areas.

entered this field of activity, and similarly the utility companies have started to set-up subsidiaries which deal with contracting solutions. However, they usually do not provide as far-reaching services as their British counterparts.

9.4.2.3 Electricity and gas prices

As mentioned before, regular electricity tariffs for franchise customers are subjected to price control. Increases of tariffs require the approval of the regulatory authorities. In practice the utility companies are in a very strong position because they have access to all the necessary data to justify higher tariffs. Furthermore, the staff situation of the regulatory authorities does not allow in-depth investigations and controls (Walz, 1994). Large customers need to negotiate electricity tariffs individually with their utility company. They are subjected to cartel regulation only. The improved opportunities for industrial cogeneration have at least allowed these large users to strengthen their negotiation position because they are now able to compare the tariffs offered with the costs of setting up an independent CHP-plant. On the other hand, the power utility companies can adjust their power prices right below the level which would make autoproduction economical.

In general, the comparatively high electricity prices in Germany (see Figure 9.7) have represented an advantage for the diffusion of autoproduction and CHP, because they make any kind of energy efficiency investment interesting. One of the reasons for these high prices was until 1995 the coal subsidy ("Kohlepfennig") which raised electricity prices by about 10%. It was abolished in response to a decision of the Federal Constitutional Court ("Bundesverfassungsgericht"), leading to a corresponding decrease of electricity prices from 1996 onwards.

The economic viability of CHP depends decisively on the price differential between natural gas (which is now the major fuel for new CHP-plants) and electricity prices. Their development is depicted in Figure 9.7. A second major determinant of the economic viability of CHP are the different export, top-up and back-up tariffs for electricity. In theory, remuneration of exported power should be based on the principle of avoided costs. As indicated in the framework regulation for electricity tariffs, long-term aspects, such as e.g. capacity terms and environmental impacts, should be integrated into the calculations, but in practice this is very difficult. Due to the lack of more specific regulations,[224] the pricing for CHP-based and industrial generation is determined by the voluntary so-called Associations' Agreement on export tariffs. It is explicitly based on the consideration of avoided costs and takes now also capacity terms into account. Updated several times since its introduction in 1979 (anon., 1979), the last time in 1994 (VIK, 1994b; VIK, 1995), it provides the basic guidelines for pricing arrangements between industry and public utilities. It applies also to other power exporters outside the official frame of reference which considers only plants between 5 and 10

---

224 In spite of many favourable claims and initiatives, cogeneration has not been included in the Electricity Feed Act ("Stromeinspeisungsgesetz") which prescribes high export prices for power generated from renewable fuels. However, the entire debate favoured higher export remunerations also for CHP.

$MW_{el.}$.[225] The agreement distinguishes between recommended tariffs for guaranteed and for non-guaranteed delivery. In practice, the export tariffs are at a level of 50-80% of the standard electricity tariffs (CEC, 1994; CHPA et al., 1995; Dufait et al., 1992; Walz, 1994), and this ratio has not changed significantly over the last years.[226] Although the dominant position of the utility companies has been challenged in the last years, their weakly regulated monopoly still enables them to control the emergence of decentralised generation capacity, not only by means of electricity pricing but also by defining technical grid conditions. The agreements achieved in the last years reflect the minimum consensus on these issues among the major stakeholders and interest groups.

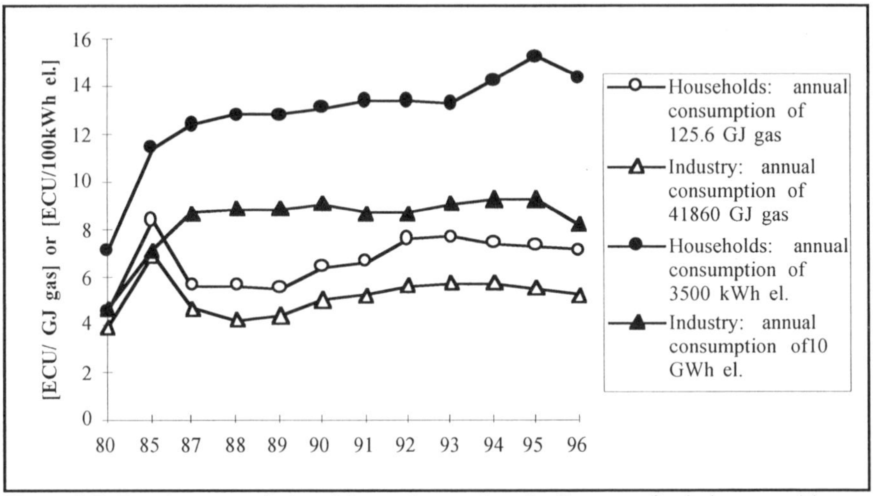

Figure 9.7: Development of electricity and gas prices in Germany for industry and households, Source: (EUROSTAT, 1997)

As pointed out by Walz (Walz, 1994), the structure of electricity prices plays also an important role as an incentive for energy efficiency, and thus for investment in technologies like CHP. Apart from the detailed specifications of differentiated tariffs (day/night, summer/winter), the electricity price digression with growing consumption is regarded as a major disincentive for energy efficiency.

The possibilities for investment in decentralised power generation is closely linked to the financial and planning competencies of the organisations responsible for such projects. The situation of public bodies like local authorities, universities, hospitals, etc. is critical in this respect. Although they are also subject to financial constraints, their possibilities are much better than in many other European countries (Gruber et al., 1993). A specific German problem is the detrimental impact of

225 In the formulation of 1994, there are no clear limits foreseen which in the end has lead to very flexible interpretations.

226 The guidelines of the association of public electricity utilities have enabled this stability (VDEW-remuneration model of 1991, updated in 1994).

the dual finance system in the health sector on CHP (Nitsch, 1994, p. 77 ff.). As the benefits of installing CHP-systems in hospitals are accrued by the operators (i.e. the health insurance companies), but the higher initial costs carried by the investors (i.e. mainly regional authorities), the interest structure is detrimental for CHP. Similar problems can be identified in the education system. As far as could be concluded from the case studies, the discretionary power of the regional authorities ("Ermessensspielraum") is decisive in this respect. By taking into account complementary selection criteria, they can often justify the additional funding needed for a CHP-scheme.

9.4.2.4 Principles of environmental regulation

The most important pieces of environmental legislation in Germany are the Federal Immissions Act ("Bundesimmissionsschutzgesetz") and its specifications, especially the Clean Air Code ("Technische Anleitung (TA) Luft") (Jarass, 1995). German environmental regulation relies on the prescription of standards to be achieved for certain types of applications. These standards are regularly updated. Reliable provisional regulations exist for the upgrading older plants, taking into account their specific requirements.[227]

Despite the prescriptive approach of the TA Luft, a key principle of environmental policy in Germany is the delegation of responsibilities for implementing environmental objectives to sectorial bodies. Especially industry and its top organisations are frequently assigned the responsibility for reaching environmental objectives, based on the conviction that they know much better than any regulator how to achieve the best possible solutions without endangering their competitiveness. Obviously, this approach requires a well-organised industrial governance system which can enforce the necessary commitments to be made by individual companies. This approach can be exemplified by looking at the case of demand-side measures for energy efficiency. While the heat insulation regulation ("Wärmeschutzverordnung") of 1994 prescribed standards for insulation of houses (BMBau, 1994), the planned heat use regulation ("Wärmenutzungsverordnung"), which foresaw similar prescriptions for industry, was replaced by a self-commitment of industry to reduce energy consumption according to the CO_2-targets of German government.

The principle of subsidiarity with respect to State and local authorities is applied in the field of environmental policy, but also in many other policy areas. For example, local authorities are in charge of implementing and monitoring environmental regulations.

---

227 Major technical specifications for CHP are the Small and Large Combustion Plant Codes ("Großfeuerungsanlagenverordnung - GFAVO" and "Kleinfeuerungsanlagenverordnung - KFAVO") where the emission limits for large and small generation and combustion plants are described in detail.

9.4.2.5 Technology supply side

Although there are several traditional engine manufacturers in Germany, especially for heavy diesel engines (e.g. MWM, MAN, KHD, MB Halberstadt and Blohm & Voss), there is no manufacturer of small gas turbines. Only at the upper end of the power range, Siemens and ABB maintain production sites in Germany. Other typical CHP-components are supplied by a large variety of firm. The most important companies involved directly in the CHP business are therefore the system packagers who provide standardised solutions to their clients. They emerged in the early Eighties with the introduction of small-scale CHP. At the beginning, this market segment was dominated by newcomers on the electrical engineering market, but in the meantime a few established engine manufacturers have come to dominate the market.[228] The first modules used standard components, which were assembled according to the scarce data available about the site conditions of typical users. Many of the early suppliers have left the market in the meantime, with some successful newcomers remaining in parallel with the established manufacturers. As shown in Figure 9.8, only a small number of utility companies have actively entered the CHP engineering business.[229] One of the reasons must be seen in their good and stable relationships with the established manufacturers which moved into the CHP-business in the mid-/late Eighties, i.e. at a time when utilities started to become interested, too. In other words, the traditional division of labour between utilities and manufacturers was maintained for quite some time.

Entry to the engine business was relatively easy because it is a standard technology which is well understood and easy to apply to different applications. Solutions based on small gas turbines are on the contrary limited to a small number of companies (mainly MWM, MAN, KHD and EES). They have exclusive contracts with one of the main gas-turbine manufacturers (EGT, Solar and Kawasaki). In addition, the scope for standardised and packaged solutions is much more limited for gas turbines, so that the entry to the market is also restricted from the technical point of view. Finally, Germany is in general a difficult market to enter for foreign companies due to the established links between suppliers and manufacturers, and the rather complex planning and approval procedures.

Especially in the Eighties, relatively close links were established in a few cases between CHP-users and packagers in order to learn about the actual requirements and build up the experience needed to improve the technology. However, the user-supplier relationships were less dense than in the UK, where most manufacturers aimed at the supply of full energy services on a contracting basis. Especially the

---

228 This becomes obvious when looking at the currently available CHP systems and their manufacturers in the small-scale CHP database (HMU, 1996). Only for mini-CHP below 50 kW, the market is still very unstable. In 1995, about 40 different companies were manufacturing mini CHP-units (Wraneschitz, 1995).

229 The active involvement of utility companies is usually restricted to project management to set up CHP-plants, but they do hardly invest directly into packaging. One of the exceptions is Ruhrgas, the major gas utility in Germany, which holds shares in EES (Erdgas Energie Systeme), one of the most successful newcomers on the small-scale CHP-market.

established German engine manufacturers have been criticised for not being responsive enough to the needs of their clients - in contrast to the newcomers. A gap was left in many cases between manufacturers and users which was closed only once the utility companies became more interested in CHP, and started to set up joint ventures and contracting solutions with their customers.

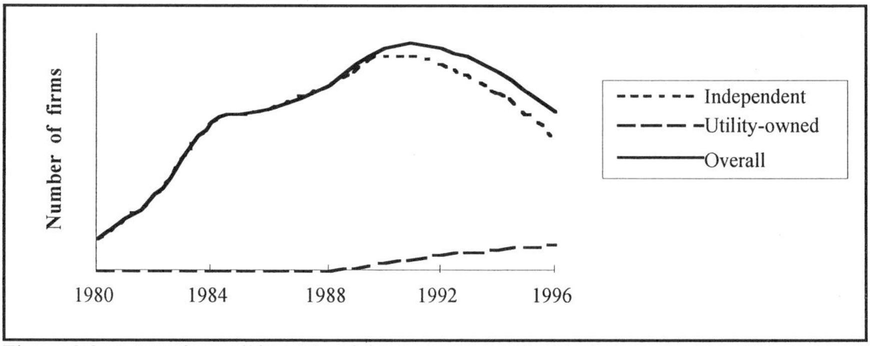

Figure 9.8: Approximate life-cycle of the German small-scale CHP industry

Basically, the economic structure on the technology supply side seems to be more stable again after the more dynamic phase in the early and late Eighties. The most important recent development is the careful attempt of foreign, especially Dutch, manufacturers to enter the German market. However, as confirmed in the interviews, they do not yet play a visible role, neither in terms of realised plants nor in terms of participation in calls for tenders.

9.4.3 Structures for the provision of technological information

In general terms, the links between users and suppliers are relatively weak in Germany. Information supply and exchange is to a large extent concentrated around technical and industrial associations. They play a critical role for the establishment of networks, especially for the dissemination of information and the definition of standards. Government institutions keep a low profile with regard to information provision and the stimulation of exchange of experiences. The main actors in information supply on cogeneration are - apart from the manufacturers themselves - the following:[230]

- District Heating Association ("Arbeitsgemeinschaft Fernwärme - AGFW") which is part of the Association of German Electricity Utilities ("Vereinigung Deutscher Elektrizitätswerke - VDEW");
- Association of Industrial Energy and Power Producers ("Verband der Industriellen Energie- und Kraftwirtschaft - VIK") which is representing mainly industrial power generators;

---

230 Actors who are only loosely linked to electricity supply have not been included in this list, like e.g. the gas and water suppliers associations.

- Association of Municipal Companies ("Verband Kommunaler Unternehmen - VKU") which is the association of public local utility companies;
- Association of Regional Energy Supply Companies ("Arbeitsgemeinschaft Regionaler Energieversorgungsunternehmen - ARE") which represents the regional utility companies, and the association for the joint operation of the grid ("Verbundgesellschaft - VBG");
- Association for the Economical and Environmentally-Friendly Consumption of Energy ("Arbeitsgemeinschaft für sparsamen und umweltfreundlichen Energieverbrauch - ASUE") which focuses on gas-based CHP with engines and gas turbines; it is supported by the gas industry;
- Association of German Machine and Plant Manufacturers ("Verband des Deutschen Maschinen- und Anlagenbaus - VDMA") where most CHP-manufacturers are organised; it plays a key role for setting standards, jointly with the
- Association of German engineers ("Verein Deutscher Ingenieure - VDI"), which is responsible for the definition of the so-called "VDI-Richtlinien", i.e. the planning and design guidelines for engineers;
- Universities and research institutes, which on the one hand carry out scientific-technical research, and on the other hand work and give advice on technical and economic aspects of energy supply;
- Federal ministries for Economic Affairs ("Bundesministerium für Wirtschaft - BMWi"), for Education, Science, Research and Technology ("Bundesministerium für Bildung, Wissenschaft, Forschung und Technologie - BMBF"), for the Environment ("Bundesministerium für Umwelt, Naturschutz und Reaktorsicherheit - BMU"), for Construction, Spatial and City Planning ("Bundesministerium für Raumordnung, Bauwesen und Städtebau - BMBau");
- State authorities which are running smaller programs for information supply, support of standardisation and monitoring of CHP plants.

Most of the associations mentioned fulfil also a political function by representing the interests of their members, e.g. the AGFW as part of the association of electricity utilities, or the VIK as the organisation of industrial generators. ASUE, which is mainly technically oriented, is complemented at the political level by the Support Association for Small-Scale Cogeneration Systems ("Fördergemeinschaft Blockheizkraftwerke - FGBHKW"). However, in spite of their political differences, they all co-operate often on technical, informational and statistical issues.

The definition of technical norms and guidelines by VDI and VDMA played not only an important role for standardisation, but also for the dissemination of information. In the early Eighties, when CHP was not yet included in the best practice guidelines of the VDI ("Stand der Technik"), it was still considered as an 'exotic' technology by many engineers and planners. Once it was recognised in these guidelines and design standards in the early Nineties, it was taken seriously and adopted.

Also university and other research institutes are well connected to the entire CHP debate; they are particularly active in advising policy-makers by providing expertise on the technical and economic implications of CHP. This indicates the fairly technical character of the entire debate on energy supply in Germany.[231]

9.4.4 Political and regulatory structures

9.4.4.1 CHP in and the political decision structures

Although Germany is usually said to be characterised by a rather closed political system (Kitschelt, 1983), it is far from being monolithic. Although it might be difficult for new actors to enter the political arenas, the decentralised structure with strong State and local authorities provides good conditions for a broadly spread representation of heterogeneous interests. The diversity of power constellations offers opportunities for decentralised energy policies by local energy utilities as well as many different entry points for lobbying.

In this context, the importance of the distribution of power between State and Federal level needs to be mentioned. It offers space for experimentation and variety within the Federal framework. On the other hand, this situation makes it difficult to implement a fully homogeneous policy approach without the support and the co-operation of the other levels of government.

Another characteristic feature should be seen in the delegation of competencies to the ESI in order to facilitate its self-organisation. Government tends to intervene only if no satisfactory arrangements can be concluded independently among utilities and users. However, the threat of legislative intervention needs to be credible in order to force the partners to achieve the necessary agreements.

The preparatory phase of policy-making on energy-related matters at national level takes place in a consultation process between the Ministries of Economic Affairs, of Environment, of Research and of Construction. While on the one hand they receive important inputs from industrial associations and other interest groups, they build on the other hand also on the more independent advice from technical associations and research institutions. The involvement of interest groups is compulsory in the formal consultation procedures for any major legislative initiative. This guarantees, that despite the greater financial and informational resources of industrial associations, also environmental and consumer interests are represented in a balanced way. Also by means of informal lobbying several interest groups in favour of CHP managed to exert influence on the political decision-making processes over the last fifteen years, e.g. in order to channel the growing importance of environmental issues in the public. Both Parliament and ministries are targets of these informal lobbying activities.

---

[231] Although it is not the focus of this study to look at cultural aspects, the case of cogeneration is also interesting with respect to the role and reputation of the academic and technical experts from science and engineering. In contrast to the UK, they are fully integrated in the political debates as more or less independent experts.

Also on the industrial side, the interest structures with regard to CHP are far from being uniform. Although the public utilities count among the most powerful lobbies in Germany, they are often confronted with opposition from a similarly influential interest group, namely the industrial generators. Although they have found a 'modus vivendi' by signing the Associations' Agreement on the remuneration of exported power, their positions on issues such as liberalisation and third party access are clearly opposed. The small-scale CHP area is rather weakly represented at the political level, but it has improved recently with the foundation of the support association for small-scale CHP ("Fördergemeinschaft Blockheizkraftwerke - FGBHKW"), which is promoted by the gas and water industry. Also the VKU, representing the interests of local utility companies, has become quite positive about small-scale CHP.[232] Moreover, the gas industry is also directly represented by the Federal Association of the Gas and Water Industry ("Bundesverband der Gas- und Wasserwirtschaft - BGW"). Other smaller support associations, e.g. the Association of Operators of Engine-based Power Stations ("Betreiberverband Motorkraftwerke"), operate as platforms for technical exchange between small CHP-users, but they also get involved in the political debates

The main interest groups involved in policy-making are clearly the industrial associations (VDEW, VIK, VKU, ASUE, ARE, VGB) and the FGBHKW. Those who are rather under-represented in the political process are the small-scale CHP users and the manufacturers/packagers.[233] Although the latter are organised in the associations of their respective industries, CHP-manufacturing or packaging represents only a small area of their activities. An important role in the political debates has also been played in the last years by research institutions. The importance of a scientific foundation for the arguments put forward by the interest groups has led to the commissioning of several studies to support their positions. This, however, has in the end rather obscured the evaluation of CHP than clarified it.[234]

This system of multiple balances lends itself to a rather evolutionary approach to political decision-making. A fast and radical reform of the ESI can hardly be imagined within the German context, and both the debates on the reform of the Energy Supply Act and on the Internal Energy Market in Europe have underlined the preference for an incremental approach and transition process.

---

232 The founders of the FGBHKW are the Federal Association of the Gas and Water Industry (BGW), a subsection of the Association of Municipal Companies (VKU), and the Association for the Economic and Environmentally Friendly Consumption of Energy (ASUE).

233 This analysis is based on two major sources, namely a) the information obtained from the different ministries on the involvement of external actors in the preparation of legislation and regulation, and b) the participation of actors in the Enquete-Commission "Schutz der Erdatmosphäre" of the German Bundestag, where CHP was one of the important topics (Enquete-Kommission, 1995).

234 Cf. similar results by Henschel and Wiedemann on the communication processes about nuclear power and renewables (Henschel & Wiedemann, 1993).

9.4.4.2 Political auto-control of decision-making

The decentralised structure of political decision-making in Germany makes it very difficult to identify clearly a centre of control. The competencies are widely distributed among different levels and branches of government. Especially during the legislative process at Federal level formal and informal consultation mechanisms operate in parallel and enable the integration of inputs from a large variety of actors. In addition, consultation processes among different ministries and among different levels of government contribute to adjusting different opinions. States ("Bundesländer") participate directly in the decision-making process at Federal level, e.g. by launching initiatives in the Upper House of Parliament ("Bundesrat"). They are also consulted during the preparatory work in the ministries, not to forget their discretionary powers in policy implementation. In other words, there is a very close inter-locking between interest groups and policy-makers at different levels.

Obviously, some interests are better organised and more influential than others. Especially the VDEW, VKU and VIK play key roles in the general energy policy debates. However, the picture of the 'corporatist state' appears to be too simple, because there is no exclusive representation and access for selected organisations. As there are manifold points of entry in the German system, a wide range of perspectives is recognised and reflected in energy and environmental policies. Political auto-control and independence of partial interests are thus guaranteed by a wide distribution of power.[235]

9.5 Behavioural aspects: Decisions, interactions and motivations

Within the three decision arenas (government policies, technology innovation and technology adoption) the decisions, interactions, perceptions and motivations of the different actors are now going to be studied. The delimitation between technology suppliers and users may sometimes appear to be arbitrary but it is due to a growing overlap of the activities of traditional manufacturers and energy supply companies.

9.5.1 Political realm

As for the UK, the decisions in the political realm are presented according to different policy areas (political auto-control, technology policy, energy policy, environmental policy, etc.) for which selected examples are then studied in detail.

---

[235] This also shows that the capture theory does not apply without modifications to the situation in Germany. It may be true that some interest groups have a very strong influence at federal level, but there are many 'checks and balances' which make it very difficult to manipulate government as a whole.

9.5.1.1 *Public policy decisions*

Measures of political auto-control

The structures of energy policy-making enable the participation of many different actors in that process, at different levels and representing different social interests. Although this balance of influences may be somewhat biased in favour of the most powerful actors, little has changed during the last two decades in terms of its set-up. The recognition of environmental drawbacks of energy supply has become increasingly important, and this is reflected in the growing integration and acceptance of environmental interest groups as partners in the political consultation processes. The structure of the ESI, as well as the main government-industry interfaces, were also replicated in East Germany in the course of reunification.

In terms of decentralisation of competencies, there have been several cases in which local authorities have tried to re-establish local control over their utility services by deciding to set up local utility companies after the end of the concession contracts with their regional suppliers. However, these have been exceptional cases only, compared to the growing influence which regional and supraregional utilities exert on their local counterparts.

Energy policy regulation

In contrast to the UK and many other European countries, the energy supply sector was not significantly reformed until 1997/98, in spite of a long-standing debate on a reform of the Energy Supply Act of 1935. The situation for CHP was improved rather gradually by changing and smoothening specific regulations.

Although a specific target for CHP capacity has never been set in Germany, support for CHP rose in the course of the Eighties and Nineties as an element of an energy strategy which is essentially based on the diversification of energy sources, on efficiency increases, on a declining role of indigenous coal, and on a freezing of current nuclear power plant capacity. The priorities of German energy policy for the Nineties are outlined in a background document of the Ministry for Economic Affairs (BMWi, 1992). On the one hand, the need to consider environmental and resource considerations in parallel with an economic, secure and diversified supply is underlined, and on the other a stronger emphasis is put on the use of economic instruments in energy and environmental policy. This is complemented by high levels of priority given to the integration of the New Länder and the adjustment of the German energy supply framework to the envisaged European Internal Energy Market. In other words, although strongly disputed, there is a long-term policy for energy supply in Germany, and cogeneration has a clear role to play in it.

The situation in Germany is still affected by the specific role of indigenous coal. This led to artificially high power prices due to the "Kohlepfennig", a special levy on electricity prices to support the national coal industry.[236] The support for in-

[236] It amounted to about 7-8% of the electricity bill, but was abolished in 1996.

digenous coal is decreasing, as reflected in a number of adjustments of the coal policy framework in the last years. The major arguments raised against the coal policy build on perceived dangers for the competitiveness of electricity-intensive industries which have to bear the costs for the coal subsidies.

The reform of the Federal Tables of Electricity Charges ("Bundestarifordnung Elektrizität") in 1989 was another regulatory decision with an impact on CHP. By defining the elements to be considered in the avoided costs of a utility purchasing independently generated power, an increase of exchange tariffs between utilities and CHP-operators became possible (Berkner et al., 1990). Tariffs had to reflect not only the fuel and capacity costs of the purchasing utility company, but also take into account the beneficial impacts of the rational and efficient use of electricity, and the protection of resources and the environment. In spite of this implicit requirement to consider external costs, it is still hardly possible to operationalise them. Consequently, they are not fully reflected in the tariffs for purchase and export of independently generated power.

A number of minor regulatory changes and decisions removed barriers to the diffusion of CHP. One of them concerns the ways of right ("Wegerechte") and the corresponding levies which can be raised by local authorities if private firms want to transfer electricity directly across public roads to other sites without using the utility company's grid. The current rules were challenged by national and European court decisions, but not yet been changed.[237] Other, informal changes have led to the practice of not applying the full licensing requirements of §12 Power Generation Act ("Verstromungsgesetz") any more, and to the facilitation of dual finance CHP-projects in the health and education sectors. The practical application of these two rules depends very much on the discretion of the State authorities. Other measures, e.g. enforce the application of least-cost-planning and demand-side-management by utility companies, were discussed but not implemented (BMWi, 1994).

A major issue in relation to renewable energies, but also of relevance to CHP, is the Electricity Feed Act ("Stromeinspeisungsgesetz"). It guarantees higher export tariffs for electricity produced from renewable energy sources such as wind, solar and waterpower, following the argument that they displace the use of fossil fuels and thus help reduce CO_2-emissions. It has been claimed frequently that these export tariffs should also be applied to CHP because of its high efficiency and comparative displacement effect on fossil generation. The power utilities which have to pay the higher tariffs argue that this would be an additional subvention which distorts the market and leads to an increase of electricity prices (Bartsch & Dingeldey, 1995). Consequently, the main guideline for export tariffs continues to be the Associations' Agreement, signed by the industrial top organisations as an attempt to reduce the high level of uncertainty associated to autoproduction.

Much support was lent to CHP by the Enquete-Commission of the German Bundestag. It recognised fully the advantages of CHP and recommended an extension

237 See for example the Almelo-decision' dealing with cross-border transfers of electricity (Hancher, 1995).

of its use. The results of the Enquete-Comission were summarised in a catalogue of recommended measures (Enquete-Kommission, 1995). Some of them have been implemented since then.

Environmental policy

In addition to the principles of environmental regulation as outlined before, several specific measures have been taken, or at least envisaged, in the last fifteen years in order to reduce environmental impacts and CO_2-emissions, the latter in response to the Rio-agreements (25-30% reduction until 2005 compared to 1990):

- The Clear Air Code ("TA Luft") with its revisions and amendments is the most important legislative instrument to achieve the different environmental objectives of German government. It prescribes emission limits for different types of plants, and outlines also the environmental licensing procedures to be followed. It is important to note, that although the strict emission limits of TA-Luft are not applied formally to small-scale applications under 1 MW, they are informally required in many cases in order to comply with "best practice" ("Stand der Technik"). Depending on the size of plant, the supervisory functions are delegated to different levels, down to the chimney-sweepers who need to inspect the heating installations of households.
- A more efficient use of energy was the target of the regulations for heat use and insulation ("Wärmenutzungsverordnung" and "Wärmeschutzverordnung"). They aimed at achieving the CO_2-emission targets. The latter one came into force in 1995 and enforces standards for insulating houses and for limiting waste heat in industry. The first one, i.e. the heat use regulation, was never implemented due to the self-commitment ("Selbstverpflichtung") of German industry to reduce CO_2-emissions by 25% until 2005.
- The fierce debate on CO_2- or carbon taxes has - for the moment - been stopped and postponed until a European-wide initiatives will be taken. Although in principle, German government favours market-based principles, no agreement could be achieved on how to reach an optimum balance and formulation of such a tax instrument. For example, it was regarded as essential to consider the situation of certain energy-intensive industries which have already made substantial efforts to reduce their energy consumption. With regard to CHP, the critical point was whether a CO_2-tax should be formulated as an "input tax", i.e. based on the amount and type of fuel used for producing energy, or as an "output tax", i.e. based on the amount of energy produced. Obviously, in the latter case, CHP would have been put at a disadvantage.

Diffusion policy

Already in the late Seventies German government set up a number of support schemes, especially for district heating. They ran out for West Germany after reunification and were concentrated on East Germany until 1994/95. CHP-investment benefited also from more general support schemes such as investment grants ("Investitionszulage") of usually 7.5% and up to 20%, or programmes to promote coal-based power generation (see Table 9.3). However, support declined in absolute terms during the Eighties (Walz, 1994).

Additionally, regional programmes were set up but they have never had a major impact due to their limited size. They rather fulfil the function of indicating political support and enabling field trials rather than widespread adoption.[238]

Information programmes have also been carried out and supported both at Federal and State level, but in general they are not regarded as important compared to those of the industrial and engineering associations. They aim to inform the wider public rather than the experienced planner and practitioner. Examples are the information about new energy technologies for citizens ("Bürgerinformationen Neue Energietechniken - BINE") or the information about municipal energy supply ("Informationen zur kommunalen Energieversorgung - KEV"). An exception is a support programme to advise small and medium-sized companies on energy-savings for which small subsidies are granted. The Future Energy Forum ("Forum für Zukunftsenergien"), founded in 1989, is another platform and network for information exchange on new energy technologies, and CHP has also been included in its activities. Energy agencies have been set up in some regions to stimulate the adoption of efficient energy technologies by providing information about contractual matters and support programmes, in some cases also complemented by small funding schemes. In general, they have developed into very successful network nodes for small-scale energy schemes.

Innovation policy

Several Federal programmes, as well as some smaller State programmes, offer possibilities to get funding for more innovative CHP-related research projects. In these programmes, particular emphasis is put on the improvement of ancillary equipment rather than of core components. For example, the improvement of pipeline technology for district heating is an important research area. In a large R&D project new concepts for heat distribution are explored (Neuffer & Witterhold, 1995; Witterhold, 1995). Other areas of interest are heat storage systems and cooling applications. CHP is not a major area for research funding in itself, but there are many possibilities within programmes which target primarily at other, though related, technologies, e.g. renewable energy, gas turbine research, efficient energy use or new power sources such as fuel cells or Stirling engines.[239]

---

238 No comprehensive overview of regional support programmes in all the sixteen federal states is available. In addition, support measures for CHP are often hidden in different budgets and sub-programmes. In Hessen, it is estimated that about 100 Mio DM were spent from 1988 -1993 to support different kinds of CHP-systems (Zinn, 1995).

239 The detailed projects which have been funded between 1974 and 1995 are accessible in a data-base provided by the project organisation BEO, working on behalf of the BMBF (BEO, 1995). An overview of the major programmes in relation to CO_2 can be found in (BMU, 1994).

Other policy measures

In addition to the aforementioned activities, there are several sectorial and State policies which affect the uptake and development of CHP. Examples are the dual financing system in the health and education sectors, State and local support programmes to stimulate CHP in certain sectors, and differences in implementing Federal regulations at State level.

Highly efficient CHP-applications also benefit from certain reductions of the mineral oil tax, if the overall efficiency exceeds 60% (BMU, 1994), while the plans for the introduction of a CO_2-tax were postponed in 1996.

Program name	Description	Time period
Federal DH-support programmes "Zukunftsinvestitionsprogramm 1 & 2"	Financial support (up to 25 %) for the extension of district heating systems	1977-1981 1981-1987
Investment grants for industry and public utilities	general minor support (typically 7.5%) for investment projects	Since 1975
Support programmes of the States	Usually smaller financial support for specific applications	Since the late 1980s
Third Power generation Act ("3. Verstromungsgesetz")	Financial support for coal-based power plants, from which also CHP-plants benefited	Since 1977
Federal District Heating Support Program East Germany	Modernisation of the East German DH-and CHP-systems (300 Mio. DM)	1992 - 1995
Federal R&D policy programs	Substantial financial support for innovative new technologies	permanent
R&D-programs at State level	Several smaller programs	since about 1990

Table 9.3: Overview of the main policy programmes to promote CHP in Germany during the Eighties and Nineties, Source: (BMU, 1994; Walz, 1994), interviews

9.5.1.2 Decision arenas and interactions for energy policy-making

Due to the multitude and complexity of political arenas dealing with energy and environmental issues in Germany, it is evident that the developments over the last years cannot be described in an all-encompassing way. Therefore only some of the interaction processes regarding important decisions are described in detail while others have to be dealt with at a more general level.

Interactions between Federal and State policy-making and the ESI

One of the fundamental principles of German energy policy is the delegation of competencies to the industries concerned, even if Federal government maintains the option to exert pressure by threatening them with direct interventions. The process which led to the Associations' Agreement on the remuneration of exported power is a good example of this approach. The partners of this agreement - VDEW, VIK and the Federal Association of German Industry ("Bundesverband der Deutschen Industrie BDI") - were urged by government to find a common solution. The initial interest in such an agreement was to define reliable principles of remuneration. Later on, the updates of the agreement were more concerned with the level of remuneration and the underlying definition of avoided costs. Especially the last agreement from 1994 needs to be interpreted against the background of the debate on the Electricity Feed Act. While government usually stood in the back and just urged the associations to negotiate, it threatened them in 1994 by considering that CHP-export tariffs would be included in the Electricity Feed Act, if the associations did not manage to find an agreement by themselves.

Another example which underlies German government's reliance on the delegation of competencies is reflected in the fate of the regulation for industrial heat use ("Wärmenutzungsverordnung"). Initially intended as a means to make sure that industry contributes to meeting the CO_2-targets, it was subsequently replaced by a self-commitment of industry. However, recent developments of industrial CO_2-emissions indicate that a reduction by 25% will be hard to achieve. The crucial argument in the debate was that it would be difficult to further substitute energy consumption by higher efficiency. Given the fact that German industry in energy-intensive branches operated already at comparatively high level of energy efficiency, there was little scope for additional reductions without endangering competitiveness. The idea behind the self-commitment approach was that the industrial organisations would be able to distribute emission reductions among their members in a more efficient way than government, knowing much better how the situation in the problematic industries was. One of the main reasons for the possible failure of this self-commitment seems to be rooted in the politics of the industrial associations. In fact, they are facing similar difficulties as government when trying to enforce commitments and balance the interests of its members.[240]

Government position

Cogeneration and energy efficiency started to become an issue in Germany in the mid-Seventies. District heating and demand side efficiency measures (insulation, end-use technologies, etc.) were the main target areas at the time. Small-scale and industrial CHP became relevant topics much later. Initially the debates at political level were driven by arguments of security of supply, but since the early Eighties

---

240 This refers to the "Initiative of German Industry for Global Climate Protection' from March 1995 by which it committed itself to a reduction of CO_2-emission by 25% by 2005, based on the level of the year 1990.

there has been a switch to environmental issues which, from the Rio-summit in 1992 onwards, have focused more and more on CO_2.

In the late phase of the security of supply debate, i.e. in the late Seventies and early Eighties, a number of technologies were promoted on the market which were highly decentralised, technically quite complex and tended to be manufactured by small young companies. The best example of this trend are heat pumps, but after the decrease of oil prices in the early/mid-Eighties the market for this highly efficient option collapsed completely. Small-scale CHP was frequently mixed up in the political debate with heat pumps, and a negative 'loser'-image was sometimes associated to it. This has changed in the meantime, as expressed in several policy documents, e.g. the report by the Enquete-Commission of the German Bundestag (Enquete-Kommission, 1995). Especially at regional level support has been strong, as shown e.g. by the initiatives of a number of States to introduce a more CHP-friendly remuneration regime for exported power (Schürmann, 1995). Both at Federal and State level, the attitudes towards CHP are far from being homogeneous even if CHP is widely supported in general terms. This becomes particularly obvious with respect to the assessment of liberalisation plans and their impact on CHP. While the BMWi is strongly in favour of a more liberal approach to energy supply, the BMU tends to be more reluctant because of the uncertainties which liberalisation would bring about for environmentally friendly technologies such as CHP. BMWi also prefers to wait for a reform of the European framework - the Internal Energy Market - before addressing the German Energy Supply Act.[241]

These pending decisions also affect the present state of debate on the Electricity Feed Act. It is a symbol of German government's commitment to renewable energies and environmental improvements. Its intention is to promote renewables by obliging utility companies to purchase electricity at a price beyond avoided costs. Until now it has been applied only to renewable energies and small energy-from-waste projects, but claims have been raised to include also CHP, at least small-scale CHP. Critique had already been raised about the additional costs of the present scheme to some regional utilities (notably in Schleswig-Holstein with its high share of wind power) and about the need to pass these costs on to the consumers. Until now the capacity which falls under the law is rather small, but this would change if CHP was considered as well. The resistance of the utility companies against an extension is therefore obvious. Neither was German government very supportive of any modification of the Electricity Feed Act, at least until the reform of the Energy Supply Act was fully operational.

Decentralisation of energy policy in Germany

The decentralised structure of the German political system facilitates the emergence of a large variety of energy supply solutions at local and regional level. There are also a number of common difficulties to overcome at these levels in order to realise CHP-projects. Especially at local level, the close links between authorities and utility companies have repeatedly led to the construction of plants

---

241 In the meantime, the reform of the Energy Supply Act is about to be implemented.

which fulfilled mainly a symbolic function of showing commitment to the environment, but they could hardly be justified on economic grounds. This problem has disappeared in recent years with the growing independence of local utilities of the authorities. On the other hand, in several cases they had to accept the involvement and financial backing of regional or supra-regional utility companies. This involvement obviously implied a growing influence of regional or supra-regional supply strategies on the decisions at local level. The possibilities for decentralised production are rather weakened by these developments.

The scattered character of competencies and the manifold interdependencies in the German system make it very difficult to implement a straightforward and coherent energy policy at Federal level. The complex balance of power is more open to slow transformation processes rather than to major reforms of the entire framework. This underlines the important role played by the political structures and the political culture in Germany on the emergence of CHP.

The role of the Enquete-Commission of the Bundestag

In the follow-up of the Rio-conference, the German Bundestag appointed an Enquete-Commission dealing with energy policy and climate change. It tried to identify and analyse options for a sustainable energy supply path. One of the technologies which attracted much attention was cogeneration. In the context of the work of the Enquete-Commission, several studies were commissioned in order to assess the benefits of CHP as compared to separate production. Most of the studies came to a positive assessment of CHP (Enquete-Kommission, 1995). One of the reactions to these studies was a further study, commissioned by PreussenElektra, one of the supraregional electricity utilities. The contractor, the Pestel-Institute for Systems Research, delivered a more critical assessment of CHP, denying substantial efficiency advantages of CHP (ISP, 1994). On the contrary, negative economic impacts were indicated for the German economy. In the follow-up of this publication, a fierce debate emerged on the reliability and quality of the results of the Pestel-study, ending with its widespread rejection. This example shows how the work of the Enquete-Commission revived interest in the different types of CHP and brought it back to the attention of policy-makers. It also highlights the role of scientific expertise in the policy debate.

New interest groups and the independence of political authorities

In spite of its corporatist elements, the German political system offers a variety of possibilities to exert influence on political decisions. This is due partly to the decentralisation of competencies, which leaves both regional and local authorities the discretion to try out new solutions and adjust the Federal framework regulations to their specific requirements. But also at Federal level the decision-processes are far from being closed. There are a many different formal and informal consultation procedures which aim to guarantee the consideration of a multitude of opinions and attitudes. Although it is hardly possible to evaluate the weight of industrial interests as compared to environmental ones in different ministries, they all dispose of formal and informal channels of influence. In addition, the powerful associations of both the ESI (VDEW) and industrial generators

(VIK) have kept a quite stable balance of interests in relation to energy issues. Decentralised local utility generation is well represented by the VKU, but small decentralised and independent generators was in a comparatively weak position until recently. The foundation of the support association for small-scale cogeneration systems (FGBHKW) aimed exactly at reducing this deficit.

Anyway, due to the fact that German ministries do not exclusively rely on the expertise from the industries affected by their policies, but also consider the positions of many other actors, they manage to get fairly well-balanced information input on the current situation. A wide spectrum of interests has access to the political process, even if some actors may benefit from some privileges, e.g. by being consulted in the formal legislative processes. Of particular interest is the role played by scientific expertise for decision-making. In Germany, the confidence in and reliance on experts from the scientific world is quite strong, though not necessarily stronger than the influence of interest groups. This is reflected not only in the composition of the Enquete-Commission, but also in the studies commissioned by the different ministries. Traditionally, industrial interests find particular attention in the context of the Ministry of Economic Affairs. Their lobbying tends to argue against a further tightening of environmental or efficiency regulations which would be problematic for the energy-intensive branches of German industry (e.g. chemical). Environmental concerns are taken more seriously by other actors, and in order to understand the political decisions made, the conflict lines among the different political actors need to be looked at in specific examples.

Controversial topics: Some recent examples

These lines of conflict become obvious when looking at some of the major topics discussed in the last few years. They have focused on the one hand on export remuneration (c.f. the Electricity Feed Act and the Associations' Agreement), and on the liberalisation issue on the other.

Despite the fairly stable application of the associations' agreement, there are still many uncertainties associated to it. Both large exporters (who are actually not concerned by the agreement), and small non-industrial users (who are under-represented in the VIK) raised the argument that avoided costs should be defined differently than foreseen in the agreement. At present, the basis for remuneration are the avoided costs of exporting to the medium voltage grid, i.e. export to regional electricity supply companies. This corresponds to the standard situation of industrial generators. Small-scale CHP exports to a local utility would justify higher remuneration, because the avoided costs of local utilities are comparable to their purchase price from regional utilities. And this purchase price is clearly higher than the avoided costs of the regional utilities, who in turn are purchasing from the supraregional generators. The FGBHKW proposed a remuneration scheme ("City-Gate model") which is based on the avoided costs of the "last supplying level", i.e. the local utilities for small-scale CHP (FGBHKW, 1995; Schürmann, 1995), but their approach did not find agreement among the other main associations. For electric utilities it would raise the spendings for imported power, and the VIK opposed it because it would maintain the monopoly position of the local utilities while they can buy on the market. In relation to the City-Gate approach it is interesting to note that it has been supported by a number of regional authorities who promoted it in the Upper Chamber of Parliament (Euler, 1996).

The broader topic of liberalisation and the Internal Energy Market has been dealt with very cautiously in Germany. The promised reforms were pending until the framework directive was accepted by the European Council in the middle of 1996. In 1997, a new Electricity Act was prepared to liberalise and re-organise the German energy and electricity supply system. Although is implementation is still pending, it seems to pursue a rather moderate approach to liberalisation which leaves the powerful position of the supra-regional utilities untouched.

9.5.2 Technology supply side

9.5.2.1 Decisions: Innovation and supply of technology

While large-scale and gas turbine CHP represent the traditional business of established manufacturing and engineering firms, small-scale CHP was initially dominated by a number of small firms. In the early Eighties, some new companies were set up which perceived the potential advantages and market opportunities of this type of CHP. In contrast to the established energy system providers they were also more innovative in terms of organisational and financial solutions to facilitate CHP. This enabled them to combine traditional manufacturing and packaging with energy utility services. However, in Germany the separation of service supply and manufacturing is still quite pronounced, with utility companies on the one side and manufacturers on the other. The vertical integration of the production and supply chain and of packaging and service supply have only started to develop recently. Apart from most of the packagers, only a few larger utility companies have set CHP-related energy service concepts.

The established engine manufacturers moved into the CHP-business quite late, namely at the end of the Eighties. Even after having recognised the growing potential for small-scale CHP in Germany, it took several years until they fully reacted to the users' new requirements. The early solutions were far from being optimally adapted to the needs, especially in terms of back-up services, financing and organisation. The young companies came up with far more intelligent offers. The situation started to change about 1990, when the supply side of CHP-equipment stabilised around a few major manufacturers.[242] With very few exceptions, the young companies either managed to keep only their regional markets or they disappeared.

The gas turbine market is also dominated by a small number of established companies. They co-operate with one of the main small gas turbine manufacturers from abroad on the basis of franchising, i.e. they provide the distribution and support network in exchange for exclusive access to their turbines.

---

242 These were mainly MWM, MAN, Jenbacher and EES, and for larger engines also Blohm and Voss. EES is the only new company in this group. After reunification also MB Halberstadt operated successfully in the large-scale engine market.

9.5.2.2 Decision arenas and interactions on innovation

Although some of the early CHP-packagers could build on earlier energy-related activities, they were often lacking competencies and experience in electrical and heat engineering as well as knowledge on the specific conditions of the energy sector. Consequently, the early Eighties were characterised by the search for reliable CHP-systems, both in technical and organisational terms. Initially, the products were of rather bad quality, and a long learning process was required until system designs became available which were fully adapted to the site conditions and met the users' expectations. A typical problem was the oversizing of plants in the Eighties, i.e. they were designed to cover more than the heat base-load of a site.[243] Due to the need either to export power or to run the units only part-time, a profitable operation was in fact impossible. Although these mis-specifications can also be traced back to a short-term profit-seeking behaviour of some manufacturers, the main problem should be seen in a lack of experience and knowledge of plant design.

The established manufacturers of energy systems and engines were initially rather conservative and reluctant to commit themselves to CHP. They did not become active in packaging CHP-systems before the second half of the Eighties when the new companies had already built up a lot of experience. The extension of their activities to energy services, contracting and financing developed very slowly, and have tended to be more limited than those of the newcomers. This strategy needs to be seen in the context of the specific energy market conditions and established relationships in Germany, where the position of local utilities does not facilitate the emergence of independent service companies outside a few niche markets. The tasks have always been clearly separated between manufacturers on the one hand and public utility companies on the other. An aggressive service strategy of the established manufacturers would have created conflicts with some of their main clients, with whom they had developed a good relationship over decades. Consequently, the responses to the interest of customers in packaged and turn-key solutions have been rather limited until recently. Nowadays, they also offer different types of full maintenance and service contracts.

One can distinguish between new companies and established ones also in terms of the technologies used. The former integrated a mixture of available components in their plants, which led to variable quality standards, but they were innovative and disposed of considerable tacit knowledge. The latter could build on their established knowledge in engine technology and disposed of substantial R&D resources to improve and stabilise the technology. The increasing sophistication of CHP made it more difficult for the newcomers to keep pace with new requirements in terms of system optimisation, environmental regulation and computer control. Especially the move from diesel to gas engines was a problem for the new compa-

243 Obviously, the planners and manufacturers may have a short-term interest in oversizing plants, but in the longer-term the resulting economic disasters would have damaged the reputation of both the planners and of CHP in general. Consequently, the reasons for planning failure should be seen primarily in incompetence and lack of knowledge rather than in the defective behaviour of the actors.

nies because there were hardly any good gas engines available on the German market. Diesel aggregates were available from several manufacturers, but gas engines for stationary applications were still in the development phase. The well-organised learning processes in the larger companies enabled them to overtake the early innovators in this crucial technology. The strategies of companies like MWM or MAN are very interesting in this respect. Initially they did not dispose of purpose-made gas engines for CHP-systems and tried to adapt their diesel engines for this purpose. The results revealed to be unsatisfactory, expensive and maintenance-intensive. Only in the early Nineties this deficit could be solved by introducing new generations of gas engines.

The quasi-compulsory application of the TA-Luft emission limits was another key issue for the manufacturers.[244] Many small firms did not dispose of the competencies to integrate highly sophisticated clean-up or lean-burn technologies into their systems. Therefore, the strict limits became not only a major marketing element in the environmentally sensitive German context, but also a key discriminatory factor for technology suppliers. Many CHP-users had more confidence in one of the established manufacturers to provide reliable, sophisticated and environmentally advanced solutions than in a small innovative firm.

Attempts were also made to standardise the design specifications for CHP-units, especially of the 50 kW scale (Meixner, 1995). This represented a major step forward for small-scale applications. For small systems, the investment and maintenance costs are comparatively high, but this disadvantage could be partly balanced by economies of scale by series production. By defining standard interfaces for the widely used 50 kW size, such economies of scale could be achieved for components, control systems and the connections between the packaged unit and the embedding system. With regard to gas turbines, the standardisation and packaging potential is more limited. As there is no domestic manufacturer of small gas turbines in Germany, the packagers can standardise their equipment and its interfaces to a limited extent only.

The initial reluctance with regard to contracting solutions has been overcome in the meantime also as a result of more favourable regulations. Especially the licences for generation and supply can be obtained much easier now, even if the extent of competition in this area is still limited. The local utility companies dominate the energy service market for both households and industry. The interviews also revealed a general conservatism regarding competitive energy service supply on the sides of both users and suppliers.

In spite of its convincing technical characteristic, CHP was initially also a risky business. Early on, the bad experiences made with heat-pumps served as a warning for many sceptics not to adopt another small system technology such as engine-based CHP - if the difference between the two technologies was understood at all. The recent developments of the regulatory framework and the opportunities also for local utilities to operate CHP provided a conducive context for developing the technology further. This is also reflected in the ongoing R&D efforts espe-

244 In practice, small-scale CHP systems below 1 MW were even required to fulfil limits of 50% of the TA-Luft values in order to keep local emissions low.

cially with regard to fuel cell CHP. They show the importance of the long-term promises behind the combination of the two technologies. In other words, one could interpret the Eighties as a learning period, not only in technical but also in organisational and institutional terms. It shows that for developing CHP successfully, a broad range of factors had to be taken into account by the technology suppliers. Among the main difficulties encountered by them are the planning and permission procedures, the conservatism of many actors at all levels from regulatory authorities to designers, planners and chimney sweepers,[245] the unclear and competitive pricing of utility companies, and the reluctance and prejudices against small-scale CHP in general.[246]

With regard to innovation and technology policies, most of the programmes were regarded as of limited help only, except for the R&D-support measures for specific components, e.g. heat storage or clean-up systems. What was essentially lacking were programmes to improve the knowledge about system operation and design. This was achieved only once the industrial and engineering associations started to deal with CHP and disseminated information for practitioners. Some of the regional programmes were helpful for small companies by facilitating the participation in demonstration projects. Monitoring and testing schemes made it possible to improve the performance and reliability of their systems. Other kinds of financial support measures were welcome as an additional source of funding, but were usually too small to have a determining impact on investment decisions.

The availability of information on CHP was also very scarce in the early Eighties. By means of some regional programmes first experiences and data could be collected on the operation of new CHP-plants. Apparently, this effect was fully underestimated by Federal government and left to industrial and engineering associations, but at that time they were not very open-minded with regard to CHP. Only in the late Eighties the situation for small-scale and industrial CHP improved with the foundation of ASUE and the cogeneration-committees in VIK, VDI and VDMA. Regarding district heating, the situation has always been different due to the role played by the AGFW which supports public electricity utilities on all district heating-related issues.

9.5.3 The user-side of CHP-technology: Adoption

As in the case study on the UK, a distinction is made between the interactions and decision processes of CHP end-users on the one hand and of utility/service-companies on the other. Especially utilities are not only technology-users, but fulfil also a gatekeeping function for the grid system and are often in charge of

[245] In Germany, chimney sweepers fulfil certain control functions on behalf of the local authorities, for example the measurement of emissions of households, and the approval of combustion sites.

[246] An interesting episode in this context concerns the high level of trust in the local utilities. In several cases, the negative (and often false) assessment of CHP by local utilities was believed by potential CHP-users. They could not even imagine that the utilities would advise them wrongly out of reasons of self-interestedness.

certain regulatory tasks. The major decisions on the user-side are described jointly for end-users and utility/service companies.

9.5.3.1 Decisions: The use of CHP-technology

The trend towards the vertical integration of energy supply chains, and especially of electricity supply, has led to a greater dependence of local utilities on regional and supra-regional companies in the last two decades. They have increased their shares in local utilities and can thus constrain the possibilities for local self-generation and CHP. Especially at local but also at regional level, various energy-related services are provided by the same company. Electricity, heat, gas, and many other activities such as swimming pools are typical task of local utilities, a situation which facilitates the use of CHP. In the last years a trend towards more separate organisational units for different services ("Spartenunternehmen") has emerged, especially among the larger local and regional companies. Competition between departments makes it more difficult to develop integrated solutions.

Principles like least-cost-planning (LCP) and demand-side-management (DSM) have started to be considered by some utility companies, but there are only very few active and innovative first movers. Their initiatives must be seen as first steps towards a service orientation in energy supply which aims to provide the customers with integrated energy supply solutions.[247] The corresponding measures may even require to perform construction works in houses or factories, e.g. for insulation or to replace inefficient machinery. Financial incentives are sometimes offered to stimulate the acceptance of DSM.

In industry, interest has raised again in self-generation, even if this is not clearly reflected in the diffusion data due to the impact of reunification. The growing penetration of contracting arrangements needs to be mentioned in this context. Due to the widespread financing problems they play a particularly important role in East Germany. But even such new contracting projects hardly involve more than one energy end-user; integrated energy systems for several customers seem to be very difficult to realise.[248] In general, many new types of applications have been explored at the user side, applications which would have been impossible to realise ten years ago for technical, economical as well as for regulatory reasons.

---

247 See for example the overview of least-cost planning measures of German energy suppliers in (BMWi, 1994, III/5-8) and (Bauerschmidt, Utermark, & Weber, 1994).

248 One of the obstacles for integrated supply solutions are the licensing requirements of § 5 Energy Supply Act (EnWG). The licence is granted on a standardised basis if only a single company is supplied with energy, while a detailed procedure needs to be followed in the case of supply to several partners.

9.5.3.2 Decision arenas and interactions I: Energy service end-users

Motivations

End-users, especially industrial firms, permanently reconsider their energy supply systems for possibilities of improvement. If CHP appears to be promising within their planning horizon, it is eventually adopted. The decision about CHP is usually part of a wider energy supply strategy which aims at integrating the different supply chains in order to exploit economies of scope, e.g. between production processes and energy supply or as part of a city development strategy. This explains also the emphasis given to the improvement of integrated design and operation of CHP.

In general terms, the expected long-term cost savings are the major arguments in favour of CHP, but they are questioned by the uncertainties of economic boundary conditions. If the power generated can be entirely used on site to replace power purchases from the grid, the conditions are particularly promising for CHP. In this case, the value of generated power corresponds to the price of imported electricity and not just to the far lower export tariff. A second important economic argument consists of the reduction of the capacity charge to be paid to the utility company. Environmental considerations have also become an important motivation, especially for public investors and local utilities. In some cases the political will to install CHP was even stronger than deterrent economic assessments. However, in the majority of cases under study, the environmental characteristics of CHP just represent an additional argument beyond economic considerations, even if the de-facto application of TA-Luft/2 emission limits to small-scale CHP below 1 MW, especially by local utilities, shows that is more than a factor of secondary order.

Since the mid-Eighties the information and experiences made with CHP have been widely disseminated in order to improve the level of awareness and knowledge of potential investors. Beyond the engineering associations (VDI, VDMA) and the associations of industry (VIK) and electricity utilities (VDEW/AGFW), especially the newly founded associations like ASUE have helped to provide sufficient information to small potential users.

Decision-making and assessments

Investment in energy supply is also a matter of priority setting. In Germany, many demand-side measures for energy efficiency had already been exploited in the Seventies so that more expensive supply-side measures such as CHP appeared on the agenda in the Eighties. This is reflected in the payback periods required for CHP-projects which are in the order of two to four years for private investors. Public utilities apply either a calculation over the plant lifetime, or aim to payback periods of up to seven years. Anyway, the role of such calculations is regarded by most users as of limited significance because investment in CHP is often strategic and cannot be measured just in terms of simple investment economics, but needs to take long-term development plans into account.

The differences between private users of CHP and public users are not only expressed in the payback-criteria applied, but also in the assumption which underlie their calculations. Local utilities need to compare CHP-investment to the supply price of their regional electricity utility, taking into account the contracted peak demand and the corresponding capacity charge. Although in principle similar considerations apply for private customers, many of them are connected to a local utility. This implies that they can calculate against the (higher) local supply tariffs. Consequently, the economic benefits of autoproduction are also higher. In general, it is more difficult to find economically viable CHP-projects at regional or even supraregional connection levels.

The uncertainties associated to prices and tariffs have been generally perceived as moderate in Germany in the Nineties, especially compared to the situation in the early and mid-Eighties when the Associations' Agreement was not yet fully operational and strongly disputed. It is still claimed frequently that exported power should be valued higher, although the tariffs have been increasing over the last years. Anyway, the agreement is only a guideline for individual negotiations, even if it fulfils an orienting function insofar as it is used as the basis for assessing the economic viability of most new plant projects.

Associated to the economics of CHP is the issue of finance. Despite increasing difficulties to find adequate resources, hardly any of the cases under study had been subjected to severe financing problems.[249] If it was the case, then this was due to the very specific situation of the project, e.g. dual finance, or the situation in East Germany. These are also the typical cases which tend to be carried out under a contracting arrangement. It may even be the only possibility to realise a project. Finance has become the strongest argument in favour of contracting in West German industry, too, but it has been seriously considered only since the middle of the Nineties, and usually in collaboration with a utility company.[250] This coincides also with the wider availability of financing schemes by energy service providers. For local utilities with a fairly solid financial basis, contracting is not a very common solution, but for urgent projects it is now considered occasionally, too. However, the comparatively healthy financial situation of many investors is only one reason why contracting is still far more underdeveloped in Germany than in the UK. Another one is the recognition that contracting and the participation of an additional partner reduces the potential economic benefits of CHP. Contracting is normally justified by the additional expertise brought in by the contractor, but many of the energy users dispose of the necessary expertise themselves. Under these circumstances, a contracting arrangement does not bring in much added-value. Finally, there is still a clear lack of familiarity and experience with contracting, as well as a lack of information.

---

249 In this respect, the fundamental bias of case study research should be kept in mind. By definition, case-studies tend to study realised examples for which financial and other barriers could be overcome.

250 The industrial associations VIK and VDEW support their members in setting up contracting arrangements, be they for full services, financing, or operation.

Perceived problems and difficulties

Given the well-established character of large-scale district heating and industrial CHP, competencies and experience were not perceived as major problems in these two application areas. The industries which adopted CHP usually dispose of the necessary background to design and operate CHP. For small-scale CHP the situation is different. Neither sufficient information on prior experiences was available, nor were planners particularly open-minded to these non-conventional systems. It was frequently mentioned, that the lack of familiarity of planners, designers and regulating authorities with small-scale CHP was a bigger obstacle than the lack of co-operation of utility companies. In the Eighties, local utilities were still quite reluctant to accept small-scale CHP as a serious technology, but once they had started themselves to install small units, the situation changed. Moreover, based on their long-standing experiences with district heating and electricity supply, they had the competencies to set up and operate plants properly. Until the middle of the Nineties, learning effects and the collection of data and experiences from existing plants have solved this knowledge problem, and it is much better known now how load profiles should look like to be able to realise good CHP-systems.

Sources	**Importance**	**Characterisation & timing**
Industrial associations (VIK, ASUE)	high	Technical, planning and legal information, contracting experiences, from late Eighties onwards
Utilities' associations (AGFW/ VDEW, VKU)	high	Technical and planning support for DH already before 1980, since early Nineties also small-scale
Engineering associations (VDI, VDMA)	high	Mainly technical and planning information, especially since early Nineties
Government	low	Little direct input for users, regional energy agencies since early Nineties
Engineering consultant offices	medium	For less competent users important inputs about technical matters and planning
Other users	high	Site visits and direct exchange about experiences with similar applications
Manufacturers and packagers	medium	Rather limited interaction with packagers and manufacturers on technical matters
Own expertise	medium/high	High level of general energy engineering expertise among utility users and some industrial ones

Table 9.4: Main information sources of CHP-users in Germany

Surprisingly, the links with the suppliers have been characterised as being of limited importance only (see Table 9.4). While the new emerging small-scale packagers had to overcome many barriers to be accepted by potential users, the more established manufacturers and packagers were regarded as not very customer-oriented. This lack of communication between users and suppliers of CHP-equipment is also expressed in the differences in perception of requirements and future needs of CHP-technology. While users were looking for reliable turn-key solutions for their specific energy supply problems, most manufacturers wanted to sell their hardware components, but without getting involved into detailed plant design, construction and operation. Although the situation has improved in the meantime especially for smaller applications, there is still a service gap between the two sides.

The problems of complexity and site-specific character of CHP have already been mentioned frequently. Although the learning processes of users and manufacturers, facilitated by the good level of basic competence of many local actors and by the density of their informal contacts, has clearly improved the planning and design of many CHP-systems, these two issues continue to be major problems.

The relationships with the utility companies

The acceptance and facilitation of independent power generation and exports are the crucial issue in the relationships between CHP-users and utility companies. Private operators who want to connect to the local or regional grid network need to find an agreement with their utility company on technical and economic conditions. While for small consumers, the general electricity tariffs apply, large electricity customers need to negotiate them individually. In general, the improvements of CHP-technology have reinforced the negotiation position of large customers because they can compare the conditions offered by the utility company for direct supply with auto-production by means of CHP. As a consequence of this, large customers have frequently been offered lower power prices by the connecting utility, once they had started to consider auto-production. To this adds the he removal of the coal levy in 1995 which has led not only to a reduction of power prices but also to a reduction of the export tariffs. According to FGBHKW the reduction of export tariffs was even overproportional, and thus reduced the incentives for auto-production and CHP.

The problem which CHP-operators had with their utility companies during the Eighties were not only due to conscious resistance of the latter, but rather due to a lack of knowledge about the characteristics of CHP. Repeatedly, it was pointed to the tendency of local utilities to reject solutions which were not generally established and known by them. This situation has changed since small-scale CHP became used more widely by local utilities themselves. In general, an improvement of the relationships and a stabilisation of the terms of exchange has been achieved, despite the fact that the connection conditions still vary widely across Germany. Some utilities are still regarded to be hostile to CHP, as expressed e.g. in the technical and economic connection conditions applied in practice (Dufait & Verbruggen, 1993; Fenzl & Pummer, 1993, p. 4/3).

In order to underline the variety of conditions for CHP in Germany, a number of conducive elements should also be mentioned as characteristic of the utility - CHP-user relationships. Personal and informal contacts between end-users in industry and utility companies have been of utmost importance for carrying individual projects through, especially at local level. This refers not only to questions of information exchange, but also to the achievement of formal or informal arrangements for export and purchase of power between (local) utilities and end-users.[251] In other words, official information on technical and economic connection conditions needs to be treated with care because in practice they may be handled in a quite different way.

Regulatory framework and public policies

The assessment and perception of the regulatory framework is dependent on the specific circumstances of an application area. For example, for hospitals and universities the dual financing system is crucial. The aforementioned informal links also play a role for the relationships between authorities and companies. Personal contacts between State authority delegation ("Regierungspräsidien"), which are responsible for licensing and financing, and local users facilitates the realisation of projects.[252]

The different financial support schemes have attracted much interest among users, but have usually not been regarded as an important argument in the decision-making process. They only tipped the balance in favour of a project if massive support was given, like e.g. in the case of the DH-programmes where up to 40 % of investment was publicly funded. Investment grants ("Investitionszulagen") of normally 7.5 % did not make a major difference for business decisions which were characterised by high uncertainty, strategic considerations and a need for very promising economic appraisals. If available, minor funding measures were welcome as an additional source of finance, but in fact they did not affect the decision for or against a project.

Consequently, the political and regulatory developments of the last years have been perceived as a far more important issue than any type of direct support measures because they induced uncertainty about future investment conditions for CHP. The abolishment of the coal levy and its effects on electricity prices was important in this respect, but also the Electricity Feed Act and the Associations' Agreement. Currently, the most important uncertainties stem from the debates about the implementation of the new Energy Supply Act and the Internal Energy Market in Europe.

---

251 Informal arrangements between industrial CHP-operator and local utility companies are not uncommon. For example, an industrial end-user agrees to contribute to the surcharge to be paid by the local to the regional utility, if the local utility exceeds its capacity limits because of a failure of the industrial plant to export power. In turn, the local utility pays export tariffs, which correspond to those for guaranteed supply.

252 This concerns e.g. the application or better non-application of §12 Energy Supply Act to CHP-plants between 10 and 50 MW, or the acceptance of CHP-projects for public funding under dual finance conditions. The discretionary possibilities of authorities and the specific interpretation of guidelines have led to the establishment of "CHP-islands" in Germany where it is administratively fairly easy to set up CHP schemes.

Expectations in past and present

Given the numerous uncertainties related to CHP, expectations and strategic perspectives play a major role for decision-making. The past improvements and future expectations with respect to CHP, as perceived by the different types of end-users, are usually quite similar, and it is interesting to compare them with those of the manufacturers. While in the mid-Eighties, there was still a clear mismatch between the needs and expectations of the users, and the opportunities and developments identified by the manufacturers, these differences have become minor in the meantime as a result of the growth of the shared knowledge-base on CHP-applications. Both sides perceived gas-based CHP with turbines and engines as the most promising technologies for self-generation of electricity in general, but there were clear differences in terms of the detailed needs for improvement and development. Users identified the integration of CHP into a wider energy supply system as the key issue, while manufacturers, dealing mostly with individual components, focused on improvements of the performance characteristics of their products. Correspondingly, system reliability and control were of major importance for users, but manufacturers and packagers started to address these issues only after having adopted a more service- and contracting-oriented approach.

Regarding future developments, both manufacturers and users set high expectations in fuel cells, especially for small and medium-sized CHP applications. However, even if their technical and economic potential may become clearer in the future, the question remains whether fuel cells can be provided as fully reliable and integral elements of an energy system. Current knowledge on small CHP-systems design may be sufficient to integrate a new power source like fuel cells into more complex systems, but what CHP users expect before adopting a new system for full-scale operation are first of all reliability, compatibility and easy operation. As demonstrated repeatedly in the New Länder, the easy upgrading of the power source while keeping the rest of the system unaltered is one of the big advantages of CHP. The adoption of fuel cells would also be facilitated by the stability of the organisational and institutional framework for CHP. Most uncertainties have been removed and enable operators to cope with new innovations. Developments of gas turbines, engines and ancillary equipment are expected to be of an incremental nature. The most significant changes will concern the packaging and integration of CHP into complex energy supply systems.

Liberalisation is not a real concern for most CHP-end users because they would mainly be affected by changes of the connection conditions and interfaces. The possibility to negotiate better tariffs and contracting solutions would be useful for CHP, but not of major importance for all those operators who auto-produce for their own consumption.

9.5.3.3 Decision arenas and interactions II: Utility and service companies

Local utilities fulfil a double role for CHP as operators of their own CHP plants and as gatekeepers for the connection of other operators' systems. They can perceive CHP as an opportunity to set up an environmentally and socially optimised supply system which integrates different supply chains, facilitated by planning tools such as integrated resource planning, least-cost planning or demand side management. Most local utility companies still see themselves as public service providers rather than as profit-making firms, but the operation of CHP-plant is also constrained by the requirement of economic efficiency and by the regulatory and economic context conditions.

This conflict between economic and social objectives is also reflected in the decision behaviour and criteria applied by utilities. Depending on the individual system, they regard small-scale CHP either as a competitor to their DH-networks, or as a complementary element. This becomes particularly obvious in comparison to gas supply. Island networks based on small-scale CHP-systems can be an excellent possibility to introduce CHP without having to build extended heat networks. An already installed gas network can be used as a complementary system rather than being a competing system. In other cases, when island networks are set up within a wider DH system by a competing service provider who just connects some major heat loads, the entire DH-network is put at a disadvantage due to the loss of the most valuable customers ("cherry-picking" effect). The example shows that one needs to be very careful in assessing technological options.

The possibility of the German local utilities to provide a wide variety of energy-related services represents a good opportunity for least-cost-planning. In fact, some local utilities have been pacemakers in this respect, offering integrated packages of active and passive measures to provide energy services to their clients (Bauerschmidt et al., 1994; Hartung, 1995; Lüschen, Sonntag, & Werner, 1995). The main problem for LCP lies in the necessity to go beyond the provision of a connection for heat, gas and electricity, but to intervene also inside the houses by suggesting insulation measures or more efficient electrical equipment. The use of waste for heat and power generation is another very interesting combination at the local utility level. These examples offer good opportunities to exploit economies of scope, but they also require to operate the corresponding activities under one roof. If this is the case, there are indeed strong driving forces in favour of highly efficient energy supply solutions. In principle, utility companies are also in a position to facilitate the interconnection of different large heat and power users in an integrated supply system. Examples from the Ruhr-area show that these utilities-industry partnerships are still feasible, but there is also widespread reluctance among private and industrial partners to become involved in large-scale and long-term contracts. In the last years, and especially in East Germany, contracting arrangements for energy supply systems have usually been set up for individual sites, involving the local/regional utilities for gas and electricity, the industrial partner, and possibly another financing partner. Large integrated multi-partner systems are rather exceptions.

Surprisingly, most local utilities interviewed do not see a particular need to prepare for the future Internal Energy Market in Europe and a new liberalised framework in Germany. It is perceived as a problem which mainly concerns the large supraregional and regional utilities which will compete at European level. They usually do not expect to be heavily affected by competitors at local level due to their long-standing relationships and experiences with their clients. Only in certain niches such as new residential areas and some industrial customers, they fear to lose their dominant role.

The behaviour and decision criteria of local utilities do not differ very much from those of any other user, except for three aspects. First of all, they tend to apply less strict criteria than private investors when comparing a CHP-plants with the costs of purchasing power from their regional supplier. Secondly, their public service function makes them more inclined to consider environmental issues. Finally, the bigger companies have been quite willing to test new technologies and thus contributed to establishing new technologies. These informal support measures for emerging technologies such as CHP or renewables have not yet been reduced over the last years. It is feared that such informal and catalysing activities will not be feasible any more once a fully competitive system is established in which the local utilities do not dispose of a protected position any more.

The interventions of local authorities in the decisions of local utility companies had double-edged impacts on CHP. In some cases, authorities have urged their utility companies to give more emphasis to environmental considerations, or even enforced the construction of CHP-plants. Some of the economic failures which could be observed afterwards had then a detrimental effect on the reputation and the further development of CHP. Since the late Eighties, the operation of utility companies has become more independent from political influence, expressed for example in their transformation into independent companies or in the outsourcing of certain tasks. In parallel, the rising influence of large regional and supraregional utilities at the local level has also led to a more efficient management.

Regional utilities play a critical role for CHP because their tariff regimes determine the tariffs of the local utilities and because they define the connection conditions for large-scale CHP-operators. A very interesting example concerns the structure of tariffs and charges they impose on local utilities.[253] The penalties to be paid for exceeding the contracted capacity limits makes it interesting for local utilities to utilise their power plants for peak optimisation rather than for energy optimisation. Economic efficiency differs in these cases clearly from energy efficiency, because plants are not operated in competition with the normal purchase price but in competition with the substantially higher peak price plus an equivalent capacity term. Beyond some large-scale district heating projects, regional utilities have also got involved in medium-scale industrial CHP-projects themselves, mostly through some kind of contracting arrangement.

The entire regional supply level has been increasingly challenged in the last years because its main function of medium voltage distribution could in principle be delegated to the supra-regional and local utilities. They are sometimes regarded as an unnecessary organisational layer which ought to be abolished in the course of the reform of the ESI. But they are also the highest level at which the integration of heat and power generation is possible because they are often involved both in power and gas distribution, and are still fairly close to the potential CHP-users. In other words, they can play a beneficial role for facilitating medium- to large-scale

---

253 In terms of avoided costs, one can distinguish four levels of assessment, namely a) avoided costs of supraregional suppliers, b) avoided costs of regional suppliers, c) avoided costs of local utilities (and possibly of service companies), and d) cost considerations of end-users.

CHP-projects which are beyond the scope of local utilities. Until now, however, they have been rather passive players in relation to CHP as reflected in the limited range of CHP-projects they have been able to set up on their own.

The supraregional utilities are mainly involved in large-scale bulk power generation and, as a by-product of this activity, also in large-scale heat supply for industrial customers and district heating. Given their parallel (though not horizontally integrated) activities in gas, heat and power supply, they dispose at least of the potential to exploit the benefits of, mainly large-scale, CHP. Small-scale, decentralised generation seems to be opposed by them, as expressed in initiatives like the "Pestel-study" (see Section 9.5.1.2). However, some of the supraregional generators own or hold shares in local and regional utility companies which they tend to use to support their general strategy of power supply. This trend towards vertical integration in supply is of a double-edged character for CHP. On the one hand, the supraregional utilities become more interested in local generation and optimised horizontally integrated services of their subsidiaries. On the other hand, this interest is constrained by their core business of large-scale power generation for which local utilities play only a subordinate role as distributors.

Due to the cost structure at this highest supply level, the scope for independent CHP is limited to large scale industrial and district heating projects. The avoided costs of a CHP plant are quite low for supraregional utilities due to the massive sunk costs and the operational characteristics of large power plants. The developments of the next years will be interesting with regard to the future structure of the German power plant park because from about 2005 onwards many power plants will have to be replaced or modernised. The crucial question will then be whether the centralised generation path will be continued or whether a switch to a locally and horizontally optimised generation will be made. At present, the supraregional utilities do not seem to be in favour of such a decentralisation of generation.

This shows that a number of strategic considerations are of major importance for the supra-regional generators. First of all, independent decentralised capacity would replace their own centralised capacity - at least to a certain extent.[254] This has as a consequence the risk of an under-utilisation of their own plants and a reduction of their power sales. In the longer term, there are also implications for their longer-term capacity planning and for the general fuel mix for power generation in Germany, because large power plants tend to run on coal or nuclear fuels.

A very supportive position has been taken in the last years by the major German gas suppliers (as far as they are not controlled by one of the electricity supply companies). Since the emergence of gas-based CHP-systems, they have promoted small- and medium-scale CHP in different ways. Initially, there were rather reluc-

---

254 It needs to be kept in mind that supraregional generators are operating power plants at different load levels (base load, medium load, peak load). They will certainly continue to generate the bulk power to the grid at base load and be responsible for system co-ordination of the other load levels. If a high availability can be guaranteed, the growing power capacities of decentralised plants could then be recognised as fully valuable replacement for centralised medium-load plants. This would also have to be reflected in future remuneration tariffs.

tant to commit themselves to CHP due to their close relationships with the ESI and the strategy of concentrating on the gas sales business. In fact, the two major companies Ruhrgas and Wintershall deliver the lion's share of their gas to public utilities. However, contracting and support for technology development in the field of gas-engines, turbines, fuel cells, system integration have contributed to the advancement of the technology. Ruhrgas is also involved in a company called Erdgas Energie Systeme (EES), one of the most successful small-scale CHP-packagers. The most important activity of the gas industry to promote CHP must be seen in the political support for modifications of the regulatory framework for CHP. Through the lobbying of their promotional organisations like ASUE on the technical and FGBHKW on the political side they have given clear signals to policy-making.

The situation for private service companies continues to be difficult in Germany. In the industrial realm, many companies prefer to set up energy supply on their own, in spite of a slow trend towards a higher acceptance of contracting. And if such arrangements are envisaged, be it in industry, in commerce or for public buildings, then local utility companies are the "natural" partners. This reflects also the experience made with many contracting projects in the past. Many things can be done very smoothly and easily if the local utility is "on board". It facilitates licensing procedures and promises reasonable grid connection conditions. For certain small-scale applications like group heating, service providers had success in competitive tenders, but in general this field of activity is still of minor importance. Liberalisation may well improve their possibilities, but for the foreseeable future local utilities will stay in a very good position.

To summarise, the situation on the user side of CHP has improved slowly, but significantly. Technology, competencies, system integration, economic boundary conditions and the regulatory framework have developed in a conducive way. Still, there is no major push for CHP, and the opportunities for the private facilitators, contractors and service-providers are uncertain. The existence of strong local utilities is generally speaking a conducive factor for CHP, in spite of the initial reluctance to adopt small-scale CHP. Their interest structure is in principle compatible with the requirements of CHP. Even at regional and supraregional level some more openness towards CHP can be identified. Strategic issues about the long-term trajectory of electricity supply come into play at these levels, and this makes the control of CHP-technology an important issue for them. Many regional companies accept independent production - be it from local authorities or industry - as long as it does not represent a massive share. The role of private energy service providers has remained marginal. They are competing with many services of the local and regional utilities, and since the latter have become more active in contracting over the last years, the perspectives for this new type of firms are less promising than in the UK.

9.6 The evolutionary dynamics of CHP-diffusion in Germany

PET suggests to apply the notion of variation, continuity and selection to the three different types of decisions made in relation to a technology. In the end, they contribute to an evolutionary process which determines the fate of the technology in question. For such a technology-oriented perspective, seven main elements have been proposed to interpret the joint operation of technological innovation, adoption and related public policies regarding CHP.

External pressure

The developments in Germany are characterised by rather incremental changes which have been induced to some extent by impulses external to the PET-system of energy supply. The *decline of energy-intensive industries* and the *stable situation of fuel supply* have not favoured the spread of energy efficient generation technologies since the Eighties. Conducive events and forces can be identified in the *widespread use and implementation of natural gas*. It provided the infrastructure needed for the adoption of new gas-based CHP-systems. Of major importance was also the awareness about and *importance of environmental concerns* which are reflected in many aspects of the preceding analysis. Regarding industry, this is expressed by its efforts to improve efficiency on the basis of economic motivations. On the technological side, *few impulses came from abroad* due to the closed and informal character of user-supplier relationships in Germany. The *completion of the Internal Market for Energy* in Europe foreshadowed changes to be introduced in the regulatory framework, but without having a major impact on CHP yet. All these external developments required both technical and organisational adjustment in the energy supply system to enable the adoption and innovation of CHP.

Variation and selection by innovation

The innovative responses to external and internal pressures, and the provision of adequate technical and non-technical solutions, were affected by several factors. Although both technology manufacturers and users disposed of a good technical knowledge, it was fairly narrowly focused on large-scale district heating and industrial applications. The new technologies for small- and medium-scale CHP were not yet available, neither in Germany nor abroad. On the technology supply side, one can therefore identify two inroads for the introduction of innovative CHP-technology. Either *new small companies developed niche markets*, or established manufacturers improved and adapted their equipment to the new needs. The *attempts by the established firms were not far-reaching* enough to meet the non-technical, systemic, organisational and service requirements of CHP-users - at least initially. This situation represented a good opportunity for newcomers, but they had to overcome the lack of confidence in the technology as well as in their companies. In the Nineties, the *more sophisticated requirements of CHP started to go beyond the skills and knowledge of the small newcomers*, and enabled the established firms and those which disposed of the support of strong partners to take the lead in the CHP-market.

From the technological side, it was crucial that the new *CHP-systems became accepted as standard technologies* for heat and power supply. Standardisation has a double meaning in this respect. First of all, it has a technical connotation of facilitating *economies of scale and compatibility*. Secondly, and even more important, it denotes the acceptance of CHP by technical associations which then spread *planning and design practices* among engineers, as well as background information among decision-makers. The technical improvements were either *incremental improvements* of components or steps towards the *optimised integration of CHP in energy supply systems*, based on learning processes and the adoption of spin-offs from control technology, car and power engineering, etc. Direct *government support measures* supported the development of new technologies for CHP, but without playing a major role.

A major variety-creating effect needs to be seen in the *diversity of local and regional regulatory and policy contexts* which provided many experimental fields for trying out different types of CHP-technology. The discretionary power left to regional and local authorities is thus a key element of the German political system which facilitates variety generation.

Selection by users in their organisational contexts

The selection conditions for CHP as determined by German users are shaped to a large extent by the *special role of local utilities*. As users they fulfil many of the essential conditions to operate CHP successfully, and as local grid operators they are the gatekeepers for the connection of other generators. Similar to the few private facilitators they can *integrate the different energy-related services* of gas, heat and electricity and exploit the corresponding economic and energy efficiency benefits. The other facilitators, such as regional utilities and gas suppliers, differ very much in the support they give to CHP. Their possibilities to get actively involved in CHP are also more limited than at the local level. While the *gas industry recognised opportunities in CHP* in order to complement its services, the regional and supraregional utilities have not been inclined to develop a supportive stance towards it. Their interest structures and plant parks reflected a logic which is very different from the one which underlies a decentralised CHP-based supply system.

The typical energy service end-users in industry and the public sector are also characterised by fairly *well-developed technological competencies*, but also by a certain degree of *conservatism* and reliance on established codified and standardised knowledge. This *knowledge is controlled by technical associations* which play a similar gate-keeping role on the information side as the local utilities on the energy supply side. Once CHP was accepted by these technical associations, it was also accepted by many of the potential users as a standard technology. Any eventual technical problems could then be overcome quite easily.

The *local adoption environment has thus developed in a fairly positive way* for CHP. Due to their organisational structure, local utility companies play an important integrating role for CHP, similar to that of Integrated Energy Service Companies in the UK. But also the *regulatory and administrative conditions have improved* to the advantage of CHP. The higher levels of the energy supply system have maintained an ambiguous stance until now, even if the connection conditions which the power utilities are willing to concede to decentralised CHP have improved. The *stability of the economic boundary conditions* for CHP, which was facilitated by the Associations' Agreement, was certainly very helpful for providing a reliable planning base for CHP-users.

Selection conditions of the regulatory and political framework

The regulatory framework of the energy supply sector represents the wider selection environment for both users and suppliers of CHP. *Energy and environmental regulation have developed in a fairly conducive way* for CHP since the early Eighties, though in a rather incremental way by *removing barriers and improving the economic principles applied to remuneration* of independently generated power. The principle of delegation of responsibilities to the top organisations revealed to have a double-edged impact on CHP. On the one hand it facilitated the *development of acceptable and pragmatic guidelines*, on the other hand these solutions reflected only those interests which are actually represented in the top organisations. Interests like those of *small- and medium-scale CHP-users have been underrepresented* in the decision-making processes, but they represent a growing population of actors.

An important characteristic of the German context is the high degree of variety it allows to emerge. This is mainly the result of the *decentralisation of competencies and responsibilities* for energy supply and political decision-making. New solutions can be tested so that niches can emerge. Due to these possibilities for creating variety, evidence could be accumulated of the advantages of CHP from a variety of perspectives which in the end led to the establishment of CHP as a topic on the political agenda. The attempt to maintain variety is also reflected in the interactions within the national political structures. *Many very different organisations participate in the preparation of political decisions*, both promoters and opponents of CHP. In this context, it is also important to underline the *importance assigned to scientific expertise* in the political process (in spite of attempts to operationalise this expertise on CHP by some actors) because it helped to analyse and assess the various experiences made.

This fairly conducive situation of the regulatory and political context is also reflected in political *targets and commitments of Federal and State governments* as well as of the Parliament. CHP has clearly become an important option for German energy policy. In practice, this development lead to a number of supportive (though not necessarily effective) *measures to stimulate the diffusion of energy efficient technologies* such as CHP. In the last years, the focus of this support has been on the modernisation of the East German energy supply system. However, *much support for CHP remains at a general and declamatory level* without triggering a substantial reorientation towards decentralised energy supply.

In terms of the basic principles governing energy supply in Germany, the situation has become rather uncertain in the last years. Despite a *general commitment to market principles* and the relaxation of several administrative and interventionist measures, the current liberalisation plans in view of the European Internal Energy Market are still under negotiation. For the future, this implies a *high level of uncertainty*, not only for CHP but for the strategies of most of the actors involved.

Stabilisation mechanisms

Despite the improvements of the selection conditions for CHP, which then provided new incentives for innovation and variation, the centralised and horizontally separated supply structure is stabilised by a number of mechanisms. Bulk power generation is operated by utilities which do not have much interest in decentralised generation. The centralised system is controlled by *well-established networks of powerful utility companies* which co-operate with the established power engineering industries. Moreover, the *spatial and size structure of power stations* represents a physical barrier to any major transformation and delays change. These technological, economic and organisational interdependencies show a strong resistance to change.

A further important stabilising mechanism must be seen in the *consensus-oriented political system* in Germany and the manifold *mutual links among the different actors*. Examples of this are the cross-ownership among utility companies, the delegation of competencies and the comprehensive preparation and consultation exercises during the policy-making process. They stabilise the current situation and make it difficult to trigger far-reaching changes. The consensus-orientation is also reflected in the *shared mental frameworks and knowledge base* which are strongly influenced by the engineering associations.

The importance of technical expertise and of independent scientific advice also opens up opportunities to overcome structural lock-in and trigger changes. In the case of CHP, technical experts who are generally independent of partial interests, *stabilised the new technology by accepting it as a standard option.* Once being accepted at this technical level, the engineering culture in the energy industries changed and developed in a more conducive way for CHP. In other words, it was possible to introduce *stabilising mechanisms in favour of CHP* and thereby prepare its more widespread acceptance.

There are other examples of such stabilising driving forces in favour of independent and decentralised generation which contributed to overcoming the current lock-in. Industrial generators, local utilities, new manufacturers, the gas industry and manifold other actors formed *implicit or sometimes even explicit coalitions of interest* and questioned the centralised energy supply trajectory. Although the entire system is largely centralised at present, there are possibilities and forces operating towards a different future trajectory.

Reinforcing and delaying mechanisms

Several reinforcement mechanisms can be also be identified in the case of Germany, which operate both in favour or to the detriment of CHP. As in the case of the UK, *new technological opportunities* as well as slowly *improving framework conditions* broadened the scope for further investment in technology development, especially by some new pioneering companies. *Successful example applications* within the new framework helped then to raise more optimistic expectations regarding CHP. They represented an *incentive to improve the technology further* and to raised interest among the established manufacturers, at least once the threshold level of their initial inertia and routinised behaviour had been overcome. Obviously, *learning effects, economies of scale and changing mental framework* also played a role for establishing CHP as an accepted technology and for triggering institutional change. This process is characterised in Germany by a particular kind of threshold effect as a result of the *standardisation of design principles and planning procedure* by engineering associations.

On the CHP supply side, the early newcomers on the market developed the first small-scale niches, but once the stabilisation of CHP was achieved, the established companies provided the expertise and sophistication to bring the technology to the level of perfection required by the wide community of energy engineers. This *professionalisation* of CHP-manufacturing was further reinforced by environmental requirements which could not be easily fulfilled by small companies without systematic R&D.

One of the early delaying mechanisms in Germany can be identified in the link which was often made between heat pump technology and small-scale CHP. The *lack of knowledge* especially in the political realm led to *misunderstandings and a precipitated refusal* of small-scale CHP, which could only partly be counterbalanced by the good reputation of DH. A second interesting delaying mechanism consisted of *the effect of failures*. Early failures of small CHP-projects confirmed many of the prejudices against the technology, and destroyed the potential markets in the vicinity of such cases. In other words, especially in early phases of diffusion it is essential to plan and design plants carefully in order to avoid a delay of market development.

Due to the comparatively slow transformation process in Germany, the problem of timing has been less crucial than in the UK, although it has certainly delayed diffusion. *Timing has been important in the case of East Germany* where attempts were made to stabilise the district heating infrastructure by modernising the generation side. In some cases this was achieved timely, in others large parts of the DH network were given up in favour of natural gas grids. With regard to industrial capacity, it is difficult to assess the question of timing given the fact that much of the industrial capacity has actually disappeared.

However, the fairly *good climate for environmentally-oriented investment* which can be expected to continue in the near future, seems to offer a *window of opportunity for CHP in Germany* in a few years time when much of Germany's generation capacity needs to be replaced. Then, the combination of a well-established technology (possibly complemented by fuel cell systems) and organisational structures which favour decentralised and horizontally integrated energy supply and service provision could facilitate the wider adoption of CHP.

System evolution

The evolutionary process of change which accompanied the diffusion of CHP can be traced back to the underlying three types of decisions, to several feedback mechanisms and to stabilising elements of continuity in the system. *The net balance of conducive and detrimental reinforcements has been quite positive in Germany.*

The process of *change has been incremental*; there was no single dominant driving force behind these developments (such as liberalisation in the UK) but *a universe of minor changes operating in favour of CHP*. The growing service orientation among utility companies, their increasing acceptance of the different types of CHP, the rising interest of different potential users and of policy-makers, the adjustments of the regulatory framework and of the underlying political processes, and the technological developments for broadening the application areas of CHP all contributed to the developments in Germany. This also means that an explanation of these changes is very complex; they cannot be traced back to any dominant driving force, but only to a *complex set of interactions* between the individual strategies of the actors concerned, and the balances between them.

Just some of the most important elements can be pointed out. The major utility companies tried to introduce pricing and tariff regimes which were detrimental to CHP-users, and they also influenced the political debates on energy supply. Although initially less well organised than the ESI, the strategies of the gas industry, of CHP manufacturers and of the scientific experts who favoured CHP due to its high efficiency counterbalanced the effect of the large utility companies. On the other hand, the decentralised structure of the German political and energy supply system opened up opportunities for experimentation. Local utilities had an organisational and interest structure which was compatible with the requirements of CHP. The Associations' Agreement was a crucial cornerstone for defining reliable economic framework conditions of CHP-operation. Industrial and engineering associations standardised design and planning procedures and made CHP a widely acceptable technology. In other words, unlike the British case, the diffusion of CHP in Germany is an example of *a co-evolutionary transformation of technology, system structures and actor positions*.

For the moment it is still unclear whether CHP will remain just a valuable but small contributor to energy supply in Germany or whether it will reach a more substantial share. Given the advanced level of development and the widespread support for it, the *prospects are good, but highly dependent on the establishment of a new regulatory framework* as part of the Internal Energy Market.

9.7 Outlook

The analysis of structural changes and underlying mechanisms has pointed out the most important factors and mechanisms of the diffusion of CHP in Germany. For the future, three basic scenarios can be outlined which are characterised by three policy strategies to frame the diffusion of CHP:

- Introduction of a moderately liberal framework compatible with CHP;
- Full application of liberal market-based principles in energy supply;
- Political enforcement and support for CHP within the current framework.

Major exogenous ruptures are not considered, so that the three scenarios are defined by different politically induced transformation processes. In the brief discussion below, four aspects will be highlighted: the basic principles of each scenario, their main elements, the implications for CHP and the positions of major actors.

9.7.1 Scenario 1: Moderate CHP-friendly liberalisation

With respect to the ongoing reform of the German Energy Supply Act, one of the most probably options is a moderate liberalisation of the current framework, which would introduce changes especially for large customers and generators. Limited third-party-access and competition between supply companies from Germany and abroad would also be possible for certain energy services such as local heat networks. However, the major principle of monopolistic local supply areas would be maintained. This scenario can be characterised in more detail by the following developments:

- Competition between generators in the large-scale market and free choice of suppliers for large customers;
- Limited competition at local level for contracting energy supply solutions to private customers;
- Concentration and further vertical integration of utility companies;
- Establishment of horizontally integrated utilities at local and regional level;
- Disintegration of informal links between utility companies and industry;
- Regulated electricity prices for small customers;
- Export remuneration based on individual negotiations, but guided by orienting agreements of the industrial associations;
- Further removal of administrative barriers;
- Widest possible removal of subsidies for specific industries like coal;
- Only very limited financial support for CHP, especially if its based on renewable fuels; it would not be financed by the utility companies any more, but from the Federal budget;
- Limited research funding for relevant basic technologies;
- Achievement of environmental objectives on the basis of voluntary agreements and by incentives for LCP/DSM rather than by taxes and specific regulations;
- Stable relationships between government and industry.

Within this framework, CHP could be expected to diffuse further, but to a limited extent only due to decreasing electricity prices. Such a price drop would be the result of more competition in the ESI and the removal of subsidies. Especially industrial CHP would be affected by these developments. No major changes are expected for DH due to the limited extent of competition at local level. The initial phase of this scenario path is characterised by a high level of uncertainty about future developments, and this would delay, if not prevent, several CHP-projects. Technological advances and R&D measures would be carried out as a response to improved opportunities for contracting in small- and medium-scale applications and for certain public service applications which had been subject to specific regulations before (e.g. in the health and education sector).

While the liberalisation approach pursued in this scenario would not go far enough for industry, it would already go beyond the positions of many utilities. Especially regional utilities would be in a difficult situation, because in contrast to the supraregional companies, they have less possibilities to compensate for fiercer competition in Germany by operating Europe-wide. Moreover, while local utilities would maintain parts of their monopoly in franchise distribution, regional utilities could not benefit from this advantage any more. The lack of strong supporters of CHP would prevent a further promotion of this technology by specific programmes and regulations. However, in general such a scenario would probably find sufficient agreement to be implemented. It matches also quite well with the incremental character of German policy-making which requires a broad consensus on major decisions.

9.7.2 Scenario 2: Full application of liberal principles

Within the framework defined by the Internal Energy Market, this scenario corresponds to the widest possible application of market-oriented principles. The liberalisation of distribution, generation and customer services are its main characteristics. More specifically, the following elements can be mentioned:

- Competition between large domestic and foreign supply companies;
- Fierce competition between local utilities and private service providers, limiting the utilities' possibilities to participate in activities which cannot be justified on short-term economic grounds (e.g. support for R&D);
- Fast spread of contracting agreements for all kinds of energy services;
- Free choice of suppliers for all customers;
- Uncertain conditions for the planning of major long-term infrastructure projects;
- Horizontal and vertical integration of energy companies as a response to competition and in order to exploit economies of scope and scale;
- Uncertainties about electricity prices as a result of very limited regulation and free negotiation of tariffs;
- Further decline of the domestic coal industry;
- No specific measures to promote CHP;
- Dominance of a short-term time horizon for investment in energy supply.
- Environmental regulation by voluntary agreements and by market-based measures, especially taxation.

The diffusion of CHP would be hampered in this scenario by uncertain economic and planning conditions. Long-term investments in integrated supply systems are difficult to justify on a reliable basis. This would affect in particular district heating because the free choice of energy service suppliers for all customers makes large district heating systems a very risky investment. The conditions for industrial and small-scale CHP would be rather double-edged. They would improve due to a more competitive contracting market, but worsen due to expectations of falling electricity prices. This last effect could be counterbalanced politically by introducing a taxation regime for energy which is conducive to CHP and energy efficiency in general.

The measures to trigger this second scenario would meet strong resistance from both regional and local utility companies, because they would be subject to competition from private companies and loose their protected area monopolies. Supraregional utilities would be less affected because as in scenario 1, they can expand abroad and integrate vertically with regional and local companies. Industry would in general support a liberalised framework because it promises lower energy bills and the free choice of their suppliers. Apart from environmental groups who would fear a neglect of environmental considerations in the liberal framework, there would be little specific support in favour of CHP. The topic would be overridden by other considerations, unless it was kept on the political agenda as a result of lobbying and scientific evidence of its advantages. However, due to the rupture with the current system which is implied by this scenario, it would be difficult to reach an agreement to implement it.

9.7.3 Scenario 3: Political enforcement and support for CHP

Scenario 3 reflects a rather narrow interpretation of the Internal Energy Market but contains a much stronger element of political intervention than the two others. Competition and third-party-access would be introduced only for the large-scale market while monopolies are maintained for transmission and distribution at regional and local level. The entire approach would be driven by the recognition that a number of politically important issues cannot be addressed adequately in a purely market-based environment. These are e.g. the coal question, environmental considerations, the long-term development of the energy supply system, and the need to provide reliable planning conditions. CHP would thus be specifically addressed as a desirable option and supported by various measures. The scenario picture can thus be complemented as follows:

- Limited scope for private contracting but good opportunities for local and regional utilities to develop energy services;
- Regulations tailored to facilitate the adoption of CHP and auto-production;
- Decentralisation trend of energy supply in order to exploit economies of scope;
- Stringent emission and efficiency regulations based on a prescriptive approach;
- Specific consideration of CHP in the tariff regulations which complement the voluntary Associations' Agreement;

- Regulatory control of utilities' investment plans which need to fulfil certain criteria to reorient long-term energy supply patterns;
- Compulsory least-cost planning and analysis of potential industry-utility company synergies in energy supply;
- Co-ordination of district heating systems with competing options like natural gas supply;
- Taxation of fuel consumption as a measure to control energy prices and to favour investment in energy efficiency;
- Higher electricity prices as a result of environmental and resource-oriented measures which favour also CHP;
- Increasing problems of complex administrative and regulatory procedures;
- Close corporatist interaction between different levels of government and industry;
- Large scope for further improvements of the technology due to the conducive framework conditions.

Scenario 3 represents the only pathway by which district heating can be expected to expand further (at least as far as other gas networks do not yet exist) because a fairly stable and economically conducive regulatory framework would be introduced. Small-scale CHP would be used where possible and reasonable, but usually in co-operation with a public utility company. Also for industry, CHP would become a very interesting option due to rising electricity prices. As auto-production would be favoured and have to be accepted by the utilities at reasonable tariffs reflecting avoided costs, it would become a viable option for many applications.

Resistance can be expected from some industries which would suffer from higher electricity prices. Also large utility companies would rather oppose the framework because the stronger intervention and regulatory control by government would lead to a loss of their independent and powerful position in the German supply system. In addition, the scenario suggests a decentralisation of supply and generation based on CHP. This is a constellation which clearly challenges the supraregional companies and implies a loss of control for them. Regional and local companies would not oppose the scenario because they would basically maintain their current positions and be just subject to a tighter control of their relationships with autoproducers. As the scenario is characterised by a decisive role of environmental and resource arguments, it would be supported by the corresponding interest groups. There is a potential conflict with the principles of the IEM, especially regarding the types of supply areas which would have to be opened to competition. In other words, the scenario would contradict the emerging paradigm of liberalisation of energy markets. However, scenario 3 would be far easier to realise in the German context than in the UK because for the moment the supply system is still based on quite well performing public monopolies. The crucial question is rather whether it could be introduced against the resistance of such important lobbies as the supraregional utilities and industry.

9.8 Excursion: CHP in the New Länder[255]

The situation in East Germany since reunification differs significantly from that in West Germany, and not only due to different levels of economic development. The rupture of 1989/90 triggered a complete re-organisation of the energy supply sector. In parallel, industrial production declined substantially, thereby also reducing the scope for CHP. In addition technologies became available in East Germany (and not only CHP) which had just not been available before reunification. They enabled the introduction of radical changes in the remaining parts of the East German energy supply system. The organisational structure of the West German system set-up was adopted with minor modifications. In other words, a process of fast change has been taking place in the New Länder since then, by which the situation of CHP was strongly affected.

The political priorities regarding the East German energy sector after reunification reflected three main objectives which were to be achieved in the shortest possible period of time:

- Achievement of a price level for electricity which should not be higher than in West Germany, in spite of the massive investment needed over the short and medium term;
- Maintaining the exploitation of indigenous brown coal for power generation;
- Modernisation of the entire supply system in order to bring it up to West European standards and connect it to the international grid system.

9.8.1 Structural developments

Regarding the modernisation of energy supply, massive amounts of capital were needed after reunification. This money was provided to a large extent by the West German utility companies, especially the supraregional firms. In order to realise the necessary investment, for instance for the modernisation and new construction of brown coal power plants, major organisational changes were required, too. Not only did RWE, PreussenElektra and Bayernwerk take over the majority of the shares of the "Vereinigte Energiewerke AG (VEAG)" in 1994, the successor of the former East German main generation and transmission company, but they also set up or participated in regional electricity and gas utilities. These, in turn, en-

---

[255] Until 1997, little material had been published on the situation in the New Länder. For some good overviews, see for example (Steiner, 1994) on major investment in power generation, (Stotz & Broßmann, 1995) on supra-regional electricity supply, (Holst, 1995) on gas supply patterns, (Muders, 1994) on small-scale CHP, and (Becker, 1995) on the political evaluation of the developments in East Germany. The results of the assessment of the district heating support programme for East Germany were not yet available, and in general the situation was still in flux. The following information is mainly taken from some specific interviews on cases in East Germany, and on selected questions asked to the West German experts and practitioners in the other interviews of this case study.

tered in partnerships with local authorities and founded new local utility companies which had enough financial backing to modernise the energy supply systems of entire cities. Usually, the supraregional utility companies did not hold direct shares in local or regional utilities in East Germany, but participated indirectly through their West German regional subsidiaries.[256] In other words, although the structural set-up of the utility system was not copied in detail, if reflected perfectly the power structures which prevailed in the West German ESI. One could even say that the West German utility companies divided the East German shares among themselves in order to dispose of the possibility to modernise the system without being challenged by competition. An important guarantee for the functioning of this system was the introduction of the so-called '30%-rule'. It means that east German local utilities may not generate more than 30% of their electricity themselves. All the capacity which goes beyond that level must be sold to the regional supplier and bought back again.[257] This issue was fiercely debated because it affects in a very fundamental way the level of independence granted to the local utility companies, and thus also to the local authorities. The influence of local utilities was further affected by the necessary re-organisation of the grid system of the former German Democratic Republic; it had to be restructured completely and its control be re-distributed among the three levels of utility companies (local, regional, supraregional). The unbundling of the grid system was very difficult, and the local utilities had to fight for obtaining the control of the local grid networks.

Beyond these general structural developments in the energy sector, the situation of CHP after reunification can be described by the following characteristics:

- Existence of many district schemes which were often interconnected in the past with industrial heat consumers.;
- Many district heating schemes have been modernised in the meantime; especially the old coal-based combustion systems were replaced by modern gas-fired turbines and engines;
- In parallel, most plants have been upgraded to comply with environmental regulation because after a phase of adaptation the same rules are now applied as in West Germany.
- Emergence of a major power generation capacity at local level, replacing several of the old brown coal-fired central electricity generation plants;
- Limited scope of small-scale CHP due to the widespread availability of heating networks;

256 An example of this is the 'Dresdner Modell'. It was based on agreements of partnership between regional suppliers and local utilities which were concluded very early after reunification. This helped to avoid many problems between the partners later on (e.g. regarding the '30%-rule'), because they were both 'in the same boat' right from the beginning and had agreed to the main objectives of their collaboration.

257 However, there was a fierce debate going on about this 30% level, up to appeals to the Constitutional Court ("Bundesverfassungsgericht"). One of the issues in question concerned the way in which the 30% level was to be applied, either for each individual local utility or for East Germany as a whole.

- While in some cases agreements could be found on the co-ordination of district heating and direct gas supply, in others the wider introduction of gas made DH-infrastructures unviable and lead to their abandonment; in addition, most of the old city gas systems are now delivering natural gas;
- Industrial decline had a direct effect on CHP by reducing its capacity and potential. It also had a negative indirect impact because it made the interconnected operation of industrial and district heating CHP from one major plant economically unviable; plants were oversized for the remaining heat load and had to be abandoned.

9.8.2 Developments at micro-level

In spite of the struggle between local authorities and West German utilities, the agreements which were concluded for power supply in the New Länder are in many cases based on a fairly co-operative basis. The local utility companies could perform their tasks of modernising energy supply due to the participation of one of the regional gas and/or electricity utilities which are at least partly owned by West German companies. These very constructive arrangements were regarded to be possible only because competition was excluded, and it is rather doubtful whether they would have been feasible within a liberalised framework.

The growing competition between district heating and gas supply has had very diverse impacts. While in some cities, the definition of priority areas ("Vorranggebiete") for district heating or direct gas supply helped avoiding destructive competition between the two technologies, in others district heating was entirely replaced by gas systems. In the first group of cases, the agreements were enabled by the horizontal integration of local utilities which co-ordinated the activities of their gas and heat supply divisions.

Since reunification, two major developments lead to the loss of a large share of the heat load of district heating schemes. First of all, industrial heat load has declined massively, and secondly, demand-side efficiency measures have reduced the heat load on the private customer side. New connections could only partly compensate for these losses, but the downward trend could be stopped by about 1992/93.

In many industrial cases, finance of new investment was a major barrier for CHP. The participation of the regional gas/electricity companies in contracting arrangements or joint ventures with private investors helped overcome this problem. A very specific East German difficulty needs to be seen in unclear property rights which could sometimes not even been solved by the middle of the Nineties. Expropriated owners did not want to invest in their former property until all doubts about receiving the property titles were cleared.

The massive financial support given to East Germany to upgrade the energy supply system went to a large extent into CHP. The CHP-investment programme (see Table 9.3) gives evidence of this as well as the active involvement of West German utility companies. In many cases the Federal support was decisive for making the investment projects viable, with subsidies of up to 35%. The fact that the district heating support was not prolonged after 1994 left still much capacity unexploited, especially in those cases which could not be addressed for reasons of unclear property rights or lack of finance.

Environmental regulation were applied in East Germany similarly to the West, but with clearly defined transition phases. However, the flexibility which was built into the regulatory approach had in some cases lead to technical specifications which went beyond the requirements of corresponding plants in West Germany. Due to the fact that the policy objectives for the development of the East German supply system gave highest priority to the modernisation of the system as whole, the emphasis given to environmental upgrading was perceived as exaggerated and endangering the economic development possibilities of the East German economy.

9.8.3 Assessment and outlook

The transition phase of the East German energy sector after reunification was a large-scale experiment. On the one hand it was successful in the sense that the energy supply system was modernised in a very short period of time and brought up to the standards of Western Europe. A significant part of the district heating infrastructure could be maintained in that process. On the other hand, due to the fact that the West German industrial structure and organisation was replicated in many respects, a major opportunity was lost to set up supply structures which were already compatible with the new framework of the European Internal Energy Market. As a consequence, the next reorganisation will take place very soon, once competition will be effectively introduced in Germany. Neither was a serious attempt made to strengthen the decentralisation of energy supply in the New Länder. On the contrary, the local level was given a rather limited degree of independence only. However, one may doubt whether a different pathway than the one actually taken would have been feasible given the fact that the co-operation of the West German utility companies was indispensable in the process, both for financial and technical reasons.

In spite of these shortcoming and the uncertainties regarding the IEM, the perspectives for CHP in the New Länder are quite positive. The lion's share of former district heating capacity could be maintained, the re-emergence of industrial CHP depends mainly on future industrial development patterns and small-scale CHP is expected to find its market in certain niche applications, though to a more limited extent than in West Germany. A solid ground for a further diffusion of this promising technology is laid.

Chapter 10: Comparison of the Developments in the UK and Germany

The two preceding chapters have given a structured account of the diffusion processes of CHP in the UK and Germany along the lines of the PET-approach. Following the same pattern, these developments are now going to be compared in order to highlight differences and similarities, especially in terms of the dynamic mechanisms and processes which have driven the developments in the two countries.

10.1 Comparison of the developments till the early Eighties

The early histories of cogeneration in the two countries laid the ground for the recent developments of CHP, not only by shaping technological and organisational structures of the ESI, but also in terms of attitudes towards the technology. They are reflected in the comparatively bad reputation which certain types of CHP had acquired after some unsatisfactory early applications, especially of district heating in the UK. Quite some time was needed to overcome widely held prejudices about a lack of reliability and control of these systems. In both countries it was important during the Eighties to raise awareness at the political level about the advantages of CHP. In Germany this debate was facilitated by the rather positive experiences made since the second world war with district heating. A large and successful infrastructure had been built up in many cities, often including industrial partners, while in the UK non-industrial CHP started to be re-considered in the late Seventies only.

Therefore, at the beginning of the Eighties the situations in the two countries were quite different. While in the UK, district heating was hardly applied, in Germany a substantial capacity had been built up and further expanded since the first oil price crisis. On the other hand, with regard to industrial applications, also in the UK a considerable capacity was in place. Small-scale applications were in an early phase of implementation at the beginning of the Eighties in both countries. Significant differences can be identified in organisational terms, especially with regard to the extent of horizontal integration of heat and power supply. While in

Germany the horizontal co-operation of public utilities, industry and users was common, it was a rare exception in the UK. The existence of strong local utilities in Germany was of particular importance in this respect.

Main differences:

- Massive exploitation of CHP and DH after the Second World War in Germany, but apart from some industrial applications hardly any CHP in the UK;
- Good reputation of CHP and DH in Germany, but a negative image of CHP/DH in the UK;
- Well developed horizontal co-ordination of energy supply chains at the local level in Germany; separate supply chains in the UK.

10.2 Differences of the context conditions

The role of the energy supply context in terms of availability and prices of fuels has been recognised as an important factor for the use of highly efficient technologies like CHP (OECD/IEA, 1983). Given the current stability of fossil fuel supply and the rather minor price differences between the two countries, the supply context does not seem to be a major factor to explain differences in recent developments. However, security of supply is still a more important issue in Germany than in the UK, where major indigenous oil and gas reserves can be exploited.

More important has been the push for CHP as a result of the environmental debate in Germany. It reinforced the recognition of the advantages of CHP and led to an improvement of the boundary conditions for its exploitation, e.g. by facilitating the planning and permission procedures. This effect was very remarkable in the small-scale area. At least until the early Nineties, environmental issues did not play an important role in the UK. On the contrary, social housing questions were more closely linked to the DH-debate than environmental issues.

In the UK, the debate on energy and environmental policy was dominated by the ideological positions taken with regard to liberalisation. It determined the re-definitions of the regulatory and organisational framework for energy supply in 1983 and 1990. This reflects also the general governance philosophy in the UK which has been characterised since 1979 by a clear preference for a market-based approach, also in response to the negative experiences made during decades of state control in utility services. In Germany, the governance approach is more pragmatic, consensus-oriented, technocratic, and less driven by ideological considerations, even if a stronger market-orientation is regarded as desirable.

The economic recession of the Eighties had a particularly strong impact in the UK where financial constraints led to the consideration of innovative contracting and service schemes, including solutions such as third-party finance, discount energy purchase and contract energy management. In terms of the restructuring of the industrial landscape, Germany underwent substantial changes earlier than the UK. This transformation was driven to a large extent by environmental imperatives. Since 1990, reunification and the developments in East Germany have represented a very specific challenge, especially with regard to the collapse of industry and the need to modernise the entire energy supply infrastructure.

Finally, the widespread supply of natural gas has affected the scope for CHP in both countries. The extension of gas networks took place at a similar pace, and it was essential for the emergence of the most advanced CHP-solutions currently available. With regard to gas prices and competition in supply, the situation in the UK has changed fast since liberalisation of gas supply in 1986; in Germany gas supply has always been a far more competitive area than electricity supply. In general, natural gas plays a double-edged role with regard to CHP because on the one hand it facilitates its use, but on the other hand direct gas heating competes with it. An important problem in this respect lies in the competition between two expensive networks for gas and heat supply, endangering the economic viability of both. In the UK, this has led to a clear dominance of gas supply while in many German cities priority areas for district heating and gas supply were formally or informally defined. This was possible due to agreements between the monopolistic local gas and heat businesses.

At European level, a number of initiatives had an impact on the conditions for CHP. Apart from triggering the use of natural gas for power generation, several environmental directives have affected regulatory practices in the Member States. The recent White Paper on the Internal Energy Market (CEC, 1995) already foreshadowed future changes and adaptations of the national energy supply frameworks, especially in Germany.

Main differences:

- Market-based governance approach in the UK since 1990, consensus-oriented system with distributed responsibilities in Germany;
- Environmental issues more important in Germany than in the UK;
- Industrial restructuring took place later in the UK than in Germany;
- Special conditions in East Germany after reunification.

10.3 Differences of technological developments

Although in principle the technical developments have been similar in both countries, a number of differences in terms of timing and focus can be observed. During the last fifteen years many technological obstacles for CHP have been addressed successfully, and this phase has been paralleled by a growth of CHP in many European countries. However, it is hard to say whether this has been the result of technological improvements or whether it has rather been a matter of framework conditions, organisational modifications and changing attitudes. Evidence from the EU Member States suggests, that these two issues have developed in parallel.[258] It is also reflected in the mutual adjustment of the manufacturers' perceptions of technological opportunities and of the improved conditions for CHP-applications, a process which could be observed in both the UK and Germany.

In technical terms, a whole bunch of improvements have been made in both countries. They opened up new fields of application for CHP. Many of the improvements of the technology were a matter of integrating it into the architecture of the energy supply systems of a site, for example by improving plant design or by adding sophisticated control systems. Component innovations need to be mentioned as well, for example core elements such as engines and gas-turbines, or ancillary equipment for chilling and heat storage.

Many of these improvements were derived from related, non-energy technological fields. With the exception of system integration, hardly any dedicated CHP-oriented R&D efforts were made. Spin-offs from component development in other areas were frequently introduced into CHP-systems in an intelligent way. The use of information technology was essential for this integration in many cases: remote monitoring and preventive maintenance, integrated system operation, grid connection, data collection for the design of integrated packaged plants, etc. Beyond these similarities, major differences between the UK and Germany concern the emphasis given to environmental improvements and standardisation of design (which were very pronounced in Germany), or metering equipment and gas-based technology (which received particular attention in the UK).

Further significant differences can be observed in terms of the timing of innovations in the markets for customised and packaged CHP-solutions in the two countries. They are characterised by differences in focus on specific types of applications. This implies that there must have been different incentives for improvements, too. They are particularly obvious in the case of remote monitoring of small CHP-units. Advanced control systems were required to reduce maintenance costs and downtimes because reliability was the crucial issue for CHP. While remote monitoring was introduced to the British market already in the middle of the Eighties, it became a key concept in Germany in the early Nineties only. This time difference was mainly due to the very conflictive relationships between regional boards and CHP-users in the UK - a situation which made a highly reliable system vital in order to avoid extremely high charges for back-up power in case of system failure. In Germany, the relationships were more constructive and supported by agreements among the associations of users and utility companies.

---

[258] For example, in the Netherlands cogeneration has been at the heart of the Dutch electricity policy strategy, and Dutch companies have also pioneered new CHP technology, which then induced changes in other countries. But also in many of the other countries the link between the organisational and the technological sides is evident and increasingly recognised. For an overview of the situation in Europe see (CHPA et al., 1995).

According to the interviews carried out, the development efforts in the CHP field were triggered by the Energy Act 1983. It made technology manufacturers perceive first opportunities for tailored CHP-systems for the British market. Other differences in timing can be identified for the development of gas engines, for grid connection and metering (an important issue in the liberalised British system) and the extent of packaging of CHP-plants. The packaging and service-oriented contracting approach, which was crucial in the UK, required formalised relationships and clearly defined products, boundaries and interfaces between the contracting partners. This points once again to the importance of organisational innovations.

There is widespread agreement in the UK and Germany about the major future technological trends. In both countries, the potential and importance of fuel cell technology and of improved system integration with chilling, heat storage, waste incineration, etc. have been underlined. The preparations in these two key technologies seem to be more advanced in Germany. Further incremental innovations are expected to reduce the costs, especially of small and micro systems by packaging and by exploiting economies of scale by series production.

Major differences between the UK and Germany can be observed with regard to the importance of different innovations. System integration, gas engine development and environmental issues are rated particularly high in Germany, whereas efficiency increases, reliability and upscaling of packaged systems were regarded as the most important areas in the UK.

Main differences:

- Less emphasis on reliability in Germany due to better co-operation between utilities and end-users; in the UK it has been one of the key issues;
- Environmental innovations played a more important role in Germany than in the UK;
- Stronger emphasis on gas-based technology development in the UK as a result of the 'dash for gas' and the expansion of small-scale CHP;
- Country-wide standardisation of design and small-scale systems in Germany, and emphasis on packaged solutions by British companies;
- Differences in terms of timing of certain innovations, e.g. remote monitoring, gas engines or metering equipment.

10.4 Comparison of structures and structural changes

10.4.1 Technological structures

The data on the general electricity supply and demand situation in the two countries show quite similar patterns. A stagnation of energy consumption has been paralleled by an increase in electricity generation. Both have a centralised generation system, though with a higher share of independent power production in Germany and a substantial generation capacity at the level of local utilities, which results from the more decentralised organisation of the German ESI.

The diffusion data of CHP show that - despite some fluctuations - there has been a slow rise of overall capacity over the last twenty years in Germany, whereas in the UK the decline was stopped in the late Eighties. Despite a fast take-up of small-scale and industrial CHP in the UK since then, capacity is still much lower than in Germany where local utilities and industry operate large CHP plant parks. These developments are superposed by the situation in East Germany where industrial capacity has declined, but most DH-schemes maintained. District heating in the UK is still of negligible importance, but a number of projects are being reconsidered at present. Small-scale CHP has emerged in parallel in the two countries during the second half of the Eighties.

It is difficult to find data on the extent of horizontal integration of heat and power supply, and especially if industry - public utility co-operations are to be included. The levels of CHP-diffusion as well as the information derived from interviews indicate rather co-operative conditions in a number of German cases by which an integration of district heating and industrial applications was favoured. In the UK, the experiences have been very diverse and are still unclear because the RECs are still in a phase of defining their attitude towards CHP. Agreement prevails that the possibilities to integrate of heat and power supply of firms and/or utility companies have improved, but without being reflected in actually realised projects.

Future diffusion of CHP is expected to continue, but to different extents for the different applications. The 5000 MW target in the UK will be approached mainly by investment in industrial CHP and some small-scale capacity, while in Germany a major contribution is also expected from DH. All these expectations and prospects are subjected to a high level of uncertainty due to the expected further changes of the regulatory frameworks during the next years.

Main differences:

- Substantially higher CHP-capacity in Germany, especially due to district heating;
- Similar development of small-scale CHP in both countries;
- Significant industrial CHP-capacities in both countries, though at a much lower level in the UK;
- Higher level of decentralised generation in Germany, although decentralised power generation in the UK is rising;
- Better conditions for horizontal integration of heat and power supply in Germany than in the UK.

10.4.2 Economic structures and regulatory framework

Regulatory framework for energy supply

Liberalisation, privatisation and the new specific regulations overturned the economic structures for electricity supply in the UK. While the Energy Act 1983 did not have a major impact on the behaviour and the positions of the main actors, it opened up at least some first opportunities for setting up new CHP-schemes in the private sector. They served as test-cases for the further improvement of the technology in some niches. The Electricity Act in 1989 was a radical step which changed many of the parameters for CHP. Competition was introduced in generation and distribution, leading to a high level of uncertainty. In principle, the new regulatory conditions facilitated the set-up of CHP-projects in the private sector, but the economic conditions (high level of uncertainty, short payback requirements and time horizons) made it very difficult to predict and assess their performance. In Germany, the organisational principles of the energy supply system, which are based on monopoly areas at three levels has of supply, have been maintained until 1997, but a number of facilitating measures were introduced. They ranged from technical and planning regulations to remuneration for exported power and financial support schemes for both R&D and investment.

The conditions for CHP in the ESI

The specification of the regulatory framework and the reliability of the planning conditions have revealed to be much more important issues for the diffusion of CHP than the question whether a liberalised or a monopolistic approach is pursued. In terms of these specific conditions, a number of important differences and similarities can be observed between the two countries. First of all, liberalisation in the UK opened up the possibilities for access to the generation business. While in Germany, generation licences and other barriers to access have been removed slowly, the liberalised settings facilitated independent generation in the UK at one blow. The main problem consists nowadays of administrative problems and the complexity of the pool system. The importance of these changes is expressed in the fast development not only of independent generation in the UK, but especially of accompanying activities: energy management, third-party financing, contracting, etc. However, their emergence was also a matter of business cultures and economic necessities of companies in the energy sector to focus on core activities. Both these forces were stronger in the UK than in Germany. Moreover, the complexity of the new liberal system was very difficult to deal with for many companies. It reinforced the trend to outsource tasks to specialised companies which were looking for investment opportunities in CHP. These were not only specific energy management companies, but also technology suppliers (especially small-scale packagers), gas supply companies, Regional Electricity Companies (RECs) and the major generators, which all tried to extend the range of their services horizontally and vertically. While the emergence of private Integrated Energy Supply Companies (IESC) in the UK has been a significant change which facilitated several CHP-projects, horizontal integration is still limited to individual sites. In Germany, this horizontally integrating function is fulfilled to a large extent by local utility companies, i.e. at a higher level of supply than that of individual sites. Although activities under the headline of CEM and public-private partnerships have also started to be considered for individual sites in Germany, they have gained importance at a comparatively slow pace and for some application areas only (e.g. health, education, East Germany) .

Despite the intention of British liberalisation and privatisation to reduce political influence on energy supply, the regulator still has a very powerful position which he also used several times in order to guarantee the 'operation of market mechanisms'. The German regulatory system may have a wider coverage, including e.g. electricity price controls and investment control, but in practice the influence is rather limited due to a lack of personnel and information, which is largely controlled by those to be regulated: the utility companies.

Of crucial importance were also the organisational structures and interests in the two countries. In Germany there were already a number of interested parties established and positioned to exploit the benefits of CHP. The horizontal integration of energy services was a key facilitating condition in this respect. For example, although power generation is dominated by large scale, power-only production, local utilities can combine the heat and power production lines horizontally. In the UK, there were hardly any local utility companies in place, and this shortcoming could only partly be compensated for by the emergence of IESC and the changing attitudes of the RECs after privatisation/liberalisation.

The RECs fulfil still several key functions for electricity supply, even in the liberalised system. At the time of the monopolistic Regional Electricity Boards, i.e. before 1990, they fully controlled the interfaces with the grid and could thus manipulate the conditions for CHP. Even during the first years after liberalisation they kept their privileged position as the only fully competent actors in the system. It took quite some time until common rules were established under the new framework which prevented the RECs from using technical or economic connection conditions to obstruct independent CHP. Due to their double role as grid operators and regional supply companies on the one hand, and as plant operators and management companies on the other, they are still gatekeepers for CHP and can obstruct projects which do not match their supply and generation interests - despite the competitive threat from other electricity and service suppliers.

Linked to the role of local utilities is also the degree of vertical integration and cross-ownership. In Germany, the web of interests and stakes is very complex, making the electricity supply system look like an inter-related block rather than a clearly defined set of competing agents. The recent trend towards the integration of supply chains in the UK, both vertically and horizontally, seems to favour the emergence of a similarly inter-related system. All major generators have got involved in energy service supply, and the RECs have started own generation businesses.

Administrative barriers in energy regulation have been observed both in the UK and Germany. The major issues in Germany have been the regional diversity of administrative practices, the complexity of administrative requirements (which were, however, not regarded as a real problem) and the issue of dual financing. In the UK, the weak financial capabilities of local authorities and the changing conditions in the health sector represented major problems.

Electricity pricing

The regulation of electricity prices is differentiated in both countries for franchise customers and for large customers. While the number of franchise customers has decreased in the UK, they still represent the largest number of customers in Germany. In 1998, the franchise limit will be entirely abolished in the UK. In both countries the franchise prices are subjected to regulatory control, but the impact on CHP depends very much on whether the regulator accepts the utilities' price regime suggestions or not. Large non-franchise customers need to negotiate their tariffs independently, in both countries subject only to monopoly control. The main difference is that British customers can choose between independent generation, participation in the electricity pool and supply from any of the energy supply companies. In Germany, there is only a choice between the local/regional monopolist and independent generation. The customers can now use the CHP-option and (at least in the UK) the threat to choose a different supplier to negotiate better tariffs and/or export conditions. In other words, the negotiation positions have changed quite significantly.

The net outcome for CHP still depends very much on the possibilities to find a co-operative arrangement between the partners. In Germany, the conflicting interests were often settled by informal agreements at local level, whereas in the UK the trend of the last years has been to set up contracting arrangements for CHP-projects. This contracting-type of solution has been emerging only recently in Germany and become fairly common in the New Länder as a response to financial constraints. For very large customers in the UK, the third option of participating in the pool system is not only an administratively complex alternative, but bears also the risks of a highly volatile and uncertain market. In contrast to this, the situation for power exports is characterised in Germany by comparatively stable and reliable export and back-up tariffs, supported by the Associations' Agreement. Although formally no obligation to purchase export power was imposed in either of the countries, the Associations' Agreement fulfils this function in practice.

Principles of environmental regulation

Apart from energy regulation, environmental regulation has also played a major role for CHP. The German TA-Luft requires strict emission limits for plants above 1 MW. Despite the fierce limits, the schedule for introducing them provided a reliable planning base for any kind of combustion plant. Even below 1 MW, it is hardly possible nowadays to install a CHP-unit which does not at least fulfil TA-Luft, if not half the limits of TA-Luft. The regulation by Her Majesty's Inspectorate (HMIP), apart from being perceived as a constraint beyond 50 MW only, is very flexible due to the application of the BATNEEC-principle,[259] but does not

---

[259] BATNEEC - Best Available Technologies Not Entailing Excessive Costs, c.f. section 8.4.2.4

provide reliable orientations. It is a rather reactive approach which depends on technological improvements achieved, but it does not stimulate new innovations in a proactive way. The rules applied in specific cases depend on the negotiation skills of the actors, and are driven forward by impulses from abroad where more advanced technology is developed under more stringent environmental regulations.

Technology supply side

While the domestic small gas turbine manufacturer EGT holds a major share on the British market, the German small turbine market is entirely supplied by foreign companies. The close interactions between EGT and its customers in the UK was perceived as an advantage for improving the technology and for having the best possible support nearby. The German gas turbine packagers were able, but not always willing, to give a similarly good support service to their customers as EGT in the UK, but there was no need seen to engage in close interactions to get feedback for turbine development. With regard to small-scale packagers, the mushrooming and dominance of new companies in the UK was not paralleled by a similarly dynamic development in Germany, mainly for reasons of informal barriers to market access in Germany. In particular, the conservatism of planners and the existence of established links and trust between utility companies and manufacturers made entry for newcomers difficult. Later on they were displaced by established manufacturers which could cope with increasingly sophisticated system requirements.

Main differences:

- Liberalised framework in the UK vs. monopolistic set-up in Germany;
- Strong position of local utility companies in Germany while the British RECs have lost their key role;
- Regionally variegated regulatory conditions in Germany, but a central regulatory body in the UK;
- Prescriptive environmental regulations in Germany and flexible guidelines in the UK;
- Trend towards integrated services supply to individual customers in the UK, but reluctance to develop a service-oriented approach in Germany;
- Onset of horizontal and vertical integration of energy supply in the UK; in Germany horizontal integration at local level by local utilities and vertical interweaving of utility companies in general;
- Higher uncertainty with respect to electricity and export/back-up prices in the UK than in Germany;
- Increasing vertical integration of supply companies and manufacturers/packagers in the UK, but only to a limited extent in Germany.

A further difference should be seen in the rising interest of the British utility companies in packaging and energy services since liberalisation. Hardly any small-scale manufacturer survived independently of the support of one of the utilities. In other words, in the UK the vertical integration of the technology production chain was achieved in parallel with the vertical integration of electricity generation and supply, and the (partial) horizontal integration of energy supply chains. This is clearly different from the German situation where most manufacturers, also small-scale, have remained independent, and the group of established engine manufacturers can be distinguished from the young and small packaging companies.

10.4.3 Structures for the provision of technological information

Reliable information on the different dimensions of CHP is crucial for promoting its wider use. Especially at the level of CHP-users in the UK it became evident that there had been a great deficit in terms of knowledge and competencies on the side of many - especially smaller - firms. While this deficit could be cleared up in Germany by the active work of a number of industrial and engineering associations which were recognised throughout industry, the associations in the UK had a more limited impact. It was not before the late Eighties, that material on the successful use of efficiency-enhancing technologies such as CHP became available due to the activities of the Energy Efficiency Office (EEO). At that time, the CHPA started to play a more significant role, too, sometimes even in co-operation with the EEO. Engineering associations were not very influential in the entire process. Due to this temporary lack of information in the UK, some of the manufacturers (especially small-scale equipment packagers) set up their own networks of users to collect feedback for product development and improvement.

German engineering associations were sufficiently influential and neutral to play a decisive role in the decision-making processes about energy-related issues by providing credible information. It needs to be pointed out that in general technical and research expertise is more highly valued in the German policy-making realm than in the UK, and goes far beyond the provision of information. In spite of their technical orientation and conservatism with regard to innovative technologies, the technical character of the CHP- and efficiency debates gave them substantial influence. In contrast to the UK, where changes were triggered by a radical transformation of the economic regime, the situation in Germany can be characterised as a slow process of incremental change. A critical threshold level for the diffusion of CHP was reached when design standards were defined by the technical associations. By focusing on design principles rather than on standard specifications (as e.g. for 50 kW_{el} -plants) they provided a guidance which had a converging and aligning impact at a broad front towards recognised best practice ("Stand der Technik"). In the UK, much more effort was put into standardisation of individual types of units by packaging schemes of individual companies. Information about their experiences was then disseminated as best practice by the EEO.

Main differences:

- Engineering and industrial associations played a decisive role for information supply in Germany, whereas in the UK government bodies and especially the EEO operated as key nodes for information provision;
- Important function of scientific and technical expertise in the political process in Germany, but only a very limited influence in the UK.
- Standardised design procedures in Germany, but standardised packaging by individual firms in the UK

10.4.4 Political and regulatory structures

It is particularly interesting to observe how the debates on energy policy in the two countries developed during the Eighties. In Germany a long and fierce debate among a great number of influential actors lead to the incremental transformation of the regulatory framework in a more CHP-friendly way. In the UK, a top-down approach broke up and restructured the established supply system. The British ESI had been immobile due to the close interconnections between ESI and government; only a massive structural shock could introduce innovation and change. In Germany, despite the strong influence of the large utility companies and their respective pressure groups, the political system was still sufficiently open and variegated to take up at least some alternative options, which were supported e.g. by industrial autoproducers, or local authorities and utilities.[260]

Political decision power is less centralised in Germany than in the UK, and offers more opportunities for decentralised energy technologies under the responsibility of local organisations. Especially the high degrees of financial autonomy and legal competencies of local authorities (which in many cases own the local utility companies) operate in favour of CHP and help create variety. Major support for CHP came from the local authorities which managed to articulate their positions at higher levels of the political system. It could be shown that new niche applications such as small-scale CHP could indeed be technically and economically successful, and this raised further interest in this option. British energy policy was (and still is) more monolithic with little possibilities to induce novelty in the political realm.

The decentralised regulatory system in Germany adds a further variety-creating element to this picture. The substantial differences among regional authorities in applying federal framework regulations enabled the testing of new solutions. Variety-creating effects were achieved in the UK by different means, namely by liberalising the entire supply system and opening it to competition. Private companies

260 Without going into detail on this aspect, it is also obvious that the majority vote in the UK has affected the policy and governance approach. While in Germany, there is always a need to negotiate widely accepted solutions, the British system is characterised by a 'winner takes all' approach which does not have to consider the balance of interests to a similar extent as in Germany.

had the possibility to provide well-adjusted solutions for different conditions, but within a fairly monolithic regulatory frame.

The comparative openness of the German energy policy[261] is also reflected in the successful and extensive lobbying in favour of CHP by several interest groups since the late Eighties. It lead to an explicit consideration of CHP in energy policy papers as well as in financial support schemes. Due to the complexity of the matter, the whole debate developed very much into a technical discussion where technical expertise was crucial for supporting the often contradictory standpoints of the interested parties involved. The preparatory political processes in and among the German ministries reflect this variety of represented interests, too. These mechanisms are far less developed in the UK where the energy policy debate was far more ideological in its model character for the British liberalisation philosophy. The lack of influential supporters of CHP (perhaps with the exception of the Combined Heat and Power Association - CHPA) adds to this. Therefore, CHP never became an important issue on the political agenda.

With regard to the relationships between government and industry, the situation in Germany is characterised by corporatist elements, which aim at sharing and delegating responsibilities to top organisations. Although this might preclude the influence of those not represented in the few top organisations, it is a very good way of making constructive use of the interest structures of industry for achieving government policy objectives. In the UK, a switch between two extremes could be observed, namely from a dirigistic to a weakly regulated liberalised system.

Main differences:

- Greater openness of German energy policy to new influences and interests, but a closed energy policy-making process in the UK until liberalisation;
- Decentralised competencies facilitated the creation of a variety of options in Germany, while the British system was more homogeneous with little possibilities for independent action at local level;
- Influence of various interest groups in the German political process, combined with corporatist elements and the delegation of responsibilities; in the centralised British political process, interest groups and other actors have not played an as active and responsible role in policy implementation;
- Importance of technical expertise in decision-making more pronounced in Germany than in the UK;
- Highly ideological debate about energy policy in the UK, and a rather pragmatic policy approach in Germany.

---

261 This remark may sound strange because usually Germany is regarded as an example of a closed political system. However, for energy policy it was more open than the former British system.

An important role was played in both countries by Parliament committees. While the German Enquete-Commission took the environmental debate about the follow-up of the Rio-conference as its starting point for debating the potential contribution of CHP and other decentralised energy options, the House of Commons Select Committees were a general forum to question government policy on energy-related issues. In both countries the discussion shaped public and political opinion about energy policy in a conducive way for CHP.

10.5 Comparison of actors' decisions, interactions and motivations

The preceding parts of the comparison have already highlighted important differences of structural factors shaping the success or failure of CHP. The differences in terms of interactions and decision-making processes at micro-level are expected to clarify in full detail the differences between the two countries. The perception of structural and exogenous developments as well as the behaviour of other actors determine jointly the decision taken.

10.5.1 Political realm

Decisions

As already mentioned, the German policy approach was rather incremental and evolutionary while liberalisation and privatisation in the UK represented a radical reversal of energy policy. The intention behind British policy was actually to reduce the direct involvement of government in energy matters. In fact, the extent of specific policy intervention has decreased since then. The incremental, though permanent attempts to adapt the German energy policy framework to new emerging requirements is expressed in initiatives such as the new Federal Tables of Electricity Charges, the non-application of §12 Power Generation Act or the support for the Associations' Agreement.

An important activity in support of CHP in the UK was the establishment of a target for CHP-capacity which was even raised after two years. However, at a formal level, hardly any specific conditions for CHP were improved to recognise its environmental benefits, although it was repeatedly stated that there was an interest in accelerating its uptake.

In terms of environmental regulations, the possibilities for CHP were affected in Germany by a number of measures to improve energy efficiency on the demand side (e.g. by the heat insulation regulation) as well as on the supply side. In combination with the TA-Luft, several complementary policy actions were taken. They underline the importance assigned to environmental issues in Germany, even if the CO_2-tax was rejected, and the planned heat use regulation for industry transformed in a voluntary agreement. The removal of the coal subsidy indicates that these policy actions were not necessarily conducive to CHP. In the UK, fewer specific environmental measures were taken, or at least they were formulated more flexibly. The BATNEEC-approach made it unnecessary to specify framework regulations in detail, because their interpretation was left to the authorities in charge of implementation.

Hardly any financial support programmes of diffusion and technology policy were set up in the UK. Instead, emphasis was given to less costly information policies about best practices, and they were indeed regarded as quite successful. In Germany, on the contrary, substantial subsidies were made available to promote especially district heating, while information measures were left to industrial organisations and - to some extent - to regional governments. In general, financial support measures had an impact only when massive support was granted (as for district heating); small investment grants did hardly affect any investment decision.

Decision-arenas and interactions

The political structures and constellations being more flexible, variegated and open in Germany, the advantages of CHP with regard to resource-related and environmental consideration were recognised early on, and transformed into a number of active support measures like e.g. the DH support programmes during the Eighties and Nineties. In addition, the use of CHP for industrial and small-scale applications was facilitated by regulatory measures. Despite the resistance of the electric utilities against independent power projects, and despite their continuing dominance within the German ESI, the regulatory conditions were improved stepwise. A rather strange coalition emerged at that time in favour of CHP, composed of environmental interest groups, local utilities and industrial firms, represented by their respective associations. Newly founded interest groups added to this mixture of political forces. They managed to move the case of CHP onto the political agenda, together with their specific CHP-related interests. In contrast to the UK, the importance of the new types of CHP[262] was recognised at the political level in Germany in the mid-Eighties already, although it was a difficult period for small innovative energy technologies due to negative experiences with other small-scale and high efficiency technologies.

In the UK, the first liberalisation step in 1983 did not specifically aim at CHP, but rather at improving the possibilities for auto-production in general. The fact that the Regional Boards still maintained their key role enabled them to control and prevent many project proposals during the Eighties. In addition, CHP was associated to the negative experiences with DH in the Fifties and Sixties. This made it difficult to raise interest at the political level. CHP has never been a real issue on the political agenda, at least since the DH-initiatives of the late 1970s which culminated in the Marshall-report (DoE, 1979). Only the second step made in 1990 fully overturned the situation, but it was driven essentially by the liberalisation philosophy of British government. Only on the basis of this strong driving force it

---

262 This refers mainly to the potential of small-scale and industrial CHP which are the two fast-growing applications. District heating appeared on the political agenda already in the seventies.

was possible to break up the link between ESI and DoE which had dominated energy policy in the UK for decades. For these reasons, the beneficial effects for CHP were only a by-product of a wider change of energy policy. Nowadays, the comparison of the two countries looks very differently than fifteen years ago, with a fairly passive role of government and the regulator in the UK, and several government interventions at different policy levels in Germany.

Moreover, while in the UK the major changes were driven by strong political motivations, a far more problem-oriented and pragmatic approach was followed in Germany, based on a consultation process which involved parliaments, experts, ministries, pressure groups, regional authorities etc. The major political topics were quite similar in both countries, namely questions of access to the grid, licences, remuneration of exports, prices for electricity and back-up, and the recognition of the environmental benefits of CHP in the calculation of avoided costs.

Main differences:

- Incremental and pragmatic policy process in Germany in contrast to a radical and politically motivated approach in the UK;
- Emergence of an influential coalition of interests in favour of CHP in Germany, but no such development in the UK;
- Strong environmental forces in the political realm in Germany, but a limited influence of environmental interest groups in the UK;
- CHP became an issue on the political agenda in Germany, in the UK it was a by-product of liberalisation in general;
- Specific support programmes for certain applications in Germany, but hardly any financial incentives for CHP in the UK;
- Emphasis on the dissemination of information and best-practice in the UK, an activity which was of very limited importance for German government.

10.5.2 Technology supply side

Decisions

On the technology supply side of CHP, the situation in the UK can be characterised by a trend towards the provision of horizontally integrated energy services to individual customers since liberalisation. It is reflected for instance in the emergence of IESCs and in DEP-initiatives. Due to the close links which developed since the middle of the Nineties between some smaller manufacturers and energy utilities, vertical integration has also been reinforced. The responses and dynamics were quite different in Germany, where the relationships between the major technology suppliers and energy suppliers have remained stable, building on informal rather than formal links. Only some of the new manufacturers of packaged CHP-equipment tried to find a niche in the supply of energy services, but were constrained by the regulatory conditions.

In terms of innovative improvements of CHP-technology, the major differences have been outlined before. It is however interesting to note that the German manufacturers adopted a more forward-looking strategy for their environmental R&D. Expecting even tighter limits in the future, they tried to improve their technology in time, while in the UK a reactive rather than a pro-active approach to R&D was adopted.

Another major difference can be identified with regard to standardisation and packaging. Whereas in Germany some industry-wide standards could be established for design and specification of CHP (especially for the small-scale 50 kW units), the British manufacturers tried to exploit standardisation potentials within their companies by packaging their systems as far as possible.

Decision arenas and interactions

In addition to utility and energy management companies, technology supply companies have also entered the field of energy services, especially in the small-scale area. Even more important, they perceived new technological opportunities and new markets in a liberalised framework very early on, and therefore started to work both on the improvement of the technology and on its adaptation to specific customer needs. Before these new initiatives started, i.e. in the early Eighties, CHP technology seemed to have reached a dead end - both in the UK and Germany. The technology was used to a some extent, be it at different levels in the two countries, but there was little scope seen for its widespread application. Consequently, the incentives for suppliers of CHP-equipment to carry out R&D work were rather low, and technological progress slow. It benefited mainly from spin-offs of turbine, engine and heat exchanger technology. Not even the R&D-support for CHP/DH in Germany lead to major impulses. The situation changed in the mid-Eighties when some new companies emerged in the small-scale area. After 1990, one could even speak of a mushrooming of small-scale CHP companies in the UK, attracted by new technological opportunities, better regulatory conditions and a growing market. The larger-scale business is more difficult to enter due to the complexity of turbine technology, the limitations of packaging and standardising units, and the site-specific character of large plants.

Although the manufacturing companies in both countries were confronted with similar technical challenges, some important differences can be discerned with regard to the focus of their development activities. Reliability being the key issue for British CHP-users, development efforts went into operation monitoring and computer control systems. This orientation was driven by the difficult relationships between regional utility companies and private operators of CHP. Due to the detrimental back-up tariffs, it was crucial to be as independent as possible of grid supply. In Germany, reliability was also perceived as important, but a large share of CHP-capacity was operated by the local utilities themselves, i.e. by those who also supplied the back-up power. The utilities thus disposed of a much higher level of autonomy than in the UK to operate their units efficiently. Consequently, the pressure to improve reliability was less strong. The fact that much effort was put into the environmental performance of all types of CHP in Germany can be explained by the strict legislation and its informal application also to plants smaller than 1 MW.

For both countries, it appears to be plausible to argue that these new technological advances preceded the improvements of the application conditions and the organisational innovations, because the technical impulses came mostly from outside the power engineering industry, but needed to be adapted to CHP. According to the interviews made, the improved regulatory conditions facilitated and stimulated the adoption and adaptation of available technological opportunities, but there was rather little original work done on CHP technology itself. The main work to be done consisted of the integration of new components into a reliable technological system. The new types of organisational and financial arrangements between energy service end-users, service providers and utilities were then necessary to facilitate and embed the diffusion process.

RTD-policies were perceived as of minor importance by the technology suppliers in both countries. In the UK, there was little support for R&D-work anyway, and information policies aimed predominantly at informing potential users. In Germany, R&D-funding was regarded as important to develop and test new components which were needed for integrated systems (e.g. heat storage, mini-motors, chilling, heat networks). The environmental upgrading of CHP systems was driven essentially by regulations, but complemented by RTD-policy measures.

Given the active role of German industrial and technical associations in enhancing information exchange between technology manufacturers and users, the learning processes were strongly influenced and controlled by these actors. In the UK, individual co-operations between individual manufacturers and their users were the prevailing means to collect information about the operation and design of CHP-plants.

Main differences:

- Orientation towards energy service provision by British manufacturers, and rather conservative product provision policy of most German companies;
- 'Mushrooming' of CHP-companies since the late Eighties in the UK; in Germany, the established manufacturers started to dominate the market at the same time;
- Design standardisation in Germany vs. packaging in the UK;
- Different foci for advancing CHP- technology in response to patterns of co-operation between end-users and utilities;
- Benefits from public R&D-funding in Germany for the development of integrated systems, but a very limited impact of British technology policy on CHP-development.

10.5.3 The user-side of CHP

Decisions

The fate of CHP depended in both countries very much on the changing positions of utility companies due to their gatekeeping function in the ESI. The extent to which the horizontal integration of their energy services could be achieved was crucial. In the UK, this integration was achieved by the establishment of Integrated Energy Service Companies, and it was a precondition to enable the uptake of CHP. It was driven both from the technology manufacturers' and from the utility and service companies' side. A similar trend can also be observed in Germany, but it was far less pronounced than in the UK where it met perfectly with the end-users' interest to outsource non-core activities. The local utilities in Germany already represented horizontally integrated companies and covered the niche of integrated energy services.

Strategic decisions of many utility companies aimed also at a vertical integration of both the energy supply and the technology supply chains. The integration of the technology supply chain has accelerated in the UK since privatisation, whereas the situation in Germany technology manufacturing and energy supply have remained separate business areas with strong informal co-operation links. Characteristic of the German ESI are, in contrast to the UK, the multiple interdependencies between the utility companies at different levels.

These trends on the utility side are reflected by complementary decisions on the end-users' side. The acceptance of a service approach has been very high in the UK , whereas in Germany most energy users tended to keep control of energy supply in house. This situation has been changing slowly since the early Nineties, and especially in East Germany since reunification.

Decision arenas and interactions of energy service end-users

For potential CHP end-users in the UK, it has become more easy since liberalisation to set up cogeneration schemes, but a number of factors can be identified which have remained rather detrimental. Among these count the short time horizons and strict payback requirements applied to investment projects. Also the uncertainties of the pool system and of electricity prices in general have made any reliable project appraisal difficult, if not impossible. The additional investment for CHP-equipment could hardly be justified on the basis of very vague and uncertain extrapolations. Uncertainty about environmental regulation according to the BATNEEC-principle further adds to this, even if until recently it was hardly perceived as a real constraint. However, stricter rules are expected to be applied by HMIP in the future. In this respect, the developments in Germany have been quite different. Environmental regulations are still among the strictest in Europe, and although they are compulsory beyond 1 MW only, they have also been applied informally to small scale units. While environmental regulations were quite well-defined and predictable, the application of energy regulations for licensing varied significantly among the Federal states. Public regulatory bodies required also longer learning process to be able to deal with CHP and specify rules for this 'new' technology.

Finally, energy supply is usually non-core investment for which many companies are not willing to spend more money than necessary. Due to these constraints, CHP has been a relevant option for many end-users only when it could meet several objectives simultaneously, e.g. in terms of the energy balance of a site, its environmental strategy, the integration of the production process, and finally also the economic balance of a firm. Moreover, if an old energy supply systems had to be replaced anyway, CHP was always an interesting option. Such an integrated view of energy supply systems, production processes and passive energy efficiency measures has been more common in Germany than in the UK where packaged solutions with clearly defined interfaces are the preferred options for CHP-plants rather than their full integration into a production site.

Economic assessments as expressed in payback periods are regarded as useful orienting guidelines for investment in CHP. However, although they must remain in an acceptable range, many experiences from both countries have shown, that strategic consideration can easily override them. This is especially the case if energy supply is a key production factor for the user in question. The criteria and time horizons applied for investment appraisal in practice are similar in both countries, though slightly stricter in the UK. German public utility companies assess the economics of their investment over the lifetime of plants, an approach which reflects their public service function.

The motivations of end-users to invest in CHP are also similar in both countries. Most important are the expected economic benefits and the independence of the utility companies. While the economic argument is evident, the independence issue needs to be seen in relation to the problem of uncertainty, because autoproduction was a way to circumvent the unpredictability of the electricity tariffs as defined by the utility companies or the electricity pool in the UK. Differences can be observed between the UK and Germany with regard to the ways in which end-users try to reduce these uncertainties. In Germany, informal contacts and local networks were important in this respect and facilitated the construction of CHP-schemes. In the UK, on the contrary, formal market relations tend to be the dominant way of interaction between utility companies and end-users.

Generally speaking, German companies were better prepared to deal in-house with the complex system character of CHP than their British counterparts. This holds particularly for local utilities which started to operate CHP themselves: they were competent, horizontally integrated and in control of the grid connection.

The problem of insufficient competencies has also a bearing on the possibilities for packaging and standardisation. Packaging for end-users was a success in the UK because it provided the end-users with a 'black box', which was remotely monitored and easy to use without knowing in detail how it worked. In Germany, the end-users tended to be involved more actively in the design and operation of plants. This lead also to the development of more site-specific solutions than in the UK. In other words, the success of energy services depended not only on the service providers, but also on the conditions on the side of end-users.

While the general level of competencies of German users was comparatively high, they often lacked specific information on CHP. This bottleneck was only removed in the late Eighties when CHP started to be fully accepted by the technical and engineering associations. In the UK, the government's information policy as implemented by EEO played an important role for the acceptance of CHP and the improvement of the knowledge base. However, it required also a generation change until engineers and board members were willing to accept its widespread application.

Regulatory and administrative procedures have been criticised frequently for inhibiting CHP-diffusion in the UK, but the situation has changed in the meantime. The first years after privatisation and liberalisation were needed to stabilise the framework. Despite the fact that CHP might still be in a disadvantaged position in some respects, the new procedures are not regarded as problematic by the users, with the exception of the complex pool system. In Germany, the changes of regulations and procedures were more gradual and depended much more on the local context than in the UK. Most of the earlier problems and uncertainties of innovative CHP-solutions have been settled in the meantime.

Decision processes and motivations of utility companies and service providers

Until liberalisation, most Electricity Boards had a detrimental stance towards CHP, but since then the situation has changed. Being constrained to supplying only electricity until 1990, there was no scope for them in developing CHP-activities, because they could only benefit from an optimisation of power generation. In the liberalised system, they suddenly had both the opportunity and the capabilities to expand their activities to other energy services, and to integrate them horizontally. After an initial phase of reluctance, all the RECs and large generators moved into the CHP-business. Although their major activities remained in distribution or bulk generation, they became very interested in CHP-projects in which they can participate themselves. Due to the threat of losing customers, a more conducive attitude is now generally held towards other types of CHP-projects, too, i.e. projects in which they do not participate themselves. This change of attitude of the utility companies has also been facilitated by a generation change and the recruitment of new staff which is more open to CHP and less biased by negative experiences of the past. Not only from the electricity side interest in CHP has grown. Due to the fact that CHP transformed it from a coal-based to a gas-based technology, British Gas as well as other gas suppliers moved also into the CHP-business. This organisational integration of gas, heat and power supply - be it at a slow pace - provides now the conditions for an easier diffusion of CHP.

In Germany, the horizontal integration of most local and some regional utilities facilitated the adoption of CHP. Several co-operative arrangements are in place between industry and utility companies. They build on relations of trust at the local level rather than on formal contracts. This does not mean that the utility companies were in general open and co-operative with regard to CHP. Several of the regional utilities and the supra-regional large-scale generators took a hostile position towards decentralised CHP. Vertical integration, i.e. the ownership of or shareholding in local utilities by regional and supra-regional companies, lead to a slowdown of self-generation at local level in order to maintain the demand for centrally generated power. However, there is still enough variety built into the German system to find many innovative examples. The recent adaptations of the regulatory framework have further facilitated CHP, and especially small-scale application. Still, in both countries a fierce debate has been going on until present times about remuneration and tariffs to be applied to CHP-plants, focusing on the issue of avoided costs, its elements and calculation.[263] The Associations' Agreement in Germany provided fairly reliable guidelines for the remuneration of export power, but the debate on its content and application has never stopped.

With regard to the internalisation of external environmental costs of power generation, CO_2-taxes were proposed in both countries as a promising option, but not been further advanced. The voluntary agreement of German industry to reduce CO_2-emissions has replaced a more interventionist legal initiative. In the UK, measures such as the CHP-target and general environmental regulations are regarded as sufficient to achieve the CO_2-emission targets.

The difficulties to deal with CHP successfully - both in technical and economic terms - have been addressed by energy management companies in the UK which sell their knowledge to less experienced users of CHP. Their specific expertise and willingness to take over the financial risk reduces the benefits which an end-user could derive from a CHP-scheme, up to the point that in some cases promising project became economically unviable. On the other hand, contracting helps to overcome the reluctance to invest additionally in non-core utility services. In other words, good projects are facilitated by energy management firms, but less promising projects can become unviable due to the additional costs of contracting. In the UK many new private companies have emerged since 1990 which address these new service needs, while in Germany the major contribution is still coming from public local and regional utilities. Their offers cover a wide range of activities and energy efficiency measures, including the operation of CHP-plants, but usually they offer less far-reaching services than British energy service companies.[264] Anyway, the diversity of German utility companies is enormous. They range from utilities which apply an extensive combination of measures (e.g. least cost planning, fully integrated supply of heat, power, gas and water, or support for

---

263 Some of the key problems concern the consideration of capacity terms, backup grid capacity provided, the level at which avoided costs apply (local or regional utilities), the connection costs and the power threshold level to which favourable conditions should be applied. With the increasing differentiation of tariffs in the UK, it has become more difficult to oversee and assess the present situation, and the confidential character of most agreements since 1990 makes this even more difficult.

264 In the UK it is quite common to hand over the entire responsibility for the CHP to the service provider (i.e. he guarantees the supply of heat and electricity) or even for the entire internal operation of energy services, including passive efficiency measures, etc. German users are less willing to accept such far-reaching interventions on their sites, and many arrangements are based on a joint energy service company to supply a site. In other words, the service user maintains control over his supply in co-operation with other investors and - very often - the local utility company.

passive efficiency measures) to rather traditional firms which do not make any particular efforts to promote efficiency and CHP.

Main differences:

- Horizontal integration now emerging in the UK due to energy service approach; in Germany horizontal integration at local level established, but in large companies a trend towards separate service division;
- Slow transformation of utilities' position in Germany in favour of CHP, fast change in the UK;
- Greater importance of local and informal contacts between end-users and utilities in Germany than in the UK;
- Vertical integration of energy supply in Germany is widely established; in the UK vertical integration was broken up in the course of privatisation but tends to become re-established by a number of take-overs;
- High uncertainty perceived as a problem in the UK, while in Germany more stable and reliable conditions prevail due to prescriptive regulatory approach and industrial agreements;
- Environmental regulation perceived as important driving force in favour of CHP in Germany, but they did not have a major impact in the UK;
- More in-house expertise of German CHP-users, also at board level, despite a lack of specific information on CHP;
- Therefore also more site-specific and integrated solutions in Germany, packaged solutions in the UK;

10.6 Differences of evolutionary dynamics

The comparison has until now focused on individual elements of the two PET-systems under study. In the following section, the evolutionary mechanisms in both countries are going to be compared in order to show how the differences in the interaction of these elements have affected the diffusion paths of CHP.

External pressure

Even when taking into account the different historical starting points, the comparison shows that the external pressures exerted on the two systems were quite similar. The technological context for CHP in terms of control systems and power source technologies developed in parallel, as well as the availability of fuels, especially natural gas. However, the supply context in the UK differs from that in Germany in at least one important respect, namely the availability of indigenous oil and gas resources. Consequently, considerations on energy efficiency, security of supply and diversification of fuel sources were a less pressing concern in the UK than in Germany.

The general awareness about environmental issues in Germany had a more significant impact in Germany than in the UK. It affected the pace of industrial restructuring, triggered energy efficient investment, and influenced also the political agendas. The re-orientation towards less energy-intensive industries and higher efficiency started already in the mid-Seventies and promoted the use of industrial CHP.

Variation and selection by innovation

With regard to innovative impulses coming from within the system, the role of the new and dynamic technology suppliers in the UK since the mid-Eighties needs to mentioned. Their activities have been complemented by organisational innovations since 1990. These pace of these developments needs to be seen in relation to the delay in CHP-diffusion in the UK until liberalisation. It was caught up rapidly since then, even if the UK has not yet reached the level of CHP-diffusion of other European countries. In Germany, only the small-scale CHP area was driven by proactive technology suppliers and service providers from the early Eighties onwards. Large-scale CHP-technology and district heating were rediscovered as R&D-areas in the early Nineties in parallel with reunification and the CO_2-debate.

In terms of the pre-selection of innovative technologies, the manufacturers in both countries pursued similar routes, though with differences in terms of timing and focus. Especially since the early Nineties, the German situation has been characterised by a trend towards the standardisation of design procedures and the integration of CHP in production and energy supply systems. Both these developments cannot be identified to a similar extent in the UK.

In other words, while in the UK variety generation was driven by technology manufacturers who explored the potential of the new regulatory framework, the situation in Germany was characterised by the existence of a multitude of application niches with specific economic and regulatory conditions. This multifaceted character provided a very variegated set of incentives for the innovative trials of manufacturers, based on a limited number of shared design principles.[265]

Selection by users in their organisational contexts

The selection conditions at the user side have fairly strict over the last years in both countries, but in Germany the payback requirements, goals and possibilities of many potential CHP-users (and especially of local utilities) have been more conducive to long-term investment than in the UK. Two main factors for CHP -

---

[265] This should not be interpreted as a conflict with standardisation but rather as complementary to it. Standardisation can be understood not only as the alignment of technological specifications for specific types of units, but also as the establishment of common design standards, i.e. of generic rules for facilitating the planning and design of CHP in different contexts.

horizontal integration of energy supply chains and the financial possibilities of the local utility companies - have been significantly better in Germany than in the UK.

Before liberalisation in 1990 the selection conditions for CHP were extremely strict in the UK. The reasons should not only be seen in the formal restrictions and the hostile stance imposed by the 'gatekeepers' of the ESI, the Electricity Boards, but also in the lack of adequate organisations to support and exploit the advantages of CHP (like the local utilities in Germany). Although the possibilities had been improved formally with the Energy Act in 1983, the informal barriers still prevented CHP to be taken seriously into consideration by many potential investors.

Two of the crucial problems of CHP have also been its complexity and its requirements in terms of knowledge and competencies. The knowledge-base in energy engineering appears to be more solid in Germany, enabling many companies to set up CHP on their own and adapting it exactly to their needs. In the UK, the lack of such competencies favoured the emergence of energy service companies as competent partners in CHP-projects.

Due to the organisational structures and the diversity of attitudes of the utility companies there was a large variety of local and regional selection niches for CHP. It lead to the development of many different types of applications in Germany already in the second half of the Eighties. However, information about these experiences was only spread to a limited extent until industrial and technical associations became more active in this respect. The British situation may have been less heterogeneous, but information dissemination was already promoted by government agencies in the early Eighties.

Selection conditions of the regulatory and political framework

In Germany, the selection conditions of the national policy framework have been evolving continuously in a positive way for CHP - at least until the mid-Nineties when the debates about the European Internal Energy Market and the coal policy required a number of modifications. The environmental policy framework was defined in a clear way by introducing the TA-Luft, and there was a rough consensus about the major lines of energy policy and the importance of energy conservation, efficiency and - since the early Nineties - of CHP.[266] This consensus was based on a problem-oriented and pragmatic approach to energy policy. The petrified energy supply framework in the UK was transformed in 1990 into a highly uncertain one, which did not provide a reliable orientation to the energy sector. These manifold regulatory and political uncertainties put CHP at a clear disadvantage and could not be balanced by 'soft' commitments such as the CHP-target. Given the abolition of clear political orientations for energy supply (which was actually a prime objective of liberalisation and privatisation), the temporary success of CHP in the UK can only be interpreted as a by-product of the wider change of the governance approach to energy policy.

---

266 This consensus did not exist with regard to nuclear power where still a longer term strategy is awaiting its formulation.

This high degree of uncertainty in the UK induced a short-term approach to all kinds of energy supply investment. Together with the fairly low importance of environmental objectives, this calls into question the longer-term prospects of CHP in the UK.

The growing support for CHP at the political level in Germany has also benefited from the structures of political influence and decision-making. They revealed to be comparatively open to new views and opinions, e.g. regarding CHP, at least once they were supported by some of the established actors. The system was sufficiently flexible to avoid a political lock-in and allowed introducing corrective measures. In the UK, on the contrary, an ESI which was closely linked to government had to be transformed radically into a self-organising industry in order to overcome the lock-in.

Stabilisation by structural interdependencies and mental frameworks

The major stabilising mechanisms in Germany must be seen in the complicated patterns of cross-ownership and in the formal and informal links between the actors in energy supply. They tie together manufacturers, utility companies, industry and the different levels of government. An important node for these network connections are the different industrial associations. Of major importance has also been the engineering culture which is strongly shaped by the corresponding associations. Although similar engineering associations exist in the UK, their influence has been much weaker. Until 1990 the main stabilising mechanism consisted in the very close co-operation between government departments and the electricity boards. Since then, stability has been broken up not only organisationally but also in terms of personnel and political-regulatory responsibilities.

The prevailing technological structures stabilised traditional patterns of development and inhibited change. For example, the investment made in the past is now 'sunk' in power and heat generation plants and grid systems. Their vintage and replacement times of up to 50 years need to be taken into account when considering the possibilities to change the current system. In order to manage and operate such large centralised systems, it is necessary to dispose of appropriate organisational structures. In the UK, these were the boards and are now the major generators. In Germany, this function is fulfilled by the supraregional and regional utility companies. Their strategies are dependent on the technological systems they operate. Generally speaking, the techno-organisational structures in both countries have contributed to the stabilisation of a centralised system and delayed the establishment of decentralised energy supply. However, in Germany there were at least some structural elements in place to support decentralisation at local level.

Not only engineering, but also political cultures in the two national systems played important roles for stabilising evolution. They are reflected both at the structural level of institutions and at the behavioural level of mental frameworks. Especially the bad reputation of district heating in the UK made it difficult to convince engineers and policy-makers of the advantages of CHP, while the strong role of the local political level in Germany was helpful to move CHP to the forefront of the political debates about energy policy.

To summarise, the extremely stable British ESI was replaced by a very dynamic framework in which several "lock-ins" could be broken up, thereby opening up possibilities for cogeneration. In Germany, stabilisation was achieved by a large variety of interdependencies which allowed only a slow introduction of novelties but did not prevent them entirely.

Reinforcing and delaying mechanisms

In both countries under study, the availability and provision of information need to be mentioned as important reinforcing elements, but they operated through different channels. They were effective not only for CHP-users, but also in the political realm where information about the advantages and success of CHP-schemes was needed in order to raise awareness and support. This political link was quite important in the German context where the comparative openness and responsiveness of the system enabled the inflow of new information. The British political system, on the contrary, was completely closed to external impulses until the late Eighties, when it lost its dominant role for steering the developments in the energy sector. However, information provision to users was promoted by government and worked effectively as a reinforcing mechanism. In contrast to this, there were a number of destabilising mechanisms in place in the UK which operated to the detriment of CHP, for example the uncertainties of the regulatory system and the complexity of the electricity pool.

Many, reinforcement mechanisms operated between the macro/structural level on the one hand, and the micro/behavioural level on the other. Changing perceptions lead to decisions and thus contribute to transforming the system structures which are then perceived again at micro-level. The crucial question is whether there are actors in place who stimulate this interplay of reinforcements between micro- and macro-level. Apparently, this role was fulfilled in Germany by two types of actors, namely by industrial/engineering associations and by political actors such as local utilities/authorities or certain interest groups. In the UK, there was hardly any influential actor able to use these levers of change. Only the Energy Efficiency Office (EEO) managed to amplify the impact of new information by its dissemination strategy.

Economies of scale and scope are also typical reinforcing mechanisms which were effective in both countries. The standardised packaged solutions provided by British small-scale manufacturers enabled especially economies of scale to operate in favour of lower unit costs as a result of series production. Although they were also exploited in Germany, the standardised design procedures and the associated learning effects represented a very powerful mechanism to exploit economies of scope in highly integrated energy systems.

The effective operation of reinforcing effects is to a great extent a matter of timing. In spite of the growth of CHP in the UK, the window of opportunity to induce a substantial switch towards CHP and decentralised power generation was missed. The structure which has emerged since 1990 is based to a small extent only on decentralised plants, while the majority of new investment since liberalisation went into fairly large combined cycle power-only stations. In Germany, it is still too early to assess questions of timing, because the transformation process following the reform of the Energy Supply Act has only started. It still offers opportunities to decentralise and co-ordinate heat and power generation.

System evolution

To summarise, the situation in Germany can be characterised as an incremental process of change at different levels and in different areas: slowly improving regulatory conditions, improvements of technology and organisational solutions, changing positions of actors, and politically driven changes of the framework, all being fuelled by information-intensive processes of interaction between the main actors. In the UK, the changes were fast and radical, starting from a reform of the regulatory structures which induced changes of the actors' interest structures, and provided opportunities for new technological solutions. A switch was made from a political deadlock in a centralised, horizontally separated trajectory of energy supply to a widely self-organising system which provides hardly any long-term co-ordinating mechanisms. The two different pathways can be labelled as an *incremental and co-evolutionary process of change in Germany* and *politically driven revolutionary overturn in the UK*, representing two basic models of change in large socio-technical systems.

<u>Main differences:</u>

- Stronger external driving forces in favour of CHP in Germany (environment and security of supply);
- Stronger technological and organisational dynamics in the UK, but manifold variety-creating context conditions for CHP in Germany;
- Less competent end-users in the UK, requiring packaged and simpler solutions than in Germany ;
- Utility companies in the UK more hostile to CHP until 1990;
- More reliable and pragmatic regulatory and policy-making selection environment in Germany;
- In the UK extreme stability was replaced by strong volatility and uncertainty, whereas in Germany several stable mechanisms stayed in place and evolution was smoother;
- Reinforcements by information provision which operates through different channels in both countries, but in the UK they played a less important role in the political realm;
- Incremental and co-evolutionary process of change in Germany vs. policy-driven revolutionary overturn in the UK.

10.7 Comparison of future perspectives

In this section, the three scenarios for CHP in the UK and Germany are going to be compared briefly. For each country three comparable future pathways have been suggested, namely a moderately liberal scenario, a scenario which is based on the widest possible application of liberal principles, and a more policy-driven scenario in which government policies and regulations exert a major influence.[267]

A comparison of the three scenarios in the two countries is interesting because it helps to show that even if roughly the same pathways in terms of approach to energy supply are followed, the trajectories of CHP would differ significantly as a result of past histories, structures and mechanisms. The second scenario offers the best possibility for converging developments in the two countries, but it is not equally probable to emerge in both countries. In view of the current situation in the UK, scenario 1 offers the most promising conditions for CHP, while in the German case scenario 3 would be very conducive due to the already existing support structure for CHP at different levels. However, scenario 3 contradicts the present liberalisation paradigm and might therefore be rather unrealistic. These considerations show that there is no single ideal model for how to design a regulatory and political framework which is favourable for CHP. It is only possible to integrate new insights into a historically grown context, taking into account the possible reactions, motivations and processes at the micro level.

10.7.1 Scenario 1: Liberal framework with special attention to CHP

The introduction of some first competitive elements into the monopolistic German system would represent a major change, while in the UK a moderately liberal framework would just mean a continuation of the present pathway. In other words, many of the experiences of the UK would still have to be made in Germany until the new framework would be stabilised. Furthermore, the differences in terms of political and business cultures imply that the future pathways would look differently in many respects, with corresponding impacts on CHP.

Among the similarities, one can certainly mention a simplification of administrative and regulatory rules and of licensing procedures, although the decentralisation of authorities in Germany would inevitably keep the situation fairly complex. In general, the importance of the local level in Germany is a characteristic feature which can not be expected to change, either. It assumed in the scenario that the local utilities in Germany would maintain a certain degree of monopolistic protection for electricity distribution. In the UK, respectively, it would be attempted to delegate certain co-ordinating competencies to local authorities. In terms of generation they would have to accept exported power from autogenerators at a price

267 In none of the scenarios, major exogenous technological impulses are expected to affect the situation of CHP. However, fuel cells might represent such an impulse, and it would also be interesting to study the potential reactions of the two energy supply systems to such a development, taking into account the scenarios outlined. Such an exploratory exercise needs to be left to future research work.

which reflects long-term avoided costs. Similarly, regional differences in Germany would persist, and contribute to maintaining a high degree of variety. Electricity prices would be split into a franchise market with price control and a spot-market or pool-system to determine prices for large customers and bulk generation. Due to the role of local utilities, the impact of a moderately liberalised framework on the emergence of private energy service companies would be rather limited in Germany, while in the UK they would play a major role and contribute to the decentralisation of energy supply down to the level of individual users. The way in which environmental issues are going to be addressed in the two countries can also be expected to differ. Target setting and individually negotiated prescriptions are expected to remain the basic principle of environmental regulation in the UK, whereas in Germany elements of prescription and voluntary agreements would be mixed. Other specific measures to support CHP would be very limited in both countries, possibly with the exception of CHP for renewable fuels. Attention would be paid to promoting it in a way which avoids distortions of market mechanisms. Voluntary agreements such as the Associations' Agreement would be more difficult to realise in Germany, and even if realised they would fulfil an orienting function only. Most export tariff agreements would have to be concluded directly between supply company and autoproducer. Financial support would be given for certain R&D activities, but there would not be any subsidisation of a technology like CHP which in principle is regarded as mature.

Due to the fact that little is expected to change at local level, the diffusion perspectives for CHP in Germany would be affected mainly for large industrial electricity users who could benefit from lower electricity prices in a competitive market. The role of district heating could be maintained due to the co-ordinating function which the local utility companies will continue to fulfil. In both countries, small-scale and industrial CHP could expand further, facilitated by the removal of administrative and regulatory barriers.

The fact that the German system is still widely monopolistic implies severe resistance against liberalisation plans from a number of actors, especially if the local level is excluded from competition. Some of the effects of liberalisation would be welcome by most actors, in particular if they go hand in hand with a removal of additional 'political' charges like the ones for the coal industry. Other elements of the new framework would almost certainly be heavily contested, especially by many of the smaller players would have to face major challenges in terms of strategy and operation. In the UK, the decisive debates took already place in the past, so that the minor adaptations required by scenario 1 would not face major resistance.

10.7.2 Scenario 2: Widest possible application of market principles

In spite of the similarities with the first scenario, a major difference consists of the extent to which market competition is introduced. As it is foreseen in the UK from 1998 onwards, competition in distribution would be effective down to the level of small individual customers. In other words, also at local level the protected position of the distribution business would be abolished, and therefore the possibilities for co-ordinating energy supply would be limited. In Germany, the increasing competitive pressure could mean that some local utility companies will their independence, implying that the vertical integration of energy supply will be accelerated. Most remaining regulatory barriers for setting up generation plants would be removed. The current guidelines for remunerating export power would be replaced by individual negotiations between generator and grid operator/distributor. To determine electricity prices, a combination of a pool system and of individual negotiations would be introduced in Germany, leading to a similar set-up as currently in the UK. In order to oversee the complex energy markets, service companies would start to compete with established utility companies and play an important role for small customers. The large customer market would be shared among the established regional/supraregional supply companies and new foreign competitors.

Without going into the details of third party access to the grid, rights of way etc., barriers to access would be removed as widely as possible, leading to converging conditions in both countries. While on the one hand the removal of certain barriers and the emergence of energy services open up new opportunities for CHP, there is on the other hand the risk of a short-term orientation in energy supply due to uncertainties and increasing difficulties to co-ordinate energy supply chains between different competing companies. In other words, the German district heating systems would be challenged by similar problems as the ones which still prevent district heating in the UK. The conditions for single-site CHP (mainly small-scale and industrial) would not change significantly or even tend to become better, but for integrated multiple-site CHP-systems it would become increasingly difficult to achieve the necessary co-ordination.

Consequently, while this second scenario can be regarded as a 'business as usual' scenario for the UK, it would break with many traditions of energy supply in Germany. Considerable resistance can be expected, especially from those who would suffer disadvantages, i.e. local utilities and to a certain extent also regional utilities. Depending on how convincingly the argument is formulated that competition would lead to 'cherry-picking, i.e. to a privileged supply of large customers and to rising costs for small electricity users, critique could also come from the public. Moreover, due to the manifold interdependencies in the German system, it is rather unlikely that such an extreme version of liberalisation could be introduced.

10.7.3 Scenario 3: Political support and enforcement of CHP

A scenario, in which strong political support is lent to CHP, and targeted policy measures enforced, may appear unrealistic for the UK, but it is not impossible. In Germany, on the contrary, it even appears to be a realistic alternative because energy supply is still understood as a public service in large parts of society, even if currently liberalisation of public services is a priority theme on the political agenda. This scenario would imply for both countries to assume a minimalist interpretation of the requirements imposed by the Internal Energy Market. Moreover, in the case of the UK any attempt to give more active support to CHP would have to respect the achievements of the present system which is based on private utility and service companies. The influence role which German government policy has maintained on the energy sector makes it easier to justify a more interventionist approach.

The two countries differ significantly in terms of the organisational structures which would be needed to support actively the uptake of CHP. Such structures are still in place in Germany where competent utility companies at all supply levels could adopt and reinforce stimulation measures at federal level. In the UK, the regional utilities have been privatised so that support measures for CHP would have to make use of financial incentive mechanisms and/or taxes. These differences also show that it would be difficult to support district heating in the UK due to the lack of organisational co-ordination at local level. Only industrial and small-scale applications could be expected to benefit from market-based support measures. The German situation, on the contrary, offers several inroads to support CHP by organisational, financial and regulatory means. Consequently, a further extension of district heating capacity would be possible. It could be reinforced by supportive environmental and tax policies. However, the impact of support measures on CHP-diffusion would depend in both countries on their timing. In the UK, several modern centralised power plants have already been built in the last years which will not be replaced for the next two decades, and in Germany the major decisions on new capacity will not be taken before about 2005. This opens up a window of opportunity in a few years time to trigger a structural change in the German energy supply system. If appropriate measures are taken in time, then this second scenario could indeed lead to an acceleration of CHP-diffusion. In other words, even if the political and organisational settings became highly conducive for CHP in both countries, the speed of diffusion would differ significantly.

In view of the opening up of energy markets in Europe, this third scenario seems to be unrealistic because it would contradict the expected principles and the development path of the Internal Energy Market in Europe. Several powerful actors in both countries have already prepared themselves for this new framework - especially the major supply and generation companies. A move towards a more proactive energy policy would have to be compatible with these boundary conditions. Otherwise, government would have to be prepared to enforce its policy against the resistance of the major energy utilities, manufacturers and industrial electricity users.

PART IV:

Concluding Remarks on Cogeneration, Political Control and the PET-Approach

In this final part, the three underlying arguments which were formulated in the introduction are going to be taken up again. Taking the applied problem of how to influence the diffusion of environmentally benign energy technologies such as CHP in large socio-technical systems by political means as a starting point, a methodological and a theoretical research question were raised. This lead to the development of the PET systems approach. The three main parts of this study addressed these three research questions by first constructing a theoretical framework, secondly by showing how to implement this framework for the problem area of CHP, and thirdly by analysing and comparing the two empirical case examples of CHP-diffusion in the UK and Germany.

In the next three sections, the discussion follows the reverse order of these arguments. First, some specific conclusions will be drawn, concerning the possibilities and limitations of political control of CHP in different political and regulatory frameworks. They will focus on the respective hypotheses formulated in Chapter 7, and on the strategies which could be followed to promote the diffusion of CHP as part of a wider approach of political, economic and technological transformation. Secondly, the general methodology which was developed to study this problem will be critically assessed, its advantages and drawbacks be highlighted. In the third section, the experiences made with the PET-approach in this study will serve to review critically its usefulness as a theoretical foundation for applied and policy-oriented research. Despite the numerous question marks and uncertainties which have inevitably remained after this first application case, it is regarded as a promising starting point for studying processes of technological change and political control. Finally, an outlook on possibilities for further research work will be given in order to indicate how the PET systems approach could be further advanced, and in which areas it could be applied as a constructive foundation for prospective analysis.

Chapter 11: Conclusions

11.1 Policy conclusions

The two following sections aim at drawing conclusions which could of interest and relevance to policy-making, both at European or national level. While the five hypotheses and the requirements of CHP can be discussed in quite neutral, value-free way, the different elements of a CHP-friendly policy framework are based on the normative assumption that the wider diffusion of CHP is a desirable objective. The issue of desirability has already been discussed in detail in Chapter 5, and it was pointed out that it depends very much on the standpoint, interests and priorities of individual actors. However, from a socio-economic perspective, it became clear that the wider diffusion of CHP would be indeed a positive development in many respects. Consequently, the framework aims to give indications of how the diffusion of CHP could be stimulated by policy means.

11.1.1 Lessons from the cases of CHP in the UK and Germany

The general hypotheses and their specifications as introduced in sections 3.8.1 and 7.1 are now taken up again. It should be kept in mind that this study does not aim at testing hypotheses but rather attempts to develop a theoretical and methodological approach to guide applied research work. However, as a part of this exercise the five hypotheses have nevertheless provided a useful orientation for research work. After the presentation and comparison of the two empirical equivalents to the PET system approach, it is now possible to review the usefulness of these hypotheses in further detail.

11.1.1.1 Co-evolution of structures and technology (H1)

The interrelationship between structural changes and the diffusion of CHP stands out as one of the key issues in this study. Especially the implications of CHP at the organisational level of the energy supply system have been identified to be of crucial importance. The two case studies have shown that these organisational issues do not only comprise technological, informational and economic aspects, but also have an important political dimension. Political organisation defines the space inside of which new alternatives for energy supply can develop. In both cases, the structure of the interfaces between public policy-making and energy industries determined whether there were possibilities for alternative technologies to emerge or not.

The structural conditions in the UK and in Germany are the result of a historical process. Similarly, technological, industrial and political cultures exist which determine the decision-making processes for innovation, adoption and policy-making. In the UK and Germany, different principles underlie decision-making, both in the techno-economic and political realms. For example, they differ in the extent of decentralisation of responsibilities, the willingness of actors at different levels to make tacit agreements, the extent of horizontal and vertical integration of energy supply, the approaches to regulation, and the range and degree of involvement of actors in political decision-making processes. Consequently, there is no single 'best' structure which ought to be implemented to promote CHP, but only country-specific pathways towards the integration of new technological options such as CHP. One of the main difficulties in this respect consists of finding a balance between avoiding a lock-in into a single pathway and providing reliable and stable incentives for long-term investment into sustainable energy supply technologies. In the end, these country-specific pathways should lead to a 'match' between the four types of structures which are central to the energy sector (technology supply, technology use, information provision and policy-making) on the one hand, and the organisational principles which underlie CHP-technology and services on the other.

11.1.1.2 Technological complementarity in large technical systems (H2)

CHP interconnects different chains and levels of energy supply systems. It needs to be integrated technically in a wider technical system, be it on an individual site or in a regional electricity supply system. In this study, these technical considerations were of minor importance because - as mentioned before - the crucial point was the shaping of non-technical structural settings to match the requirements of the technical system rather than the shaping of new types of technical artefacts. The technical integration of CHP into both heat and power supply systems has nevertheless been an important factor in the process of defining its application niches. CHP needs to meet technological complementarities at different supply levels, and this implies also different levels of application niches. This has become apparent in the complementarity and competition with gas supply, in the initial technical problems of grid connection, and in the need to fit CHP into the energy supply system of an individual site.

11.1.1.3 Behavioural hypothesis (H3)

Decisions on CHP are dependent on a large number of factors and constraints, ranging from the mental frameworks of decision-makers to expectations concerning structural developments. Structural changes seem to be more important than for other types of investment due to the long planning horizon of CHP-systems. Different forms of uncertainty turned out to be of crucial importance for technological choices and the planning horizons applied. In many cases, uncertainty made all rational arguments and data irrelevant, because future conditions could not be appraised in a reliable way. For a technology such as CHP higher initial investment is required than for simpler alternatives, but it promises to pay off in the longer term. Under conditions of high uncertainty, potential projects can be carried out only if they either promise very short payback periods or are important for strategic reasons. CHP has also been subject to major technological uncertainties in both countries, but these could be reduced in the course of time by learning, by building up a knowledge base, and by disseminating information on experiences made. Uncertainty and information access are also important problems for policy-makers, but the experiences from the two countries have shown that ideological factors can easily push out more rational arguments. The political realm is a typical example of a decision arena where decisions are made as a result of long interaction processes between a large number of actors. Especially the German case confirms this. CHP users and suppliers had to overcome also their initial knowledge and experience gaps by consulting many other actors. The need for these interactions diminished once the new types of applications had been fully established and standardised.

Decision-making processes are therefore the key to understanding the process of CHP-diffusion. The study has made clear that one needs to go beyond the description of structural changes and decisions, but look at the underlying motivations and expectations. Key mechanisms of reinforcements and delays over time can only be understood on the basis of such a micro-foundation. In this regard, it is also important to note that the individual actors showed quite complex behavioural patterns and responses which can not be captured in highly simplified actor models. Even a model based on the notions of routines and satisficing behaviour can not replace a detailed empirical description of strategies and objectives.

11.1.1.4 Dynamic hypothesis: integrating strategies over time (H4)

The dynamics of change of innovation diffusion processes do not follow regular patterns. The fast transformation process in the UK underlines the importance of sudden events and of the timing of decisions for breaking up lock-in situations. However, the 'window of opportunity' to establish a more decentralised supply path based on CHP was very short, and the necessary incentives, structures and political pressures were not in place at the right time to trigger a wider diffusion of CHP. The developments in Germany highlight the importance of structural and behavioural interdependencies for the dynamics of change. It was possible to observe the detailed operation of reinforcing and delaying mechanisms, and how they influenced the diffusion path.

Reinforcing mechanisms could be observed in both countries, but the variety of mechanisms of reinforcements and inertia differed. Information provision and learning have been of crucial importance, together with the model function of successful application and the integration of CHP into the standardised design procedures of the engineering communities. These mechanisms represented also key links between innovation and diffusion elements of CHP-technology. As a classical example of a reinforcing mechanism, economies of scale by series production need to be mentioned as well. Finally, indirect network externalities in terms of costs can be identified in the case of district heating once the initial investment into the network has been made.

Before these reinforcing mechanisms could operate, the lock-ins into the power-only, centralised supply system had to be broken up. This provides further evidence for the non-linearity of diffusion processes which are due to structural interdependencies between technological, economic and political organisation. In order to disentangle these interdependencies, the behavioural perspective has been essential. Individual decision-making processes have been the node through which the complex temporal interplay of technological, economic, and political factors could be studied.

The two case studies provide good examples of very different models of change in large socio-technical systems. While the British approach can be characterised as a politically driven revolutionary overturn of the ESI, the German case represents an incremental and co-evolutionary process of change.

11.1.1.5 Control hypothesis: Liberalisation, monopolies and policies (H5)

Perhaps the most important lesson to be learnt from the case of cogeneration is that there is no single best structure and framework for CHP, and consequently neither a single best policy approach. The dynamics of change which interconnect different areas and mechanisms are highly individual for each country. While liberalisation as it was implemented in the UK can be regarded as conducive for CHP under the specific British circumstances, it would be dangerous to expect to similar benefits in Germany because the starting point is completely different. Rather than the general framework, the specific implementation of measures and regulations determines whether CHP diffuses or not. Within a liberalised framework, the emergence of a service-oriented approach seems to be a conducive force, but it needs to be balanced with the need for co-ordination and a long-term planning horizon. Within a monopolistic framework, variety creation and selection pressures for economic efficiency may be limited, but co-ordination can be achieved more easily. The German case shows that variety can also be created by the decentralisation of competencies because it leads to differences in the implementation of the regulatory and policy framework.

The empirical cases have indicated a large variety of possible intervention points for policy. They show that one needs to think beyond the traditional means of control (regulation, finance and information), but take also cultural and structural aspects into account. For example, attempts to change engineering cultures can be more effective than financial incentives. Structural measures should not only target the sector to be controlled, but also the political decision-making process itself and its interfaces with society and economy. The example of the UK shows convincingly that a lock-in situation can be based on the structure of political interfaces.

In addition to defining a framework of control, it is also important to keep this framework open to further evolution. The German political system may have been slow in reacting to new requirements and opportunities, but it enabled at least the taking up of some new ideas and their integration into the policy framework. Similarly, the new liberalised British system seems to be quite successful in complementing and adjusting regulations when it has been recognised that this is necessary and useful.

Specific political measures appear to be most effective when they build on reinforcement mechanisms and incentives which are already in place in the system. For CHP, the information interface was one of these crucial mechanisms, and it was used successfully in both countries, be it in different ways. One could call such an approach the targeted management of technological niches. In order to achieve the diffusion of a certain technology which has been recognised as beneficial, it is not enough to define a rough framework. The objective should rather be to define specific conditions within the framework which facilitate the improvement of infant technologies and their adjustment to the dominant structural conditions. At the same time, the framework conditions should be modified to accommodate the new technologies.[268] In other words, the targeted management of technological niches is complementary rather than an alternative to the definition of a comprehensive regulatory framework. It is based on an incremental model of change, whereas the structural shock which was triggered in the UK reflects a more radical approach. In most European countries the energy supply systems perform quite well, so that an incremental strategy seems to be more appropriate. The strong interdependencies in energy supply and especially for CHP would make a radical transformation a very risky endeavour, especially in countries where - in contrast to the UK - already a large capacity of technologically and organisationally integrated systems has been built up.

11.1.2 Requirements of a CHP-oriented framework

Rather than formulating detailed recommendations which would have to be differentiated for individual countries, it appears to be more appropriate to formulate requirements for a policy framework which is compatible with CHP. The aim is to highlight principles which should be taken into account in such a policy framework, independently of whether a liberal or a monopolistic approach is followed. In view of the contingencies of each specific case, such generalised principles regarding a CHP-oriented energy supply framework have clear limitations, but they are supported by the findings of this study. The principles reflect that targeted control is hardly possible due to the complexity of the technology and the energy sector. Three groups of principles are distinguished, which reflect the two main levels of analysis (structural and micro) and the political orientation of the research work.

---

268 See also the concept of strategic niche management which follows a similar line of argument, though less focused on political control (Kemp et al., 1996).

1. Organisational and structural principles:

- Decentralisation of competencies for energy supply in order to enable the compatibility of organisational structures and interests of actors with the requirements of CHP and district heating;
- Facilitation of the horizontal integration of energy supply at local and regional level, i.e. at the levels at which CHP can actually be applied;
- A limited degree of vertical integration between local energy suppliers and centralised generation companies in order to create interest in CHP at the higher levels of supply;
- Clear and reliable regulatory framework conditions which take into account the environmental benefits of CHP;
- Definition of a clear and fair interaction regime at the interface between utility companies and auto-producers in order to reduce uncertainties about tariffs and remuneration; it should be based on the recognition of long-term avoided costs;
- Opening of the structure of the policy-economy interface in order to enable the inflow of novel ideas and the creation of variety in the political realm;
- Decentralisation of political responsibilities in order to establish competent actors at local level which can support and co-ordinate energy-related tasks, and contribute to experimentation and variation.

2. Principles at micro-level:

- Based on the preceding principles, facilitation of a service orientation in energy supply, of least-cost planning and demand-side management, and of efficiency-orientation in generation; it should be facilitated by the internalisation of costs and benefits by a single horizontally integrated organisation;
- Internalisation of external costs to reflect the efficiency advantages of CHP in investment decisions without endangering economic viability, e.g. by means of incorporating environmental considerations in investment appraisal procedures;
- Preservation of variety-creating effects by shaping diversified application niches;
- Preservation of a balance between liberal elements to avoid unnecessary barriers and facilitate competitive services on the one hand, and sufficient co-ordination to enable integrated energy service supply and avoid detrimental side-effects of competition on the other;
- Facilitation of information exchange to improve learning processes and raise awareness about CHP among decision-makers.

3. Principles for the formulation of specific policy measures:

- Intelligent use of reinforcing mechanisms in the design of policy measures, e.g. by supporting new application areas, by establishing generic standards, or by facilitating information exchange;

- Exploitation of self-organisational mechanisms in the energy industry, e.g. by stimulating the provision of information;
- Targeted financial support to promote specific promising innovations or model application rather than widely spread small grants to provide just some additional funding.

11.1.3 Policy strategies for promoting CHP

Strategic foundations

While the preceding list of principles has been formulated in general terms, the development of a strategy for a decision-maker requires a number of further specifications which are summarised in the following questions:

- For whom should a strategy be developed? Who is the 'client'?
- What are the problems addressed and the objectives pursued by this decision-maker?
- Which options for intervention does the decision-maker have at his disposal to influence the system?
- How can these options be combined in a smart way to achieve a system development in line with the decision-maker's objectives?

In the case of this study on CHP, the focus is on policy-makers, but the approach is not limited to this perspective. In principle, any other actor in the system could be the 'client', too. Below, strategies both for the national and the European policy level are going to be outlined. It is assumed that the objective is to promote the diffusion of CHP within the constraints of the Internal Energy Market, taking into account the historically grown conditions of the energy supply system. In order to address the other three questions, it is necessary to dispose of a solid understanding of both the macro- and micro-mechanisms of cogeneration diffusion. At the micro-level, the main decision criteria and problems with which the different actors are confronted need to be known. At the macro level, the changes of structural conditions are of major importance. Consequently, most of the work on this study has been dedicated to the identification of mechanisms which operate at these two levels of the system.

As mentioned before, political control is not just a matter of the general framework (which can be liberal or monopolistic) but also of the specific regulations and balances built into the system. They can facilitate or inhibit CHP along the lines of the principles mentioned above. From a policy perspective, a number of intervention levels can be identified, which represent potential entry points for political measures. These levels are:

- Structures of political control;
- Regulatory principles;
- Specific regulations;

- Targeted adoption support measures;
- Technology and innovation policy;
- Information provision;
- Raising awareness.

Beyond these seven levels, the problem of timing needs to be taken into account as well. However, this problem can not be discussed here in detail, but only with respect to individual cases.

Strategies for promoting CHP at national level

National governments have a wide variety of control and intervention means at their disposal by which they could affect the diffusion of CHP. Despite the fact that the complexity of both the technology and the energy sector make it difficult to derive clear and simple policy indications, a number of potential strategies can be derived from the two case-studies on the UK and Germany.

<u>1. Structures of political control</u>

The structure of the policy-economy interface fulfils an important function to enable policy changes, and consequently also changes in the energy supply system. This interface needs to be sufficiently open and diversified to take up novel ideas and implement them through appropriate policies. The decentralisation of political competencies to the local level is linked to this issues, because it facilitates the implementation of decentralised energy supply systems and utilities. It offers the possibility to formulate a clear framework for energy service companies to develop systems which are well-adapted to local conditions and requirements.

<u>2. Regulatory principles</u>

A liberalised and privatised energy supply framework can be conducive for CHP, but it can also have detrimental effect. In any case, it is not a universal remedy. The decentralisation of responsibilities for energy supply and horizontal integration is an important principle; it ought to be respected in order to reinforce interest structures which are compatible with CHP. At regional, and even more so at local level, it is of major importance to support the co-ordination of supply chains in order to avoid duplication of infrastructure construction (e.g. gas and heat networks). Regulations, be it for energy or environmental issues, should be reliable and aim to reduce uncertainty for customers. On the other hand, this reliability needs to be balanced with the need for flexibility to be able to react to new developments. Finally, a supportive framework for CHP would have to incorporate principles which facilitate the consideration of long-term interests, e.g. in terms of externalities or long-term avoided costs of investment in CHP.

<u>3. Specific regulations</u>

Rather than the general framework, the specific implementation decides whether CHP can diffuse or not. Electricity tariffs and export prices are just one example of this. Simple, reliable and appropriate rules need to be established which allow taking the longer-term benefits of CHP into account. This could be achieved by delegating the responsibility for negotiating these tariffs to the organisations of the main actors concerned, while defining the guiding principles politically. Obviously, such a strategy requires that a governance culture prevails which is compatible with such an approach.

In order to intervene as little as possible, it would be promising to make effective use of reinforcing mechanisms which are already built into the system. Examples of such mechanisms are the exchange of information on best practices of CHP-plants, the controlled delegation of responsibilities to the actors in the energy supply system, or the definition of generalised incentives for energy efficiency such as Least-Cost Planning. Such incentives could then be translated by energy users into different technical solutions, among which CHP would be just one example. In addition, compulsory least-cost-planning would be a means of investment control to oblige generators to justify the construction of additional generation capacity.

4. Targeted support measures

The direct stimulation of the adoption of CHP by financial support measures seems to be effective only when it is applied massively. Small grants just fulfil the function of additional funding and - as shown in the case-studies - hardly ever tip the balance for or against CHP. Especially for innovative solutions which would otherwise not be taken up, it would be promising to develop dedicated strategies composed of a whole set of support measures. Such 'integrated niche development programmes' could have different forms. In some European countries fairly strong measures have been taken (e.g. in Denmark), and in terms of the performance of the energy supply system they have shown good results. Such enforcement-type strategies might not be acceptable in other countries, but they can be implemented also in a less interventionist way. The combination of financial support, environmental commitments, R&D-support and advanced planning and finance schemes (like the ones pioneered by several German, and also some British, cities) seems to be a useful strategy.

5. Technology and Innovation policy

Due to the fairly advanced character of CHP-technology and its dependence on local conditions, technology policy may appear to be of limited importance, but there are a number of generic components and solutions which should be supported by targeted measures. Among these generic technologies, fuel cells stand out, but they are already funded by both national funding and industrial research. Other more specific issues are pipeline technologies for DH, chilling equipment or heat storage systems.

As part of an innovation policy, the stimulation of links between users and suppliers should be considered. During the phase of stabilisation of a technology, these co-operations play a major role. They can take many different forms: direct co-operation between user and supply side are as useful as indirect ones through associations. Standardisation policies, despite their negative impact on variety, are particularly helpful for CHP when they are applied not to narrow technical solutions but to the design principles for energy supply systems with CHP.

Finally, public procurement can play an important role, especially if public or semi-public institutions, e.g. local utilities, control a large part of the energy market. In this case, their purchasing power affects not only the size of the market, but also the technological developments envisaged by the manufacturers.

6. Information provision

Closely associated to RTD-policies are also information programmes of government. Obviously, these are particularly important if insufficient information is provided within the available settings of industrial associations and other initiatives in industry. Standardisation measures at European level could be useful to a limited extent as well, but there is still a lack of organisations to do this. National organisations, like the German VDI, play a more important role in this respect. In other words, information dissemination does not necessarily have to be done by government itself, but it should stimulate the provision of such information where it is missing.

Associated to the provision of information, it is necessary to improve the users' capabilities to exploit the information provided. This could be achieved by supporting the establishment of design standards and their spread to the people actually in charge of energy planning. In addition, efforts should be made to introduce energy efficiency issues into the training and education programmes not only for engineers, but also of planners, managers and administrators.

7. Raising awareness

Raising awareness and concern about a new and promising technological principle is decisive to establish it in the major decision arenas, be they at board level or in the political realm. Only once it figures prominently on the political agenda, there is a possibility to trigger change. In the UK, the push for CHP was a by-product of a major political driving force, namely the liberalisation debate. In Germany it took years of discussion to establish it in all its variants. In addition, authorities can promote the awareness about CHP in several ways, e.g. by defining targets and objectives, or by committing themselves to a CHP-oriented procurement policy.

Strategies for promoting CHP at European level

Building on this discussion of strategies and control measures at national level, a number of conclusions can also be drawn on the possibilities to stimulate CHP-diffusion by policies at European level, even if the options available are rather limited. In this context, the question needs to be raised what the strategic political possibilities at this level are, given the variety of national systems and structures, the different options available, and the controversial opinions about experiences made with cogeneration in different countries. In addition, European policy is constrained in its possibilities to affect energy supply by the principle of subsidiarity, and by its limited competencies in the field of energy policy. However, basically the same seven levels of entry points for policy can be distinguished as for the Member States, even if they tend to be addressed by 'softer' measures.

1. Structures of political control

CHP being a local technology, it is obvious that the most adequate control structures are located far downstream of the EU-level. Formally, this is expressed in the principle of subsidiarity which ought to be respected, but the completion of the Internal Energy Market shows that the influence of the European policy level is growing. However, this is only insufficiently reflected in the policy - industry interfaces. Currently, a few major players dominate the representation of interests in Brussels, with the effect of biasing policies in favour of large companies. In many respects, it appears to be necessary to rethink the institutional set-up (also in terms of scientific advice) in order to develop more balanced policies and, in the field of energy policy, take the concerns of the local level into account.

2. Regulatory principles

The EU can intervene in energy supply only in certain areas as specified for reasons outlined in the Single European Act and the Maastricht Treaty. Competition policy and the completion of a Single European Market are the two most relevant arguments in relation to energy policy. They are at the root of the plans for a comprehensive liberalisation of energy supply in Europe (despite fierce resistance from some Member States). In the last years, the debate about the Internal Energy Market has caused a lot of uncertainty. Consequently, it would be helpful for CHP as well as for many other technologies to settle the dispute and provide reliable framework conditions as soon as possible. A major step in this direction was made with the adoption of a White Paper on energy policy (CEC, 1995) but it still awaits full implementation in many Member States. The White Paper argues clearly in favour of a liberalisation of energy markets, in spite of doubts about the environmental implications of such a policy framework. Based on the experiences with liberalisation in the UK, it would be important to take into account that it would be conducive for CHP only if it was complemented by appropriate support measures which respect the specific historically grown conditions of the national energy supply systems. European environmental regulation as a second policy component is also limited in its specific influence, but it defines at least minimum standards to be achieved throughout the Union. Rather than harmonising specific environmental legislation, it provides shared guidelines which can be implemented in a compatible way within the national frameworks. Consequently, the possibility is maintained to define conditions which take the local requirements of CHP into account.

3. Specific regulations

As mentioned before, the specific regulations designed within a regulatory framework can be more decisive for the possibilities of CHP than the framework itself. The EU hardly has the competencies to impose specific regulations, but can adopt framework directives which need to be implemented at national or even subnational level. Due to the variety of energy policy frameworks in the EU it would certainly not be desirable to impose detailed European regulations because they could not take the specific conditions and constraints at national, regional and local level into account.

4. Targeted support measures

Direct support for the uptake of individual technologies would not be a promising element of a European CHP-policy either, not only for reasons of disagreement about the need for such a policy, but also because funds are too limited to have a real impact. Moreover, past programmes had a comparatively small impact, and only in very specific areas. Applying for specific support measures from the EU is usually a too cumbersome and time-consuming procedure to be justified for local investors. The same argument applies to the establishment of an integrated niche development strategy for CHP, which could only be implemented locally; it would clearly go beyond the scope of EU-policy.

5. Technology and innovation policy

European technology and innovation policy for CHP is another area of rather limited importance. In view of currently available funds, only a very small number of projects could be funded by the EU. Secondly, CHP-manufacturing is still an industry with a national basis and rather limited intra-European co-operation for technology development. This is due to the sophisticated character of national regulations and administrations, the informal access barriers, and differences in standards for equipment and design. The last two points might become relevant for EU-policies in the future. By establishing compatible design standards throughout Europe the informal barriers to access to the national energy supply markets could be removed as part of the internal market policy. Currently, standards for technologies such as CHP are still a national domain.

6. Information provision

Information provision is one of the means by which the EU can facilitate the uptake of CHP, and a number of recent initiatives have taken up this opportunity. Such an information policy should be directed in the first place towards decision-makers, not towards users and manufacturers of CHP who are already quite well informed. The exchange of experiences across Europe, e.g. about experiences with niche applications would also be of value, even if the transferability of these experiences is limited due to the differences between national frameworks. For the foreseeable future, these differences are expected to remain within the Internal Energy Market.

7. Raising awareness

The EU is an important and prominent actor, and its activities and objectives are being reflected in many ways in the decisions at other levels and in other areas. It can thus play a major role in raising awareness about the benefits CHP. Examples of such highly visible and often symbolic measures could be capacity targets, but also the emphasis given to energy efficiency in the formulation of policy objectives.

The limits of political control of large socio-technical systems

The preceding discussion has shown that the possibilities for political control of CHP-diffusion at European level are still very limited. At national level full, a wider range of potential measures can be applied but targeted control cannot be exerted by any of the actors, either. They mutually adjust their decisions and strategies to each other and jointly shape the development path at technological and structural level through interactive decision-making processes in different realms. Governments are just one among several other actors, even if they have in principle the power to define the frameworks and rules of the game for the interactions of the others.

The inherent complexity and uncertainty of large socio-technical systems such as energy supply is one reason for this limited influence. The constraints which result from the technological structures are particularly strong in these cases: The technical system is extended, interrelated in a complex way, covering different levels with long replacement times. They are reinforced by similarly complex networks of actors. Moreover, socio-economic and technical realms are strongly pervaded by political interventions. This points to one of the main hypothesis of this study, namely that for a large socio-technical system like energy supply, it is necessary to endogenise the political realm into the analysis of technological change.

A further limitation to control is rooted in the feedback effects and reinforcement mechanisms which could be identified also in the case-studies on CHP. In the absence of dominant single driving forces, the interplay between different elements of the system can lead to very powerful dynamics which are very hard to predict. Although several of these dynamic mechanisms could be carved out, it is extremely difficult to derive clear recommendations for an appropriate timing of interventions to reinforce (or slow down) their operation. As shown for the case of the UK, there was only a short period of time when active measures to stimulate the diffusion of CHP could have changed the trajectory of energy supply before the new liberalised structures stabilised again.

11.2 Methodological conclusions

11.2.1 The constructive character of PET

An important research interest of this study consists of the development of a methodological approach to investigate processes of innovation diffusion and political control, building on the corresponding theoretical perspective of the PET-system. A methodology of nine basic steps has been formulated, but these steps need to be specified in more detail for each individual investigation. A broad range of specific research methods can be used to collect data on the different aspects touched upon. It has been shown in this study how such a specification process can be implemented, as a model for further research work along similar lines. In this interpretation, PET should be seen as a constructive tool for analysis, which helps to structure research work and interpret its findings.

The PET-methodology forces the researcher to be highly explicit and define his work very well, but it is also difficult to use due to the complexity of the underlying approach. A great amount of data needs to be collected to fill the pattern of analysis for the diffusion period under study, ranging from detailed technological information to the operation of the political system, and from highly aggregated diffusion data to individual interest structures and motivations. This is obviously a very time-consuming approach, but the result is a comprehensive and structured analysis which offers the possibility to explore future policy strategies. The main strength of the PET-approach should therefore be seen in the possibility it offers to analyse actor strategies from a system's perspective, taking into account the behaviour and interactions of all relevant actors within a broad, but carefully delimited, context. For this study of CHP, policy-makers have been targeted as the fictive 'clients', but the PET system approach could be used equally well from the perspective of other actor, too.

11.2.2 Theory-based prospective analysis: requirements and possibilities

Although this study was not meant in the first instance to be a prospective exercise, an outlook has been given on how it could be applied to develop scenarios and explore future developments. Prospective work requires a different approach than 'normal' science which is usually based on retrospection. The objective of prospective work is not to test theories and hypotheses, but to explore the future evolution of systems. The methodological challenge consists of making *constructive use of theory* in this endeavour.

This has a number of important implications. Due to the fact that futures research cannot exclude any variable ex ante, an encompassing and multi-disciplinary approach needs to be pursued. It must be taken into account that a large variety of factors could potentially have an impact on future development paths. Theory can fulfil very important functions in this respect. It can provide a foundation for the construction of a basic pattern of analysis, to be used for exploring the impacts of future changes, decisions and external determinants on the evolution of the system. It is not sufficient to know the general impact of certain factors on system evolution; one needs to understand the interdependencies and mechanisms which underlie the operation the system under study. This knowledge is available in a highly condensed form in the theories and concepts developed in the (retrospective) research work on social, economic, political and technological systems. The PET system approach is at attempt to integrate these different theoretical inputs into a coherent framework for exploratory analysis.

The analysis of trends and past developments is an important input to the understanding of the aggregate development of a system, but it does not provide much information on how to affect and control a system's development path. Trends are only useful from a control perspective if they are brought together with information on actor strategies and interest structures by means of micro-level analysis. It reveals structural changes and trends are translated into decisions on future developments. In other words, prospective analysis requires to analyse both micro and macro level, and PET provides a pattern to facilitate such an analysis.

While in principle the PET-approach seems to have some clear methodological strengths, its qualitative character makes it still very difficult to deal with system dynamics and control. The conclusions which can be drawn remain at a fairly general level, even if this might reflect also some of the fundamental limitations to which system control is subjected. The quality of findings on control issue could certainly be improved by a more refined prospective analysis, going far beyond the short prospective exploration presented in this study. While the basic tools for a PET-based analysis have been made available, further methodological refinements would still be desirable, e.g. by developing a formalised model which would allow carrying out a more refined dynamic analysis.[269]

11.3 Theoretical conclusions

11.3.1 On the usefulness of the PET-approach for structuring processes of change in socio-technical systems

PET represents a structured systems approach to describe the evolution of a socio-technical system. As far as it could be developed in this study, it represents a first step towards a comprehensive analytical and prospective framework which integrates potential determinants from several different areas.

The approach has been developed in parallel lines to an application in the field of energy supply. Therefore, some actors (e.g. end-users and private customers) have played a less prominent role than others, who are at the core of innovation and adoption decisions of CHP. In other application areas (e.g. transport or telecommunications) these restrictions would have to be taken into account. It points also to the need to extend and generalise the approach further.

The PET-system is based on the assumption that technological change is driven by at least three different types of logic: economic, socio-political and technological. They are regarded as interrelated, driving an evolutionary process of change. In addition, the PET-system interrelates two levels of analysis: micro/behavioural level, and macro/structural level, which are both embedded in a changing exogenous system environment. Due to this comprehensive character, it represents in many respects a more realistic foundation for exploratory and applied research than many other, more theory-oriented approaches.

An important element of this applied and realistic orientation is the consideration of a 'client' in PET-based research work. Although PET has been developed with the primary objective in mind to give advice to policy-makers, it can in principle

---

[269] In order to achieve this, a promising path appears to be the approach of fuzzy modelling and scenarios which make use of expert knowledge (which is fuzzy by nature) in single areas and aggregate it by the rule system identified (Canarelli, 1996). In contrast to system dynamics based on differential equations, fuzzy modelling is more flexible and better adapted to future exploration because it does not require to translate existing knowledge into a rigid system of equations which is far more demanding and sophisticated than what can actually be obtained in terms of accuracy of data.

also be applied from the perspective of any other actor/client in the system. The decisive point is that PET provides a picture of the world which comprises this client - be it a policy-maker or an industrial firm - as part of the system under study. He/she is endogenised, and the conclusions can be tailored to the perspective of this client.

11.3.2 Theory, hypotheses and their limitations: Arguments for a different perspective on theory

Rather than for testing theoretical statements and hypotheses, PET is meant as a realistic and comprehensive theoretical foundation to guide applied research. It reflects the methodological needs of prospective technological analysis, which are different from standard scientific approaches. For the purposes of prospective analysis it is essential to dispose of a comprehensive view of the problem area to be studied, covering all possible intervening factors and their interrelationships. While individual disciplines aim at controlling as many variables as possible by keeping them invariant, prospective analysis does rather the opposite. It keeps many variables open while recognising the need to be as simple as acceptable for practical reasons. This leads inevitably to quite complicated representations of the systems under study, and makes any analysis very difficult.

Maybe even more important, prospective research runs counter to the conventional scientific approach of generating testable hypotheses. While the advancement of theory can be said to be the final objective of traditional scientific research, it has a completely different function in prospective work. It becomes the knowledge base from which to start exploring the future. In other words, theory is used as a tool to guide and structure prospective analysis. The exploratory and open character of prospective empirical work makes it even impossible to test and check theoretical statements. Scenario development is a means to deal with this openness. Instead of requiring the rigorous testing of hypotheses, the criteria of consistency, plausibility and relevance three are applied to evaluate the quality of scenarios.

The preceding arguments indicate that the formulation of hypotheses is of secondary importance for PET-based research because they cannot be tested anyway. What is essential, on the contrary, is a transparent, comprehensive and well-structured pattern of analysis in order to provide a convincing picture of potential future developments. However, hypotheses can still have an important function as orienting and focusing devices of research, and this is the way in which they have been used in this study.

11.3.3 On the needs and problems of multi-disciplinary research

Multi-disciplinary work seems to be imperative for prospective analysis in order to consider the wide range of factors which can have an impact on system evolution. A further claim in favour of multi-disciplinary research can be made from a control perspective. In order to be able to give useful inputs for political control, it is necessary to know about the impacts which political measures would have on a complex socio-technical system. These impacts comprise both intended and non-intended effects. Therefore, the research question can not be restricted to the impact of an independent variable X on a dependent variable Y. Instead, it needs to include the impact of individual decisions on the evolution of a system as a whole, because each individual decision is itself integrated into the system as a dependent variable. In other words, multi-disciplinarity is essential to become aware of potential feedback loops and reinforcing mechanisms in the system under study.

In order to meet these requirements of prospective analysis, the PET system approach aims at bridging between holism and reductionism. It is based on a broad perspective, which is nevertheless kept manageable by being very explicit with regard to underlying assumptions and explanations. In other words, the PET-approach reflects the interrelated and multi-disciplinary character of real-world problems, but in a reproducible and explicit way.

A frequent obstacle to multi-disciplinary research consists of communication problems between the disciplines. Each discipline has its own established conceptual and methodological vocabulary. Any attempt to integrate them can fail easily due to misunderstandings and a lack of openness on the side of the individual disciplines. It seems to be very hard to replace seemingly clear, but often unrealistic, disciplinary certainties by a more complex, but also more realistic, new framework. The fact that the PET approach is explicitly based on a number of research fields could be helpful for bridging between disciplines while avoiding misunderstandings. It could thus be used as common conceptual ground for multi-disciplinary research groups in the field of technology studies, which are composed of engineers, scientists, economists, and social scientists.

11.4 Research needs

In the course of this study it was not possible to explore all the research opportunities opened up by the PET system perspective. There are several pathways left improve, refine and extend the theoretical and methodological approach, as well as the empirical basis of PET. Five of the most promising lines for future research can be outlined as follows:

1. Refinement of PET as a research tool

The PET-approach should be further refined in order to make it more easily and generally applicable. This is not meant to be done by formulating more refined hypotheses, but by advancing it as a research tool. Part of this refinement could consist of the application of different mathematical modelling techniques, especially those which allow using qualitative expert knowledge as a complementary input to quantitative statistical data.

2. Generalisation of PET on the basis of further case-studies

The further generalisation of the PET-approach requires inputs from other empirical applications, especially from outside the field of energy supply. Transport and communication systems could be interesting areas for investigation. They show a number of similar characteristics to energy supply (e.g. in terms of complexity, involvement of public policy, importance of local level, several parallel technology chains, etc.) but also very interesting differences (e.g. a much stronger involvement of individual citizens in the decision-making processes about the technologies in question). Further country studies on CHP would be instructive in order to check the applicability of the PET-system to different national contexts. Such case studies would also expand the knowledge base on the underlying mechanisms of innovation diffusion processes and their political control.

3. Full-scale prospective applications of PET

While this study on CHP could only sketch briefly the prospective application of PET, it would be important to use the approach for a full-scale prospective study. This would also allow refining the prospective elements of the methodology.

4. Applications of PET from an industrial perspective

In order to demonstrate the applicability of PET from a different perspective than that of public policy makers, studies which focus on the exploration of technology strategies of other 'clients', such as manufacturing firms or large users, would be very valuable.

5. Expansion of the theoretical foundation of PET

Finally, the theoretical approach was build on a vast, but still limited, amount of conceptual and theoretical literature from the different fields of innovation economics, industrial sociology and political control. It should be linked also to areas and approaches which have not yet been taken fully into account like e.g. the managerial and organisational literature. Due to the specific character of the example technology, research findings from decision theory on the behaviour of individuals (e.g. consumers) were not considered prominently. However, it could be relevant to other application areas and should be taken into account in a review of the theoretical foundation.

These five issues show that there is scope for further work on the basis of the PET system approach. It opens up a full research agenda which would require bringing together contributors from a wide range of social science and technological disciplines. Such an effort could be worthwhile because it promises to generate results which are not only scientifically interesting but also relevant and useful to society.

References

Abernathy, W., & Clark, K. (1985). Innovation: mapping the winds of creative destruction. *Research Policy*, 14, 3-22.

Abernathy, W. J., & Utterback, J. (1978). Patterns of Industrial Innovation. *Technology Review*, 50, 41-47.

ADEME (Ed.). (1989). *La Cogénération dans les reseaux de chaleur français. Situation actuelle et developpement, Research Report*. Paris: Agence Française pour la Maitrise de l'Energie.

ADEME. (1994). Les politiques de maîtrise de l'énergie selon les secteurs. *Energie et Environment*, 54-56.

AGFW (Ed.). (1995). *Hauptbericht der Fernwärmeversorgung für das Jahr 1993*. Frankfurt a.M.: Arbeitsgemeinschaft für Fernwärme (AGFW).

AGFW (Ed.). (1997). *Hauptbericht der Fernwärmeversorgung für das Jahr 1996*. Frankfurt a.M.: Arbeitsgemeinschaft Fernwärme (AGFW).

AGFW/UNICHAL (Ed.). (1993). *Jahrbuch Fernwärme International 1993*. Frankfurt: Arbeitsgemeinschaft Fernwärme (AGFW)/International Union of Heat Distributors (UNICHAL).

Albisu, F. (1990). Cogeneration Technologies: Present and Future Developments. In J. Sirchis (Ed.), *Combined Production of Heat and Power (Cogeneration)* (pp. 10-19). Barking: Elsevier.

Alemann, U. v. (Ed.). (1981). *Neokorporatismus*. Frankfurt: Campus.

Alemann, U. v. (1989). *Organisierte Interessen in der Bundesrepublik Deutschland*. Opladen: Westdeutscher Verlag.

Andersen, E. S. (1994). *Evolutionary Economics. Post-Schumpeterian contributions*. London: Pinter.

anon. (1979). Energiepolitische Einigungserklärung. Grundsätze über die Intensivierung der stromwirtschaftlichen Zusammenarbeit zwischen öffentlicher Elektrizitätsversorgung und industrieller Kraftwirtschaft. *VIK-Mitteilungen*(4), 71-73.

anon. (1995). Comeback im Kleinen. Stirling-Motoren als Mini-BHKW. *Energie Spektrum*, 9(3), 50-53.

Archibugi, D. (1988). In Search of a Useful Measure of Technological Innovation (to Make Economists Happy without Discontenting Technologists). *Technological Forecasting and Social Change*, 34, 253-277.

Archibugi, D. (1992). Patenting as an indicator of technological innovation: a review. *Science and Public Policy*, 19(6), 357-368.

Ardone, A., Fichtner, W., Tsai, W., Wietschel, M., & Rentz, O. (1995). Beitrag der Kraft-Wärme-Kopplung zur CO2-Minderung im Vergleich mit anderen Optionen. *BWK*, 47(6), 247-254.

Arthur, B. W. (1988a). Competing technologies: an overview. In G. Dosi & al. (Eds.), *Technical Change and Economic Theory* (pp. 590-607). London: Pinter.

Arthur, B. W. (1988b). Self-Reinforcing Mechanisms in economics. In P. W. Anderson, K. J. Arrow, & D. Pines (Eds.), *The Economy as an Evolving Complex System* (pp. 9-32). Redwood City (CA): Addison-Wesley.

Arthur, B. W. (1989). Competing Technologies, Increasing Returns and Lock-In by Historical Events. *The Economic Journal*, 99(March), 116-131.

Arthur, W. B., Landesmann, M., & Scazzieri, R. (1991). Introduction: Dynamics and Structures. *Structural Change and Economic Dynamics*, 2(1), 1-7.

ASUE (Ed.). (1992). *Genehmigungsverfahren für Kraft-Wärme-Kopplungs-Anlagen*. Hamburg: Arbeitsgemeinschaft für sparsamen und umweltfreundlichen Energieverbrauch (ASUE).

ASUE (Ed.). (1996). *BHKW-Marktübersicht*. Hamburg: Arbeitsgemeinschaft für sparsamen und umweltfreundlichen Energieverbrauch (ASUE).

ASUE (Ed.). (1997). *Tätigkeitsbericht 1996*. Hamburg: Arbeitsgemeinschaft für sparsamen und umweltfreundlichen Energieverbrauch (ASUE).

Auer, W.P. (1985). Entwicklung der Kraftwerks-Gasturbine. *BWK,* 37(5), 97-163

Ayres, R. (1991). Evolutionary Economics and Environmental Imperatives. *Structural Change and Economic Dynamics*, 2(2), 255-273.

Babus'Haq, R. F., Probert, S. D., & Shilston, M. J. (1986). The Total Energy Approach: Evolution of combined heat and power for district heating and/or cooling. *Applied Energy*, 25, 97-163.

Bartsch, M., & Dingeldey, T. (1995). Rechtsprobleme der Einspeisevergütung. *Energiewirtschaftliche Tagesfragen*, 45(4), 249-254.

Bass, F. M. (1969). A New Product Growth Model for Consumer Durables. *Management Science*, 15, 215-227.

Bauer, C. (1994). Praktische Erfahrungen mit dem Einsatz der KWK in den neuen Bundesländern. *VIK-Mitteilungen*(1), 2-8.

Bauerschmidt, R., Utermark, D., & Weber, W. (1994). LCP und DSM-Maßnahmen bundesdeutscher EVU. *Elektrizitätswirtschaft*, 93(18), 1044-1052.

Baumgartner, T., & Midttun, A. (Eds.). (1987). *The politics of energy forecasting. A comparative study of energy forecasting in Western Europe and North America*. Oxford: Clarendon Press.

Beaver, R. (1993). Structural comparison of the models in EMF-12. *Energy Policy*, 21(3), 238-248.

Becker, E. (1995). Energiepolitik in den neuen Bundesländern. *Energiewir t-schaftliche Tagesfragen*, 45(9), 555-558.

Beitz, W., & Küttner, K.-H. (Eds.). (1986). *Dubbel. Handbuch des Maschinenbaus*. Berlin: Springer.

Bell, G., & Callon, M. (1994). Techno-Economic Networks and Science and Technology Policy. *STI Review*(14), 59-118.

BEO (Ed.). (1995). *Zusammenstellung vom BMFT geförderter Projekte 1974-1995*. Jülich: Projektträger Biologie, Energie, Ökologie des BMFT (BEO).

Berkner, U., Herrmann, H. P., & Schmitz, E. (1990). Die neue Bundestarifordnung Elektrizität (BTO Elt) - Veränderungen und Fortführungen des bisher geltenden Strompreisrechts im Tarifabnehmerbereich. *Recht der Elektrizitätswirtschaft*, 51(1), 2-17.

Bijker, W. (1987). The Social Construction of Bakelite: Toward a Theory of Invention. In W. Bijker, T. P. Hughes, & T. Pinch (Eds.), *The Social Construction of Technological Systems. New Directions in the Sociology and History of Technology* (pp. 159-189). Cambridge (MA): MIT Press.

Bijker, W., Hughes, T. P., & Pinch, T. (1987). General Introduction. In W. Bijker, T. P. Hughes, & T. Pinch (Eds.), *The Construction of Technological Systems. New Directions in the Sociology and History of Technology* (pp. 9-16). Cambridge (MA): MIT Press.

Bijker, W. E., & Law, J. (1992a). Postscript: Technology, Stability, and Social Theory. In W. E. Bijker & J. Law (Eds.), *Shaping Technology/Building Society. Studies in Sociotechnical Change* (pp. 290-308). Cambridge (MA): MIT Press.

Bijker, W. E., & Law, J. (Eds.). (1992b). *Shaping Technology/Building Society. Studies in Sociotechnical Change*. Cambridge (MA): MIT Press.

Birchenhall, C., Kastrinos, N., & Metcalfe, J. S. (1994). *Genetic Algorithms in Evolutionary Modelling*. Proceedings of the EUNETIC Conference on "Evolutionary Economics of Technological Change: Assessment of results and new frontiers", Strasbourg, 6-8 October 1994.

Blauwhof, G. (1994). Non-equilibria Dynamics and the Sociology of Technology. In L. Leydesdorff & P. v. d. Besselaar (Eds.), *Evolutionary Economics and*

Chaos Theory. New Directions in Technology Studies (pp. 152-166). London: Pinter.

Blok, K. (1991). *On the reduction of carbon dioxide emissions*. Unpublished PhD-thesis, University of Utrecht, Utrecht.

Blok, K. (1993). The development of industrial CHP in the Netherlands. *Energy Policy*, 21(2), 158-175.

BMBau (Ed.). (1994). *Verbesserter Wärmeschutz als Gemeinschaftsaufgabe, mit der neuen Wärmeschutzverordnung und der neuen Heizungsanlagen-Verordnung*. Bonn: Bundesministerium für Raumordnung, Bauwesen und Städtebau (BMBau).

BMU (Ed.). (1994). *Climate Protection in Germany. First Report of the Government of the Federal Republic of Germany Pursuant to the United Nations Framework Convention on Climate Change*. Bonn: Bundesministerium für Umwelt (BMU).

BMWi (Ed.). (1992). *Energiepolitik für das vereinte Deutschland.* Bonn: Bundesministerium für Wirtschaft (BMWi).

BMWi (Ed.). (1994). *Energieeinsparung und erneuerbare Energien. Berichte aus den energiepolitischen Gesprächszirkeln beim Bundesministerium für Wirtschaft*. Bonn: Bundesministerium für Wirtschaft (BMWi).

BMWi (Ed.). (1995). *Die Energiemärkte Deutschlands im zusammenwachsenden Europa - Perspektiven bis zum Jahr 2020. Kurzfassung*. Bonn: Bundesministerium für Wirtschaft (BMWi).

BMWi (Ed.). (1996). *Energiedaten '96. Nationale und internationale Entwicklung.* Bonn: Bundesministerium für Wirtschaft (BMWi).

Boehmer-Christiansen, S. A. (1990). Energy policy and public opinion. Manipulation of environmental threats by vested interests in the UK and West Germany. *Energy Policy*, 18(11), 828-837.

Böhret, C., & Franz, P. (1986). Die Technologiefolgenabschätzung (technology assessment) als Instrument der politischen Steuerung des technischen Wandels? In W. Bruder (Ed.), *Forschungs- und Technologiepolitik in der Bundesrepublik Deutschland* (pp. 349-390). Opladen: Westdeutscher Verlag.

Brand, M., & Jochem, E. (1990). The internal energy market. The new coalition against energy efficiency and environmental concerns? *Energy Policy*, 18(10), 694-701.

Bräunling, G. (1986). Ansätze, Konzepte und Instrumente staatlicher Technologiepolitik. In H. H. Hartwich (Ed.), *Politik und die Macht der Technik. Tagungsbericht des 16. wissenschaftlichen Kongresses der Deutschen Vereinigung für Politische Wissenschaft, Bochum 7. bis 10. Oktober 1985* (pp. 264-277). Opladen: Westdeutscher Verlag.

Bräunling, G., & Harmsen, D. M. (1975). *Die Förderungsprinzipien und Instrumente der Forschungs- und Technologiepolitik. Ein Analyse ihrer Wirksamkeit.* Göttingen: Schwartz.

Bruckner, E., & al. (1994). Hyperselection and Innovation Described by a Stochastic Model of Technological Evolution. In L. Leydesdorff & P. van den Besselaar (Eds.), *Evolutionary Economics and Chaos Theory. New Directions in Technology Studies* (pp. 79-90). London: Pinter.

Burness, H. S., Cummings, R. G., & Paik, I. (1981). Environmental Trade-Offs with the Adoption of Cogeneration Technologies. *Journal of Environmental Economics and Management*, 8(1), 45-58.

Cabe, R. (1991). Equilibrium Diffusion of Technological Change Through Multiple Processes. *Technological Forecasting and Social Change*, 39, 265-290.

Callon, M. (1980). The state and technical innovation: A case study of the electric vehicle in France. *Research Policy*, 9, 358-376.

Callon, M. (1987). Society in the Making: The Study of Technology as a Tool for Sociological Analysis. In W. Bijker, T. P. Hughes, & T. Pinch (Eds.), *The Social Construction of Technological Systems. New Directions in the Sociology and History of Technology* (pp. 83-106). Cambridge (MA): MIT Press.

Callon, M. (1992). The dynamics of techno-economic networks. In R. Coombs, P. P. Saviotti, & V. Walsh (Eds.), *Technological change and company strategy. Economic and sociological perspectives* (pp. 72-102). London: Academic Press.

Cameron, H. M., & Metcalfe, J. S. (1987). On the Economics of Technological Substitution. *Technological Forecasting and Social Change*, 32, 147-162.

Campbell, J. L., Hollingsworth, J. R., & Lindberg, L. N. (Eds.). (1991). *Governance of the American Economy.* Cambridge: Cambridge University Press.

Canarelli, P. (1994). *Exploratory study of the technologies to lessen urban transport drawbacks* (Research Report EUR 15947 EN). Sevilla: Joint Research Centre, IPTS.

Canarelli, P. (1995a). Analyzing the Past and Managing the Future Using Neural Networks. *Futures*, 27, 325-338.

Canarelli, P. (1995b). *Use of Genetic Algorithms as a Tool for Prospective Studies with Application to the Assessment of the Future of Urban Transport* (Research Report EUR 16296 EN). Sevilla: Joint Research Centre, IPTS.

Canarelli, P. (1996). *Fuzzy Scenarios of the Future of Congestion in Urban Centres* (Research Report EUR 16444 EN). Sevilla: Joint Research Centre, IPTS.

Carlsson, B., & Stankiewicz, R. (1991). On the nature, function and composition of technological systems. *Journal of Evolutionary Economics*, 1(1), 93-118.

CEC (Ed.). (1994). *Attitudes of the Electric Utilities towards CHP in the EC Member States*. Brussels: European Commission (CEC).

CEC (Ed.). (1995). *An Energy Policy for the European Union. White Paper*. Brussels: European Commission (CEC).

Chadburn, R. (1994). *Opportunities for Combined Heat and Power in Europe*. London: MDIS Publications.

CHPA, PWK, & COGEN (Eds.). (1995). *The Barriers to Combined Heat and Power in Europe*. Brussels: COGEN Europe.

Clark, N. (1991). Organisation and Information in the Evolution of Economic Systems. In P. P. Saviotti & J. S. Metcalfe (Eds.), *Evolutionary Theories of Economic and Technological Change. Present Status and Future Prospects* (pp. 88-107). Chur: Harwood Academic Publishers.

Clark, N., & Juma, C. (1987). *Long-Run Economics. An Evolutionary Approach to Economic Growth*. London: Pinter.

Collinson, S., & Molina, A. (1995). *Reorganising for Knowledge Integration and Constituency-Building in the Era of Multimedia: Product Development at Sony and Philips*. Proceedings of the Third International ASEAT-Conference "Managing Technology into the 21st Century", Manchester, 6-8 September 1995.

Conrads, F. (1992). Zur CO2-Vermeidung durch Kraft-Wärme-Kopplung. Energiepolitische Metamorphose: Von der Ölablösung zur Klimaschonung. *Energiewirtschaftliche Tagesfragen*, 42(12), 816-821.

Coombs, R., Green, K., Richards, V., & Walsh, V. (Eds.). (1998). *Technological Change and Organization*. Aldershot: Edward Elgar.

Coombs, R., Savioti, P., & Walsh, V. (1987). *Economics and Technological Change*. Houndsmill: Macmillan.

Coombs, R., Saviotti, P., & Walsh, V. (1992). Technology and the firm: the convergence of economic and sociological approaches? In R. Coombs, P. Saviotti, & V. Walsh (Eds.), *Technological Change and Company Strategies: Economic and sociological perspectives* (pp. 1-24). London: Academic Press.

Coutard, O. (1994). Economics of Grid Systems in Reconfiguration: Competition in the Electricity Supply Industry. In J. Summerton (Ed.), *Changing Large Technical Systems* (pp. 163-192). Boulder/CO: Westview.

Cyert, R. M., & March, J. G. (1963). *A Behavioral Theory of the Firm*. Eaglewood Cliffs (NJ): Prentice Hall.

David, P. (1975). *Technical Choice, Innovation and Economic Growth*. Cambridge: Cambridge University Press.

David, P. (1992). Heroes, Herds and Hysteresis in Technological History: Thomas Edison and 'The Battle of the Systems' Reconsidered. *Industrial and Corporate Change*, 1(1).

David, P., & Bunn, J. (1988). The Economics of Gateway Technologies and Network Evolution. *Information Economics and Policy*, 3, 165-202.

David, P. A. (1985). Clio and the Economics of QWERTY. *American Economic Review*, 75(2), 332-337.

Davies, S. (1979). *The Diffusion of Process Innovations*. Cambridge: Cambridge University Press.

Davison, P., & al. (1997). *Appraisal of Impact Evaluation Methods* (Research Report of the EC-funded project FANTASIE D 6). Harwell: ETSU.

DBDH (Ed.). (1982). *Facts about Danish District Heating*. Odense: Danish Board of District Heating (DBDH).

DEA (Ed.). (1993a). *District Heating in Denmark. Research and Technological Development*. Copenhagen: Danish Energy Agency (DEA).

DEA (Ed.). (1993b). *Small-scale Combined Heat and Power in Denmark*. Copenhagen: Danish Energy Agengy (DEA).

DeCanio, S. J. (1993). Barriers within firms to energy efficient investment. *Energy Policy*, 21(9), 906-914.

Dienhart, H., & Nitsch, J. (1995). Ökologische Bewertung der Kraft-Wärme-Kopplung. *Energiewirtschaftliche Tagesfragen*, 45(8), 500-505.

DoE (Ed.). (1979). *Combined heat and electrical power generation in the United Kingdom. Energy Papers No. 35*. London: HMSO/Dept. of Energy (DoE).

DoE (Ed.). (1984). *Combined Heat and Power District heating Feasibility Programme: Stage 1. Summary Report and Recommendation. Energy Papers No. 53*. London: HMSO/Dept. of Energy (DoE).

DoE (Ed.). (1988). *Industrial Combined Heat and Power: The Potential for New Users, Energy Efficiency Series No. 7*. London: HMSO/Dept. of Energy (DoE).

DoE (Ed.). (1990). *Co-generation and electricity production in British industry 1988, Energy efficiency Series No. 8*. London: HMSO/Dept. of Energy (DoE).

DoEnv (Ed.). (1990). *This Common Inheritance. Britain's Environmental Strategy White Paper. Dept of Environment Command Paper No. CM 1200*. London: HMSO/Dept. of Environment (DoEnv).

DoEnv (Ed.). (1991). *This common inheritance - the first year report, Dept. of Environment Command Paper No. CM 1655*. London: HMSO/Dept. of Environment (DoEnv).

DoEnv (Ed.). (1996). *A UK Strategy for Combined Heat and Power*. London: HMSO/Dept. of Environment (DoEnv).

Dosi, G. (1982). Technological Paradigms and Technological Trajectories: A Suggested Interpretation of the Determinants and Directions of Technological Chagne. *Research Policy*, 11, 147-162.

Dosi, G. (1988a). The nature of the innovative process. In G. Dosi & al. (Eds.), *Technical Change and Economic Theory* (pp. 221-238). London: Pinter.

Dosi, G. (1988b). Preface to Part III: How well does established theory work. In G. Dosi & al. (Eds.), *Technical Change and Economic Theory* (pp. 120-123). London: Pinter.

Dosi, G. (1988c). Sources, procedures and microeconomic effects of innovations. *Journal of Economic Literature*, 26, 1120-1171.

Dosi, G. (1991a). The Research on Innovation Diffusion: An Assessment. In N. Nakicenovic & A. Grübler (Eds.), *Diffusion of Technologies and Social Behaviour* (pp. 179-208). Berlin: Springer.

Dosi, G. (1991b). Some thoughts on the promises, challenges and dangers of an "evolutionary perspective" in ecnomics. *Journal of Evolutionary Economics*, 1(1), 5-7.

Downs, A. (1957). *An Economic Theory of Democracy*. New York: Harper & Row.

Dror, I. (1988). Forecasting Technologies Within Their Socioeconomic Framework. *Technological Forecasting and Social Change*, 34(1), 69-80.

Druwe, U. (1989). Rekonstruktion der "Theorie der Autopoiese" als Gesellschafts- und Steuerungsmodell. In A. Görlitz (Ed.), *Politische Steuerung sozialer Systeme* . Pfaffenweiler: Centaurus.

DTI (Ed.). (1996). *Digest of United Kingdom Energy Statistics 1996*. London: HMSO/Department of Trade and Industry (DTI).

DTI (Ed.). (1997). *Digest of United Kingdom Energy Statistics 1997*. London: HMSO/Department of Trade and Industry (DTI).

Dufait, N., Martens, A., Guilmot, J. F., & Verbruggen, A. (1992). *Industrial Cogeneration in Europe* (Research Report ENE.RB9208). Mol: Vlaamse Instelling voor Technologisch Onderzoek (VITO).

Dufait, N., & Verbruggen, A. (1993). *Attitudes of the Electric Utilities Towards CHP in the EC-Member States*. Proceedings of the "PowerGen Europe '93 Conference", Paris, 25-27 Mai 1993.

Easton, D. (1965). *A Systems Analysis of Political Life*. New York: J. Wiley & Sons.

Eden, R. J. (1993). World energy to 2050: outline scenarios for energy and electricity. *Energy Policy*, 21(3), 231-237.

Edquist, C. (Ed.). (1997). *Systems of Innvoation. Technology, Institutions and Organizations*. London: Pinter.

EEO (Ed.). (1990). *Cogeneration and electricity production in British industry 1988. Energy Efficiency Series No. 12*. London: HMSO.

EEO (Ed.). (1992a). *The Application of Combined Heat and Power in the UK Health Service*. London: Energy Efficiency Office (EEO).

EEO (Ed.). (1992b). *Introduction to Large-Scale Combined Heat and Power. Good Practice Guide No. 43*. London: Energy Efficiency Office (EEO).

Elzen, B., & Mulder, K. (1995). *What can Technology Assessment (not) learn from Technology Studies? An Analysis using Case Studies on Electric Vehicles and Biodegradable Plastics*. Proceedings of the Third International ASEAT-Conference "Managing Technology into the 21st Century", Manchester, 6-8 September 1995.

ENERDATA (Ed.). (1996). *CD-ROM ENERDATA. World energy database*. Grenoble: ENERDATA.

Energiereferat (Ed.). (1995). *Richtpreisübersicht Motor-Heizkraftanlagen. Leistungsbereich 4.5- 2400 kWel.* Frankfurt a.M.: Energiereferat des Dezernats für Umwelt, Energie und Brandschutz der Stadt Frankfurt a.M.

Englmann, F. C. (1989). *Technischer Fortschritt: Diffusion, Erträge und Beschäftigung*. Tübingen: Mohr.

Enquete-Kommission (Ed.). (1995). *Mehr Zukunft für die Erde. Nachhaltige Energiepolitik für dauerhaften Klimaschutz. Schlußbericht der Enquete-Kommission "Schutz der Erdatmosphäre" des12. Deutschen Bundestages*. Bonn: Economica.

Euler, H. (1996). *Einspeisevergütung für Blockheizkraftwerke: Brauchen wir eine neue Regelung?* Proceedings of the workshop "Blockheizkraftwerke - Option für die Energie- und Umweltpolitik", Bonn, 6 March 1996.

EUROSTAT (Ed.). (1997). *Energiepreise. Eine Übersicht über die Verbraucherpreise für Gas Elektrizität und Erdölprodukte in der Europäischen Union bis 1996*. Luxembourg: EUROSTAT.

Fenzl, J., & Pummer, B. (1993). *Evaluation of the market potential of micro CHP-units in Germany*. Regensburg: Zweckverband regionale Entwicklung.

FGBHKW (Ed.). (1995). *Der "City-Gate-Ansatz". Ein Vorschlag der Fördergemeinschaft "Blockheizkraftwerke" zur angemessenen Vergütung von Stromeinspeisungen*. Bonn: Fördergemeinschaft Blockheizkraftwerke (FGBHKW).

Foray, D., & Grübler, A. (1990). Morphological analysis, diffusion and lock-out of technologies: Ferrous casting in France and the FRG. *Research Policy*, 19, 535-550.

Foster, N. (1994). *Government support for CHP*. Proceedings of the conference on "Small-scale CHP", London, 22/23 November 1994.

Foster, V., Cooper, N., & McCulloch, N. (1994). *Guide to the Environmental Regulation of the Water and Electricity Industries*. Oxford: OXERA.

Frances, J., & al. (1991). Introduction. In G. Thompson & al. (Eds.), *Markets, Hierarchies and Networks. The Coordination of Social Life* (pp. 1-20). London: Sage.

Frances, J., Levacic, R., Mitchell, J., & Thompson, G. (1991). Introduction. In J. Thompson, J. Frances, R. Levacic, & J. Mitchell (Eds.), *Markets, Hierarchies and Networks. The Coordination of Social Life* (pp. 1-20). London: Sage.

Francis, J. G. (1993). *The Politics of Regulation. A Comparative Perspective.* Oxford: Blackwell.

Freeman, C. (1988). Introduction. In G. Dosi & al. (Eds.), *Technical Change and Economic Theory* (pp. 1-8). London: Pinter.

Freeman, C. (1991). Networks of Innovators: A Synthesis of Research Issues. *Research Policy*, 20, 499-514.

Freeman, C., & Perez, C. (1988). Structural Crisis of Adjustment, Business Cycles and Investment Behaviour. In G. Dosi & al. (Eds.), *Technical Change and Economic Theory* (pp. 38-66). London: Pinter.

Fulton, K. (1994). Small radial designs promise low emissions and low costs. *Gas Turbine World*, 20(7), 24-26.

Georghiou, L., & al. (1986). *Post-Innovation Performance: Technological Development and Competition*. London: MacMillan.

Glagow, M., & Willke, H. (Eds.). (1987). *Dezentrale Gesellschaftssteuerung. Probleme der Integration polyzentrischer Gesellschaft.* Pfaffenweiler: Centaurus.

Godet, M. (1993). *From Anticipation to Action. A Handbook of Strategic Prospective*. Paris: UNESCO.

Goldthorpe, J. H. (Ed.). (1984). *Order and Conflict in Contemporary Capitalism. Studies in the Political Economy of Western European Nations*. New York: Oxford University Press.

Goodland, R., & Ledec, G. (1987). Neoclassical Economics and Sustainable Development. *Ecological Modelling*, 38(1), 19-46.

Görlitz, A. (Ed.). (1989). *Politische Steuerung sozialer Systeme.* Pfaffenweiler: Centaurus.

Görlitz, A. (Ed.). (1994). *Umweltpolitische Steuerung*. Baden-Baden: Nomos.

Görlitz, A. (1995). *Politische Steuerung. Ein Studienbuch. Unter Mitarbeit von H.-P. Burth u. Ulrich Druwe*. Opladen: Leske & Budrich.

Görlitz, A., & Druwe, U. (Eds.). (1990). *Politische Steuerung und Systemumwelt*. Pfaffenweiler: Centaurus.

Gort, M., & Klepper, S. (1982). Time Paths in the Diffusion of Product Innovations. *The Economic Journal*, 92(July), 630-653.

Grohnheit, P. E. (1993). Modelling CHP within a national power system. *Energy Policy*, 21(4), 418-429.

Grohnheit, P. E., & Verbruggen, A. (1990). *Cogeneration and the Internal Market for Electricity*. Proceedings of the conference of the International Association for Energy Economics "Integrated Energy Markets and Energy Systems-Lessons and Perspectives", Copenhagen, 28-30 June 1990.

Gröner, H., & Smeets, H.-D. (1988). Regulierung der leitungsgebundenen Energiewirtschaft. In M. Krakowski (Ed.), *Regulierung in der Bundesrepublik Deutschland. Die Ausnahmebereiche des Gesetzes gegen Wettbewerbsbeschränkungen* (pp. 117-194). Hamburg: Verlag Weltarchiv.

Grubb, M., Walker, J., & al. (1992). *Emerging Energy Technologies. Impacts and Policy Implications*. Dartmouth: Royal Institute for International Affairs (RIIA).

Gruber, E. (1992). Efficiency of Energy Conservation Programmes in European Countries. *Energy and Environment*, 3(2), 122-132.

Gruber, E., Mannsbart, W., & Walz, R. (1993). *Institutional Obstacles to a Local Energy Policy and its Management in the Context of the Internal Energy Market* (Research Report CEC 4.1040/E/92-02). Karlsruhe: Fraunhofer-Institut für Systemtechnik und Innovationsforschung (ISI).

Grübler, A. (1990). *The Rise and Fall of Infrastructures. Dynamics of Evolution and Technological Change in Transport*. Heidelberg: Springer.

Grübler, A., & Nakicenovic, N. (1987). *The Dynamic Evolution of Methane Technology* (Research Report WP-87-2). Laxenburg: IIASA.

Grupp, H. (Ed.). (1987). *Problems of Measuring Technological Change. Seminar Report*. Köln: Verlag TÜV Rheinland.

Grupp, H. (1994). The measurement of technical performance of innovations by technometrics and its impact on established technology indicators. *Research Policy*, 23, 175-193.

Hagedoorn, J., & Schakenraad, J. (1990). Inter-firm partnerships adn co-operative strategies in core technologies. In C. Freeman & L. Soete (Eds.), *New Explorations in the Economics of Technical Change* (pp. 3-37). London: Pinter.

Haken, H. (1983). *Synergetics. An Introduction. Nonequilibrium Phase Transitions and Self-Organization in Physics, Chemistry and Biology*. Berlin: Springer.

Hancher, L. (1995). Die Almelo-Entscheidung des Europäischen Gerichtshofs. Hintergrund, Bewertung und mögliche Auswirkungen. *Energiewirtschaftliche Tagesfragen*, 45(1/2), 68-76.

Hård, M. (1991). *The Diffusion of Cogeneration: Technology Commanding and Organizational Culture*. Proceedings of the "International Workshop on Constructive Technology Assessment", University of Twente, 20-22 September 1991.

Hård, M., & Olsson, S. O. (1995). Reluctant Marriage: How District Heating and Electricity Systems Have Been Combined. In A. Kaijser (Ed.), *Nordic Energy Systems: Historical Perspectives and Current Issues* . Canton (MA): Science History Publications.

Hartmann, J. (1985). *Verbände in der westlichen Industriegesellscahft. Ein international vergleichendes Handbuch.* Frankfurt/New York: Campus.

Hartung, R. (1995). Zukunftsstrategien kommunaler EVU. Entwicklung neuer Produkte und Dienstleistungen. *Energie Spektrum*, 9(1), 10-17.

Henderson, R., & Clark, K. (1990). Architectural Innovation: The Reconfiguration of Existing Product Technologies and the Failure of the Established Firm. *Administrative Science Quarterly*, 35, 9-30.

Henschel, C., & Wiedemann, P. W. (1993). Facing the Greenhouse Effect. Communication about Energy in Germany. *Technology in Society*, 15(3), 301-310.

Hippel, E. v. (1988). *The Sources of Innovation.* Oxford: Oxford University Press.

HMU (Ed.). (1996). *Datenbank Blockheizkraftwerke. Stand 1995.* Wiesbaden: Hessisches Ministerium für Umwelt, Energie, Jugend , Familie und Gesundheit (HMU).

HoC (Ed.). (1991). *House of Commons Energy Select Committee's Third Report "Energy Efficiency", Report No. HC 91, 90/91.* London: HMSO.

HoC (Ed.). (1992). *House of Commons Energy Commmittee's Second Report "Consequences of Electricity Privatisation".* London: HMSO.

HoC (Ed.). (1993). *House of Commons Environment Committee's 4th Report "Energy Efficiency in Buildings".* London: HMSO.

Hodgson, G. M. (1988). *Economics and Institutions. A Manifesto for a Modern Institutional Economics.* Oxford: Polity Press.

Hodgson, G. M. (1993). *Economics and Evolution. Bringing Life Back into Economics.* Oxford: Polity Press.

Hollingsworth, J. R., Schmitter, P. C., & Streeck, W. (Eds.). (1994). *Governing Capitalist Economies. Performance and Control of Economic Sectors.* New York/Oxford: Oxford University Press.

Holst, K.-E. (1995). Entwicklung der Erdgasversorgung in den neuen Bundesländern. *Energiewirtschaftliche Tagesfragen*, 45(9), 563-567.

Horlock, J. H. (1987). *Cogeneration - Combined Heat and Power (CHP). Thermodynamics and Economics.* Oxford: Pergamon Press.

Horst, H. (1988). Regulierung, Deregulierung und Innovation. In P. J. J. Welfens & L. Balcerowicz (Eds.), *Innovationsdynamik im Systemvergleich. Theorie und Praxis unternehmerischer, gesamtwirtschaftlihcer und politischer Neuerung* (pp. 141-158). Heidelberg: Physica.

Hourcade, J.-C. (1993). Modelling long-run scenarios: methodology lessons from a prospective study on a low CO2 intensive country. *Energy Policy*, 21(3), 309-326.

Hughes, T. P. (1983). *Networks of Power. Electrification in Western Industry, 1880-1930*. Baltimore: John Hopkins University Press.

Hughes, T. P. (1987). The Evolution of Large Technological Systems. In W. Bijker, T. P. Hughes, & T. Pinch (Eds.), *The Social Consruction of Technological Systems. New Directions in the Sociology and History of Technology* (pp. 51-82). Cambridge (MA): MIT Press.

ICIE/OPET (Ed.). (1992). *Small scale cogeneration in non residential buildings*. Brussels: European Commission (CEC).

IEA/Novem (Ed.). (1993). *IEA District Heating. Promotional Manual for District Energy Systems*. Paris/Sittard: International Energy Agency (IEA)/Novem.

Imai, K., & Baba, Y. (1991). Systemic Innovation and Cross-Border Networks: Transcending Markets and Hierarchies to Create a New Techno-Economic System. In OECD (Ed.), *Technology and Productivity: The Challenges for Economic Policy* . Paris: OECD.

Irwin, A., & Vergragt, P. (1989). Re-thinking the Relationship between Environmental Regulation and Industrial Innovation: The Social Negotiation of Technical Change. *Technology Analysis & Strategic Management*, 1(1), 57-69.

ISP (Ed.). (1994). *Vergleich der Strom- und Heizenergieerzeugung in gekoppelten und ungekoppelten Anlagen vor dem Hintergrund der Einsparmöglichkeiten durch Wärmedämmung*. Hannover: Eduard Pestel Institut für Systemforschung (ISP).

IVE (Ed.). (1993). *A Review of Cogeneration Equipment and Selected Installations in Europe*. Valencia: Institut Valencia de l'Energia (IVE).

Janes, M. C. (1996). A Review of the Development of Technology Assessment. *International Journal of Technology Management*, 11(5/6), 507-522.

Jansen, D. (1991). Policy Networks and Change: The Case of High-Tc Superconductors. In B. Marin & M. R. (Eds.), *Policy Networks. Empirical Evidence and Theoretical Considerations* (pp. 137-175). Frankfurt/Boulder: Campus/Westview.

Jarass, H. D. (Ed.). (1995). *Bundes-Immissionsschutzgesetz mit Durchführungsverordnungen, TA Luft und TA Lärm*. München: Beck.

Jasanoff, S., Markle, G. E., Petersen, J. C., & Pinch, T. (Eds.). (1995). *Handbook of Science and Technology Studies*. Thousand Oaks/London: Sage.

Jochem, E. (1991). Volkswirtschaftliche und energiewirtschaftliche Aspekte der KWK-Anwendung. In K.-H. Suttor & W. Suttor (Eds.), *Handbuch Kraft-Wärme-Kopplung. Für Planer, Betreiber, Ausbildung* (pp. 143-169). Karlsruhe: C.F. Müller.

Jochem, E., & al. (1986). *Zum Einfluß technisch-wirtschaftlicher Rahmenbedingungen auf die Anwendungspotentiale neuer Technologien Rationeller Energienutzung - Ein internationaler Vergleich am Beispiel der Energieumwandlung für Strom und Wärme* (Research Report BMFT-FB-T 86-063). Karlsruhe: Fraunhofer-Institut für Systemtechnik und Innovationsforschung (ISI).

Jochem, E., & Gruber, E. (1990). Obstacles to rational electricity use and measures to alleviate them. *Energy Policy*, 18(5), 340-350.

Joerges, B. (1988). Large technical systems: Concepts and issues. In R. Maynz & T. P. Hughes (Eds.), *The development of large technical systems* (pp. 9-36). Frankfurt a.M.: Campus.

Katz, M., & Shapiro, C. (1994). Systems Competition and Network Effects. *Journal of Economic Perspectives*, 8(2), 93-115.

Katz, M. L., & Shapiro, C. (1985). Network Externalities, Competition, and Compatibility. *The American Economic Review*, 75(3), 424-440.

Katz, M. L., & Shapiro, C. (1986). Technology Adoption in the Presence of Network Externalities. *Journal of Political Economy*, 94(4), 822-841.

Keck, O. (1984). *Der Schnelle Brüter. Eine Fallstudie über Entscheidungsprozesse in der Großtechnik*. Frankfurt a.M.: Campus.

Keck, O. (1986). Die gesellschaftliche Steuerung der Technik - ein institutioneller Ansatz. In G. Bechmann & F. Meyer-Krahmer (Eds.), *Technologiepolitik und Sozialwissenschaft* (pp. 17-41). Frankfurt a.M.: Campus.

Kemp, R., Miles, I., Smith, K., & al. (1994). *Technology and the Transition to Environmental Stability. Continuity and Change in Technological Systems. Final Report from project "Technological Paradigms and Transition Paths: The Case of Energy Technologies"* (Research Report PL910282). Maastricht: MERIT.

Kemp, R., Schot, J., & Hoogma, R. (1996). *Regime Shifts through Processes of Niche Formation. The Case of Transport*. Proceedings of the "Greening of Industry Conference", Heidelberg, 24-27 November 1996.

Kemp, R., & Soete, L. (1990). Inside the 'green box': on the economics of technological change and the environment. In C. Freeman & L. Soete (Eds.), *New Explorations in the Economics of Technical Change* (pp. 245-258). London: Pinter.

Kemp, R., & Soete, L. (1992). The Greening of Technological Progress. An Evolutionary Perspective. *Futures*, 24, 437-457.

Kemp, R. P. M., Olsthoorn, A. A., Oosterhuis, F. H., & Verbruggen, H. (1991). *An Economic Analysis of Cleaner Technology: Theory and Evidence*. Proceedings of the "Greening of Industry Conference", Noordwijk aan Zee, 17-19 November 1991.

Kenis, P. (1991). The Preconditions for Policy Networks: Some Findings fro a Three-Country Studyon Industrial Restructuring. In B. Marin & M. R. (Eds.), *Policy Networks. Empirical Evidence and Theoretical Considerations* (pp. 297-330). Frankfurt/Boulder: Campus/Westview.

Kitschelt, H. (1983). *Politik und Energie. Energie-Technologiepolitiken in den USA, der Bundesrepublik Deutschland, Frankreich und Schweden*. Frankfurt/New York: Campus.

Klein, J. (1983). *Investitionsverhalten bei Prozeßinvestitionen. Ein Beitrag zur mikroökonomischen Diffusionsforschung*. Frankfurt/NewYork: Campus.

Kline, S. J., & Rosenberg, N. (1989). An overview of innovation. In R. Landau & N. Rosenberg (Eds.), *The Positive Sum Strategy. Harnessing Technology for Economic Growth* (pp. 275-306). Washington: National Academy Press.

Knie, A. (1994). Gemachte Technik. Zur Bedeutung von "Fahnenträgern", "Promotoren" und "Definitionsmacht" in der Technikgenese. In W. Rammert & G. Bechmann (Eds.), *Technik und Gesellschaft. Jahrbuch 7* (pp. 41-66). Frankfurt/NewYork: Campus.

Kohler-Koch, B. (Ed.). (1989). *Regime in den internationalen Beziehungen*. Baden-Baden: Nomos.

Krasner, S. D. (1982). Structural Causes and Regime Consequences: Regimes as Intervening Variables. *International Organisation*, 36(2), 185-205.

Krause, F., Koomey, J., & Olivier, D. (1995). *Energy Policy in the Greenhouse, Vol. 2/3D: Renewable Power. The Cost and Potential of Low-Carbon Resource Options in Western Europe*. El Cerrito (CA): International Project for Sustainable Energy Paths (IPSEP).

Krupp, H. (1992). *Energy politics and Schumpeter Dynamics. Japan's Policy Between Short-Term Wealth and Long-Term Global Welfare*. Tokyo: Springer.

Kuhn, T. (1970). *The Structure of Scientific Revolutions*. (2nd edition ed.). Chicago: The University of Chicago Press.

Law, J., & Callon, M. (1988). Engineering and sociology in a military aircrft project: A network analysis of technological change. *Social Problems*, 35(3), 284-297.

Lehmbruch, G. (1984). Concertation and the Structure of Corporatist Networks. In J. H. Goldthorpe (Ed.), *Order and Conflict in Contemporary Capitalism. Studies in the Political Economy of Western European Nations* (pp. 60-80). New York: Oxford University Press.

Lehner, F. (1981). *Einführung in die Neue Politische Ökonomie*. Königstein/Ts.: Athenäum.

Leyten, J., & Smits, R. (1996). The role of technology assessment in technology policy. *International Journal of Technology Assessment*, 11(5/6), 688-702.

Lezuo, A., Knappstein, H., Langnickel, U., & Nymoen, H. (1995). Brennstoffzelleneinsatz im Mehr-MW-Heizkraftwerk. *Energiewirtschaftliche Tagesfragen*, 45(9), 585-590.

Liso, N. d., & Metcalfe, J. S. (1994). *On Technological Systems and Technological Paradigms. Some Recent Developments in the Understanding of Technological Change*. Proceedings of the Fifth Conference of the International Joseph A. Schumpeter Society on "Economic Dynamism: Analysis and Policy", Münster, 17-20 August 1994.

Lotka, A. (1924). *Elements of Physical Biology*. New York: Dover Publications.

Lucas, K., Rebstock, F., Roth, H., & Solfrian, W. (1992). *Analyse der Hemmnisse für die Nutzung industrieller Abwärme zur Fernwärmeversorgung*. Duisburg: Research Report, Institut für Umwelttechnologie und Umweltanalytik.

Luhmann, N. (1984). *Soziale Systeme: Grundriß einer allgemeinen Theorie*. Frankfurt: Suhrkamp.

Luhmann, N. (1988). *Die Wirtschaft der Gesellschaft*. Frankfurt: Suhrkamp.

Lundvall, B. A. (1988). Innovation as an interactive process: from user-producer interaction to the national system of innovation. In G. Dosi & al. (Eds.), *Technical Change and Economic Theory* (pp. 349-369). London: Pinter.

Lundvall, B. A. (1992a). Introduction. In B. A. Lundvall (Ed.), *National Systems of Innovation. Towards as Theory of Innovation and Interactive Learning* (pp. 1-22). London: Pinter.

Lundvall, B. A. (Ed.). (1992b). *National Systems of Innovation. Towards as Theory of Innovation and Interactive Learning*. London: Pinter.

Lüschen, H., Sonntag, J., & Werner, R. (1995). Least-Cost Planning im Dienstleistungskonzept der EVS. *Energiewirtschaftliche Tagesfragen*, 45(3), 128-133.

Lutzenhiser, L. (1994). Barriers to energy efficiency in the US housing industry. *Energy Policy*, 22(10), 867-876.

MacKenzie, D. (1987). Missile Accuracy: A Case Study in the Social Processes of Technological Change. In W. Bijker, T. P. Hughes, & T. Pinch (Eds.), *The Social Construction of Technological Systems. New Directions in the Sociology and History of Technology* (pp. 195-222). Cambridge (MA): MIT Press.

MacKenzie, D. (1992). Economic and sociological explanation of technical change. In R. Coombs, P. P. Saviotti, & V. Walsh (Eds.), *Technological change and company strategy. Economic and sociological perspectives* (pp. 25-48). London: Academic Press.

Mahajan, V., Muller, E., & Bass, F. M. (1990). New Product Diffusion Models in Marketing: A Review and Direction for Research. *Journal of Marketing*, 54(January), 1-26.

Mahajan, V., & Peterson, R. A. (1985). *Models for Innovation Diffusion*. Beverly Hills: Sage.

Mansfield, E. (1961). Technical change and the rate of imitation. *Econometrica*, 29, 741-766.

Mansfield, E. (1968). *Industrial Research and Technological Innovation. An Econometric Analysis*. New York: W.W. Norton.

Marchetti, C., & Nakicenovic, N. (1979). *The dynamics of Energy Systems and the Logistic Substitution Model* (Research Report RR-79-13). Laxenburg: IIASA Laxenburg.

Marin, B., & Mayntz, R. (1991a). Introduction: Studying Policy Networks. In B. Marin & R. Mayntz (Eds.), *Policy Networks. Empirical Evidence and Theoretical Considerations* (pp. 11-24). Frankfurt/Boulder (CO): Campus/Westview.

Marin, B., & Mayntz, R. (Eds.). (1991b). *Policy Networks. Empirical Evidence and Theoretical Considerations*. Frankfurt/Boulder (CO): Campus/Westview.

Marvin, S. (1991a). Combined Heat and Power/District Heating: A Review. *Planning Practice and Research*, 6(2), 31-35.

Marvin, S. J. (1991b). *Localisation of Labour Party CHP/DH Policy: 1977-1987*. Unpublished PhD-Thesis, Open University, Milton Keynes.

Maturana, H. R., & Varela, F. J. (1980). *Autopoiesis and cognition. The Realization of the Living*. Dordrecht: Reidel.

Mayntz, R. (1987). Politische Steuerung und gesellschaftliche Steuerungsprobleme - Anmerkungen zu einem theoretischen Paradigma. In T. Ellwein & al. (Eds.), *Jahrbuch zur Staats- und Verwaltungswissenschaft, Band, 1/1987* (pp. 89-110). Baden-Baden: Nomos.

Mayntz, R., & Hughes, T. P. (Eds.). (1988). *The development of large technical systems*. Frankfurt a.M.: Campus.

Mayntz, R., & Scharpf, F. W. (1995). Steuerung und Selbstorganisation in staatsnahen Sektoren. In R. Mayntz & F. W. Scharpf (Eds.), *Gesellschaftliche Selbstregelung und politische Steuerung* (pp. 9-38). Frankfurt/New York: Campus/Westview.

Mayntz, R., & Schneider, V. (1995). Die Entwicklung technischer Infrastruktursysteme zwischen Steuerung und Selbstorganisation. In R. Mayntz & F. W. Scharpf (Eds.), *Gesellschaftliche Selbstregelung und politische Steuerung* (pp. 73-100). Frankfurt/New York: Campus/Westview.

Meckel, B. (1996). *National Perspectives and Market Prospects Germany*. Proceedings of the COGEN Europe conference on "Europe's Cogeneration Markets of the Future: Prospects and Challenges", Brussels, 10-11 October 1996.

Meixner, H. (1995). Neue Strategien für Kraft-Wärme-Kopplung in Hessen. In HMU (Ed.), *Kraft-Wärme-Kopplung in Hessen. Konzepte, Anlagen, Erfahrungen, Strategien und Perspektiven* (pp. 113-137). Wiesbaden: Hessisches Ministerium für Umwelt, Energie, Jugend, Familie und Gesundheit (HMU).

Menzler, K. (1996). Stand der Stirlingmotorentwicklung in der BRD : Personal communication: 19 February 1996.

Merton, R. K. (1973). *The Sociology of Science. Theoretical and Empirical Investigations*. Chicago: The University of Chicago Press.

Metcalfe, J. S. (1981). Impulse and Diffusion in the Study of Technical Change. *Futures*, 13, 347-359.

Metcalfe, J. S., & Boden, M. (1992). Evolutionary epistemology and the nature of technology strategy. In R. Coombs, P. P. Saviotti, & V. Walsh (Eds.), *Technological change and company strategy. Economic and sociological perspectives* (pp. 49-71). London: Academic Press.

Meyer-Krahmer, F. (1989). *Der Einfluß staatlicher Technologiepolitik auf industrielle Innovationen*. Baden-Baden: Nomos.

Midttun, A., & Baumgartner, T. (1986). Negotiating energy futures: the politics of energy forecasting. *Energy Policy*, 14(3), 219-241.

Mokyr, J. (1990). *The Lever of Riches*. Oxford: Oxford University Press.

Molina, A. (1993). In Search of Insights into the Generation of Techno-Economic Trends: Micro- and Macro-Constituencies in the Microprocessor Industry. *Research Policy*, 22, 479-506.

Moncada Paterno Castello, P. (1996). Incentives for Innovation in Electric Power Utilities. *The IPTS-Report*, 1(4), 13-17.

Muders, H. (1994). Entwicklung der Blockheizkraftwerke in den neuen Bundesländern, Stand 1994. *Elektrizitätswirtschaft*, 93(14), 772-774.

Müller, J., & Vogelsang, I. (1979). *Staatliche Regulierung. Regulated Industries in den USA und Gemeinwohlbindung in wettbewerblichen Ausnahmebereichen in der Bundesrepublik Deutschland*. Baden-Baden: Nomos.

Nadiri, M. I. (1987). Joint Production. In J. Eatwell, M. Milgate, & P. Newman (Eds.), *The New Palgrave. A Dictionary of Economics* (Vol. 2, pp. 1028-1030). London/Basingstoke: Macmillan.

Nagel, K., Rasmussen, S., & Barrett, C. L. (1997). Network Traffic as a Self-Organised Critical Phenomenon. In F. Schweitzer (Ed.), *Self-Organization of Complex Structures. From Individual to Collective Dynamics* (pp. 579-592). Amsterdam: Gordon and Breach.

Nakicenovic, N. (1986). The Automobile Road to Technological Change. Diffusion of the Automobile as a Process of Technological Substitution. *Technological Forecasting and Social Change*, 29, 309-340.

Nakicenovic, N. (1991). Diffusion of Pervasive Systems. A Case of Transport Infrastructure. *Technological Forecasting and Social Change*, 39, 181-200.

Nakicenovic, N., & al. (1993). Long-term strategies for mitigating global warming. Special issue. *Energy Policy*, 18(5).

Nakicenovic, N., & John, A. (1991). CO2 Reduction and Removal: Measures for the Next Century. *Energy Policy*, 16(11/12), 1347-1377.

Nelson, R. R. (1987). *Understanding Technological Change as an Evolutionary Process*. Amsterdam: North Holland.

Nelson, R. R. (Ed.). (1993). *National Innovation Systems. A Comparative Analysis*. New York: Oxford University Press.

Nelson, R. R. (1995). Recent Evolutionary Theorizing About Economic Change. *Journal of Economic Literature*, 33, 48-90.

Nelson, R. R., & Winter, S. G. (1982). *An Evolutionary Theory of Economic Change*. Cambridge/MA: Harvard University Press.

Neuffer, H., & Witterhold, F. G. (1995). Kostensenkung durch Fortschritte im Fernwärmeleitungsbau. Ergebnisse aus F&E-Vorhaben in Deutschland. In AGFW (Ed.), *Aus Forschung und Entwicklung. Mitteilungen der Forschungsstelle und des Forschungsbeirates der AGFW*. Frankfurt a.M.: Arbeitsgemeinschaft Fernwärme (AGFW).

Niosi, J., & al. (1993). National Systems of Innovation: In Search of a Workable Concept. *Technology in Society*, 15(2), 207-228.

Nitsch, J. (1994). *Wirtschaftliches und ausschöpfbares Potential der Kraft-Wärme-Kopplung in Baden-Württemberg, Untersuchung im Auftrag des Wirtschaftsministeriums Baden Württemberg*. Stuttgart: Deutsche Forschungsanstalt für Luft- und Raumfahrt (DLR-STB).

OECD/IEA (Ed.). (1983). *District heating and combined heat and power systems. A technology review*. Paris: OECD.

OECD/IEA (Ed.). (1994). *Combined Heat and Power Generation in IEA Countries*. Paris: OECD.

Olson, M. (1968). *Die Logik des kollektiven Handelns. Kollective Güter und die Theorie der Gruppen*. Tübingen: Mohr.

Pasinetti, L. L. (Ed.). (1980). *Essays on the Theory of Joint Production*. London: Macmillan.

Patel, P., & Pavitt, K. (1994). The Nature and Economic Importance of National Innovations Systems. *STI Review*(14), 9-32.

Pavitt, K. (1984). Sectoral patterns of technical change: Towards a taxonomy and a theory. *Research Policy*, 13, 343-373.

Pearce, D., & Warford, J. J. (1993). *World without End. Economics, Enviroment, and Sustainable Development*. New York: Oxford University Press.

Pfeifer, T., Fahl, U., & Voß, A. (1991). *Comparison of Centralized and Decentralized Energy Supply Systems*. Proceedings of the Second International Conference on Energy Consulting "Closing the Gaps: Efficiency - Inefficiency in East and West", Graz, September 1991.

Pinch, T. J., & Bijker, W. E. (1987). The Social Construction of Facts and Artifacts: Or How the Sociology of Science and the Sociology of Technology Might Benefit Each Other. In W. E. Bijker, T. P. Hughes, & T. J. Pinch (Eds.), *The Social Construction of Technological Systems. New Directions in the Sociology and History of Technology* (pp. 17-50). Cambridge (MA): MIT Press.

Przeworski, A., & Teune, H. (1970). *The Logic of Comparative Social Inquiry*. New York: John Wiley & Sons.

Rammert, W. (1988). *Das Innovationsdilemma. Technikentwicklung im Unternehmen*. Opladen: Westdeutscher Verlag.

Reddy, A. K. N. (1991). Barriers to improvements in energy efficiency. *Energy Policy*, 19(10), 953-961.

Rickards, T. (1985). *Stimulating Innovation. A Systems Approach*. London: Pinter.

Rip, A., & Kemp, R. (1996). Towards a Theory of Socio-Technical Change, *to be published in the book "Human Choice and Climate Change. An International Assessment"* . New York: Battelle Press.

Rip, A., Misa, T., & Schot, T. (Eds.). (1995). *Managing Technology in Society. The Approach of Constructive Technology Assessment*. London: Pinter.

Rogers, M. E. (1983). *Diffusion of innovations*. (revised 2nd ed.). New York: Free Press.

Rosenberg, N. (1982). *Inside the Black Box. Technology and Economics*. Cambridge: Cambridge University Press.

Ruddy, T. (1994). *German Utilities as Potential Partners*. Proceedings of the "Power-Gen Europe '94 Conference", Köln, 17-19 May 1994.

Rüdig, W. (1986). Energy conservation and electricity utilties: a comparative analysis of organizational obstacles to CHP/DH. *Energy Policy*, 14(No. 2), 104-116.

Rüdig, W. (1990). *Towards a "New" Political Science of Technology* (Strathclyde Papers on Government and Politics 63): University of Strathclyde, Dept. of Government.

Russell, S. (1986a). *The Political Shaping of Energy Technology: Combined Heat and Power in Britain*. Unpublished PhD-thesis, Aston University, Birmingham.

Russell, S. (1986b). The social construction of artefacts: A response to Pinch and Bijker. *Social Studies of Science*, 16, 331-346.

Russell, S. (1993). Writing energy history: explaning the neglect of CHP/DH in Britain. *British Journal for the History of Science*, 26(1), 33-54.

Russell, S. (1994). Combined Heat and Power in Britain. In M. Hård & S.-O. Olsson (Eds.), *The Combined Generation of Heat and Power in Great Britain and the Netherlands* . Stockholm: NUTEK.

Rydén, B., Johnsson, J., & Wene, C.-O. (1993). CHP production in integrated energy systems. Examples from five Swedish communities. *Energy Policy*, 21(2), 176-190.

Sahal, D. (1985). Technology Guide-Posts and Innovation Avenues. *Research Policy*, 14, 61-82.

Saviotti, P., & Trickett, A. (1992). The Evolution of Helicopter Technology, 1940-1986. *Economics of Innovation and New Technology*, 2(2), 111-130.

Saviotti, P. P., & Metcalfe, J. S. (1984). A theoretical approach to the construction of technological output indicators. *Research Policy*, 13, 141-151.

Scharpf, F. (1978). Die Theorie der Politikverflechtung: ein kurzgefaßter Leitfaden. In J. J. Hesse (Ed.), *Politikverflechtung im föderativen Staat. Studien zum Planungs- und Finanzverbund zwischen Bund, Ländern und Gemeinden* . Baden-Baden: Nomos.

Scharpf, F. W., & Mayntz, R. (1995). Steuerung und Selbstorganisation in staatsnahen Sektoren. In R. Mayntz & F. W. Scharpf (Eds.), *Gesellschaftliche Selbstregelung und politische Steuerung* (pp. 9-38). Frankfurt a.M.: Campus.

Schefold, B. (1987). Joint Production in Linear Models. In J. Eatwell, M. Milgate, & P. Newman (Eds.), *The New Palgrave. A Dictionary of Economics* (Vol. 2, pp. 1030-1034). London/Basingstoke: Macmillan.

Schimank, U. (1987). Evolution, Selbstreferenz udn Steuerung komplexer Systeme. In M. Glagow & H. Willke (Eds.), *Dezentrale Gesellschaftssteuerung: Probleme der Integration polyzentrischer Gesellschaften.* Pfaffenweiler: Centaurus.

Schmookler, J. (1966). *Invention and Economic Growth.* Cambridge (MA): Harvard University Press.

Schneider, V. (1989). *Technikentwicklung zwischen Politik und Markt: Der Fall Bildschirmtext.* Frankfurt: Campus.

Schneider, V., & Werle, R. (1991). Policy Networks in the German Telecommunications Domain. In B. Marin & R. Mayntz (Eds.), *Policy Networks. Empirical Evidence and Theoretical Considerations* (pp. 97-136). Frankfurt/Boulder (CO): Campus/Westview.

Schot, J. (1992). Constructive Technology Assessment and Technology Dynamics: The Case of Clean Technologies. *Science, Technology, & Human Values*, 17(1), 36-56.

Schubert, K. (1989). *Interessenvermittlung und staatliche Regulation*. Opladen: Westdeutscher Verlag.

Schubert, K. (Ed.). (1992). *Leistungen Grenzen politisch-ökonomischer Theorie. Eine kritische Bestandsaufnahme zu Mancur Olson*. Darmstadt: Wissenschaftliche Buchgesellschaft.

Schumpeter, J. (1964). *Theorie der wirtschaftlichen Entwicklung*. (6th edition ed.). Berlin: Duncker & Humblodt.

Schumpter, J. (1946). *Kapitalismus, Sozialismus und Demokratie*. Bern: Francke.

Schürmann, H. J. (1995, 24 July). Strittige Eckdaten für Kraft-Wärme-Kopplung. *Handelsblatt*, pp. 3-4.

Schwartz, P. (1991). *The Art of the Long View. Planning for the Future in an Uncertain World*. New York: Doubleday.

Schweitzer, F. (Ed.). (1997). *Self-Organization of Complex Structures. From Individual to Collective Dynamics*. Amsterdam: Gordon and Breach.

Schweitzer, F., & Steinbrink, J. (1997). Urban Cluster Growth: Analysis and Computer Simulations of Urban Aggregations. In F. Schweitzer (Ed.), *Self-Organization of Complex Structures. From Individual to Collective Dynamics* (pp. 501-518). Amsterdam: Gordon and Breach.

Seidel, M. (1994). Tendenzen bei Entwicklung und Einsatz von Blockheizkraftwerken. *BWK*, 46(9), 373-376.

Sigurdson, J. (Ed.). (1990). *Measuring the dynamics of technological change*. London: Sage.

Silverberg, G. (1991). Adoption and Diffusion of Technology as a Collective Evolutionary Process. In N. Nakicenovic & A. Grübler (Eds.), *Diffusion of Technologies and Social Behavior* (pp. 209-230). Berlin: Springer.

Silverberg, G., Dosi, G., & Orsenigo, L. (1988). Innovation, Diversity and Diffusion: A Self-Organisation Model. *The Economic Journal*, 98(December), 1032-1054.

Simon, H. (1955). A Behavioral Model of Rational Choice. *Quarterly Journal of Economics*, 69, 99-108.

Simon, H. (1959). Theories of Decision Making in Economics. *American Economic Review*, 49, 253-283.

Sirchis, J. (Ed.). (1990). *Combined Production of Heat and Power (Cogeneration)*. London: Elsevier.

Smith, K. (1992). Technological innovation indicators: experience and prospects. *Science and Public Policy*, 19(6), 383-392.

Smits, R., Leyten, A., & Geurts, J. (1987). *The Possibilities and Limitations of Technology Assessment. Report for the Dutch Ministry of Education and*

Science in co-operation with the Commission of the European Communities (DG XII/FAST). Den Haag: TNO-STB.

Solow, R. (1957). Technical Change and the Aggregate Production Function. *Review of Economics and Statistics*, 39, 312-320.

Starr, C., Searl, M. F., & Alpert, S. (1992). Energy Sources: A Realistic Outlook. *Science*, 256.

Steiner, K. (1994). Strom-Investitionen in Ostdeutschland. *Elektrizitätswirtschaft*, 93(1/2), 22-24.

Stoneman, P. (1983). *The Economic Analysis of Technological Change*. Oxford: Oxford University Press.

Stotz, J., & Broßmann, E. (1995). Entwicklung der überregionalen Stromversorgung in den neuen Bundesländern. *Energiewirtschaftliche Tagesfragen*, 45(9), 559-562.

Streeck, W. (1984). Neo-Corporatist Industrial Relations and the Economic Crisis in West Germany. In J. H. Goldthorpe (Ed.), *Order and Conflict in Contemporary Capitalism. Studies in the Political Economy of Western European Nations* (pp. 291-314). New York: Oxford University Press.

Street, J. (1992). *Politics and Technology*. NewYork: Guildford Press.

Summerton, J. (1992). *District Heating Comes to Town: The Social Shaping of an Energy System, Linköping Studies in Arts and Sciences No. 80*. Linköping: Linköping University.

Summerton, J. (Ed.). (1994). *Changing Large Technical Systems*. Boulder (CO): Westview.

Thompson, G., & al. (Eds.). (1991). *Markets, Hierarchies and Networks. The Coordination of Social Life*. London: Sage.

Topp, A. (1993). Das Recht der Fernwärmeversorgung. In AGFW/UNICHAL (Ed.), *Fernwärme International* (pp. 47-53). Frankfurt a.M.: Arbeitsgemeinschaft Fernwärme (AGFW)/International Union of Heat Distributors.

Traube, K., & Schulz, W. (1995). *Ökologische und ökonomische Wirkungen des Zubaus von Kraft-Wärme-Kopplungsanlagen, insbesondere von Blockheizkraftwerken, in der Bundesrepublik Deutschland*. Research Report, Bremen: Bremer Energie Institut.

Tushman, M. L., & Rosenkopf, L. (1992). Organizational Determinants of Technological Change: Towards a Sociology of Technological Evolution. *Research in Organizational Behavior*, 14(4), 311-347.

Ulrich, G. (1994). *Politische Steuerung. Staatliche Intervention aus systemtheoretischer Sicht*. Opladen: Leske + Budrich.

Utterback, J. (1979). The Dynamics of Product and Process Innovation in Industry. In C. Hill & J. Utterback (Eds.), *Technological Innovation for A Dynamic Economy* : Pergamon Press.

Utterback, J. (1994). *Mastering the Dynamics of Innovation. How Companies Can Seize Opportunities in the Face of Technological Change*. Boston (MA): Harvard Business School Press.

Utterback, J., & Suarez, F. (1993). Innovation, Competition and Industry Structure. *Research Policy*, 22, 1-21.

vanderDoes, T., Brown, M., & Verhagen, B. (1996). The role of promotional organisationas in the development of cogeneration. *International Journal of Global Energy Issues*, 8(4), 342-348.

vanLente, H. (1993). *Promising Technology. The Dynamics of Expectations in Technological Development*. Delft: Eburon.

Vasko, T. (1987). *The Long-Wave Debate. Selected papers from an IIASA International Meeting, held in Weimar (GDR), June 1985*. Berlin: Springer.

Verbruggen, A. (1992a). Combined heat and power. A real alternative when carefully implemented. *Energy Policy*, 20(9), 884-900.

Verbruggen, A. (1992b). *Constraints on the development of decentral power generation*. Proceedings of the Conference "Lausitz wohin? Schöpferische Vielfalt - Stukturen für eine tragfähige Zukunft", Cottbus, 14-16 October 1992.

Verbruggen, A. (1992c). Pricing independent power production. *International Journal of Global Energy Issues*, 2(1), 41-49.

Verbruggen, A. (1994). *The Electricity Industry Structure in Europe*. Proceedings of the "Power-Gen Europe '94 Conference", Cologne, 17-19 May 1994.

Verbruggen, A., Dufait, N., & Martens, A. (1993). Economic evaluation of independent CHP projects. *Energy Policy*, 21(4), 408-417.

Verbruggen, A., & Marcelis, C. (1986). *Institutional Constraints on District Heating and Cogeneration Feasibility*. (Research Report 86/197). University of Antwerp: Studiecentrum voor Economisch en Sociaal Onderzoek (SESO).

Verbruggen, A., Wiggin, M., Dufait, N., & Martens, A. (1992). The impact of CHP generation on CO2 emissions. *Energy Policy*, 20(12), 1207-1214.

VIK (Ed.). (1991). *Statistik der Energiewirtschaft 1990/91*. Essen: Verband der Industriellen Energie- und Kraftwirtschaft (VIK).

VIK (Ed.). (1994a). *Statistik der Energiewirtschaft 1993/94*. Essen: Verband der Industriellen Energie- und Kraftwirtschaft (VIK).

VIK (Ed.). (1994b). *Verband der Industriellen Energie- und Kraftwirtschaft. Tätigkeitsbericht 1993/94*. Essen: Verband der Industriellen Energie- und Kraftwirtschaft (VIK).

VIK (Ed.). (1995). *Verband der Industriellen Energie- und Kraftwirtschaft. Tätigkeitsbericht 1994/95*. Essen: Verband der Industriellen Energie- und Kraftwirtschaft (VIK).

VIK (Ed.). (1996). *Statistik der Energiewirtschaft 1995/96*. Essen: Verband der Industriellen Energie- und Kraftwirtschaft (VIK).

VIK (Ed.). (1997). *Statistik der Energiewirtschaft 1996/97*. Essen: Verband der Industriellen Energie- und Kraftwirtschaft (VIK).

Vogel, D. (1986). *National Styles of Regulation. Environmental Policy in Great Britain and the United States*. London: Cornell University Press.

Volterra, V. (1931). *Leçons sur la théorie mathématique de la lutte pour la vie*. Paris.

Walsh, V. (1984). Invention and innovation in the chemical industry: Demand-pull or dicovery-push? *Research Policy*, 13, 211-234.

Walz, R. (1994). *Die Elektrizitätswirtschaft in den USA und der BRD. Vergleich unter Berücksichtigung der Kraft-Wärme-Kopplung und der rationellen Energienutzung*. Heidelberg: Physica.

Waterson, M. (1988). *Regulation of the Firm and Natural Monopoly*. Oxford: Basil Blackwell.

Weber, H. (1986). Technokorporatismus. Die Steuerung des technologischen Wandels durch Staat, Wirtschaftsverbände und Gewerkschaften. In H. H. Hartwich (Ed.), *Politik und die Macht der Technik* (pp. 278-297). Opladen: Westdeutscher Verlag.

WEC (Ed.). (1991). *District Heating/Combined Heat and Power. Decisive factors for a sucessful use - as learnt from experiences*. London: World Energy Council.

Weidlich, W. (1991). Modelling Concepts of Synergetics for Dynamic Processes in the Society. In E. W., M. Peschel, & W. Weidlich (Eds.), *Models of Selforganization in Complex Systems* (pp. 9-28). Berlin: Akademie Verlag.

Weidlich, W. (1992). Das Modellierungskonzept der Synergetik für dynamische sozioökonomische Prozesse. In U. Witt (Ed.), *Studien zur evolutorischen Ökonomik II* (pp. 39-64). Berlin: Duncker & Humblodt.

Welfens, P. J. J. (1988). Innovationstheorie, -politik und -dynamik im Systemvergleich. In P. J. J. Welfens & L. Balcerowicz (Eds.), *Innovationsdynamik im Systemvergleich. Theorie und Praxis unternehmerischer, gesamtwirtschaftlihcer und politischer Neuerung* (pp. 1-27). Heidelberg: Physica.

Willke, H. (1987). Kontextsteuerung und Re-Integration der Ökonomie. Zum Einbau gesellschaftlicher Kriterien in ökonomische Rationalität. In M. Glagow & H. Willke (Eds.), *Dezentrale Gesellschaftssteuerung: Probleme der Integration polyzentrischer Gesellschaft* (pp. 155-172). Pfaffenweiler: Centaurus.

Witt, U. (1987). *Individualistische Grundlagen der evolutorischen Ökonomik*. Tübingen: Mohr.

Witterhold, F. G. (1995). Forschung und Entwicklung für die Fernwärmeversorgung in Deutschland. In AGFW (Ed.), *Aus Forschung und Entwicklung.*

Mitteilungen der Forschungsstelle und des Forschungsbeireates der AGFW . Frankfurt a.M.: Arbeitsgemeinschaft Fernwärme (AGFW).

Worldbank (Ed.). (1994). *World Development Report 1994. Infrastructure for Development*. New York: Oxford University Press.

Wraneschitz, H. (1995). Kleine ganz groß. Marktübersicht Klein-BHKW. *Betrieb & Energie Sonderheft zu Energie Spektrum*, 9(2), 12-16.

Zimmermann, K. (1988). Umweltpolitik und integrierte Technologien: Entwicklungen und Determinanten in empirischer Analyse. *Konjunkturpolitik*, 34, 327-351.

Zinn, H. (1995). Strategien der Landesregierung zum Ausbau der Kraft-Wärme-Kopplung. In HMU (Ed.), *Kraft-Wärme-Kopplung in Hessen. Konzepte, Anlagen, Erfahrunge, Strategien und Perspektiven* (pp. 1-12). Wiesbaden: Hessisches Ministerium für Umwelt, Energie, Jugend, Familie und Gesundheit (HMU).

Zipper, A. (1988). Öffentliche Elektrizitätsversorgungsunternehmen in der Bundesrepublik Deutschland als potentielle Schrittmacher umwelttechnologischer Innovationen. In P. J. J. Welfens & L. Balcerowicz (Eds.), *Innovationsdynamik im Systemvergleich. Theorie und Praxis unternehmerischer, gesamtwirtschaftlicher und politischer Neuerung* (pp. 206-223). Heidelberg: Physica.

Abbreviations

AGFW	Arbeitsgemeinschaft Fernwärme bei der Vereinigung Deutscher Elektrizitätswerke
AIEP	Association of Independent Electricity Producers
ARE	Arbeitsgemeinschaft Regionaler Energieversorgungsunternehmen
ASUE	Arbeitsgemeinschaft für Sparsamen und Umweltfreundlichen Energieverbrauch
BATNEEC	Best Available Technologies Not Entailing Excessive Costs
BGW	Bundesverband der Gas- und Wasserwirtschaft
BMBF	Bundesministerium für Bildung, Wissenschaft, Forschung und Technologie
BMFT	Bundesministerium für Forschung und Technologie
BMU	Bundesministerium für Umwelt, Naturschutz und Reaktorsicherheit
BMWi	Bundesministerium für Wirtschaft
CCGT	Combined Cycle Gas Turbines
CEC	Commission of the European Communities
CEGB	Central Electricity Generating Board
CEM	Contract Energy Management
CHP	Combined heat and power
CHPA	Combined Heat and Power Association
DEP	Discount Energy Purchase
DH	District Heating

DoE	Department of Energy
DoEnv	Department of Environment
DSM	Demand-Side Management
DTI	Department of Trade and Industry
EEO	Energy Efficiency Office
ESI	Electricity Supply Industry
EU	European Union
EVU	Energieversorgungsunternehmen
FGBHKW	Fördergemeinschaft Blockheizkraftwerke
HMIP	Her Majesty's Inspectorate of Pollution
HMSO	Her Majesty's Stationary Office
HMU	Hessisches Ministerium für Umwelt, Energie, Jugend, Familie und Gesundheit
HoC	House of Commons
IEA	International Energy Agency
IESC	Integrated Energy Service Companies
LCP	Least-Cost Planning
MEUC	Major Energy Users' Council
NFFO	Non-Fossil Fuels Obligation
OECD	Organisation for Economic Co-operation and Development
OFFER	Office of Electricity Regulation
OFGAS	Office of Gas Supply
PET	Politics - Economics - Technology
REC	Regional Electricity Companies
TA Luft	Technische Anleitung Luft
TPA	Third Party Access
VDEW	Vereinigung Deutscher Elektrizitätswerke
VDI	Verein Deutscher Ingenieure
VIK	Verband der Industriellen Energie- und Kraftwirtschaft
VKU	Verband Kommunaler Unternehmen

Appendices

Appendix 1: Deutschsprachige Zusammenfassung

Innovationsdiffusion und politische Steuerung von Energietechnologien. Eine vergleichende Untersuchung zur Kraft-Wärme-Kopplung in Großbritannien und Deutschland

Einleitung

Angesichts der Notwendigkeit, eine nachhaltigere Energieversorgung aufzubauen, stellt sich die Frage, ob und wie ein solcher Prozeß politisch gesteuert werden, oder zumindest politisch gefördert werden könnte. Eine Antwort auf diese Fragestellung läßt sich nur entwickeln, wenn die förderlichen oder hinderlichen Faktoren für die Diffusion von Energietechnologie untersucht werden, näherhin deren technologische, ökonomische, organisatorische, institutionelle und politische Bedingungen.

Die anwendungsbezogene Problemstellung

In den letzten Jahren sind die Liberalisierung und die Stärkung des Wettbewerbs zu den zentralen Themen der Debatte über die anzustrebenden Grundsätze der Energieversorgung geworden. Jedoch ist die Kompatibilität von Wettbewerbs- und Nachhaltigkeitsprinzipien auch wiederholt in Frage gestellt worden, vor allem im Hinblick auf den Anstieg der CO_2-Emissionen und den Resourcenverbrauch. Auch nach der Verabschiedung des Weißbuchs zur Europäischen Energiepolitik im Jahre 1995 besteht Uneinigkeit darüber, bis zu welchen Grad staatliche Intervention und Steuerung im Bereich der Energieversorgung nötig ist und ob dafür monopolistische Elemente beibehalten werden sollen. Aufschlüsse bezüglich dieser Frage lassen sich von einem Vergleich der Auswirkungen erhoffen, die unterschiedliche nationale Steuerungsansätze im Energiesektor auf die Verbreitung der als vergleichsweise "nachhaltig" geltenden Technologie der Kraft-Wärme-Kopplunghatten. Zu diesem Zweck ist eine vergleichende Betrachtung Großbritanniens und Deutschlands besonders geeignet, weil diese beiden Länder stellvertretend für die beiden wesentlichen Steuerungsphilosophien stehen können: Großbritannien für die die weitestgehende Anwendung von Wettbewerbs-prinzipien und Deutschland für die Beibehaltung monopolistischer Strukturen.

Kraft-Wärme-Kopplung (KWK) gilt als eine der Technologien oder, besser gesagt, als eines der technologischen *Prinzipien*, auf das große Hoffnungen im Hinblick auf eine effizientere Energieversorgung gesetzt werden. Was die Innovations- und Diffusionsprozesse in den verschiedenen europäischen Ländern angeht, zeigt sich ein sehr unterschiedliches Bild, das sich exemplarisch in den beiden ausgewählten Ländern widerspiegelt. Während in Deutschland in der Vergangenheit eine bedeutende KWK-Kapazität aufgebaut worden ist, hat diese Technologie in Großbritannien bislang nur eine untergeordnete Rolle gespielt. Seit der Liberalisierung und Privatisierung der britischen Elektrizitätswirtschaft im Jahre 1990 hat sich die Situation jedoch geändert. Großbritannienstellt daher ein sehr interessantes Beispiel für das Verhältnis von Nachhaltigkeit und Wettbewerb dar.

Die entscheidende, politisch ausgerichtete Forschungsfrage dieser Arbeit besteht somit darin, ob und wie ein Übergang zu einer stärker KWK-basierten Energieversorgung ermöglicht werden könnte. Dabei gilt den Hindernissen, die sich der Innovation und Diffusion der KWK entgegenstellen, sowie den Möglichkeiten und Grenzen, darauf politisch Einfluß zu nehmen, ein besonderes Interesse. Das Forschungsinteresse richtet sich jedoch nicht nur auf das Entwickeln und Testen von Hypothesen, sondern besonders auch auf die Erforschung zukunftsgestaltender Strategien. Eine vorausschauende Analyse, z.B. im Sinne von Szenarien, , auf deren Basis dann politische Handlungsstrategien entwickelt werden könnten wird daher im Mittelpunkt der Arbeit stehen. Um diese Handlungsorientierung angemessen berücksichtigen zu können, wird von fiktiven Auftraggebern der Studie ausgegangen. Diese sind im vorliegenden Fall in erster Linie politische Entscheidungsträger auf nationaler oder europäischer Ebene.

Die methodologische Problemstellung

Unter einer methodologischen Perspektive wirft diese Fragestellung ein großes Problem auf: Wie kann man überhaupt das Zusammenspiel von technologischen Entwicklungen, Innovations- und Adoptionsentscheidungen, strukturell-organisatorischen Veränderungen und politischen Steuerungsmechanismen systematisch und nachvollziehbar untersuchen?. Weder existieren umfassende theoretische Ansätze für eine solche Analyse noch geeignete spezifische Methodologien. Die Formulierung eines verallgemeinerbaren methodologischen Werkzeugs wäre auch deshalb wünschenswert, weil sich das Problem der integrierten Analyse von Innovation, Diffusion und politischer Steuerung nicht nur in Bezug auf den konkreten Fall der KWK stellt, sondern ebenso in anderen Technologiefeldern. Beispielsweise existieren in den Bereichen der Telekommunikation oder dem Transportwesen ebenfalls sehr enge Verknüpfungen zwischen politischen Institutionen und Akteuren im technisch-wirtschaftlichen Umfeld eines großtechnischen Systems.

Die theoretische Problemstellung

Ein methodologischer Ansatz benötigt eine solide Fundierung, und die in den vergangenen rund zwanzig Jahren geleistete Arbeit zu Fragen technologischen Wandels hat eine Fülle empirischer und theoretisch-konzeptioneller Ergebnisse geliefert. Die Herausforderung für die vorliegende Arbeit besteht darin, die relevanten Konzepte unterschiedlicher, bislang weitgehend unverbunden nebeneinander stehender Disziplinen in einen konsistenten theoretisch-konzeptionellen Ansatz zu integrieren. Erst auf der Basis einer solchen Synthese kann dann der Versuch unternommen werden, eine geeignete Methodologie für einen solchen interdisziplinären Ansatz zu entwickeln.

Unter einer wissenschaftstheoretischen Perspektive zielt diese Vorgehensweise darauf ab, Theorie konstruktiv als Fundierung für die anwendungs- und zukunftsorientierte Forschung zu nutzen. Während der traditionelle Ansatz sozialwissenschaftlicher Forschung auf die Weiterentwicklung der Theorie ausgerichtet ist, soll hier der umgekehrte Weg gegangen werden. Aufbauend auf einer breiten, integrativen theoretischen Basis soll ein methodologischer Ansatz formuliert werden, der für eine voraussschauende Analyse geeignet ist. Besonderer Wert wird dabei darauf gelegt, daß alle Arbeitsschritte explizit und nachvollziehbar dargestellt werden. Dadurch soll eine Übertragung des Forschungsansatzes auf andere Problemstellungen erleichtert werden.

Teil I: PET - ein theoretischer und methodologischer Ansatz zur Analyse von Prozessen der Innovationsdiffusion und politischen Steuerung

Die Theorie im Überblick

Die formulierten Problemstellungen haben bereits angedeutet, daß aufgrund der Ausrichtung auf Fragen politischer Steuerung und Steuerbarkeit eine Analyse nur aus einer disziplinenübergreifenden Perspektive sinnvoll sein wird, da nur dadurch eine möglichst umfassende Berücksichtigung von Mechanismen und Faktoren der Innovationsdiffusion gewährleistet werden kann. Zunächst werden einige der grundlegenden Begrifflichkeiten geklärt, z.B. hinsichtlich der verschiedenen Dimensionen des Begriffs "Innovation" und der Abgrenzung von so gängigen Konzepten wie "Regulierung" und "politische Steuerung". Dem folgt ein Überblick über den Problemkomplex "technologischer Wandel", der die zu berücksichtigenden Analyseebenen und disziplinären Perspektiven mit drei grundlegenden Modellansätzen - linear, rückgekoppelt systemisch - verknüpft.

Aufbauend auf dieser Kategorisierung werden die wesentlichen Ansätze der für die Problemstellung relevanten Disziplinen vorgestellt und diskutiert. Diese sind die Innovationsökonomie, die Techniksoziologie und die Politikwissenschaft. Diesem Überblick vorangestellt sind Bemerkungen zur Messung und quantitativen Darstellung des technologischen Wandels als aggregiertem Phänomen, z.B. mit Hilfe von Methoden wie der Patentanalyse, der Technometrie, der morphologischen Analyse oder einfacher Wachstumsmodelle.

Die wirtschaftswissenschaftliche Debatte zur Fragen des technologischen Wandels hat sich in den vergangenen Jahren vornehmlich auf die Auseinandersetzung zwischen neoklassischen und evolutorischen Ansätzen konzentriert. Neoklassisch orientierte Modelle der Innovation und Diffusion ermöglichen eine mathematisch geschlossene Darstellung, aber ihnen liegen Annahmen zugrunde, die häufig als unrealistisch und somit als inadäquat für die Analyse des technologischen Wandels angesehen werden. Zudem ist es kaum möglich, neoklassische Modelle mit einer detaillierten technischen Problemanalyse zu verbinden, wie es im Sinne eines multidisziplinären Ansatzes unumgänglich wäre. Die evolutorische Forschungsrichtung scheint hingegen ein vielversprechenderer Ausgangspunkt zu sein, weil sie historisch gewachsene soziale Strukturen und technologiedynamische Aspekte explizit berücksichtigt. So stellen einige ihrer Elemente, z.B. technologische Entwicklungspfade, Paradigmen und Regime, aber auch die institutionellen Rahmenbedingungen von Innovationssystemen gute Anknüpfungspunkte für die die Integration mit technischen, kognitiven, soziologischen und politischen und technischen Ansätzen dar.

In der soziologischen Debatte läßt sich ebenfalls eine Abkehr von der simplen Analyse von “Faktoren”, die einen Einfluß auf Innovation und Diffusion ausüben, feststellen. Stattdessen geschieht eine Hinwendung zu stärker systemorientierten Ansätzen. Stichworte wie Netzwerkanalyse, akteurszentrierter Institutionalismus oder die Analyse großtechnischer Systeme weisen auf Parallelen zu den vorgenannten evolutorischen Ansätzen hin. Neuere Richtungen der Technikfolgenbewertung, wie z.B. das "Constructive Technology Assessment" gehen über diese analytischen Ansätze noch hinaus, indem sie auf eine engere Verbindung zwischen Forschungs- und Entscheidungsprozessen abzielen und damit auch auf einen konstruktiveren Beitrag der Forschung zur konkreten Problemlösung.

Politikwissenschaftliche Beiträge zur Rolle der Politik für den technologischen Wandel sind eher rar. Stattdessen gibt es eine reiche Literatur über die Bewertung konkreter technologiepolitischer Programme. Diese konzentriert sich aber in erster Linie auf deren Einzelwirkung und nicht auf den Gesamtprozeß der Innovationsdiffusion bestimmter Technologien. Darüber hinaus hat die politikwissenschaftliche Steuerungsdebatte jedoch Erkenntnisse über die Möglichkeiten und Grenzen der politischen Steuerung im allgemeinen bereitgestellt, die die Aussage untermauern, daß eine gezielte und unabhängige Steuerung durch einzelne Akteure kaum realisiert werden kann. Dies liegt an einer

Reihe von Problemen, wie z.B. den strukturellen und kognitiven Interdependenzen zwischen Steuerungsobjekt und -subjekt, dem Mangel an Informationen und Informationsverarbeitungskapazitäten und letztlich auch dem teilautonomen Steuerungsverhalten der verschiedenen Akteure und Subsysteme. Während Netzwerkanalysen im wesentlichen noch einem mechanistischen, wenn auch inzwischen dynamisierten Modell des technologischen Wandels anhängen, greifen die jüngeren selbstreferentiellen Ansätze ("Autopoiese") auf biologische Systemanalogien zurück, mit deren Hilfe eine realitätsnähere Beschreibung von politischen Steuerungsprozessen möglich ist. Sie bestätigen eine eher kritische Einschätzung der Möglichkeiten politischer Steuerung. Von besonderem Interesse am Autopoieseansatz ist das zugrundeliegende Akteursmodell, das detaillierte Kenntnis der internen Strukturen, Routinen und Erwartungshaltungen der verschiedenen Akteure verlangt, um Steuerungsstrategien entwerfen zu können.

Während eine Reihe von Verbindunglinien zwischen den Ansätzen der betrachteten Disziplinen ausgemacht werden können, fehlt es bislang an einem übergreifenden Ansatz, der die verschiedenen Perspektiven systematisch in einen konzeptionellen Rahmen integriert. Darüber hinaus ist allen Ansätzen gemeinsam, daß technologische Aspekte nur in einer sehr allgemeinen und unspezifischen Form berücksichtigt werden können, was in einem integrierten Ansatz korrigiert werden müßte. Der Überblick hat ferner ergeben, daß ein in Systemsprache gefaßter Ansatz am ehesten geeignet erscheint, eine übersichtliche und klar strukturierte Integration der verschiedenen Perspektiven zu erzielen.

PET als theoretisch-konzeptioneller Ansatz

Der hier entwickelte "PET-Systemansatz", der die Integration politischer ("Politics"), ökonomischer ("Economics"), technologischer ("Technology") und im weiteren Sinne auch soziologischer Aspekte leisten soll, greift auf Elemente der verschiedenen diskutierten Ansätze zurück, um die Dynamik und die Mechanismen technolgischen Wandels darzustellen. Ein wesentliches Problem besteht dabei in der Abgrenzung des zu untersuchenden Systems, d.h. des Bereichs, innerhalb dessen Interaktionen und Rückkopplungen betrachtet werden. Zwar werden Impulse aus der Systemumwelt auf das System berücksichtigt, nicht jedoch Rückkopplungen vom System auf dessen Umwelt. Im konkreten Anwendungsfall ist es daher entscheidend, die Abgrenzung so zu wählen, daß diese Rückkopplungen über die Systemgrenze tatsächlich vernachlässigt werden können. Die Analyse der Systemdynamik baut auf einer evolutorischen Interpretation auf, welche die Interdependenzen und Rückkopplungen zwischen Mikro-/Akteursebene und Makro-/struktureller Ebene im System berücksichtigt. Mit anderen Worten, Transformationsprozesse im PET-System werden als das Ergebnis von Entscheidungen der verschiedenen Akteure betrachtet. Deren Haltungen werder wiederum von den Erwartungen hinsichtlich struktureller und technologischer Veränderungen geprägt sind, sowie von der Wahrnehmung des Verhaltens anderer Akteure. Vorgeformte Denk- und Entscheidungsmuster ("technologische Paradigmen") üben zusätzlich einen stabilisierenden Einfluß aus.

Für das Entscheidungsverhalten der Akteure wird von einem autopoietischen Modellverständnis ausgegangen, das als kompatibel mit dem vorgenannten Konzept technologischer Paradigmen angesehen wird, aber in seiner Betonung der Selbstreferenz noch darüber hinausgeht.

Vereinfacht lassen sich somit die folgenden Elemente des PET-Systemansatzes unterscheiden, die sich auch in der nachfolgenden Abbildung wiederspiegeln:

- Die Ebene der untersuchten Technologie, die wiederum eingebettet sein kann in ein umfassenderes großtechnisches System. Die möglichst objektive und vollständige Erfassung ihrer verschiedenen Bewertungsdimensionen stellt dabei eine wichtige Voraussetzung für die weitere Analyse dar;
- Die Ebene der Akteure, die unterteilt wird in eine technologieanbietende Seite und eine technologienachfragende, sowie in den Bereich politischer Akteure;
- Die Ebene der Interaktionen in Entscheidungsarenen, in denen die Entscheidungen einzelner Akteure vorbereitet und ausgehandelt werden. Innovations-, Adoptions- und politische Entscheidungen werden hierbei als die drei wichtigsten Handlungskategorien angesehen. Ferner wird zwischen verschiedenen Organisationformen (Märkte, Hierarchien und Netzwerke) und Inhalten/Medien (Geld, Information, Macht/Recht, Artefakte) von Interaktionen unterschieden.
- Die strukturelle Ebene oder system-endogene Umwelt der Akteure und der betrachteten Technologie. Sie wird aufgegliedert in technologische, ökonomische, informationsbezogene und politische Strukturen.
- Die exogene Systemumwelt, deren Veränderungen die Prozesse im System beeinflussen, aber selber weitgehend vom System unabhängig sind.

Die Dynamik des PET-Systems wird erfaßt durch eine Reihe von grundlegenden Mechanismen, die teilweise aus der ökonomischen Evolutionstheorie entlehnt sind, teilweise aber auch darüber hinaus gehen. Im wesentlichen handelt es sich dabei um die folgenden Analyseelemente:

- Externer Anpassungsdruck;
- Variation und Vorselektion von Technologien im Rahmen von Innovationsprozessen auf der Anbieterseite;
- Selektionsmechanismen auf der Seite der Technologieanwender in ihrem organisatorischen Umfeld;
- Regulatorische und politische Selektionsbedingungen;
- Stabilisierungsmechanismen;
- Verstärkungsmechanismen;
- Systemwandel.

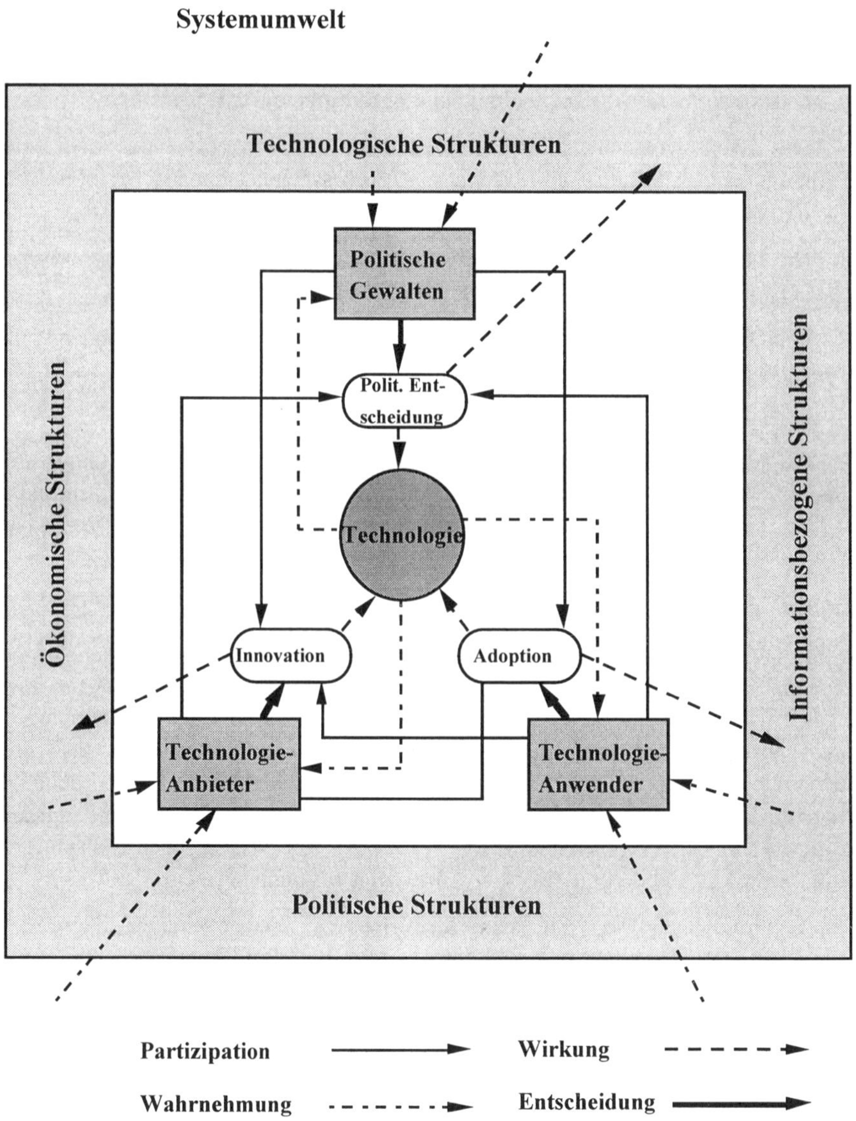

Abbildung 1: Das PET-System

Das PET-System dient weniger der Aufstellung detaillierter Hypothesen als vielmehr der Strukturierung und Anleitung des explorativen Forschungsprozesses im Bereich der systemischen Steuerung von Innovations- und Diffusionsprozessen. Es stellt ein Raster zur Verfügung, mit dessen Hilfe konkrete Analysen durchgeführt und Steuerungsstrategien entwickelt werden können. Wenn trotzdem fünf Arbeitshypothesen formuliert werden, dann in erster Linie um die PET-Perspektive in vereinfachter Form forschungsleitend einsetzen zu können. Die fünf Kernhypothesen zielen dabei auf folgende Aspekte ab:

- Die Koevolution von Technologie und Strukturen im PET-System;
- Die Komplementarität von untersuchter Technologie und dem sie umgebenden technischen System als Voraussetzung für eine erfolgreiche Diffusion;
- Das Entscheidungsverhalten der Akteure, das auf Wahrnehmungen der Umwelt, Erwartungshaltungen und vorgeprägten Positionen aufbaut;
- Die Dynamik des Systemwandels und der zugrundeliegenden Mechanismen;
- Die Möglichkeiten und Grenzen der politischen Steuerung in interdependenten sozio-technischen Systemen.

Über diese allgemeinen Hypothesen hinaus können konkretere Aussagen auf sinnvolle Weise nur in Bezug auf einen bestimmten Anwendungsfall formuliert werden, und zwar erst dann, wenn detaillierte Informationen über die Technologie und ihre wichtigsten Bewertungsdimensionen zur Verfügung stehen.

Eine allgemeine Methodologie zur Umsetzung von PET

Aufgrund des umfassenden Charakters des PET-Systemansatzes erscheint es geboten, ihn in eine relativ einfach handhabbare Methodologie zu übersetzen, mit deren Hilfe dann konkrete Fragestellungen bearbeitet werden können. Dies wäre auch im Hinblick auf das Bestreben, Theorie konstruktiv für die anwendungsbezogene Forschung einzusetzen, wünschenswert. Eine möglichst explizite Darlegung der Methodologie würde darüber hinaus die Anwendung des Ansatzes auf neue Problemstellungen erleichtern. Die wesentlichen Schritte der entwickelten Methodologie lassen sich wie folgt zusammenfassen:

- Technologieanalyse, d.h. die Identifizierung der Charakteristiken der untersuchten Technologie und ihres technologischen Kontexts anhand der wesentlichen Bewertungsdimensionen;
- Analyse technologischer Veränderungen, d.h. die Untersuchungen von Verbesserungen der Technologie in Vergleich zu anderen Optionen;
- Abgrenzung des Systems, die u.U. im Laufe der Forschungsarbeit überprüft werden muß;
- Analyse der Systemstrukturen und der Transformationsprozesse, denen sie unterworfen sind;
- Identifizierung der Akteure im technologieanbietenden, technologienachfragenden und politischen Bereich;
- Ereignisanalyse, d.h. die Identifizierung der wesentlichen Entscheidungen und Ereignisse während des Untersuchungszeitraums;

- Analyse der Interaktionen, Motivationen und Erwartungen, die zu diesen Entscheidungen geführt haben;
- Analyse der Systemdynamik und der ihr zugrundliegenden Mechanismen;
- Prospektive Analyse, Szenarioentwicklung und Strategieformulierung.

Die prospektive Analyse stellt den Endpunkt der Methodologie dar und ermöglicht es, handlungsleitende Strategien für einen bestimmten Akteur oder Auftraggeber zu entwickeln. Im Falle der politischen Steuerung wird es sich dabei häufig um Akteur aus dem Bereich nationaler Regierungen handeln, auch wenn im Prinzip für jeden andern Systemakteur Strategien entwickelt werden könnten. Entscheidend ist hierbei, daß der Akteur, für den Strategien entwickelt werden sollen, explizit als Teil des zu untersuchenden Systems aufgefaßt wird. Mit anderen Worten, auch der politische Bereich als *interdependenter* Bestandteil des Systems aufgefaßt und nicht als autonome Steuerungsinstanz.

Teil II: Einführung in das Problemfeld und Spezifizierung der Forschungsmethodologie

Das Ziel dieses zweiten Teils besteht darin zu zeigen, wie der im ersten Teil entwickelte PET-Ansatz in einer konkreten empirischen Untersuchung angewendet werden kann. Als Beispiel ist die Technologie derKraft-Wärme-Kopplung gewählt worden. Aufgrund ihrer Einbettung in das Energieversorgungssystem kann sie als Teil eines großtechnischen Systems aufgefaßt werden.

Einführung in die Probleme der Energieversorgung und Kraft-Wärme-Kopplung

Die Entwicklungen der vergangenen Jahre haben zu großen Veränderungen im Bereich der Energieversorgung geführt, durch die viele der über Jahrzehnte gewachsenen Strukturen neu bewertet werden mußten. Monopolstrukturen, die über lange Zeit als "natürliche" Basis für eine effiziente Energie- und speziell Elektrizitätsversorgung galten, sind durch eine Reihe technologischer Entwicklungen in Frage gestellt worden. So nehmen z.B. die Vorteile von Skaleneffekten durch immer größere Kraftwerkseinheiten ab. Entsprechendes läßt sich von zentralisierten Organisationsstrukturen sagen. Darüber hinaus haben Aspekte wie die Versorgungssicherheit, der Umweltschutz, die Resourcenknappheit oder auch die strategische Bedeutung der Energieversorgung eine Neubewertung erfahren und weltweit ein Nachdenken über wesentliche Reformen im Energiebereich eingeleitet.

Die KWK wird weithin - trotz zahlreicher Auseinandersetzungen über ihre umweltbezogenen Vor- und Nachteile - als eine der Technologien eingestuft, die als wünschenswert und ausbauwürdig gelten. Ihre Brisanz besteht darin, daß sie als dezentrale Erzeugungstechnologie eine strukturelle Bedrohung für die etablierten Akteure in den zentralisierten Elektrizitätsversorgungssystemen darstellt. Darüber hinaus ist unklar, welche Auswirkungen eine weitreichende Liberalisierung der Elektizitätsversorgung auf die Anwendung und Verbreitung der KWK hätte.

Die Kraft-Wärme-Kopplung kann sehr einfach definiert werden als technologisches Prinzip zur gleichzeitigen Erzeugung von Elektrizität und Wärme im selben Prozeß. Um jedoch zu verstehen, warum die verschiedenen Akteure zu so unterschiedlichen Bewertungen der KWK kommen, müssen die Implikationen der Technologie anhand einer ganzen Reihe von Dimensionen analysiert werden. Dies ist eine zwingende Voraussetzung für die spätere empirische Analyse der Entscheidungsprozesse und Akteursstrategien. Diese Technologieanalyse kann zunächst noch unabhängig von den konkreten Länderfallstudien durchgeführt werden und konzentriert sich auf folgende Dimensionen:

- Die technologische Dimension, die zentral für das Verständnis aller weiteren Aspekte ist. Hierbei muß unterschieden werden zwischen dem Aufbau des technischen Systems KWK einerseits und seinen Eigenschaften als Teil eines Energieversorgungssystems andererseits. Auch die drei wesentlichen Anwendungsfälle der KWK (Fernwärme, industrielle Anwendungen, Blockheizkraftwerke) müssen in diesem Zusammenhang unterschieden werden.

- Die umweltbezogene Dimension, die in den letzten Jahren im Zentrum vieler Debatten um die KWK gestanden hat. Das wesentliche Charakteristikum der KWK besteht darin, daß aufgrund der Kopplung ein deutlich höherer energetischer Wirkungsgrad erzielt werden kann als in herkömmlichen Kraftwerken. Wirkungsgraden von bis zu 90% für die Erzeugung von Wärme und Strom in KWK-Systemen stehen Werte im Bereich von 60-75% für die Kombination von konventionellen Kondensationskraftwerken mit individuellen Heizungsanlagen gegenüber. Je nach konkretem Anwendungsfall kann somit der relative Wirkungsgradgewinn der KWK im Bereich von 10-30% liegen, was auch einen entsprechend niedrigeren Brennstoffverbrauch zur Folge hat.

- Die mikroökonomische Dimension, die sowohl für die Herstellerseite untersucht werden muß (Innovation) als auch für die verschiedenen Akteure auf der Anwenderseite (Adoption). Letztere müssen in zwei Gruppen unterteilt werden. Erstens handelt es sich um Endanwender und -verbraucher wie Industriebetriebe, Kommunen, Krankenhäuser und andere öffentliche und private Einrichtungen. Zweitens müssen auch die verschiedenen intermediären Organisationen des Energieversorgungssystems, d.h. in erster Linie die EVUs (Elektrizitätsversorgungsunternehmen), untersucht werden. Sie kontrollieren nicht nur die wesentlichen Schnittstellen des Systems, sondern betreiben auch eigene Anlagen. Darüber hinaus sind im Zuge der Liberalisierung auch noch private Energiedienstleistungsunternehmen entstanden, die ebenfalls berücksichtigt werden müssen.

- Die makro- und sozioökonomische Dimension, bei der es vor allem um die Frage nach einer volkswirtschaftlich möglichst effizienten Energieversorgungung geht, aber z.B. auch um Implikationen für die Versorgungssicherheit oder die sozialen Aspekte der Energieversorgung.
- Organisatorische und Machtdimension, die sich mit den Auswirkungen der KWK auf die langfristigen Organisationsstrukturen der Energieversorgungswirtschaft befaßt, sowie den sich daraus ergebenden Konsequenzen für die Macht- und Einflußpositionen der Hauptakteure.

Die Kraft-Wärme-Kopplung steht zudem in Konkurrenz mit einer Reihe anderer Technologieoptionen, und zwar auf mehreren Ebenen: zwischen verschiedenen KWK-Varianten, zwischen gekoppelten und nicht-gekoppelten Technologien, sowie zwischen zentralen und dezentralen Versorgungskonzepten. Aufgrund von technologischen Verbesserungen und sich wandelnden Rahmenbedingungen ändern sich diese Konkurrenzverhältnisse im Laufe der Zeit. Die mehrdimensionale Betrachtung dieser verschiedenen Optionen macht deutlich, daß eine simple Technologiebewertung in Kategorien wie Über- oder Unterlegenheit nicht sinnvoll ist. Vielmehr muß stärker differenziert werden, zumal die fünf aufgeführten Dimension zu unterschiedlich sind, als daß man sie aggregierenkönnte. Von Über- oder Unterlegenheit kann man letztlich nur sprechen, wenn man sich auf einen bestimmten Akteur und eine einzelne Bewertungsdimension zu einem konkreten Zeitpunkt bezieht.

Überblick über bisherige Forschungsergebnisse

Im Rahmen der empirischen Umsetzung der PET-Methodologie für den Fall der KWK ist es notwendig, eine Vielzahl von Daten zu den verschiedenen Elementen des Ansatzes zu sammeln. Um spezifizieren zu können, zu welchen dieser Aspekte neues Datenmaterial generiert werden muß, bedarf es zunächst einer Aufarbeitung des aktuellen Forschungsstandes. Außerdem ist der Aufbau einer solchen Datenbasis notwendig, um die Auswahl der im Detail zu untersuchenden Länder begründen zu können. Wie bereits angedeutet, werden dies Großbritannien und Deutschland sein.

Die bisherigen Diffusionsverläufe der KWK in den verschiedenen europäischen Ländern zeigen ein sehr heterogenes Bild. Während z.B. in Frankreich die KWK bislang kaum eine Rolle gespielt hat, werden in den Niederlanden und Dänemark mehr als 30% des Stroms in Kraft-Wärme-Kopplungs-Anlagen erzeugt. Die Faktoren und Hemmnisse, auf die diese Heterogenität in den verschiedenen Forschungsbeiträgen zurückgeführt wird, lassen sich grob in exogene, strukturelle und akteursbezogene unterscheiden. Aufbauend auf einer Analyse dieser Faktoren werden dann in den meisten Arbeiten Empfehlungen für die Politik formuliert.

Zusammenfassend läßt sich festhalten, daß die bislang zur KWK durchgeführten Studien zwar viele für die PET-Perspektive brauchbare Daten bereitstellen, daß jedoch auch in einigen wichtigen Bereichen Lücken existieren, die durch neues empirisches Material geschlossen werden müssen. Dies läßt sich vor allem im Hinblick auf die folgenden Aspekte feststellen:

- Beschränkung auf eine statische Sichtweise des Diffusionsproblems;
- Vernachlässigung zahlreicher relevanter Akteure;
- Vernachlässigung der politischen Dimension der KWK-Diffusion und speziell der zugrundeliegenden politischen Strukturen und Prozesse;
- Vernachlässigung der Rolle innovativer Entwicklungen der KWK-Technologie für die Diffusion;
- Genereller Mangel an Aussagen über den Ablauf von Entscheidungsprozessen auf der Ebene der Akteure.

Spezifizierung der empirischen Forschungsarbeit

Um das Zusammenspiel von politischer Steuerung und Innovationsdiffusion für den Fall der KWK zu untersuchen, werden zwei Länder verglichen, die bei ähnlichen Grundvoraussetzungen unterschiedliche Diffusionsmuster und Strukturen des Energieversorgungssystems aufweisen. Im Hinblick auf die gegenwärtige Liberalisierungsdebatte wurden Großbritannien und Deutschland ausgewählt. Wie bereits erwähnt wurde, repräsentieren sie die beiden auf Wettbewerb bzw. auf Monopolen basierenden Grundmodelle der Steuerung des Elektrizitätssektors. Das britische System ist seit 1990 der europäische Modellfall für die Liberalisierung im Elektrizitätssektor. Parallel zur Liberalisierung hat sich die KWK vergleichweise rasch verbreitet, auch wenn dabei berücksichtigt werden muß, daß die Ausgangskapazität Ende der achtziger Jahre sehr niedrig war. Im Gegensatz dazu bestand zu diesem Zeitpunkt in Deutschland bereits eine beträchtliche KWK-Kapazität, die aber innerhalb eines monopolistischen Systems aufgebaut worden war. Die Themenbereiche, auf die in beiden Ländern bei der empirischen Arbeit besonderer Wert gelegt wurde, sind zum einen die Rolle technologischer und struktureller Veränderungen, und zum anderen die Entscheidungsprozesse auf der Akteursebene, speziell auch auf Seiten der KWK-Anwender. Neben den Positionen der wesentlichen Akteure im Bereich der Energieversorgung und Energiepolitik werden daher in beiden Ländern rund fünfzehn vergleichbare KWK-Anwendungen untersucht, welche die drei Hauptanwendungsfälle abdecken (Fernwärme, industrielle Anwendungen, Blockheizkraftwerke). Da die KWK in beiden Ländern Anfang der achtziger Jahre neue technologische und politische Impulse erhalten hat, werden Fälle herangezogen, bei denen die Investitionsentscheidungen in den Zeitraum von damals bis in die frühen neunziger Jahre fallen. Aufgrund des Schwerpunkts, der auf die Entscheidungen der verschiedenen Akteure gelegt wurde, wurden halbstrukturierte Interviews durchgeführt, die diesen Aspekten nachgehen.

Die Datenquellen und Erhebungsmethoden, die im Rahmen der empirischen Arbeit verwendet werden sollen, werden explizit dargelegt, um die Umsetzung der Methodologie demonstrieren und gleichzeitig Hinweise für zukünftige Anwendungen geben zu können. Ferner wird auf die Grenzen der empirischen Forschungsmethoden hingewiesen, die sich zum einen aus der Qualität der verschiedenen Datenquellen und zum anderen aus den gewählten Untersuchungstechniken ergeben. Speziell die Qualität des knappen statistischen Datenmaterials zur KWK hat sich als Problem erwiesen und eine weitestgehend qualitative Analyse notwendig gemacht.

Teil III: Die empirische Darstellung: KWK in Großbritannien und Deutschland vom Anfang der 80er bis zur Mitte der 90er Jahre

Liberalisierung der Energieversorgung in Großbritannien

Das entscheidende Jahr für die Elektrizitätsversorgung in Großbritannien war 1990, als nicht nur weite Bereiche des Sektors liberalisiert und damit dem Wettbewerb unterworfen wurden, sondern auch die bis dahin staatlichen Versorgungsunternehmen privatisiert wurden. Allerdings hatte es schon 1983 einen Versuch gegeben, die Bedingungen für die Stromerzeugung durch nicht-EVUs zu verbessern. Diese Initiative scheiterte jedoch daran, daß die EVUs weiterhin die Möglichkeit hatten, die technischen und ökonomischen Randbedingungen für die Eigenerzeugung und damit auch für die KWK zu definieren. Gleichzeitig profitierte die KWK auch von Neuerungen auf technologischem Gebiet, so z.B. durch die Einführung von Gasturbinen und -motoren, die Entwicklung von automatischen Steuerungs- und Kontrollsystemen oder die Vorfertigung von kompletten Systemen ("packaging"). Diese Entwicklungen hatten positive Auswirkungen auf die Zuverlässigkeit und Wirtschaftlichkeit von KWK-Anlagen. Es mußten jedoch noch eine Reihe weiterer Barrieren auf Seiten der verschiedenen Akteure überwunden werden, um eine Akzeptanz der KWK auf breiter Basis zu erzielen. So haben Vorurteile und strategische Interessen eine schnellere Einführung ebenso verzögert wie der eher schlechte Wissensstand von Anwendern und Planern über die Möglichkeiten der Technologie. Bis 1990 war die KWK zudem inkompatibel mit den vorherrschenden technischen, wirtschaftlichen und politischen Strukturen, die durch einen hohen Grad an Zentralisierung und einen Mangel an horizontaler Integration der Energieversorgungsketten für Wärme und Elektrizität gekennzeichnet waren.

Die Liberalisierung und Privatisierung von 1990 brach die bis dahin enge Verzahnung von Energiepolitik und Elektrizitätswirtschaft auf. Die nun privaten EVUs begannen sich langsam in integrierte Energiedienstleistungsunternehmen zu verwandeln; neue private Unternehmen entstanden, die die Energieversorgung ebenfalls als integrierte Dienstleistung verstanden. Der Wettbewerbsdruck zwang die Unternehmen, wirtschaftlich effizienter zu arbeiten, wozu eine höhere

Energieeffizienz beitragen konnte. Während bereits in den Achtzigern einige Hersteller die sich verbessernden Möglichkeiten für die KWK wahrgenommen hatten und ihre Entwicklungsbemühungen verstärkt hatten, wuchs das Interesse ab 1990 weiter. In der Zwischenzeit wurden zahlreiche der ursprünglich unabhängigen Hersteller von den EVUs übernommen, und auch die Gaswirtschaft stieg seit der Freigabe der Verwendung von Erdgas für die Stromerzeugung aktiv in die KWK ein. Die horizontale Integration der Energieversorgungsketten hat dadurch wichtige Impulse erhalten. Andere für die KWK wichtige Randbedingungen haben sich jedoch nicht wesentlich verändert, wie z.B. die geringe Bedeutung kommunaler Energieversorger und Verwaltungen. Dadurch sind die Aussichten gerade für die Fernwärme nach wie vor eher düster, während die Situation für die industrielle KWK und die Blockheizkraftwerke weiterhin sehr vielversprechend ist.

Unter dem Blickwinkel der politischen Steuerung muß diese Expansion in erster Linie als Nebenprodukt der ideologisch motivierten Liberalisierung gesehen werden. Allerdings war sie sicherlich nur das auslösende Moment für die schnelle Verbreitung einer Technologie, die technisch und wirtschaftlich bereits seit einigen Jahren bereit stand, aber innerhalb des alten energiewirtschaftlichen Rahmens nicht eingesetzt werden konnte. Obwohl sich einige der Informationsprogramme der britischen Regierung als sehr erfolgreich erwiesen haben und auch ein Kapazitätsziel für die KWK formuliert wurde (5000 MW im Jahr 2000), gab es nur sehr wenig aktive und direkte Unterstützung für die KWK, und wenn, dann nur für Nischenanwendungen. Darüber hinaus zeigten sich eine Reihe von Anpassungen der neuen Regulierungen seit 1990 als vorteilhaft für die KWK. Es sollte jedoch klar herausgestellt werden, daß das liberalisierte System auch einige grundlegende Probleme für die KWK aufwirft. Diese sind vor allem in den Unsicherheiten in Bezug auf die Strompreise und Planungsvorgaben zu sehen, sowie in der nach wie vor unzureichenden Koordination der Energieversorgungsketten auf lokaler Ebene. Dies hat u.a. dazu geführt, daß Gas- und Dampfturbinen-Anlagen zur reinen Stromerzeugung die derzeit am schnellsten wachsende Kraftwerkslinie darstellen, weil sie leichter mit dem kurzfristigeren Planungshorizont im liberalisierten System vereinbar sind. Die Verbreitung der KWK bleibt daher noch weit hinter ihren Möglichkeiten zurück, und dies wird sich auch während der nächsten fünfzehn Jahre kaum ändern.

Die Konstellation der wesentlichen Akteure im liberalisierten System hat sich inzwischen stabilisiert, und dies ermöglicht es, einen Blick in die Zukunft zu werfen. Die weitere Entwicklung der KWK in Großbritannien ist natürlich von einer Reihe von Bedingungen abhängig, die mit Hilfe verschiedener Szenarien abgebilded werde können. Zweifelsohne hat die Liberalisierung im Falle Großbritanniens dazu beigetragen, die Anwendungsmöglichkeiten für die KWK zu verbessern. Allerdings heißt dies nicht, daß diese nun besonders günstig wären, sondern lediglich, daß ein über Jahre hinweg blockiertes Potential teilweise ausgefüllt werden konnte. Speziell für die Fernwärme bestehen weiterhin nur geringe Aussichen auf einer breitere Anwendung, und auch die anderen Anwendungsfälle sind durch die spezifischen Charakteristiken des liberalisierten britischen Systems nur begrenzt expansionsfähig. Die Vorteile, die der Wettbewerb für die KWK mit sich gebracht hat, werden durch zahlreiche neue Mechanismen im System wieder aufgehoben, so daß die eher moderate Zielmarke der britischen Regierung von 5000 MW KWK-Kapazität im Jahr 2000 als durchaus realistisch angesehen werden kann.

Transformierung der Monopole in Deutschland

Die KWK spielt für die Energieversorgung in Deutschland eine wesentliche Rolle, und das bereits seit vielen Jahren. Alle drei Anwendungsbereiche - Fernwärme, Industrie- und Kleinanlagen - sind in nennenswertem Maße vertreten. Bis 1998 war das auf Versorgungsmonopolen aufbauende Energiewirtschaftsgesetz von 1935 in Kraft, das gegenwärtig im Rahmen der Umsetzung des europäischen Binnenmarkts für Energie refomiert wird. Seit den frühen achtziger Jahren kam es zu ähnlichen technologischen Entwicklungen wie in Großbritannien, wenn auch mit entscheidenden Unterschieden im Detail, die häufig auf die spezifischen Bedingungen des Energiewirtschaftsrechts zurückgeführt werden können. Ein wesentliches Charakeristikum der Situation in Deutschland besteht in der Variationsbreite der Randbedingungen in den verschiedenen Bundesländern und Kommunen. Obwohl es von Seiten der EVU Widerstände gegen dezentrale und unabhängige KWK-Anlagen gegeben hat, ist gerade das Verhältnis zwischen den kommunalen Energieversorgern (Stadtwerken) und den privaten KWK-Anwendern im großen und ganzen durch kooperative und informelle Beziehungen gekennzeichnet. Darüber hinaus sind die Stadtwerke selbst eine der wichtigsten Betreibergruppen von KWK-Systemen. Sie verfügen auch über das notwendige technische Fachwissen, um neue Entwicklungen aufgreifen zu können. Generell muß die grundlegende technische Kompetenz der Anwender in Deutschland als sehr hoch eingeschätzt werden. Während sowohl die Fernwärme als auch industrielle Anwendungen weithin eine hohe Akzeptanz genießen, bestanden anfangs erhebliche Vorurteile gegenüber Blockheizkraftwerken. Diese konnten erst dann aus dem Weg geräumt werden, als die Technologie von den wesentlichen technischen Verbänden akzeptiert worden war, und zwar im Rahmen der Vereinfachung und Standardisierung der Auslegungs- und Planungsrichtlinien.

In den letzten Jahren hat sich die Diskussion über die Bedingungen für die KWK vor allem auf die Frage der Vergütung von Stromeinspeisungen der privaten Betreiber in das öffentliche Netz zugespitzt. Obwohl die sogenannte Verbändevereinbarung zwischen den Vertretern der Energiewirtschaft und der industriellen Stromerzeuger als Richtlinie für Stromexporte weitgehend anerkannt wird, hat es regelmäßig Auseinandersetzungen über deren konkrete Anwendung und Fortschreibung gegeben. Trotz dieser Unsicherheiten haben sich die ökonomischen Randbedingungen für die KWK schrittweise verbessert. Ebenso sind zahlreiche administrative Hindernisse und Unsicherheiten ausgeräumt worden. Das wachsende Interesse der Anwerder hat zu gezielten Entwicklungsbemühungen geführt, durch die technologische Verbesserungen, eine Ausweitung des Spektrums verfügbarer Lösungen und erhebliche Kostensenkungen erzielt werden konnten.

Im Gegensatz zu Großbritannien hat es in Deutschland neben der Forschungsförderung auch eine Reihe wichtiger Förderprogramme für den Ausbau der Fernwärme gegeben, die seit der Wiedervereinigung vor allem der Modernisierung der Fernwärmesysteme Ostdeutschlands zugute kamen. Hingegen zählt die Bereitstellung und Verbreitung von Information für Anwender nicht zu den wesentlichen Aktivitätsbereichen der Politik, sieht man von der Arbeit regionaler Einrichtungen wie den Energieagenturen einmal ab. Die zentrale und wichtige Rolle der industriellen und technischen Verbände als Informationsquellen in den Netzwerke der Energiewirtschaft und Energietechnik machte die Bereitstellung von Information durch die Politik allerdings auch nicht notwendig. Ein weiteres Charakteristikum der deutschen Energiewirtschaft und -politik besteht in der häufigen Delegierung von Verantwortlichkeiten an die industriellen Verbände, wofür die Verbändevereinbarung zur Stromeinspeisevergütung nur eines von vielen Beispielen ist. Die vielschichtige Verflechtung von Verantwortungsbereichen im politischen System ermöglicht die breit abgestimmte Umsetzung inkrementaler Veränderungen. Im Hinblick auf grundlegende Reformen ist sie aber eher ein Innovationshemmnis.

Die Situation in Ostdeutschland ab 1990 stellte insofern eine besondere Herausforderung dar, als ein massiver Modernisierungsbedarf bestand, speziell für die großen Fernwärmesysteme in vielen Städten. Die industrielle KWK-Kapazität war indes weitgehend zusammengebrochen. Um nach der Wiedervereinigung die Modernisierung der Energieversorgung in die Wege zu leiten und die bestehenden Versorgungskapazitäten zu sichern, sind weitestgehend die gleichen industriellen Strukturen wie in Westdeutschland eingeführt worden, und zwar unter Federführung der westdeutschen EVUs. Aufgrund der massiven finanziellen Unterstützung für die Fernwärme konnten die Kapazitäten weitgehend erhalten werden. Auch bei der Modernisierung der verbliebenen Industriebetriebe kommt häufig die KWK zum Zuge, wobei aufgrund der Finanzierungsprobleme in den Neuen Ländern verstärkt Contracting-Lösungen gewählt werden.

Abhängig vom Grad der zu erwartenden Liberalisierung der Energiewirtschaft in Deutschland schwanken die Aussichten für die KWK stark. Als entscheidend für das Schicksal der Fernwärme wird das Weiterbestehen der lokalen Koordination der Versorgungsketten durch die Stadtwerke angesehen, das eng mit der Existenz kommunaler Gebietsmonopole verknüpft ist. Die anderen Anwendungsgebiete - Industrie und Kleinanlagen - könnten von den erwarteten Veränderungen profitieren, vor allem wenn sich eine stärkere Dienstleistungsorientierung in Energiebereich durchsetzen sollte, und zwar sowohl auf Seiten kommunaler EVUs als auch durch das Entstehen neuer privater Energiedienstleistungsunternehmen. Ein großer Unsicherheitsfaktor ist da zukünftige Strompreisniveau. Bei einem starken Verfall der Strompreise würde die Eigenerzeugung in KWK-Anlagen in vielen Fällen wirtschaflich uninteressant werden.

Vergleich der Entwicklung der KWK in Großbritannien und Deutschland

Während sich die Technologie der KWK auf den ersten Blick in beiden Ländern recht ähnlich entwickelt zu haben scheint, legt eine tiefergehende Analyse doch einige Unterschiede frei. Beispiele dafür sind die Steuerungssysteme und die Strom- und Wärme-Meßtechnik, die in Großbritannien eine zentrale Rolle gespielt haben, während in Deutschland die Systemintegration und umweltbezogene Innovationen eine größere Bedeutung hatten. Darüber hinaus gibt es Differenzen im Hinblick auf die Zeitpunkte der Einführung von Neuerungen. Ursachen für diese Unterschiede sind zum einen in den organisatorischen und regulatorischen Randbedingungen zu sehen, aber auch in den Kompetenzen und Strategien der Unternehmen im Energiesektor.

Der Hauptunterschied zwischen den beiden Ländern liegen jedoch auf der strukturellen Ebene. Neben den schon genannten Diffusionsmustern und den damit zusammenhängenden technologischen Strukturen fallen auch die Differenzen in den anderen untersuchten Strukturelementen (ökonomisch, politisch, informationsbezogen) deutlich aus. Auf der Informationsseite ist hier vor allem die große Bedeutung industrieller und technischer Verbände in Deutschland zu nennen. Sie tragen entscheidend zur Formung der kognitiven und planerischen Standards für die KWK bei. Ähnliche Funktionen werden in Großbritannien in begrenztem Maße von staatlichen bzw. halbstaatlichen Organisationen wahrgenommen.

Im Hinblick auf die anstehenden Liberalisierungsbemühungen spielen die anzustrebenden elektrizitätswirtschaftlichen Strukturen eine entscheidende Rolle. Bezüglich einer weiteren Verbreitung der KWK zeigt die Detailanalyse der beiden Länder, daß eine eindeutige Bewertung zugunsten eines liberalisierten oder eines monopolistischen Ansatzes nicht möglich ist. Beide Ansätze weisen spezifische Vor- und Nachteile auf, die auf unterschiedlichen Weise in den beiden Ländern zur Geltung kommen. Einige Mechanismen im liberalisierten britischen System haben sich als sehr vorteilhaft für die KWK erwiesen, wie z.B. das Entstehen von integrierten Energiedienstleistungsunternehmen, das erst durch die Einführung des Wettbewerbsprinzips möglich war. Im Zusammenhang damit ist es im Bereich der KWK auch zu einer Annäherung von Technologieanbietern und Dienstleistern wie den EVU gekommen. Deren Aktivitäten überlappen sich in Großbritannien deutlich stärker als in Deutschland. Andere Mechanismen im britischen System sind hingegen sehr problematisch für eine Technologie, die einen längerfristigen Planungshorizont und höheren Koordinationsaufwand erfordert. Ähnliche Beispiele lassen sich auch für Deutschland finden, wo einige Elemente wie die kommunalen Monopole sich als wichtige Voraussetzungen für die Planung von Fernwärme-Systemen erwiesen haben. Andererseits hat der Mangel an Wettbewerb auf allen Versorgungsebenen bislang die Umsetzung innovativer Lösungen eher behindert. Die Rolle dieser Mechanismen kann im Detail aber nur im Kontext der historisch gewachsenen Strukturen des jeweiligen Landes bewertet werden.

Die Inflexibilität des Energieversorgungssystems und die enge Verquickung mit der Energiepolitik waren sicher die Hauptgründe dafür, daß in Großbritannien nur ein revolutionärer Umbruch wie die Liberalisierung und Privatisierung im Jahre 1990 eine Veränderung einleiten konnte. Im Gegensatz zum deutschen System, das durch die Dezentralisierung von Verantwortlichkeiten zahlreiche Nischen für das Testen und Weiterentwickeln der KWK bot, war das britische System viel monolithischer und abgeschlossener gegenüber innovativen Impulsen. Einer der entscheidenden Schwachpunkte war und ist nach wie vor die vernachlässigbare Rolle der Kommunen für die Energieversogung, durch die ein Ausbau der KWK - sei es als Fernwärme oder in Form von Kleinanlagen - erschwert wird. Ein zweiter wesentlicher Unterschied zwischen beiden Ländern manifestiert sich in der Anwendung des Prinzips der Delegierung von Verantwortlichkeiten an industrielle und technische Verbände in Deutschland. Dies hat im Falle der KWK zu Lösungen geführt hat, die von den betroffenen Akteuren zumindest akzeptiert werden konnten. Dies darf jedoch nicht über das Problem der einseitigen Einflußnahme auf die Energiepolitik hinwegtäuschen, das z.B. dazu führt, daß sich starke und gut organisierte Interessengruppen erheblich besser in den politischen Entscheidungsprozeß einbringen können.

Die verschiedenen Akteursstrategien in Bezug auf die Innovation, Adoption und Politik der KWK müssen in diesen Selektionsrahmen eingeordnet werden. Die Akteure sind in ihren Handlungsmöglichkeiten durch strukturelle Gegebenheiten beschränkt, aber andererseits prägen sie durch ihre Handlungen diese Strukturen mit. Derartige Rückkopplungen sind entscheidend für das Verständnis der Evolutionsdynamik des Systems. In Form von Verstärkungs- oder Verzögerungsmechanismen formen sie die technologischen Entwicklungspfade der KWK. Dies läßt sich für die beiden untersuchten Länder zeigen. Die Lernprozesse, die durch die Verbreitung von Informationen und anhand erfolgreicher (und erfolgloser) Anwendungsfälle angestoßen werden konnten, sind weitere gute Beispiele für solche dynamischen Mechanismen, ebenso wie die Abhängkeiten zwischen technologischen, ökonomischen, informationsbezogenen und politischen Strukturen.

Zusammenfassend läßt der Vergleich sicher keine pauschale Aussage über die Überlegenheit eines liberalisierten oder eines monopolistischen Systems im Hinblick auf die Verbreitung der KWK zu. Die historisch gewachsenen Strukturen und Randbedingungen in beiden Ländern erfordern differenzierte Aussagen darüber, welcher der beiden Ansätze in der jeweils gegebenen Situation geeigneter wäre, um die Diffusion einer gesellschaftlich und ökologisch wünschenswerten Technologie wie der KWK zu ermöglichen. Eine der wesentlichen Schlußfolgerungen dieser Untersuchung besteht sicher auch darin, daß der gewählte Steuerungsansatz (liberalisiert oder monopolistisch) weniger wichtig zu sein scheint als dessen detaillierte Implementierung und Regulierung. Das deutsche, auf Monopolen basierende System stellt trotz aller Kritik recht gute Bedingungen für die KWK bereit, die zudem durch das Beseitigen von administrativen und organisatorischen Hemmnissen noch weiter verbessert werden konnten. In einer solchen Situation bietet es sich an, einen inkrementalen Politikansatz zu verfolgen, wie er auch in der Vergangenheit bereits eingesetzt wurde. Im Gegensatz dazu war in Großbritannien der revolutionäre Umbau der Energie- und Elektrizitätswirtschaft eine Vorraussetzung dafür, daß die KWK überhaupt expandieren konnte. Die schnelle Diffusion der letzten Jahre ist daher auch auf einen starken Nachholbedarf zurückzuführen und nicht nur auf den positiven Einfluß des liberalisierten Systems per se. Einige der identifizierten Mechanismen weisen außerdem darauf hin, daß eine Liberalisierung nicht zwangsläufig einen Vorteil für die KWK mit sich bringt. So hätte die Einführung des britischen Systems in Deutschland vermutlich fatale Folgen für zahlreiche typische Anwendungsfälle der KWK.

Die drei Szenarien, die für beide Länder skizziert werden können, unterstreichen den begrenzten Einfluß des Steuerungsansatzes für die Diffusion der KWK. Erstens hängen die Aussichten für die KWK sehr stark von den gewachsenen Strukturen ab. Zweitens sind das Ausmaß und die Bereiche einer Liberalisierung zwar entscheidend für das weitere Diffusions*potential* der KWK, nicht jedoch auch zwangsläufig für die tatsächliche Verbreitung. Drittens sind letztlich die konkreten und detaillierten Elemente des Regulierungsrahmens von zentraler Bedeutung dafür, ob die relevanten Akteure Anreize für die Weiterentwicklung, Anwendung und Unterstützung der Technologie erkennen können oder nicht.

Teil IV: Schlußfolgerungen

Schlußfolgerungen zur politischen Steuerung der Innovationsdiffusion von KWK-Systemen

Die oben angeführten fünf Arbeitshypthesen stellen den Ausgangspunkt für eine Reihe allgemeiner Schlußfolgerungen zur Diffusion und Innovation von KWK-systemen und ihrer politische Steuerung dar. Die Aussagen zur Interdependenz von Technologie und Systemstrukturen, sowie zur Komplementarität von spezifischer Technologie und dem sie umgebenden großtechnischen System können für den Fall der KWK weitestgehend bestätigt werden. Auch das beobachtete Entscheidungsverhalten der Akteure in beiden Ländern spiegelt die entsprechenden Hypothesen wider. In den Fallstudien ließen sich auch einige wichtige Mechanismen identifizieren, die für das Verständnis der Systemdynamik entscheidend sind. Hinsichtlich der Steuerungshypothese konnten zumindest einige interessante Interventionspunkte identifiziert werden, durch die die Innovationsdiffusion der KWK positiv beeinflußt werden kann.

Aufbauend auf den Ergebnissen des Vergleichs lassen sich zudem eine Reihe von Anforderungen herausarbeiten, die generell respektiert werden sollten, um ein für die KWK günstiges Umfeld zu schaffen. So sollten die organisatorischen und strukturellen Prinzipien, auf denen das Zusammenwirken von politischem, energiewirtschaftlichem und energietechnologischem Bereich basiert, die wesentlichen technischen Charakteristika der KWK unterstützen, z.B. durch die horizontale Integration von Energieversorgungsketten oder die Förderung langfristiger Investitionen in den Aufbau von Fernwärmesystemen. Darüber hinaus gibt es eine Reihe von Ansatzpunkten auf der Akteursebene, die zur Unterstützung von Entscheidungen zugunsten der KWK eingesetzt werden können. Die Förderung des Informationsaustauschs oder der Kooperation zwischen Energieversorgern und Eigenerzeugern sind hierfür nur zwei Beispiele. Schließlich lassen sich auch einige grundlegende Prinzipien angeben, die jedem Steuerungprozeß zugrunde liegen sollten, z.B. der gezielte Einsatz von Verstärkungseffekten und die selbstorganisatorischen Prozesse im Energiesektor. Diese Überlegungen spiegeln erstens wider, daß ideologische Positionen pro oder contra Liberalisierung nicht dem komplexen Charakter der Steuerung der Innovationsdiffusion gerecht werden. Zweitens können sie als Beleg dafür dienen, daß aufgrund des selbstorganisatorischen und selbstreferentiellen Charakters der untersuchten Systeme sehr konkrete Steuerungsempfehlungen weder möglich noch sinnvoll sind. Die Komplexität, Unsicherheit, Pfadabhängigkeit und Autonomie der in den beiden Länderstudien untersuchten Innovations- und Diffusionsprozesse zeigen deutlich, daß den Möglichkeiten gezielter Steuerung klare Grenzen gesetzt sind.

Dementsprechend gilt, daß zwar einige allgemeine Anhaltspunkte für politische Steuerungsstrategien auf nationaler oder europäischer Ebene angegeben werden können, daß diese aber erst anhand konkreter Fallbeispiele spezifiziert werden können. Diese allgemeinen Ansatzpunkte für die politische Steuerung der Innovationsdiffusion der KWK sind:

- Die politischen Steuerungsstrukturen und Schnittstellen zwischen den Akteuren im energiewirtschaftlichen und politischen Bereich;
- Die Grundsätze der Regulierung im Umwelt- Energie- und Baubereich;
- Die konkrete Ausformulierung spezifischer Regulierungsmaßnahmen;
- Die gezielten Programme zur Investitionsförderung für die KWK;
- Die Innovations- und Forschungsprogramme;
- Die Bereitstellung und Verbreitung von Informationen für die konkrete Umsetzung von KWK-Projekten;
- Die Schärfung des Bewußtseins für die Möglichkeiten und Grenzen der KWK.

Methodologische Schlußfolgerungen

Im Rahmen dieser Studie ist versucht worden, die entwickelte PET-Methodologie schrittweise und explizit umzusetzen. Ziel war es zu zeigen, wie aufbauend auf einer multi-disziplinären theoretischen Basis ein konstruktiver Beitrag zur Entwicklung politischer Handlungsstrategien geleistet werden kann. Die Informationsmengen, die dafür gesammelt und aufgearbeitet werden mußten, haben deutlich gemacht, wie aufwendig bereits der empirisch-retrospektive Teil einer solchen Untersuchung ist. Die vorausschauende Anwendung konnte daher nur skizziert werden, aber sie hat zumindest eine Einsichtnahme diesbezüglich ermöglicht, wie der entwickelte Ansatz umgesetzt werden kann. Die neun Schritte der Methodologie haben sich dabei als sehr hilfreiche Orientierung für die empirische Forschungsarbeit erwiesen.

Darüber hinaus zeugen die Ergebnisse von den Schwierigkeiten, zu konkreten Schlußfolgerungen für die politische Steuerung zu kommen. Als unabdingbar hat sich die detaillierte Kenntnis des zu steuernden Bereichs erwiesen, und zwar sowohl in technischer und ökonomischer als auch in politischer Hinsicht. Partielle Einsichten in Teilbereiche des Systems sind für die Steuerungsanalyse unzureichend, da das Entscheidungsverhalten der Akteure, das für die Steuerung entscheidend ist, von einer großen Bandbreite von Faktoren und Mechanismen abhängt.

Theoretisch-konzeptionelle Schlußfolgerungen

Das PET-System ist bewußt als disziplinenübergreifender theoretischer Ansatz formuliert worden, um das Problemfeld der Steuerung technologischen Wandels zu untersuchen. Die Systemsprache, in die der Ansatz gefaßt ist, hat es ermöglicht, die zugrundeliegenden Prozesse und Mechanismen umfassend darzustellen, sie aber gleichzeitig übersichtlich und nachvollziehbar zu strukturieren. Existierende Forschungsansätze aus den Sozial- und Wirtschaftswissenschaften sind dafür mit einer detaillierten technologischen Analyse verknüpft worden.

Aufbauend auf den Bausteinen des PET-Ansatzes sind sieben Merkmale zur qualitativen Beschreibung der Systemdynamik vorgeschlagen worden. Dies kann sicherlich nur als ein erster Schritt verstanden werden, der der weiteren Verbesserung bedarf. Zwar konnten unter den sieben Merkmalen alle wesentlichen Elemente behandelt und durch die Betonung von Verstärkungs- und Verzögerungsmechanismen eine Interpretation der Innovationsdiffusion als Prozeß erreicht werden, aber diese Interpretation blieb im Rahmen der vorliegenden Arbeit weitgehend retrospektiv. Für eine umfassende prospektive Analyse müßten verbesserte Forschungsmethoden entwickelt und eingesetzt werden, die neuere Entwicklungen der Ssystemsimulation aufgreifen.

Forschungsausblick

Aufbauend auf den Erfahrungen und Ergebnissen der vorliegenden Arbeit lassen sich einige weitere Bereiche identifizieren, in denen weiterführende Forschungsanstrengungen notwendig erscheinen. Zum einen wären empirische Studien auf der Basis des PET-Systemansatzes für eine Verallgemeinerung des Ansatzes notwendig. Sie sollten sowohl andere Länder als auch andere Technologiebereiche abdecken. Vor allem weitere großtechnische Systeme wie Transport oder Telekommunikation kämen hierfür in Frage. Zum anderenkonnte die prospektive Anwendung des PET-Ansatzes im Rahmen der vorliegenden Arbeit nur knapp behandelt werden und sollte zunächst im Energiebereich für einige Beispieltechnologien in vollem Umfang durchgeführt werden. Hierfür wäre es wünschenswert, bessere qualitative oder quantitative Analysemethoden zu entwickeln und einzusetzen. Da generell auch die Möglichkeit besteht, eine andere Perspektive als die der *politischen* Steuerung im Rahmen von PET einzunehmen, wären Fallstudien für die Entwicklung von Strategien für industrielle oder technologische Akteure von großem Interesse. In theoretischer Hinsicht könnte die PET-Perspektive sicher noch besser mit den verfügbaren theoretischen Ansätzen der Innovations-, Diffusions- und Steuerungsforschung verzahnt werden. So wurden die Management- und organisationstheoretische Literatur nur am Rande gestreift; sie könnte aber weitere Impulse geben, um die multi-disziplinäre Basis des PET-Ansatzes zu verbreitern und dadurch dessen Einordnung in die wissenschaftliche Literatur verbessern.

Appendix 2: Questionaire outline

The set-up of the following questionaire, or better series of questionaires, has been used as a guideline for the interviews carried out with the different actors. As similar though different questions had to be asked to the different actors, it is structured into seventeen "issues". It is indicated for each issue for which types of actors it has been raised. The numbers indicate the respective type of actor:

(1) Technology user;

(2) Technology supplier and manufacturer;

(3) Energy utility or service company;

(4) Industrial associations and interest groups;

(5) Political authorities.

The seventeen blocks are not further specified into individual questions but only into the issues raised because depending on the individual interviewee the questions had to be modified slightly.

A) ORGANISATION
1. Organisational Profile (1, 2, 3, 4, 5)
• Type of organisation (public utility, auto-producer, association, manufacturer, political organisation, etc.)
• Main areas of activity
B) CHP-SPECIFIC ISSUES
2. CHP-related activities (1, 2, 3, 4)
• General technical and economic characteristics of the plant operated or planned
• Specification of the type of CHP-related services/products provided (including involvement in RD&D-work)
• Specification of other types of activities in relation to CHP
3. Energy supply strategy and motivations for dealing with CHP (1)
• Objectives associated to energy supply
• General position to and confidence in CHP-technology
• Competing technical and service options considered instead of the CHP

• Complementary energy efficiency measures and activities
• Importance and institutionalisation of energy efficiency management
4. Information sources and cooperation patterns (1, 2, 3, 4, 5)
• Major information sources on CHP and assessment of their effectiveness and importance
• Characterisation of the quality and importance of cooperation patterns between users and manufacturers/packagers/utility companies/other actors
• Identification of other actors and activities which had an impact on the state of information and knowledge on CHP
5. Investment decisions and associated expectations (1, 3)
• Reasons for the initial interest in CHP
• Criteria applied for selecting the technology/projects
• Expectations with regard to the investment made
• Responsibilities and debates in the course of the decision process
• Arguments raised in favour of and against CHP
• Main uncertainties perceived with regard to CHP and ways to deal with them
• Involvement of external expertise in the course of the project decision
6. Realisation of CHP-projects (1, 3)
• Main obstacles to the realisation of the CHP-project (technical, economic, administrative, institutional)
• Opportunities and problems of coordination of heat and power supply systems with other actors
7. Experiences with CHP-systems (1, 2, 3)
• Reasons for success and failure of projects/development
• Comparison of initial expectations and realised outcomes (technical and economic)
C) PERCEPTION AND ASSESSMENT OF TECHNOLOGICAL AND STRUCTURAL CHANGES
8. Perception and assessment of technological developments and R&D on CHP (1, 2, 3, 4, 5)
• Improvements of components (technical, economic, environmental)
• Improvements of system characteristics (reliability, control, efficiency, others)

• Assessment of the comparative importance of these improvements
• Identification of the most important actors on the technology supply side
• Major expected future improvements and breakthroughs
9. Perception and assessment of economic and regulatory changes in the energy sector (1, 2, 3, 4, 5)
• Impacts of and experiences with the structural organisation of the ESI with regard to CHP
• Identification of the most important actors in the energy supply sector, including regulatory bodies
• Positions of these actors with regard to CHP and changes of these position
• Major areas of uncertainty for CHP in the ESI
10. Perception and assessment of energy and energy technology policies (1, 2, 3, 4, 5)
• Identification of the most important actors in these policy areas associated to energy and CHP
• Positions of these actors with regard to CHP and changes of these position
• Identification and assessment of specific technology, innovation or diffusion policy measures
• Experiences, reliability and uncertainties associated to these policies at different policy levels
11. Perception and assessment of environmental policies (1, 2, 3, 4, 5)
• Identification of the most important actors in these policy areas associated to energy and CHP
• Positions of these actors with regard to CHP and changes of these position
• Identification and assessment of specific environmental policy measures
• Experiences, reliability and uncertainties associated to environmental policy
D) POLITICAL INTERACTIONS AND FEEDBACKS
12. Own position with regard to the role of CHP in the ESI (3, 4, 5)
• Characterisation of the own attitude towards CHP
• Positions with regard to CHP-related policy issues
13. Direct and indirect participation and role in the political decision-making process (1, 2, 3, 4, 5)
• Participation in major interest groups

• Participation in industrial/technical associations
• Direct involvement in political processes
14. Assessment of the political decision-making processes (3, 4, 5)
• Assessment of possibilities to access and participate in the preparation ofpolicies
• Identification of levels of policy-making interactions (local, regional, national)
• Assessment of the distribution of responsibilities in the policy-making processes
• Reliability of the outcomes of policy-making
E) OTHER ISSUES
15. Other important aspects to be mentioned (1, 2, 3, 4, 5)
16. Final assessment of the situation and prospects of CHP (1, 2, 3, 4, 5)
17. Further contacts (1, 2, 3, 4, 5)

Appendix 3: List of interviews

1. Interviews in the UK (December 1994 - March 1995)

Name	**Organisation**	**Place**
Mr. M. Doble	Energy Technology Support Unit (ETSU)	Harwell
Mr. R. Potter	Dept. of Trade & Industry	by telephone
Mr. A. Christy	Dept. of Environment	by telephone
Mr. N. Foster	Dept. of Environment	London
Mr. M. Gylee	Office of Electricity Regulation (OFFER)	Birmingham
Mr. A. Jackson	Office of Electricity Regulation (OFFER)	Birmingham
Mr. M. Odedra	Office of Electricity Regulation (OFFER)	Birmingham
Mr. M. Brown	Combined Heat and Power Association, London & COGEN Europe	London & Brussels
Mr. D. Porter	Association of Independent Power Producers (AIEP)	London
Mr. P. Rost	Major Energy Users' Council (MEUC) & former MP	London
Mr. R. Chadburn	Rodney Chadburn Associates & former Eastern Electricity	Ipswich
Mr. B. Ingham	Eastern Group plc.	Ipswich
Mr. A. Halstead	PowerGen CHP Ltd.	Coventry
Mr. J. Pearson	British Gas	London
Mr. I. Parker & Mr. P. North	East Midlands Electricity plc.	Caythorpe
Mr. S. Booth	Yorkshire Electricity	Scarcroft
Mr. E. Wilson	AHS Emstar	Coventry
Mr. W. Stean & Mr. M. Robinson	Balfour Beatty	Sidcup
Mr. A. Melrose	Manchester Airport	Manchester

Mr. M Barnard	European Gas Turbines	Lincoln
Mr. R. Weston	Combined Power Systems	Manchester
Mr. R. D'Rane	Nedalo b.v.	Horsham
Mr. A.Widdowson	Walsall Metropolitan Borrow Council	Walsall
Mr. M. Sprackling	Shropshire County Council	Shrewsbury
Mr. T. Whitehead	Central Power Ltd.	Birmingham
Mr. R. Edmonds	UML Ltd.	Birkenhead
Mr. Atkins	Wandsworth Borough Council	London
Mr. S. Lewis & Mr. S. Lamch	Addenbrookes Hospital Cambridge	Cambridge
Mr. P. Birch	University of Liverpool	Liverpool
Mr. G. Leech	University of Keele	Newcastle-under-Lyme
Mr. J. Forte	Forte Hotel Group	Nottingham
Mr. H. Tschoepe	Manchester Ramada Hotel	Manchester
Mr. J. Buckhan	Newcastle City Council	Newcastle
Mr. J. Buxton	Stockport Metropolitan Borough Council	Stockport
Mr. A. Hall	Doverstrand Ltd.	Grimsby
Mr. J. Watson	Ciba-Geigy Chemicals Ltd	Grimsby
Mr. D. Bowman	Tioxide Europe Ltd.	Grimsby
Mr. P. Ellen	Rover Ltd.	Solihull

2. Interviews in Germany (October 1995 - February 1996)

Name	**Organisation**	**Place**
Dr. F.-G. Witterhold, Dr. H.P. Winkens & Mr. H. Neuffer	Arbeitsgemeinschaft Fernwärme (AGFW) beim Verband Deutscher Elektrizitätswerke	Frankfurt a.M.
Dr. G. Sauer, Mr. W. Niebisch & Mr. K. Menzler	Verband der Industriellen Kraftwirtschaft (VIK)	Essen
Mr. K. Telges	Arbeitsgemeinschaft für Sparsamen & Umweltfreund-lichen Energieverbrauch (ASUE)	Hamburg

Dr. M. Seidel	Verband Kommunaler Unternehmen (VKU)	Köln
Dr. H. Pick	Fördergemeinschaft Blockheizkraftwerke (FGBHKW)	Bonn
Mr. V. Mund & Dr. H. Menz	Bundesministerium für Wirtschaft (BMWi)	Bonn
Dr. N. Salomon	Bundesministerium für Umwelt, Naturschutz und Reaktorsicherheit (BMU)	Bonn
Dr. H. Meixner	Hessenenergie	Wiesbaden
Mr. J. Maaß	HASTRA	Hannover
Dr. N. Scheirle	Energieversorgung Schwaben AG (EVS)	Stuttgart
Mrs. C. Wacker	Maingas AG	Frankfurt a.M.
Mr. W.G. Paulick	Fichtel & Sachs	Schweinfurt
Mr. L. Höfler	Lothar Höfler BHKW	Lindau
Dr. F. Hirschbichler	M.A.N.	Augsburg
Mr. A.-I. Franz	Erdgas Energie Systeme GmbH (EES)	Essen
Mr. K. Hettel	Dr. Schwabe Arzneimittel	Durlach
Mr. W. Schwab	Carl Bardusch GmbH	Ettlingen
Dr. H. Heinisch	Stadtwerke Tübingen	Tübingen
Mr. P. Küppers & Mr. T. Banholzer	Stadtwerke Rottweil	Rottweil
Mr. T. Maier & Mr. W. Scholl	Stadtwerke Ingolstadt	Ingolstadt
Mr. K.H. Krömmelbein & Mr. Röchling	Stadtwerke Gelsenkirchen	Gelsenkirchen
Mr. R. Hörl	Flughafen München GmbH	München
Mr. K. Weireter	C.F. Ploucquet GmbH & Co.	Heidenheim
Dr. K. Schilf	TU Hamburg-Harburg	Hamburg
Mr. N. Ernst	Klinikum Bayreuth	Bayreuth
Mr. K. Niethammer & Mr. H. Böttcher	Kübler & Niethammer	Kriebethal
Mr. T. Dechert	Dresden Elektrizität und Fernwärme	Dresden

3. Interviews with independent experts (February 1994 - August 1996)

Name	Organisation	Place
Prof. Dr. A. Verbruggen & Ms. N. Dufait	University of Antwerp	Antwerp
Dr. R. Walz	Fraunhofer Institut für System-technik und Innovationsforschung	Karlsruhe
Dr. S. Marvin	University of Newcastle-upon-Tyne	Newcastle
Dr. S. Russell	University of Wollongong (Australia)	London
Mr. A. Brown	DG XVII (Energy), European Commission	Brussels
Mr. A.M. Klom	DG XVII (Energy), European Commission	Brussels

Appendix 4: Complementary data

1. Cost digression of CHP-units up to 500 $kW_{el.}$

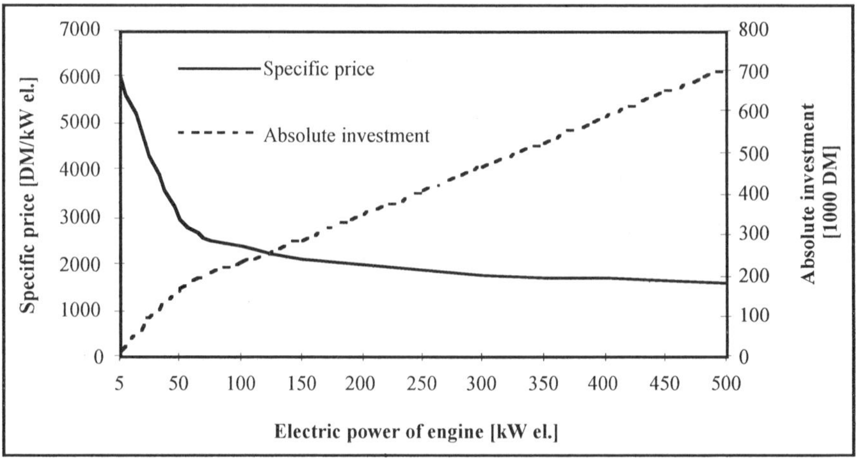

Source: (Energiereferat, 1995)

2. Technical performance characteristics of gas-turbines by time

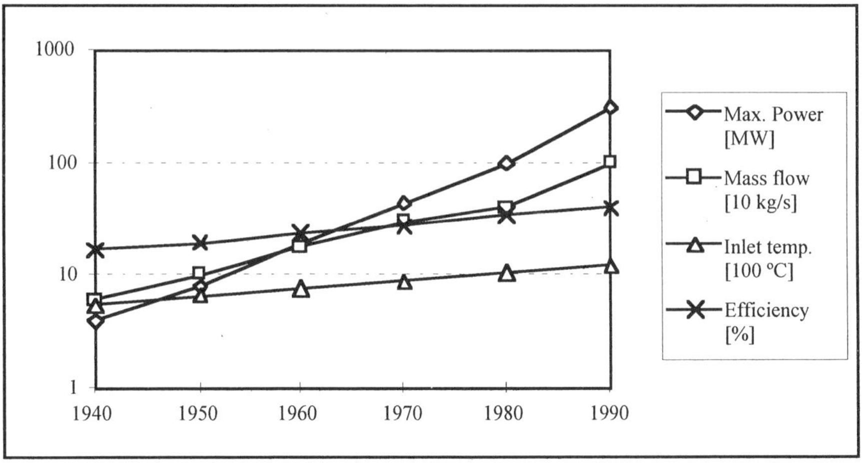

Source: (Auer, 1985) and manufacturers' brochures, sourcedata interpolated

3. Capacity development of CHP in the UK by sectors in MW_{el}.

	1977	1983	1988	1991	1993	1994	1995	1996
Total industry	2792	2353	1793	2033	2593	2771	3093	3069
• *Chemicals*	972	962	654	949	918	964	1228	1147
• *Refineries*	463	504	400	414	433	440	460	495
• *Paper and board*	550	311	210	262	211	277	331	337
• *Food and drink*	210	205	171	221	201	211	196	208
• *Iron and steel*	243	241	131	135	130	130	130	130
• *Other industrial*	354	130	227	52	702*	750*	750*	752*
Transport, commerce and administration	n.a.	n.a.	n.a.	together	122	157	179	185
Other small-scale and district heating	n.a.	n.a.	n.a.	279	174	210	213	309

Source: (DTI, 1996; DTI 1997) ; (*) The CHP capacity of the public electricity supply system is included

4. Capacity development of CHP in Germany by sectors in MW_{el}.

	1975	1983	1988	1991	1993	1994	1995	1996
Total industry	8770	9164	8133	11154	11520	11121	10906	9960
• *Mining*	1251	1272	930	2847	3576	3129	2748	2481
• *Chemical*	3058	3601	3556	4126	3893	3922	3816	3524
• *Refineries*	781	475	441	682	634	570	682	651
• *Iron and steel*	1388	1278	730	765	712	736	494	368
• *Paper & board*	1260	1264	1253	1362	1347	1363	1475	1354
• *Food and drink*	532	770	733	865	860	915	891	875
• *Other industry*	500	504	490	507	498	486	800	707
District heating	5100	7000	7100	8200	8700	9300	9710	10020
Small-scale	10	180	430	700	850	1100	1450	1600

Source: (AGFW, 1997, AGFW, 1995; VIK, 1997; VIK, 1994a; ASUE, 1996, ASUE 1997); data for East Germany included from 1991 onwards, for small scale CHP from 1983 onwards

Appendix 5: Comparative summary overview

United Kingdom	Germany
1. Preceding history	
• Hardly any district heating; • Some industrial capacities; • Weak local authorities without own utility companies;	• Construction of DH infrastructures after 2nd World War; • Major industrial CHP-capacity; • Horizontally integrated local energy supply companies;
2. Exogenous developments	
a) Fuel supply • Indigenous reserves of coal, gas and oil; • Relatively low fuel prices; • Wide use of gas, also for power generation since late 1980s;	*a) Fuel supply* • Few indigenous fuel resources apart from coal, and high mportance of a balanced and secure supply; • Major financial support to support indigenous coal, therefore comparatively high power prices; • Extended gas supply;
b) EU-policy • Uncertainties about Internal Energy Market, CO2-taxation and stricter environmental regulations; • Model role of other countries with regard to CHP;	*b) EU-policy* • Uncertainty about future IEM, but first preparations are made by the major companies; • CO2-tax postponed or probably abandoned;

c) Economic development • Tecession limited investment in utility services; • Restructuring of energy-intensive industries since late 80s;	*c) Economic development* • Moderate economic growth; • Early restructuring of industry towards high efficiency; • Special situation and needs in East Germany;
d) Public opinion and governance • Low but growing importance priority of environmental issues; • Bad reputation of CHP/DH due to negative early experiences; • Introduction of a liberal governance approach;	*d) Public opinion and governance* • Strong impact of the environmental discussion; • Negative experience with heat pumps created reluctance towards small CHP; • Consensus-oriented governance approach;
3. Technology changes	
a) Components • Spin-offs from other technologies, especially microelectronics for remote control and grid connection (late 80s); • Higher efficiency of gas-turbines, new stationary applications; • Environmental improvements as far a needed; • Reliability and cost-reducing innovations of major importance; • Further cost-reductions and reliability major objectives	*a) Components* • Spin-offs from other technologies, especially microelectronics for remote control and grid connection (early 90s); • Particular need to improve gas-engines; • Ongoing development of micro-CHP systems; • Environmental performance an important area of research; • Improvement of reliability important but not priority; • Now incremental improvements for cost reductions and further system integration;

b) Architecture • Integrated system design since early 90s; • Remote control & online monitoring for preventive maintenance in the late 80s; • Packaging of individual systems since mid 80s; • Major organisational innovations towads integrated service supply;	*b) Architecture* • Integrated system design since late 80s; • Online monitoring and control, and preventive maintenance introduced in the early 90s; • Packaging and standardisation of design procedures and guidelines since late-80s; • Limited uptake of new organisational solutions like CEM or service orientation;
c) Future developments • Some minor efforts for new systems based on Stirling and fuel cells; • Integration with chilling and industrial processes;	*c) Future developments* • New development and demonstration projects for fuel cells and Stirling engines; • Futher system integration of CHP with other components for chilling, incineration, etc.;
4. Structural changes	
a) Technological structures • Fast diffusion of industrial and small-scale CHP over the last ten years; • Reconsideration of some DH-projects and group heating networks;	*a) Technological structures* • Fast diffusion of small-scale and gas-turbine systems since the late 80s; • Slow diffusion of district heating at a comparatively high level and upgrading of generation sets to CHP in old DH-systems, especially in Eastern Germany;

• Widest possible use of gas for CHP; • Fast growth of CCGT-based power-only capacity; • Highly centralised generation structure is slowly decentralising; • High overcapacity and stagnation of demand;	• Wider use of group heating schemes in cities; • Extension of gas-based CHP; • Centralised generation structure with a substantial share (15%) of independent, mainly industrial generation capacity; • Modest overcapacity and stagnation of demand;
b) Economic structures • Liberalisation & privatisation opened up a long-blocked potential; • Emergence of IESC and merging of technology supply and demand functions; • Fast growth of CEM and DEP-schemes; • Re-organisation of the ESI with possibilities for horizontal integration but still a weak local level; • Area monopolies broken up, competition in supply; • Changing relationships between users and suppliers of technology; • Entry of new technology suppliers (mainly small ones) and Involvement of power generators and regional supply companies in technology packaging;	*b) Economic structures* • Adjustments of the monopolistic regulatory framework with regarde to CHP; • Negotiated export prices between utilities' association and autogenerators; • Slowly growing importance of CEM and similar arrangements; • Continuity of the industy structure with powerful local utilities, regional distributors and supra-regional generators, mutually intertwined, both horizontally and vertically; • Area monopolies maintained; • Informal but rather limited relationships between technology manufacturers and users; • Mushrooming of small technology suppliers, with onset of shakeout in the mid-90s, dominance of some established manufacturers;

• Complex regulatory structure and pool system to determine electricity prices, controlled franchise area until 1998; • BATNEEC-principle as flexible environmental regulation;	• Franchise electricity prices subjected to control, and indivually negotiated prices for large costumers; • Reliable prescriptions as principle of environmental policy;
c) Informational structures • Emergence of an information network for CHP around government institutions; • Close cooperation users/suppliers in the first years of diffusion to stablise the technology and develop packaged systems; • Minor influence of experts as compared to interest groups;	*c) Informational structures* • Information-networks centered around the main industrial associations, especially for industrial CHP and utilities; • Engineering associations involved in standardisation and design;limited exchanges between manufacturers and users; • Important role of technical expertise in decision-making;
d) Political structures • Losening of links ESI-government; • Delegation of responsibility to the regulators and dissolution of Dept. of Energy; • Ideologically driven policy approach in favour of liberalisation; • No decentralisation of political decision power and means to the local level and industry;	*d) Political structures* • Many different inroads for political negotiation at local, regional and federal level, possibilities for variation; • Integration and consideration of a variety of interest groups in the political decision process; • Pragmatic and incremental policy approach, at present favouring limited liberalisation; • Strong local policy level, also in terms of energy planning; delegation of regulatory responsibility to industry

5. Decisions, interactions and motivations	
a) Innovation • Recognition of new opportunities for suppliers from mid 80s onwards, reinforced in the early 90s; • Major learning processes were achieved since the early Eighties; • Combination of technical and non-technical solutions; • Adaptation to changing needs of users; • Packaging as a standardisation step, but no standardisation of planning and design;	*a) Innovation* • Recognition of business opportunities from the mid 80s onwards in the small-scale area, large-scale established and extended; • Learning process during the late 80s and early 90s: introduction of an integrated systems view and changing design philosophies; • Standardisation of planning procedures and technical specifications by industrial associations;
b) Adoption end-users: • Perception of high uncertainty about regulation, long-term energy policy and electricity prices; • CHP as additional investment, therefore short payback times;	*b) Adoption* end-users: • Uncertainty about tariffs and reliability of CHP was limited due to Associations' Agreement, but remaining uncertainty about long-term liberalisation plans; • Payback-times for industry shorter than for public enterprises, but often overriden by strategic considerations;

• Trend towards outcontracting of utility services;	• Frequent facilitation of adoption by informal cooperation between local utilities and users;
• High uncertainty is a major barrier to CHP investment;	• Recognition of a need to build up in-house competencies, specificity of site conditions very important;
• Lack of competencies for operating CHP;	• Stabilisation of boundary and planning conditions;
• Specific conditions apply to some CHP-areas (local authorities & health service funding)	• In specific areas (hospitals, some public) crucial problems with financing, therefore opportunities for CEM;
<u>service and utility companies</u>:	<u>service and utility companies</u>:
• Changing roles and attitudes of ex-utilities due to competitive pressure, move towards horizontal and vertical integration;	• Dominance of utility companies, slow onset of service orientation and local LCP, horizontal integration;
• Utilities set up their own CHP-activities since 1990, fast growth of IESC;	• Limited importance of private service companies in CEM;
• Provision of new financing and maintenance solutions;	• Especially local utilities consider non-economic criteria in their decisions, like e.g. environmental ones;
• Recognition of economic opportunities in CHP, but reluctance to accept independent CHP;	• Recognition of independent and small-scale CHP as an additional element in the supply strategy;
c) Political decisions	*c) Political decisions*
• Very limited specific support for CHP, mainly information supply,	• Specific financial support for CHP during the 70 and 80s, and until 1994 for the Neue Länder only;
• Recognition of the advantages of CHP in the late Eighties but still a weak impact of environmental regulation;	• Strong impact of environmental regulation (TA Luft) on product technology, against the fierce resistence of industry

• Giving up of an active energy policy, supported by a strong liberal philosophy;	• Pragmatic policy-approach, based more on technical expertise than on a strong will to liberalise;
• International commitments triggered environmental policies;	• Rio conference and CO2-targets used as arguments in favour of CHP;
• Parliament Committees as main arenas of critique;	• Enquete Commission in favour of CHP;
• Some efficiency measures, but mainly information policy;	• Further legislation plans to improve energy efficiency;
6. Evolutionary dynamics	
• Limited external pressure in favour of efficiency and CHP;	• Several external forces in favour of CHP;
• Innovative impulses from technological and organisational side, but little institutional variety;	• Limited technological and service innovations to create variety, but high institutional diversity to facilitate it;
• Fast transition to more open service-providers and utilities, end-users with strict requirements and limited competencies;	• Utilities slowly accepted CHP, moderately strict requirements of fairly competent CHP end-users
• Switch from a closed system to an uncertain and liberal regulatory context with little political intervention;	• Slowly improving and reliable regulatory framework with many intervention points
• Strong interdependencies between government, ESI and technology supply broken up;	• Manifold interdependencies between different levels of economic, technological and political decision-making, both horizontally and vertically;
• Few reinforcing effects, mainly based on information provision on best practice;	• Different reinforcements due to information provision, the role of associations and informal contacts;
• Policy- and ideology-driven revolutionary overturn	• Incremental and co-evolutionary change;

7. Future perspectives	
a) Scenario 1: CHP-friendly liberalisation • Minor adjustments of the present system needed; • Stronger recognition of the benefits of CHP; • Some delegation of competencies to the local level; • Electricity pricing as in present system; • Orienting environmental regulations; • Limited research funding; • Further opportunities for small-scale and industrial CHP;	*a) Scenario 1: CHP-friendly liberalisation* • Major change of the existing system; • Major administrative simplifications; • Maintenance of monopolistic local supply; • Introduction of a pool-like system for large costumers; • Lower importance of voluntary agreements; • Prescriptive approach to environmental regulation; • Financial support for R&D only; • DH, small-scale and industrial can further expand;
b) Scenario 2: Full-scale liberalisation • Certain reinforcements of the present regulatory system; • Further expansion of private service companies; • Very limited regulatory interventions; • Electricity pricing as foreseen for 1998; • Short-term orientation for investment in energy supply • Expansion of small-scale and industrial CHP, but not DH;	*b) Scenario 2: Full-scale liberalisation* • Major problems for utility companies due to competition, also at local level; • Emergence of new private service companies; • Very limited regulatory interventions and access barriers; • Fully individual negotiation of electricity tariffs; • Difficulties for long-term coordination of supply chains; • Further opportunities for single-site CHP like small-scale and industrial, but decline of DH;

c) Scenario 3: Political enforcement of CHP	*c) Scenario 3: Political enforcement of CHP*
• Financial incentives for local level CHP/DH, but no carrying organisations;	• Reinforcement of the local supply level to facilitate horizontal service coordination;
• Taxes and price control in favour of CHP;	• Financial incentives and taxes in favour of CHP-adoption;
• Strong incentives to improve small-scale and industrial CHP-technology;	• Investment control to increase energy efficiency;
• The only way for DH in the UK, but of limited importance due to already achieved lock-in into gas-based power only generation;	• Special tariffs for CHP power exports;
• Expansion of small-scale and industrial CHP;	• Major administrative rigidities; • Strong incentives to advance the technology of CHP; • Expansion of all types of CHP-applications;

Contributions to Economics

Hagen Bobzin
Indivisibilities
1998. ISBN 3-7908-1123-8

Helmut Wagner (Ed.)
Current Issues in Monetary Economics
1998. ISBN 3-7908-1127-0

Peter Michaelis/Frank Stähler (Eds.)
Recent Policy Issues in Environmental and Resource Economics
1998. ISBN 3-7908-1137-8

Jessica de Wolff
The Political Economy of Fiscal Decisions
1998. ISBN 3-7908-1130-0

Georg Bol/Gholamreza Nakhaeizadeh/
Karl-Heinz Vollmer (Eds.)
Risk Measurements, Econometrics and Neural Networks
1998. ISBN 3-7908-1152-1

Joachim Winter
Investment and Exit Decisions at the Plant Level
1998. ISBN 3-7908-1154-8

Bernd Meyer
Intertemporal Asset Pricing
1999. ISBN 3-7908-1159-9

Uwe Walz
Dynamics of Regional Integration
1999. ISBN 3-7908-1185-8

Michael Carlberg
European Monetary Union
1999. ISBN 3-7908-1191-2

Giovanni Galizzi/Luciano Venturini (Eds.)
Vertical Relationships and Coordination in the Food System
1999. ISBN 3-7908-1192-0

Gustav A. Horn/Wolfgang Scheremet/
Rudolf Zwiener
Wages and the Euro
1999. ISBN 3-7908-1199-8

Dirk Willer
The Development of Equity Capital Markets in Transition Economies
1999. ISBN 3-7908-1198-X

Springer and the environment

Druck: Strauss Offsetdruck, Mörlenbach
Verarbeitung: Schäffer, Grünstadt